T0178057

The NEUROBIOLOGICAL BASIS of SUICIDE

FRONTIERS IN NEUROSCIENCE

Series Editor

Sidney A. Simon, Ph.D.

Published Titles

Apoptosis in Neurobiology

Yusuf A. Hannun, M.D., Professor of Biomedical Research and Chairman, Department of Biochemistry and Molecular Biology, Medical University of South Carolina, Charleston, South Carolina

Rose-Mary Boustany, M.D., tenured Associate Professor of Pediatrics and Neurobiology, Duke University Medical Center, Durham, North Carolina

Neural Prostheses for Restoration of Sensory and Motor Function

John K. Chapin, Ph.D., Professor of Physiology and Pharmacology, State University of New York Health Science Center, Brooklyn, New York

Karen A. Moxon, Ph.D., Assistant Professor, School of Biomedical Engineering, Science, and Health Systems, Drexel University, Philadelphia, Pennsylvania

Computational Neuroscience: Realistic Modeling for Experimentalists

Eric DeSchutter, M.D., Ph.D., Professor, Department of Medicine, University of Antwerp, Antwerp, Belgium

Methods in Pain Research

Lawrence Kruger, Ph.D., Professor of Neurobiology (Emeritus), UCLA School of Medicine and Brain Research Institute, Los Angeles, California

Motor Neurobiology of the Spinal Cord

Timothy C. Cope, Ph.D., Professor of Physiology, Wright State University, Dayton, Ohio

Nicotinic Receptors in the Nervous System

Edward D. Levin, Ph.D., Associate Professor, Department of Psychiatry and Pharmacology and Molecular Cancer Biology and Department of Psychiatry and Behavioral Sciences, Duke University School of Medicine, Durham, North Carolina

Methods in Genomic Neuroscience

Helmin R. Chin, Ph.D., Genetics Research Branch, NIMH, NIH, Bethesda, Maryland

Steven O. Moldin, Ph.D., University of Southern California, Washington, D.C.

Methods in Chemosensory Research

Sidney A. Simon, Ph.D., Professor of Neurobiology, Biomedical Engineering, and Anesthesiology, Duke University, Durham, North Carolina

Miguel A.L. Nicolelis, M.D., Ph.D., Professor of Neurobiology and Biomedical Engineering, Duke University, Durham, North Carolina

The Somatosensory System: Deciphering the Brain's Own Body Image

Randall J. Nelson, Ph.D., Professor of Anatomy and Neurobiology, University of Tennessee Health Sciences Center, Memphis, Tennessee

The Superior Colliculus: New Approaches for Studying Sensorimotor Integration
William C. Hall, Ph.D., Department of Neuroscience, Duke University, Durham, North Carolina
Adonis Moschovakis, Ph.D., Department of Basic Sciences, University of Crete, Heraklion, Greece

New Concepts in Cerebral Ischemia
Rick C.S. Lin, Ph.D., Professor of Anatomy, University of Mississippi Medical Center, Jackson, Mississippi

DNA Arrays: Technologies and Experimental Strategies
Elena Grigorenko, Ph.D., Technology Development Group, Millennium Pharmaceuticals, Cambridge, Massachusetts

Methods for Alcohol-Related Neuroscience Research
Yuan Liu, Ph.D., National Institute of Neurological Disorders and Stroke, National Institutes of Health, Bethesda, Maryland
David M. Lovinger, Ph.D., Laboratory of Integrative Neuroscience, NIAAA, Nashville, Tennessee

Primate Audition: Behavior and Neurobiology
Asif A. Ghazanfar, Ph.D., Princeton University, Princeton, New Jersey

Methods in Drug Abuse Research: Cellular and Circuit Level Analyses
Barry D. Waterhouse, Ph.D., MCP-Hahnemann University, Philadelphia, Pennsylvania

Functional and Neural Mechanisms of Interval Timing
Warren H. Meck, Ph.D., Professor of Psychology, Duke University, Durham, North Carolina

Biomedical Imaging in Experimental Neuroscience
Nick Van Bruggen, Ph.D., Department of Neuroscience Genentech, Inc.
Timothy P.L. Roberts, Ph.D., Associate Professor, University of Toronto, Canada

The Primate Visual System
John H. Kaas, Department of Psychology, Vanderbilt University, Nashville, Tennessee
Christine Collins, Department of Psychology, Vanderbilt University, Nashville, Tennessee

Neurosteroid Effects in the Central Nervous System
Sheryl S. Smith, Ph.D., Department of Physiology, SUNY Health Science Center, Brooklyn, New York

Modern Neurosurgery: Clinical Translation of Neuroscience Advances
Dennis A. Turner, Department of Surgery, Division of Neurosurgery, Duke University Medical Center, Durham, North Carolina

Sleep: Circuits and Functions
Pierre-Hervé Luppi, Université Claude Bernard, Lyon, France

Methods in Insect Sensory Neuroscience
Thomas A. Christensen, Arizona Research Laboratories, Division of Neurobiology, University of Arizona, Tuscon, Arizona

Motor Cortex in Voluntary Movements
Alexa Riehle, INCM-CNRS, Marseille, France
Eilon Vaadia, The Hebrew University, Jerusalem, Israel

Neural Plasticity in Adult Somatic Sensory-Motor Systems
Ford F. Ebner, Vanderbilt University, Nashville, Tennessee

Advances in Vagal Afferent Neurobiology
Bradley J. Undem, Johns Hopkins Asthma Center, Baltimore, Maryland
Daniel Weinreich, University of Maryland, Baltimore, Maryland

The Dynamic Synapse: Molecular Methods in Ionotropic Receptor Biology
Josef T. Kittler, University College, London, England
Stephen J. Moss, University College, London, England

Animal Models of Cognitive Impairment
Edward D. Levin, Duke University Medical Center, Durham, North Carolina
Jerry J. Buccafusco, Medical College of Georgia, Augusta, Georgia

The Role of the Nucleus of the Solitary Tract in Gustatory Processing
Robert M. Bradley, University of Michigan, Ann Arbor, Michigan

Brain Aging: Models, Methods, and Mechanisms
David R. Riddle, Wake Forest University, Winston-Salem, North Carolina

Neural Plasticity and Memory: From Genes to Brain Imaging
Frederico Bermudez-Rattoni, National University of Mexico, Mexico City, Mexico

Serotonin Receptors in Neurobiology
Amitabha Chattopadhyay, Center for Cellular and Molecular Biology, Hyderabad, India

TRP Ion Channel Function in Sensory Transduction and Cellular Signaling Cascades
Wolfgang B. Liedtke, M.D., Ph.D., Duke University Medical Center, Durham, North Carolina
Stefan Heller, Ph.D., Stanford University School of Medicine, Stanford, California

Methods for Neural Ensemble Recordings, Second Edition
Miguel A.L. Nicolelis, M.D., Ph.D., Professor of Neurobiology and Biomedical Engineering, Duke University Medical Center, Durham, North Carolina

Biology of the NMDA Receptor
Antonius M. VanDongen, Duke University Medical Center, Durham, North Carolina

Methods of Behavioral Analysis in Neuroscience
Jerry J. Buccafusco, Ph.D., Alzheimer's Research Center, Professor of Pharmacology and Toxicology, Professor of Psychiatry and Health Behavior, Medical College of Georgia, Augusta, Georgia

In Vivo Optical Imaging of Brain Function, Second Edition
Ron Frostig, Ph.D., Professor, Department of Neurobiology, University of California, Irvine, California

Fat Detection: Taste, Texture, and Post Ingestive Effects
Jean-Pierre Montmayeur, Ph.D., Centre National de la Recherche Scientifique, Dijon, France
Johannes le Coutre, Ph.D., Nestlé Research Center, Lausanne, Switzerland

The Neurobiology of Olfaction
Anna Menini, Ph.D., Neurobiology Sector International School for Advanced Studies, (S.I.S.S.A.), Trieste, Italy

Neuroproteomics
Oscar Alzate, Ph.D., Department of Cell and Developmental Biology, University of North Carolina, Chapel Hill, North Carolina

Translational Pain Research: From Mouse to Man
Lawrence Kruger, Ph.D., Department of Neurobiology, UCLA School of Medicine, Los Angeles, California
Alan R. Light, Ph.D., Department of Anesthesiology, University of Utah, Salt Lake City, Utah

Advances in the Neuroscience of Addiction
Cynthia M. Kuhn, Duke University Medical Center, Durham, North Carolina
George F. Koob, The Scripps Research Institute, La Jolla, California

Neurobiology of Huntington's Disease: Applications to Drug Discovery
Donald C. Lo, Duke University Medical Center, Durham, North Carolina
Robert E. Hughes, Buck Institute for Age Research, Novato, California

Neurobiology of Sensation and Reward
Jay A. Gottfried, Northwestern University, Chicago, Illinois

The Neural Bases of Multisensory Processes
Micah M. Murray, CIBM, Lausanne, Switzerland
Mark T. Wallace, Vanderbilt Brain Institute, Nashville, Tennessee

Neurobiology of Depression
Francisco López-Muñoz, University of Alcalá, Madrid, Spain
Cecilio Álamo, University of Alcalá, Madrid, Spain

Astrocytes: Wiring the Brain
Eliana Scemes, Albert Einstein College of Medicine, Bronx, New York
David C. Spray, Albert Einstein College of Medicine, Bronx, New York

Dopamine–Glutamate Interactions in the Basal Ganglia
Susan Jones, University of Cambridge, United Kingdom

Alzheimer's Disease: Targets for New Clinical Diagnostic and Therapeutic Strategies
Renee D. Wegrzyn, Booz Allen Hamilton, Arlington, Virginia
Alan S. Rudolph, Duke Center for Neuroengineering, Potomac, Maryland

The Neurobiological Basis of Suicide
Yogesh Dwivedi, University of Illinois at Chicago

The NEUROBIOLOGICAL BASIS of SUICIDE

Edited by
Yogesh Dwivedi
University of Illinois at Chicago

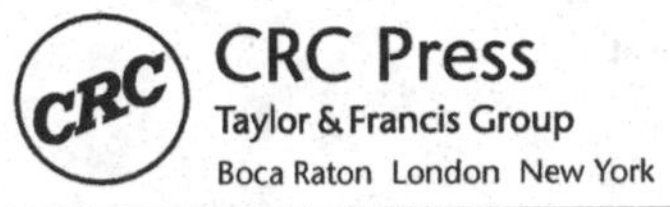

CRC Press is an imprint of the
Taylor & Francis Group, an **informa** business

CRC Press
Taylor & Francis Group
6000 Broken Sound Parkway NW, Suite 300
Boca Raton, FL 33487-2742

First issued in paperback 2019

ISBN-13: 978-1-4398-3881-5 (hbk)
ISBN-13: 978-0-367-38119-6 (pbk)

Library of Congress Cataloging-in-Publication Data

The neurobiological basis of suicide / editor, Yogesh Dwivedi.
p. ; cm. -- (Frontiers in neuroscience)
Includes bibliographical references and index.
ISBN 978-1-4398-3881-5 (alk. paper)
I. Dwivedi, Yogesh. II. Series: Frontiers in neuroscience.
[DNLM: 1. Suicide--psychology. 2. Adrenergic Neurons--physiology. 3. Serotonin--physiology. WM 165]

616.85'8445--dc23 2012006499

Visit the Taylor & Francis Web site at
http://www.taylorandfrancis.com

and the CRC Press Web site at
http://www.crcpress.com

Contents

Series Preface

The Frontiers in Neuroscience Series presents the insights of experts on emerging fields and theoretical concepts that are, or will be, at the vanguard of neuroscience.

The books cover new and exciting multidisciplinary areas of brain research and describe breakthroughs in fields like visual, gustatory, auditory, olfactory neuroscience, as well as aging and biomedical imaging. Recent books cover the rapidly evolving fields of multisensory processing, glia, depression, and different aspects of reward.

Each book is edited by experts and consists of chapters written by leaders in a particular field. The books are richly illustrated and contain comprehensive bibliographies. The chapters provide substantial background material relevant to the particular subject.

The goal is for these books to be the references every neuroscientist uses in order to acquaint themselves with new information and methodologies in brain research. I view my task as series editor to produce outstanding products that contribute to the broad field of neuroscience. Now that the chapters are available online, the effort put in by me, the publisher, and the book editors hopefully will contribute to the further development of brain research. To the extent that you learn from these books, we will have succeeded.

Sidney A. Simon
Duke University

Preface

The act of suicide has been defined as a "fatal self-inflicted destructive act with explicit or inferred intent to die" (Institute of Medicine of the National Academies, 2002). Suicide is a major public health concern not only in the United States but throughout the world. According to World Health Organization studies, approximately 2% of people commit suicide worldwide every year. Recent data released in 2011 from the Centers for Disease Control and Prevention state that suicide is the 10th leading cause of death in the United States. When accounting for all age groups, 36,035 people committed suicide in 2008. When different age groups were factored, suicide was the second leading cause of death among those aged 25 to 34 years and the third leading cause of death among those aged 10–24 years. Although approximately 90% of suicides are associated with mental disorders, several studies suggest that suicidal behavior has distinct psychiatric phenotypes and that suicidal patients have a certain predisposition that is different from other psychiatric illnesses. This is based on the argument that 10% of people who commit suicide do not have any diagnosed psychiatric illness and that 95% of people who have some form of psychiatric illness do not commit suicide. Moreover, the therapeutic approaches and their outcomes appear to be different between suicidal patients and patients with associated mental disorders. Psychological, psychosocial, and cultural factors are important in determining the risk factors for suicide; however, they offer weak prediction and can be of less clinical use. Also, the clinical history of a patient provides low specificity in predicting suicide. Interestingly, cognitive characteristics are different among depressed suicidal and depressed nonsuicidal subjects, and could be involved in the development of suicidal behavior. Thus, characterizing neurobiological basis will be key in delineating the risk factors associated with suicide. In recent years, research has been initiated to develop the neurobiological basis of suicide. This book aims to focus on the recent neurobiological findings associated with suicide.

Suicide is not predictable in the individual and results from a complex series of factors that may differ across individuals; yet 30%–70% of suicides occur in patients who are receiving some treatment. Learning more about the interacting biological, clinical, and situational factors that lead to suicide can help the clinician to recognize risk factors and initiate clinical interventions to reduce suicide risk. In Chapter 1, Fawcett describes diagnosis, traits, states, and co-morbidity in suicide.

Because of its key role in aggression and impulsivity, the role of the serotonergic system in suicidal behavior has been the focus of attention for a long time. Several clinical and postmortem brain studies have consistently shown that an abnormal serotonergic system is critical in the development of suicidal behavior. In Chapter 2, Bach and Arango discuss the neuroanatomical basis of serotonergic abnormalities that are specific to suicide. This is consistent with a homeostatic brain response, in both source 5-hydroxytryptamine synthesizing neurons in the raphe nuclei and in postsynaptic target neurons in the cortex, to deficits in serotonergic neurotransmission.

In addition to the serotonergic system, evidence indicates a dysfunctional central noradrenergic system in suicide and suggests that deficits in noradrenergic signaling may contribute to suicide. In Chapter 3, Chandley and Ordway review the neurobiological features and functional output of the brain noradrenergic system in relation to the potential involvement of the noradrenergic system in suicide.

Although γ-aminobutyric acid (GABA) was previously implicated in depression and suicidal behavior, there has not been much research on this aspect. In recent years, interest has renewed regarding the role of GABA in suicide. In Chapter 4, Anisman and colleagues review how GABA is involved in depressive behavior and provide evidence that, in depressive illness, it works in collaboration with serotonin, and corticotropin-releasing hormone and that these actions are moderated by neurosteroids. They further discuss that the coordination in the appearance of the subunits that comprise $GABA_A$ receptors may be fundamental in the timing and synchronization of neuronal activity and might, thus, influence depressive illness.

The brain endocannabinoid (eCB) system modulates several neurobiological processes, and its dysfunction is suggested to be involved in the pathophysiological characteristics of mood and drug use disorders. CB1 receptor–mediated signaling, in particular, has played a critical role in the neural circuitry that mediates mood, motivation, and emotional behaviors. The association of the eCB system with alcohol addiction and the existence of a high incidence rate of suicide in cannabis and alcohol abusers indicate that dysfunction of the eCB system might be one of the contributing factors for suicidal behavior. In Chapter 5, by using preclinical and clinical evidence, Vinod describes in detail how the eCB system may be involved in suicidal behavior.

Stress is one of the major risk factors in suicide. In Chapter 6, by using the stress-diathesis model, van Heeringen reviews how stress can play a role in suicidal behavior. He also discusses the advantages of using stress-diathesis models for treating and preventing suicide risk. In Chapter 7, on a related issue, Coryell discusses the association of hypothalamic–pituitary–adrenal axis hyperactivity and a low serum cholesterol level to a heightened risk for suicide and provides evidence showing that serum cholesterol concentrations and dexamethasone suppression test results are clinically useful tools for the assessment of risks in suicide.

An emerging hypothesis suggests that the pathogenesis of suicidal behavior and depression involves altered neural plasticity, resulting in the inability of the brain to make appropriate adaptive responses to environmental stimuli. Furthermore, stress, a major factor in suicide, hinders performance on hippocampal-dependent memory tasks and impairs induction of hippocampal long-term potentiation. The brain-derived neurotrophic factor plays a critical role in regulating structural, synaptic, and morphological plasticity and in modulating the strength and number of synaptic connections and neurotransmission. In Chapter 8, Dwivedi provides a detailed account of how a brain-derived neurotrophic factor and its cognate receptors may be involved in the development of suicidal behavior.

Neuroimaging methods provide a great opportunity to improve our understanding of the pathway between adverse environmental conditions and suicide. Because neuroimaging is completed *in vivo*, it is tailored to investigate the intermediary links between neuropathological characteristics and the symptoms or traits associated with suicide.

Such symptoms and traits may be measured at brain scanning. In Chapter 9, Meyer describes neuroimaging studies of 5-HT_{2A} receptors, serotonin transporters, monoamine oxidase A, dopamine D_2 receptor binding, and μ-opioid receptors in suicide. He also discusses how environmental conditions lead to alterations in brain neurochemistry that can increase the risk for suicide with neuroimaging studies of monoamine oxidase A in the early postpartum period.

The etiology of suicide has been considered to be a complex combination of genetic and environmental components. Studies have increasingly shown that gene–environment interactions are crucial in the development of suicidal behavior. Approximately 30%–40% of the variance in suicidal behavior is related to genetic risk factors. Given the importance of genetics and gene–environment interactions, several chapters discuss this issue. In Chapter 10, Roy elaborates on this aspect in much greater detail. In this chapter, he describes the importance of both the environment and genes in suicidal behavior and reviews clinical studies of gene–environment interactions in relation to suicide attempts. In Chapter 11, using a similar approach, Zai and colleagues discuss genetic findings in suicidal behavior in greater detail. They discuss the twin, family, and adoption studies that establish a genetic basis for suicidal behavior. They review the results from studies of candidate genes in the neurotransmitter systems, neurotrophin pathways, and hypothalamic–pituitary–adrenal axis. They further discuss recent developments in whole genome association studies, epigenetics, and gene–environment interactions.

To expand its focus, suicide research has moved toward high-throughput gene expression microarrays in an effort to identify novel biological pathways and molecular mechanisms associated with suicide. By analyzing tissues obtained from suicide completers and examining the expression of many genes in parallel, these technologies allow researchers to obtain a functional profile of gene expression, thus providing valuable insight into the overall biological processes underlying suicide. In Chapter 12, Fiori and Turecki discuss the methods that have been used to profile gene expression alterations and examine several of the neurobiological mechanisms implicated in suicide. They also discuss the future uses of gene expression profiling technologies in suicide research.

Gene expression is highly regulated through an epigenetic mechanism. Epigenetic changes regulate gene function via alternative mechanisms to the coding DNA sequence. Until recently, it was believed that only physical and chemical environmental factors altered epigenetic markings; more recently, studies have indicated that the social environment can also induce epigenetic changes. Suicide is frequently associated with a history of early-life adversity, and recent studies have suggested that epigenetic changes are present in specific areas of the brain from individuals who died by suicide. In Chapter 13, Labonté and Turecki discuss this aspect and review the growing body of literature examining epigenetic alterations associated with early-life environment and suicide.

The neurobiological characteristics of suicide in adolescents may be different than in adults because adolescent suicide is more associated with impulsive and aggressive behavior. Suicide in adolescents has been studied much less than in adults. Three chapters are devoted to adolescent suicide. In Chapter 14, Zalsman provides a detailed account of various aspects of adolescent suicide, including peripheral blood and postmortem brain studies; however, the main focus of this chapter is the genetics and gene–environment interaction studies linked to adolescent suicidality. In another chapter (Chapter 15) devoted to adolescent suicide, Pandey and Dwivedi discuss the findings of postmortem brain studies, with emphasis on neurotransmitter receptors, cellular signaling, and neurotrophins. With a similar theme, in Chapter 16, Nanayakkara and colleagues discuss suicidal behavior in the pediatric population, assessing risk among pediatric patients with bipolar disorder.

Another important age group that is vulnerable to suicide is the elderly population. Suicide rates are highest in elderly persons in most countries around the world, and suicide has a more severe impact in this demographic group. In Chapter 17, Ajilore and Kumar discuss the clinical and neurobiological risk factors among the elderly suicidal population. They also discuss how cognitive neuroscience can be used to understand impaired decision making in elderly suicidal patients.

The identification of personality features related to suicide attempts may be useful to prevent suicide mortality. This is because of their strong correlation with attempted and completed suicide. In Chapter 18, Rujescu and Giegling devote a chapter on the issue and discuss that personality trait is a crucial intermediate phenotype of suicidal behavior. They also discuss that aggression, impulsivity, anger, temperament, and neuroticism are important risk factors in suicidal behavior and show how a genetic component is involved in suicidal behavior in relation to personality.

Okusaga and Postolache have been studying triggers and vulnerabilities for suicide originating in the natural environment. More specifically, they have shown that during certain seasons there is a consistent peak of suicide. In Chapter 19, Okusaga and Postolache review the evidence connecting immune activation to *Toxoplasma gondii* infection and suicidal behavior.

Because of the inaccessibility of the human brain, peripheral tissues such as serum, blood cells, saliva, or urine may provide information regarding biological abnormalities. Pandey and Dwivedi discuss this approach in Chapter 20 and critically assess whether abnormalities in peripheral tissues can serve as biomarkers for suicidal behavior.

Finally, in Chapter 21, Mann and Currier outline the main alterations in neurobiological functions in suicide attempters and in individuals who died by suicide and describe the putative mechanisms of action by which different classes of medications may help prevent suicidal behavior.

In summary, this book covers a wide array of neurobiological abnormalities associated with suicide in a comprehensive manner. From various studies described in this book, it can be inferred that neurobiological correlates are important substrates in identifying the risk factors associated with suicidal behavior and that these neurobiological correlates may provide a promising approach to improve old approaches and develop new treatment modalities in the prevention of suicidal behavior.

Acknowledgments

I would like to dedicate this book to the people who have been victims of one of the most devastating mental disorders, i.e., suicide. Their family members and caregivers deserve equal recognition for the immense care they provide to them. My sincere thanks go to all the authors for their invaluable contributions and ideas that could be of great help in understanding the biological basis of suicide. CRC Press, Barbara Norwitz, and Joselyn Banks-Kyle deserve much credit for their support in publishing this book. Dr. Hui Zhang, Michele Agustin, and Mary Kingzette have been of immense help in editing several of the book chapters. With deep gratitude, I would like to thank Professor Ghanshayam N. Pandey, who introduced me to this important area of research. Finally, I would like to thank my wife, Aparna, and son, Tushar, whose constant enthusiasm and support have been an invaluable source to sustain my research career.

Editor

Yogesh Dwivedi received his PhD in biochemistry in 1992 from the Central Drug Research Institute, Lucknow, India, and did his postdoctoral research training in neuroscience at the Illinois State Psychiatric Institute, Chicago, Illinois, United States. Currently, he is a tenured professor of psychiatry in the Department of Psychiatry, University of Illinois at Chicago. He is also a tenured professor in the Department of Pharmacology, University of Illinois at Chicago, and faculty at the Graduate College, University of Illinois at Chicago.

Dr. Dwivedi has pursued a productive and useful line of investigation based on innovative hypotheses addressing a topic of great importance and concern in the area of public health, i.e., the neurobiology of suicide and depression. Dr. Dwivedi focuses on investigating the molecular and cellular nature of events in the brain that may lead to depressive and suicidal behavior. His research has shown that abnormalities in cellular signaling may serve as critical vulnerability factors predisposing a person to such behavior and that abnormalities in trophic/growth factors and cell survival pathways may lead to abnormalities in brain structure and, thus, to depressive/suicidal behavior. Furthermore, he has demonstrated that abnormal functioning of the hypothalamic–pituitary–adrenal axis and stress, which are closely associated with suicidal and depressive behavior, dysregulate the signal transduction molecules, causing such abnormalities. By using various investigative approaches, namely, human postmortem brain studies of suicide victims, peripheral blood cell studies of suicidal/depressed patients, and preclinical animal models, Dr. Dwivedi has successfully integrated basic and clinical neuroscience. More recently, by using nocoding RNA, epigenetic, and gene knockout approaches, his group is attempting to unravel the molecular mechanisms that may be responsible for gene regulation in depressed and suicidal brains.

Dr. Dwivedi has published more than 90 peer-reviewed articles in high-impact journals. He is the author of many book chapters and reviews. He has organized and chaired several national and international symposia on mood disorders and suicide. He has also received many prestigious awards in recognition of his achievements, including Young Investigator Awards from the American Foundation for Suicide Prevention, the National Alliance for Schizophrenia and Depression, and the International Society of Neurochemistry; a Memorial Fellowship Award from the American College of Neuropsychopharmacology; and a Rafaelson Fellowship Award from the International College of Neuropsychopharmacology. He is well funded by the National Institutes of Health, the American Foundation for Suicide Prevention, the National Alliance for Research on Schizophrenia and Depression, the Stanley Foundation, and the Marshall-Reynolds Foundation. He is a reviewer for several grant committees, including the National Institute of Mental Health, the Canadian Institute of Health Research, the American Foundation for Suicide Prevention, the Ontario Mental Health Foundation, and the Department of Biotechnology, Government of India. He is also a member of many prestigious societies, including the American College of Neuropsychopharmacology, the Society of Biological Psychiatry, the International Association of Suicide Prevention, the International Association of Suicide Research, the International College of Neuropsychopharmacology, the Society of Neuroscience, the International Society of Bipolar Disorder, the International Neuropsychiatric Association, and the International Society of Neurochemistry. Dr. Dwivedi is also a member of the Society of Biological Psychiatry Program Committee, the Advisory Board of the American Foundation for Suicide Prevention, and the International Association of Suicide Prevention Task Force on Genetics.

Contributors

Olusola A. Ajilore
Department of Psychiatry
University of Illinois at Chicago
Chicago, Illinois

Hymie Anisman
Institute of Neuroscience
Carleton University
Ottawa, Ontario, Canada

Victoria Arango
Department of Psychiatry
New York State Psychiatric Institute
Columbia University
New York, New York

Helene Bach
Department of Psychiatry
New York State Psychiatric Institute
Columbia University
New York, New York

Michelle J. Chandley
Department of Pharmacology
Quillen College of Medicine
East Tennessee State University
Johnson City, Tennessee

William Coryell
Department of Psychiatry
Carver College of Medicine
University of Iowa
Iowa City, Iowa

Dianne Currier
Department of Psychiatry
College of Physicians and Surgeons
Columbia University
New York, New York

Vincenzo de Luca
Department of Psychiatry
University of Toronto
and
Neurogenetics Section
and
Neuroscience Research Department
Centre for Addiction and Mental Health
Toronto, Ontario, Canada

Yogesh Dwivedi
Department of Psychiatry
Psychiatric Institute
College of Medicine
University of Illinois at Chicago
Chicago, Illinois

Jan Fawcett
Department of Psychiatry
School of Medicine
University of New Mexico
Albuquerque, New Mexico

Laura M. Fiori
Douglas Mental Health University Institute
McGill University
Verdun, Quebec, Canada

Ina Giegling
Department of Psychiatry
University of Munich
Munich, Germany

James L. Kennedy
Department of Psychiatry
University of Toronto
and
Neurogenetics Section
and
Neuroscience Research Department
Centre for Addiction and Mental Health
Toronto, Ontario, Canada

Anand Kumar
Department of Psychiatry
University of Illinois at Chicago
Chicago, Illinois

Benoit Labonté
Douglas Mental Health University Institute
McGill University
Verdun, Quebec, Canada

J. John Mann
New York State Psychiatric Institute
and
Department of Psychiatry
College of Physicians and Surgeons
Columbia University
New York, New York

Zul Merali
Institute of Mental Health Research
University of Ottawa
Ottawa, Ontario, Canada

Jeffrey H. Meyer
Department of Psychiatry
University of Toronto
and
Research Imaging Center
Centre for Addiction and Mental Health
Toronto, Ontario, Canada

Sonali Nanayakkara
Department of Psychiatry
Institute for Juvenile Research
University of Illinois at Chicago
Chicago, Illinois

Olaoluwa Okusaga
Department of Psychiatry
School of Medicine
University of Maryland
Baltimore, Maryland

and

St. Elizabeths Hospital Residency Training Program
Washington, District of Columbia

Gregory A. Ordway
Department of Pharmacology
Quillen College of Medicine
East Tennessee State University
Johnson City, Tennessee

Ghanshyam N. Pandey
Department of Psychiatry
University of Illinois at Chicago
Chicago, Illinois

Mani Pavuluri
Department of Psychiatry
Institute for Juvenile Research
University of Illinois at Chicago
Chicago, Illinois

Teodor T. Postolache
Department of Psychiatry
School of Medicine
University of Maryland
and
Suicide Prevention Service
Baltimore VA Medical Center
Baltimore, Maryland

Michael O. Poulter
Faculty of Medicine
Department of Physiology and Pharmacology
Robarts Research Institute
University of Western Ontario
London, Ontario, Canada

Kiran Pullagurla
Department of Psychiatry
Institute for Juvenile Research
University of Illinois at Chicago
Chicago, Illinois

Alec Roy
Psychiatry Service
New Jersey Health Care System
East Orange, New Jersey

Dan Rujescu
Department of Psychiatry
University of Munich
Munich, Germany

Isaac Sakinofsky
Department of Psychiatry
and
Dalla Lana School of Public Health
University of Toronto
and
High Risk Consultation Clinic
Centre for Addiction and Mental Health
Toronto, Ontario, Canada

John Strauss
Department of Psychiatry
University of Toronto
and
Medical Informatics
Centre for Addiction and Mental Health
Toronto, Ontario, Canada

Ryan P. Tong
Neurogenetics Section
and
Neuroscience Research Department
Centre for Addiction and Mental Health
Toronto, Ontario, Canada

Gustavo Turecki
Douglas Mental Health University Institute
McGill University
Verdun, Quebec, Canada

Kees van Heeringen
Unit of Suicide Research
Department of Psychiatry and Medical Psychology
University Hospital
University of Gent
Gent, Belgium

K. Yaragudri Vinod
Division of Analytical Psychopharmacology
Nathan Kline Institute for Psychiatric Research
Orangeburg, New York

and

Department of Child and Adolescent Psychiatry
Langone Medical Center
New York University
New York, New York

Clement C. Zai
Department of Psychiatry
University of Toronto
and
Neurogenetics Section
and
Neuroscience Research Department
Centre for Addiction and Mental Health
Toronto, Ontario, Canada

Gil Zalsman
Sackler Faculty of Medicine
Child and Adolescent Division
Geha Mental Health Center and Psychiatry Department
Tel Aviv University
Tel Aviv, Israel

and

Department of Psychiatry
Columbia University
New York, New York

1 Diagnosis, Traits, States, and Comorbidity in Suicide

Jan Fawcett

CONTENTS

Suicide is one of the most tragic outcomes in clinical practice. It is not predictable in the individual and results from a complex series of factors that may differ across individuals, yet 30%–70% of suicides occur in patients who are receiving some treatment [1–3]. Learning more about the interacting biological, clinical, and situational factors that lead to suicide can help the clinician to recognize risk factors and initiate clinical interventions to reduce suicide risk.

1.1 DIAGNOSIS

Suicide occurs in the presence of any psychiatric diagnosis, but studies repeatedly show that suicide is most common in the mood disorders, major depressive disorder and bipolar disorder [4]. Many other disorders have elevated rates of suicide, such as mixed drug abuse, alcohol and opioid abuse, eating disorders, schizophrenia and personality disorders, and even acute stress disorders [4,5]. It may be that suicide is related to the occurrence of mood depression, severe anxiety, and increased trait impulsivity that occur in the course of the entire range of psychiatric disorders [6]. The increased risk of suicide has particularly been emphasized in bipolar disorder [7]. More recently, the role of early child abuse in suicide has been recognized as a factor in elevated suicide risk [8]. Child abuse has been considered to be associated with both early onset mood disorders and the trait of impulsivity [8].

1.2 TRAITS

The presence of behavioral traits that mediate the risk of suicidal behavior has been a recent theme in studies of suicidal behavior. There is strong evidence for familial-genetic transmission of suicidal behavior and increasing evidence for certain behavioral traits mediating this risk [8–12].

The trait of angry impulsivity has been repeatedly identified as a risk factor for suicidal behavior [6,8–12]. The trait of impulsivity is distributed across the range of diagnoses but is known to be highly associated with bipolar disorder, substance abuse, and cluster B personality disorders as well as a history of early child abuse [6,9]. One study reported elevated scores for neuroticism and hostility as well as impulsivity in prisoners with a positive family history of suicide [13].

Suicide risk factors can be grouped into chronic high-risk factors and immediate or acute high-risk factors. A prior suicide attempt is the most commonly found risk factor for suicide by many studies [4,14], while the rate of actual suicide in subjects with prior suicide attempts is about 5%–10%. A prior suicide attempt could be an indicator that the risk factor of impulsivity is present. Past suicidal behavior is sometimes not helpful in a suicide risk assessment, since Isometsa and Lonnqvist [15] have reported from a large group of studied suicides that about 62% of male suicides and 38% of female suicides died on their first suicide attempt.

While suicidal ideation and particularly a suicide plan is considered a strong predictor of suicide risk, but as it is often erroneously assumed, the opposite, a denial of suicidal ideation or plan, is a very poor predictor by itself that the risk of suicide is low. This is aptly demonstrated by Isometsa et al.'s [16] report on 100 cases of suicide that occurred on the day that the patients had seen their psychiatrist. It was reported that only 22% of the patients who committed suicide endorsed suicidal thoughts at their last visit. In 76 cases of inpatient suicide reported by Busch et al. [17], it was reported that 76% of inpatients had a nursing note quoting the patient as denying suicidal ideation as the last noted communication from the patient prior to their suicide. It is quite humane to assume that if a positive indicates risk, a negative indicates low risk. Such does not appear to be the case for suicide, since a patient at high risk for suicide may be subject to suicidal impulses subsequent to denying suicidal intent or may be lying since they have decided to end their life and do not want to be interfered with. A denial of suicidal intent or plan should be taken as a neutral factor to be considered along with the patient's past history, life situation, and current clinical state.

Angry impulsivity has been repeatedly found to be chronic high suicide risk factor, which in the presence of certain situations [6,8,18–20], mood states, or anxiety can become a precipitant of suicidal behavior. Other chronic risk factors for suicide include being of male gender, living alone, owning a hand gun, or a history of significant chronic pain [21–24].

Therefore, in assessing a patient for suicide risk, it is not uncommon that the patient could be considered a high chronic risk for suicide but not at acute risk for suicide at the time of the assessment. The presence of chronic risk factors may color the clinician's assessment as to whether a patient is currently in a high-risk group. The term high-risk group indicates the fact that suicide is not individually predictable, but certain risk factors may lead the clinician to consider the patient in a high-risk group and manage the patient as such. While chronic risk factors for suicide may be useful in an actuarial prediction of a group of individuals more likely to manifest suicidal behaviors at some future point, the patient's clinical state and current life situation at the time of assessment will (along with a consideration of chronic risk factors) be the clinician's best source of information in assessing whether a patient should be managed like a patient in an acute high-risk group for suicide.

The patient's clinical state at the time of clinical assessment in the context of the clinical state over the recent past as well as knowledge of recent losses or stresses occurring in their life situation is a crucial factor in assigning risk group status and planning treatment.

1.3 CLINICAL STATE

A careful assessment of the patient's current clinical state and life situation is the most important source of information concerning whether or not the patient should be managed as patient is at high acute risk for suicide. The occurrence of recent clinical worsening of symptoms of depression or anxiety is of great importance and merits a full suicide assessment. The presence of a mixed state (or mixed features with depressive symptoms co-occurring with manic or even hypomanic features) is often associated with increased activity, impulsiveness, and severe anxiety/agitation resulting in increased risk of suicidal behavior.

Critical occurrences associated with high risk include being recently admitted or discharged to/from a psychiatric inpatient facility and the risk is elevated by the recentness of discharge up to a year following the event [15,25–27]. The recent occurrence of a loss of a loved one or divorce, a major financial setback, job loss, serious medical diagnosis (e.g., a recent diagnosis of cancer), or legal problem can precipitate a suicidal state, particularly in the presence of depression or in a highly impulsive individual [5,28–30]. It is important to assess how a patient is coping with the "bludgeonings of chance" when such events have occurred.

An assessment of the current clinical state should take into account how they are coping with life stress as well as whether their negative traits (such as negative affect) have recently increased. Has the patient increased the use of alcohol, which can further increase angry impulsivity, for instance [25,31]? Is the patient manifesting increased comorbid anxiety, agitation, or substance abuse [25,31]?

1.4 COMORBID ANXIETY: DYSPHORIC AROUSAL

It is not yet determined whether "comorbid" anxiety in mood disorders is really a comorbid anxiety disorder or is an intrinsic aspect of the symptomatology found in mood disorders. Our current diagnostic system separates symptoms of mood disorders such as major depression and bipolar depression from anxiety disorder. One study using the schedule for affective disorders and schizophrenia, current (SADS-C), a scale that rates both the presence and severity of each symptom in a major depression population, found that anxiety was present to a moderate degree in 62% of patients and panic attacks occurred in 29% [32]. A study by Clayton et al. [33] reported anxiety levels in over 300 patients with primary (no prior diagnosis) depression and showed a high frequency of anxiety symptoms and a wide severity level. The frequency of both anxiety occurring after the onset of a depression and the frequency of depression occurring in a patient already diagnosed as having a generalized anxiety disorder (GAD) (Diagnostic and Statistical Manual-IV [DSM-IV] requires 6 months of criteria symptoms for GAD and 2 weeks of criteria symptoms for major depressive episode) raises the question whether anxiety should not be considered a criterion symptom for major depression and bipolar depression with or without a mixed state.

In 1990, Fawcett et al. [25] reported a prospective study of 13 suicides over the first year and 34 suicides over 10 years in a largely hospitalized sample of patients with major affective disorders ($N = 954$) followed for 10 years. Prior suicide attempts (recent and past), severity of suicidal ideation, and severity of hopelessness were not significantly greater in 13 suicides over the first year of follow-up compared to the majority of non-suicides, but were significantly associated with suicide over a 2- to 10-year follow-up period [25]. However, the levels of psychic anxiety and panic attacks were significantly more severe or frequent at baseline in the 13 suicide patients compared with the majority of non-suicides. A subsequent study by Hall et al. [34] found elevated levels of psychic anxiety in 90% of a sample of 100 patients hospitalized for suicide attempts and then interviewed, as measured by their SADS-C psychic anxiety scores reflecting the month prior to their suicide attempt.

In 2003, Busch et al. [17] reported a review of 76 cases of inpatient suicide. Charts of these cases were stripped of identity and reviewed over the period of 1 week prior to their suicide. Seventy-nine percent of these cases showed the presence of severe anxiety and/or agitation for at least 3–7 days before their suicide (while the last communication noted by nursing staff prior to their suicide conveyed a denial of suicide thoughts of intent in 76% of cases) [17]. In 2007, Simon et al. [35] published a review of 32,000 cases of bipolar disorder from the combination of two managed care databases showing that the presence of comorbid diagnoses of GAD was associated with an elevated rate of suicide (odds ratio [OR] = 1.8) and suicide attempts (OR = 1.4). It was also found that while the suicide attempt rate was elevated with the presence of comorbid substance abuse, there was no increased risk of suicide found in this sample [35].

In 2008, a study by Stordal et al. [36] showed that 60,995 subjects in Norway, rating themselves monthly (except for July) on the Hospital Anxiety and Depression scale from 1995 to 1997, who committed suicide ($N = 10{,}670$ males and 3933 females) showed a simultaneous peak in the severity of depression and anxiety ratings during the month of their suicide ($r = 0.72$, $p = .01$), which occurred in the spring and early fall ($p = .01$). Finally, a study by Pfeiffer et al. [37] of over 887,000 veterans treated for depression showed that suicide was significantly associated with diagnoses of comorbid GAD, anxiety disorder not otherwise specified (NOS), and panic disorder, but not with post traumatic stress disorder (PTSD) or other anxiety disorders (OR = 1.8). In addition, a significant increase in suicide associated with taking antianxiety medication (OR = 1.8) was reported along with a further elevation in suicides in patients taking high-dose antianxiety medications (OR = 2.2). This would also suggest that anxiety severity, rather than anxiety disorder presence, might be associated with suicide as was found by the Fawcett et al. [32] study reviewed earlier.

While a study of suicidality in bipolar patients as opposed to suicide has not found the incidence of prior suicide attempts to be associated with the presence of comorbid GAD, it is argued that suicide attempts may not be totally equivalent to suicide and that these studies did not measure anxiety severity at the time of a suicide attempt, only a past history of suicide attempts, and therefore do not address the issues presented earlier [37].

A recent cross-national study by Nock et al. [6] of suicide attempts in a sample of over 100,000 subjects found suicide associated with the co-occurrence of anxiety. Nock postulated in this report that while depression was associated with suicidal thoughts, disorders characterized by anxiety and poor impulse control lead people to carry out a suicidal behavior [6]. This was reported by Brown et al. [38] in personality disorders as well as in mood disorders and found to be associated with low cerebrospinal fluid (CSF) 5-hydroxy indolacetic acid (5-HIAA) levels, suggesting a decreased turnover of serotonin in brain associated with impulsivity. Indeed, Swann et al. [20] have reported increased impulsivity associated with a history of more frequent suicide attempts and, more recently, Taylor et al. [39] reported that increased anxiety leads to increased impulsivity in bipolar patients.

In 1965, Bunney and Fawcett [40] reported three cases of suicide, which were found to have elevated excretion of 17-hydroxycorticosteroids (a urinary metabolite of cortisol in plasma) in 24 h urine collections made the days leading up to their suicide (they were patients on a research unit at National Institute of Mental Health [NIMH]). This was followed by a second paper of several more such cases [41]. Subsequently, the dexamethasone suppression test, which measured an overactive hypophyseal pituitary adrenal system (HPA), was found to show HPA overactivity in patients who committed suicide [42–46]. None of these reports, except for a schizophrenic patient who committed suicide with elevated 17-hydroxycorticoid excretion reported by Sachar et al. [47], manifest anxiety or dysphoric hyperarousal. Sachar reported that his patient was not depressed but exhibiting "ego disintegration." The question raised is whether patients who are on the verge of suicide are manifesting state anxiety or dysphoric hyperarousal as a clinical symptom, which correlates with HPA axis overactivity.

It may turn out that increased anxiety symptoms in patients with major depression or bipolar disorder may not be secondary to a comorbid anxiety disorder, but a feature of the primary mood disorder. It is hoped that an anxiety severity dimension will be added to all mood disorder diagnoses in DSM-V. This would draw attention to the significance of the role of anxiety in the outcome of mood disorders and focus more attention on the more successful treatment of severe anxiety symptoms in mood disorder patients.

It has been reported that in the Sequenced Treatment Alternatives to Relieve Depression (STAR*D) study severe anxiety predicted poor response to antidepressant treatment [48,49]. Recent analyses by Coryell and coworkers [50,51] have demonstrated that elevated anxiety severity at base line predicts significantly more time spent in depression in a sample of major mood disorder patients followed over 16–20 years.

The presence of severe anxiety or dysphoric hyperarousal is one of the clinical state variables that should be assessed and addressed in the management of suicide risk. A state of increased impulsiveness related to increased anxiety/arousal in response to negative events should also be considered and addressed in the management of suicide risk. While this factor is certainly not present in all patients prior to suicide, the previously cited inpatient suicide study found it present in 76% of cases studied [17]. The authors' experience from reviewing ~100 additional outpatient suicide cases suggests that severe anxiety/dysphoric hyperarousal was present in more than half the cases. The presence of severe anxiety/dysphoric hyperarousal may be one of the most common clinical indicators of imminent suicide risk available to the clinician. A subgroup of patients who carefully plan their suicide over a period of days or weeks may show no signs of severe anxiety/hyperarousal and in fact appear calm and convey no signs of ambivalence, having made their decision. These patients will not indicate their plan to a clinician or significant other and are very difficult, if not impossible, to intervene with prior to their suicide.

1.5 COMORBID SUBSTANCE ABUSE

Another comorbid clinical factor that also requires assessment is the presence of increased substance abuse. It is well known that alcohol and other substance abuse are associated with suicidal behavior [31]. In the collaborative study, it was found that the recent onset of moderate alcohol abuse was seen in the weeks/days prior to suicide. It appeared that these patients were using alcohol as self-treatment of untreated severe anxiety and insomnia in these cases. The behavioral disinhibiting effects of alcohol and other substances are well known. Such substance abuse tends to increase impulsiveness and impair judgment, thereby increasing the risk of suicidal behavior.

The review of Harris and Baraclough [4] found elevated rates of suicide in patients suffering with mixed drug abuse, opioid abuse, and alcohol abuse. Increases in the severity of abuse may portend increased risk of suicidal behavior.

1.6 SHOULD THERE BE A DIAGNOSIS OF SUICIDAL BEHAVIOR?

Since suicide/suicidal behavior occurs across the entire spectrum of psychiatric diagnosis and seems to be transmitted independently in families, with a significant proportion of genetic risk, and can be predicted statistically by prior suicidal behavior, the question arises whether suicidal behavior should be a separate diagnostic category.

A diagnosis of suicidal behavior disorder would increase the clinical focus on interventions to prevent suicidal behavior in vulnerable patients. Currently, evidence suggests that patients with a history of suicidal behavior receive inadequate pharmacologic treatment, which is similar to treatment of other depressed patients [52].

There is a significant amount of scientific data supporting a suicidal behavior diagnosis including genetic factors in transmission found in twin studies, familial transmission independent of diagnosis, biological risk markers, and prediction of subsequent suicidal behavior from past suicidal behavior [53].

On the other hand, it could be argued that this diagnosis may have a stigmatizing effect and undermine the doctor–patient relationship. Also, since a significant proportion of suicides occur on the first attempt as noted earlier [14], such a diagnosis may add a modest amount to the clinical assessment of current suicide risk. Would the effect of a separate diagnosis of suicidal behavior disorder have the net effect of improving clinical care of suicidal patients? Such a category may help with research efforts to better detect and treat acute suicide risk. What would be the appropriate criteria for this disorder: prior suicide attempts, suicide plans or rehearsals, or immediate/chronic suicide ideation?

From the clinical practice point of view, the greatest lack is the relative dearth of clinical or biological markers to detect an individual at acute risk of suicide. Given the difficulty in predicting behavior in an individual, this may always be a limited area of knowledge. However, every piece of information relevant to detecting acute high-risk states for suicide is valuable in our efforts to prevent suicide in our patients.

1.7 SUMMARY AND CONCLUSIONS

It appears that suicide occurs across diagnoses and stems from mood depression, hopelessness, severe anxiety, and increased impulsivity, often but not always related to histories of early abuse and a past history of suicidal behavior, as well as situational factors such as clinical worsening of symptoms, not infrequently in the context of a real or anticipated major loss. Chronic risk factors such as early childhood abuse, impulsivity, a past history of substance abuse, living alone, and a history of past or recent suicide attempts are important to elicit and take into consideration. Acute risk factors such as severe anxiety, insomnia, evidence of increased impulsivity, clinical worsening of symptoms, and an admission of suicide plan or preparation for a suicide attempt and recognition of situational factors such as a recent or anticipated major loss may allow for an intervention prior to a lethal attempt. Biological trait factors such as impulsivity often relating to a history of early childhood abuse, substance abuse and changes in HPA, adrenergic response, and serotonin systems seem to occur across diagnoses. In the presence of mood depression, which is seen across diagnoses and always present in mood disorders, suicidal ideas may occur that are translated into suicide attempts in the presence of the aforementioned factors, often triggered by adverse events or symptom worsening.

Further development of understanding of these pathways to suicide will both increase our clinical knowledge of when to intervene to prevent suicide and enhance our capacity to intervene in ways to prevent the development of acute suicidal risk states in vulnerable individuals.

REFERENCES

1. Luoma JB, Martin CE, Oearson JL. 2002. Contact with mental health and primary care providers before suicide: A review of the evidence. *Am J Psychiatry* 159:908–916.
2. Baraclough B, Bunch J, Nelson B, Sainsbury P. 1974. A hundred cases of suicide: Clinical aspects. *Br J Psychiatry* 125:355–373.
3. Robins E. 1981. *The Final Months*. Oxford University Press, Oxford, U.K., p. 47.
4. Harris EC, Baraclough B. 1997. Suicide as an outcome for mental disorders. A meta-analysis. *Br J Psychiatry* 170:205–228.

5. Gradus JL, Qin P, Lincoln AK, Miller M, Lawler E, Sørensen HT, Lash TL. 2010. Acute stress reaction and completed suicide. *Int J Epidemiol* 39:1478–1484.
6. Nock MK, Hwang I, Sampson N, Kessler RC, Angermeyer M, Beautrais A, Borges G, Bromet E, Bruffaerts R, de Girolamo G, de Graaf R, Florescu S, Gureje O, Haro JM, Hu C, Huang Y, Karam EG, Kawakami N, Kovess V, Levinson D, Posada-Villa J, Sagar R, Tomov T, Viana MC, Williams DR. 2009. Cross-national analysis of the associations among mental disorders and suicidal behavior: Findings from the WHO World Mental Health Surveys. *PLoS Med* 6:1–17.
7. Baldessarini RJ, Pompili M, Tondo L. 2006. Suicide in bipolar disorder: Risks and management. *CNS Spectr* 11:466–471.
8. Brent D. 2010. What family studies teach us about suicidal behavior: Implications for research, treatment, prevention. *Eur Psychiatry* 25:260–263.
9. Baldessarini RJ, Hennen J. 2004. Genetics of suicide: An overview. *Harv Rev Psychiatry* 12:1–13.
10. Roy A. 1983. Family history of suicide. *Arch Gen Psychiatry* 40:971–974.
11. Mann JJ, Bortinger J, Oquendo MA, Currier D, Li S, Brent DA. 2005. Family history of suicidal behavior and mood disorders in probands with mood disorders. *Am J Psychiatry* 162:1672–1679.
12. Roy A. 2006. Family history of suicide and impulsivity. *Arch Suicide Res* 10940:347–352.
13. Sarchiapone M, Carli V, Janiri L, Marchetti M, Cesaro C, Roy A. 2009. Family history of suicide and personality. *Arch Suicide Res* 13:178–184.
14. Coryell W, Young EA. 2005. Clinical predictors of suicide in primary major depressive disorder. *J Clin Psychiatry* 66:412–417.
15. Isometsa ET, Lonnqvist JK. 1998. Suicide attempts preceding completed suicide. *Br J Psychiatry* 173:531–535.
16. Isometsa ET, Heikkinen ME, Marttunen MJ, Henriksson MM, Aro HM, Lönnqvist JK. 1995. The last appointment before suicide: Is suicide intent communicated? *Am J Psychiatry* 152:919–922.
17. Busch KA, Fawcett J, Jacobs D. 2003. Clinical correlates of inpatient suicide. *J Clin Psychiatry* 64:14–19.
18. Zhang J, Wieczorek W, Conwell Y, Tu XM, Wu BY, Xiao S, Jia C. 2010. Characteristics of young rural Chinese suicides: A psychological autopsy study. *Psychol Med* 40:581–589.
19. Dumais A, Lesage AD, Alda M, Rouleau G, Dumont M, Chawky N, Roy M, Mann JJ, Benkelfat C, Turecki G. 2005. Risk factors for suicide in major depression: A case control study of impulsive and aggressive behaviors in men. *Am J Psychiatry* 162:116–124.
20. Swann AC, Dougherty DM, Pazzaglia PJ, Pham M, Steinberg JL, Moeller FG. 2005. Increased impulsivity associated with severity of suicide attempt history in patients with bipolar disorder. *Am J Psychiatry* 162:1680–1687.
21. Ligen MA, Zivin K, Austin KL, Bohnert AS, Czyz EK, Valenstein M, Kilbourne AM. 2010. Severe pain predicts greater likelihood of subsequent suicide. *Suicide Life Threat Behav* 40:597–608.
22. Lofman S, Rasanen P, Hakko H. 2011. Suicide among persons with back pain: A population-based study of 2310 suicide victims in Northern Finland. *Spine* 36:541–548.
23. Scott KM, Hwang I, Chiu WT, Kessler RC, Sampson NA, Angermeyer M, Beautrais A, Borges G, Bruffaerts R, de Graaf R, Florescu S, Fukao A, Haro JM, Hu C, Kovess V, Levinson D, Posada-Villa J, Scocco P, Nock MK. 2010. Chronic physical conditions and their association with first onset of suicidal behavior in the world mental health surveys. *Psychosom Med* 72:712–719.
24. Kessler RC, Borges G, Walters EE. 1999. Prevalence of risk factors for lifetime suicide attempts in the National Comorbidity Survey. *Arch Gen Psychiatry* 56:617–626.

25. Fawcett J, Scheftner WA, Fogg L, Clark DC, Young MA, Hedeker D, Gibbons R. 1990. Time-related predictors of suicide in major affective disorder. *Am J Psychiatry* 147:1189–1194.
26. Kan CK, Ho TP, Dong JY, Dunn EL. 2007. Risk factors for suicide in the immediate post-discharge period. *Soc Psychiatry Psychiatr Epidemiol* 42:208–214.
27. Goldacre M, Seagroat V, Hawton K. 1993. Suicide after discharge from psychiatric inpatient care. *Lancet* 342:283–286.
28. Peteet JR, Maytal G, Rokni K. 2010. Inimaginable loss: Contingent suicidal ideation in family members of oncology patients. *Psychosomatics* 51:166–170.
29. Ahn E, Shin DW, Cho SI, Park S, Won YJ, Yun YH. 2010. Suicide rates and risk factors among Korean cancer patients, 1993–2005. *Cancer Epidemiol Biomarkers Prev* 19:2097–2105.
30. Fang F, Keating NL, Mucci LA, Adami HO, Stampfer MJ, Valdimarsdóttir U, Fall K. 2010. Immediate risk of suicide and cardiovascular death after a prostate cancer diagnosis: Cohort study in the United States. *J Natl Cancer Inst* 102:307–314.
31. Flensborg-Madsen T, Knop J, Mortensen EL, Becker U, Sher L, Grønbaek M. 2009. Alcohol use disorders increase the risk of completed suicide-irrespective of other psychiatric disorders. A longitudinal cohort study. *Psychiatry Res* 167:123–130.
32. Fawcett J, Kravitz H. 1983. Anxiety syndromes and their relationship to depressive illness. *J Clin Psychiatry* 44:8–11.
33. Clayton P, Grove WM, Coryell W. 1991. Follow-up and family study of anxious depression. *Am J Psychiatry* 148:1512–1517.
34. Hall RC, Platt DE, Hall RC. 1999. Suicide risk assessment: A review of risk factors for suicide in 100 outpatients who make severe suicide attempts. *Psychosomatics* 40:18–27.
35. Simon GE, Hunkeler E, Fireman B, Lee JY, Savarino J. 2007. Risk of suicide attempt and suicide death in patients treated for bipolar disorder. *Bipolar Disord* 9:526–530.
36. Stordal E, Morken G, Mykletun A, Neckelmann D, Dahl AA. 2008. Monthly variation in rates of comorbid depression and anxiety in the general population at 63–65 degrees North: The HUNT study. *J Affect Disord* 106:273–278.
37. Pfeiffer PN, Ganoczy D, Ilgen M, Zivin K, Valenstein M. 2009. Comorbid anxiety as a suicide risk factor among depressed veterans. *Depress Anxiety* 26:752–757.
38. Brown GL, Ebert MH, Goyer PF, Jimerson DC, Klein WJ, Bunney WE, Goodwin FK. 1982. Aggression, suicide, and serotonin: Relationships to CSF amine metabolites. *Am J Psychiatry* 139:741–748.
39. Taylor CT, Hirshfeld-Becker DR, Ostacher MJ, Chow CW, LeBeau RT, Pollack MH, Nierenberg AA, Simon NM. 2008. Anxiety is associated with impulsivity in bipolar disorder. *J Anxiety Disord* 22:868–876.
40. Bunney WE Jr, Fawcett JA. 1965. Possibility of a biochemical test for suicidal potential: An analysis of endocrine findings prior to three suicides. *Arch Gen Psychiatry* 13:232–239.
41. Bunney WE, Fawcett JA, Davis JM, Gifford S. 1969. Further evaluation of urinary 17-hydroxycortocosteroids in suicidal patients. *Arch Gen Psychiatry* 21:138–150.
42. Coryell W, Schlesser MA. 1981. Suicide and the dexamethasone suppression test in unipolar depression. *Am J Psychiatry* 138:1120–1121.
43. Targum SD, Rosen L, Capodanno AE. 1983. The dexamethasone suppression test in suicidal patients with unipolar depression. *Am J Psychiatry* 140:877–879.
44. Yerevanian Bl, Olafsdottir H, Milanese E, Russotto J, Mallon P, Baciewicz G, Sagi E. 1983. Normalization of the dexamethasone suppression test at discharge from hospital. Its prognostic value. *J Affect Disord* 5:191–197.
45. Jokinen J, Nordström AL, Nordström P. 2009. CSF 5-HIAA and DST non-suppression-orthogonal biologic risk factors for suicide in male mood disorder inpatients. *Psychiatry Res* 165:96–102.

46. Coryell W, Schlesser M. 2007. Combined biological tests for suicide prediction. *Psychiatry Res* 150:187–191.
47. Sachar EJ, Kanter SS, Buie D, Engle R, Mehlman R. 1970. Psychoendocrinology of ego disintegration. *Am J Psychiatry* 126:1067–1068.
48. Fava M, Alpert JE, Carmin CN, Wisniewski SR, Trivedi MH, Biggs MM, Shores-Wilson K, Morgan D, Schwartz T, Balasubramani GK, Rush AJ. 2004. Clinical correlates and symptom patterns or anxious depression among patients with major depressive disorder in STAR*D. *Psychol Med* 34:1299–1308.
49. Fava M, Rush AJ, Alpert JE, Carmin CN, Balasubramani GK, Wisniewski SR, Trivedi MH, Biggs MM, Shores-Wilson K. 2006. What clinical and symptom features and comorbid disorders characterize outpatients with anxious major depressive disorder: A replication and extension. *Can J Psychiatry* 51:823–835.
50. Coryell W, Solomon DA, Fiedorowicz JG, Endicott J, Schettler PJ, Judd LL. 2009. Anxiety and outcome in bipolar disorder. *Am J Psychiatry* 166:1238–1243.
51. Coryell W. Fiedorowicz JG, Solomon D, Leon AC, Rice JP, Keller MB. 2012. Effects of anxiety on the long-term course of depressive disorders. *Br J Psychiatry* 200:210–215.
52. Oquendo MA, Kamali M, Ellis SP, Grunebaum MF, Malone KM, Brodsky BS, Sackeim HA, Mann JJ. 2002. Adequacy of antidepressant treatment after discharge and the occurrence of suicidal acts in major depression: A prospective study. *Am J Psychiatry* 159:1746–1751.
53. Oquendo MA, Baca-Garcia E, Mann JJ, Giner J. 2008. Issues for DSM-V: Suicidal behavior as a separate diagnosis on a separate axis. *Am J Psychiatry* 165:1383–1384.

2 Neuroanatomy of Serotonergic Abnormalities in Suicide

Helene Bach and Victoria Arango

CONTENTS

2.1 INTRODUCTION

The serotonergic system has been shown to be affected in a number of psychiatric illnesses in the last 50 years. A large body of work has focused on serotonin (5-HT) deficits in major depressive disorder (MDD) and suicide. A great deal has been learned about the anatomy, development, and functional organization of the 5-HT system and the alterations in this system that are present within the suicide brain. Historically, evidence for the involvement of 5-HT in suicide stemmed from findings of low cerebral spinal fluid (CSF) 5-hydroxyindoleacetic acid (5-HIAA) in depressed suicide attempters and in the brain stems of completed suicides (Åsberg 1976; Åsberg et al. 1976; Banki et al. 1984; Carlsson et al. 1980; Mann and Malone 1997; Placidi et al. 2001; Roy et al. 1986; Träskman et al. 1981). Suicide attempters also exhibit a blunted release of prolactin in response to administration of fenfluramine, a measure of 5-HT activity (Dulchin et al. 2001; Duval et al. 2001; Malone et al. 1996; Mann et al. 1995; Pandey 1997; Weiss and Coccaro 1997). These studies provided evidence for deficits in serotonergic neurotransmission in the brain stem or serotonergic targets in the forebrain of suicidal individuals. 5-HT is produced by neurons embedded in the midline raphe nuclei in the brain stem with widespread targets that appear to be topographically organized. In this chapter, we will discuss data that shed light into the contribution of these serotonergic neurons to brain diseases such as MDD and suicide.

2.2 ANATOMY OF THE RAPHE NUCLEI

The raphe nuclei in the brain stem are a collection of cytoarchitectonically ill-defined aggregates of neurons that flank the midline and contain all the 5-HT-synthesizing neurons in the brain. The word raphe is derived from the Greek word "suture" and indicates a seam between two halves. In general, the rostral raphe nuclear group, which is contained in the mesencephalon and rostral pons, projects to the forebrain, while the caudal group, extending from the caudal pons to the caudal medulla, has projections to the spinal cord and medulla (Hornung 2003; Törk and Hornung 1990).

The rostral group of raphe nuclei consists of the *caudal linear nucleus* (CLi), *dorsal* (DRN), and *median* (MRN) raphe nuclei. The CLi (Halliday and Törk 1986) is a nucleus of the ventral mesencephalic tegmentum that contains dopaminergic pigmented and substance P neurons as well as serotonergic neurons. The 5-HT neurons in the CLi are small to medium neurons with dendrites running parallel to the midline.

In nonhuman primates, serotonergic innervation of the cerebral cortex and much of the forebrain is derived from 5-HT-synthesizing neurons in the DRN and in the MRN of the brain stem. In the human, the DRN is a large group of neurons embedded in the ventral part of the central gray matter of caudal mesencephalon and rostral pons (Figure 2.1A). Based on topographic and cytoarchitectonic characteristics in

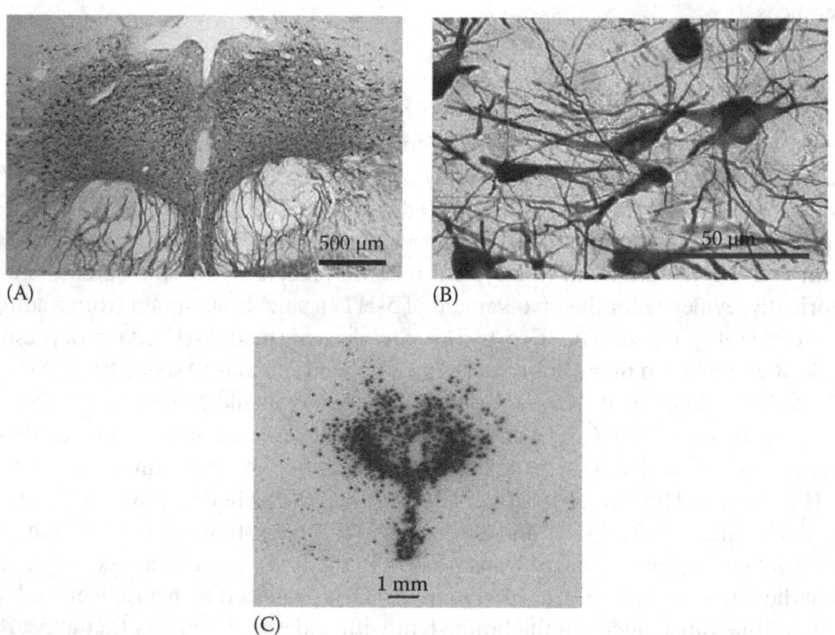

FIGURE 2.1 (A) A low-power (10×) photomicrograph of a rostral section of human brain stem DRN reacted for PH-8 (TPH2) by immunocytochemistry. (B) A higher power (40×) image of the rostral section shown in (A) demonstrating the specific expression of PH-8 in large multipolar neurons in the DRN. (C) A comparable rostral section reacted by *in situ* hybridization with a riboprobe specific for TPH2 mRNA.

Nissl-stained material, the DRN is subdivided into distinct subnuclei (Baker et al. 1990). These subdivisions correspond to those observed in tissue immunoreacted with anti-phenylalanine hydroxylase (anti-PH-8) sera (Hornung 2003; Törk 1990; Törk and Hornung 1990), which also revealed an additional component (the ventral subnucleus), not previously recognized in Nissl material. The subnuclei of the DRN are median (or interfascicular), ventrolateral, dorsal, lateral, and caudal. The *median* subnucleus is long and dense; its cells are oriented parallel to the midline and the processes extend within the medial longitudinal fasciculi. Its rostrocaudal extent goes from the anterior pole of the DRN to the rostral appearance of the median sulcus of the fourth ventricle. Caudal to this point, all DRN neurons are on either side of the floor of the fourth ventricle and none are on the midline. The *ventrolateral* subnucleus is made up of small, multipolar neurons extending caudally from the central gray to a position just dorsal to the trochlear nuclei and the medial longitudinal fasciculus. Olszewski and Baxter (1954) referred to this part of the DRN as nucleus supratrochlearis to indicate its appearance surrounding the nucleus of the trochlear nerve and extending dorsally and laterally. This subdivision has the highest density of PH-8-immunoreactive neurons, has no midline component, and extends further caudally than the median subnucleus in the rostral pons. A characteristic of this subnucleus in the nonhuman primate is that it contains the greatest density of neurons (Azmitia and Gannon 1986; Hornung and Fritschy 1988). The *dorsal* subnucleus has loosely arranged medium-sized neurons dorsomedially flanking the dense ventrolateral subgroup. The two wings of this subnucleus are joined in the midline. The *caudal* subnucleus is made of two dense strips of PH-8-immunoreactive neurons lateral to the midline and dorsal to the medial longitudinal fasciculus. Its cells are small- to medium-sized and processes are oriented parallel to the floor of the fourth ventricle. The *lateral* subdivision cannot be recognized in Nissl-stained material. It is very loosely organized and has the largest multipolar neurons of any of the DRN subdivisions. Neuronal processes extend across large fields in the central gray and axons are characteristically coiled close to the soma. The *ventral* subdivision is made up of round neurons located between the dorsal and ventrolateral subnucleus.

Although it has been known for many decades that there are massive projections from the raphe to the telencephalon (Brodal et al. 1960), at present it is not possible to verify the cortical targets of the various DRN nuclear subdivisions in the human. The projection from the DRN to cortical targets in the monkey exhibits a coarse rostrocaudal topographic relationship, as opposed to the MRN projections that are not separated rostrocaudally (Wilson and Molliver 1991a,b). The serotonergic projection to the prefrontal cortex (PFC) has a very heavy component arising from cells in the rostral part of the DRN. Regarding cortical innervation in the primate, the density is highest in layer 1, except in sensory areas where the highest density is in layer IV. The serotonergic target cells in the cortex are GAD-IR, indicating that they are GABAergic, inhibitory neurons.

The ascending dual serotonergic projection system described in the rat (Mamounas et al. 1991) has also been demonstrated in nonhuman primates (Wilson and Molliver 1991a,b) as arising from both the DRN and the MRN. The fine varicose axon system (D-fibers) originates in the DRN and branches profusely in the target area. It has been difficult to estimate the incidence of synaptic contacts from this system. The second system has large round varicosities (M-fibers), more divergent innervation and originates in the MRN. The fine, DRN nucleus system is more susceptible to degeneration by amphetamine derivatives, while the median raphe fiber system appears to be spared (Wilson et al. 1989). Both the DRN and the MRN project to the cortex and they differ in their distribution within the cortex (Kosofsky and Molliver 1987). Delineation of projections to and from DRN subnuclei has also been described in the rat. Using retrograde tracers into the dorsal subnucleus, afferent projections from the bed nucleus of the stria terminalis were found to be selectively labeled (Peyron et al. 1998). Efferent projections from the rat dorsal subnucleus have been shown to densely innervate the central amygdaloid nucleus and the dorsal hypothalamic area (Commons et al. 2003). Serotonergic innervation of the medial PFC and the nucleus accumbens originates almost exclusively from the dorsal subnucleus in rats (Van Bockstaele et al. 1993). All these regions are involved in regulation of stress-induced behavioral responses and anxiety-related behaviors.

Recent rodent studies have evaluated the role of specific DRN subnuclei in mediating 5-HT-related stress responses in behavioral paradigms. The lateral wing subregion of the mouse DRN is specifically involved in mediating the stress response in the forced swim test. Further, electrophysiological membrane studies of the serotonergic neurons of the lateral wing subregion show that these neurons have elevated intrinsic excitability compared to serotonergic neurons in the ventromedial subregion of the DRN (Crawford et al. 2010). This suggests that there are fundamental differences in serotonergic neuronal physiology that regulate stress-related behaviors and are regionally specific within the DRN.

2.3 5-HT INDICES AND NEURON COUNTS IN THE DRN OF DEPRESSED SUICIDES

Brain serotonergic neurotransmission is regulated by a network of pre- and postsynaptic receptors and the 5-HT transporter (SERT). The available clinical and postmortem data suggest that reduced serotonergic input constitutes a critical element in the vulnerability to suicidal behavior, regardless of the associated psychiatric illness. Postmortem studies using SERT autoradiography with the specific 5-HT transporter ligand, cyanoimipramine, demonstrate reduced 5-HT transporter binding in the PFC of suicides (Arango et al. 1995, 2002; Laruelle et al. 1993; Mann et al. 2000) and raise the possibility that there is reduced serotonergic innervation. The functional capacity of serotonergic neurons may be reduced because of inadequate innervation of the target brain region or a reduction in the number of SERT sites synthesized. In 1999 (Underwood et al. 1999), we sought to estimate the total number and the morphometric characteristics of serotonergic neurons in the DRN from a group of suicides and nonpsychiatric controls. Tissue was sectioned, stained for Nissl, and processed with an antiserum (PH-8) (Cotton et al. 1988), which cross-reacts with tryptophan hydroxylase (referred to as TPH) (Figure 2.1A and B). All DRN neurons were identified, counted, and analyzed every 1000 μm. We found that suicide victims had 35% greater density and number of 5-HT neurons in the DRN compared to non-suicide controls (see Figure 2.2). The total volume of the DRN did not differ in the two groups, suggesting that there is a difference in

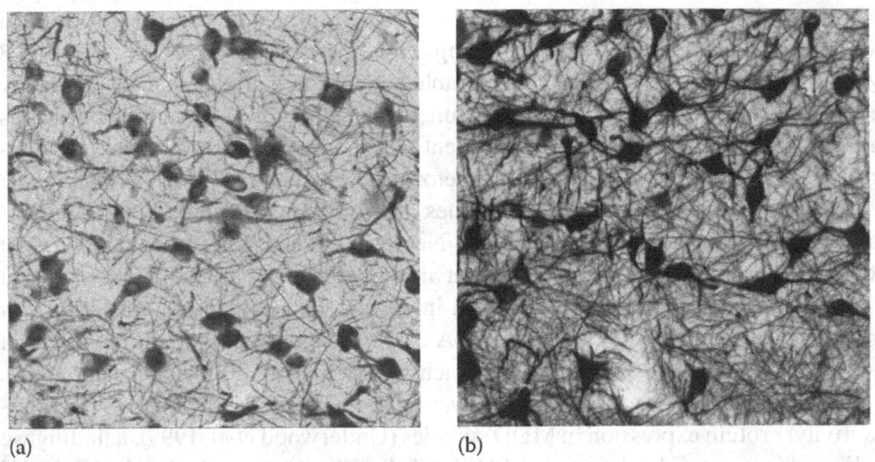

FIGURE 2.2 A high-power photomicrograph of PH-8-immunoreacted sections of human brain stem from an age- and sex-matched control (a) and suicide (b). Note the increased density of TPH-immunoreactive neurons and neuropil in the suicide case compared to the control.

the absolute number in DRN neurons. Moreover, the mean density of serotonergic neurons is also significantly greater in the suicide victims.

We used stereology to count Nissl-stained neurons in adjacent sections to those stained for TPH in 1999 (Underwood et al. 1999). The number of neurons was determined using the fractionator method (Gundersen et al. 1988; West 1993). Suicides did not differ from controls in the total number of Nissl-stained DRN neurons. However, the percent of DRN neurons that were also serotonergic, as measured in adjacent immunostained sections, was 79.9% in the suicides and 57.7% in the controls. The phenotype of the non-serotonergic DRN neurons is unknown in humans.

2.4 NEURONAL TRYPTOPHAN HYDROXYLASE IN SUICIDE

Recent studies have provided further evidence of 5-HT alterations involving the neuronal isoform of the rate-limiting serotonergic biosynthetic enzyme, TPH2 (Bach-Mizrachi et al. 2006, 2008; Boldrini et al. 2005). TPH2 converts the amino acid tryptophan to 5-hydroxytryptophan (5-HTP) en route to subsequent decarboxylation into 5-hydroxytryptamine (5-HT, see Mockus and Vrana 1998 for review). TPH2 is a critical component in the determination of the amount of brain 5-HT synthesized *in vivo* (Patel et al. 2004; Zhang et al. 2004). Deficits in TPH2 amount or catalytic activity may result in aberrant 5-HT production and subsequent behavioral changes. Until the discovery of TPH2 (Walther et al. 2003), quantitative studies of *TPH* gene expression in brain had been hampered by the almost undetectable level of transcript expression in the raphe (Austin and O'Donnell 1999; Clark and Russo 1997). These studies were measuring peripheral TPH (TPH1), the nonneuronal isoform of the enzyme (Walther et al. 2003) predominantly expressed in the pineal gland and the gut and responsible for the production of melatonin.

TPH is a member of a superfamily of structurally and functionally related monoamine biosynthetic enzymes along with tyrosine hydroxylase and PH-8 (Mockus and Vrana 1998). The considerable amount of shared sequence homology between the enzymes conveyed similar functional regulatory and catalytic domains and provided the basis for the development of PH-8, an antibody for the immunocytochemical detection of this family of enzymes (Haan et al. 1987). TPH, tyrosine hydroxylase, and the other related enzymes are highly regulated by both transcription and posttranslational modification. Neuronal TPH is the product of a gene located on chromosome 12q15 (Walther et al. 2003; Zhang et al. 2004) and is present in quantities large enough to be detected in postmortem human brain using *in situ* hybridization (Figure 2.1C). TPH2 mRNA expression is specific to all raphe nuclei tested thus far in rodents and primates (Bach-Mizrachi et al. 2006; Patel et al. 2004). In the earlier studies described previously, we found elevated TPH2 (PH-8 immunoreactivity) protein expression in MDD suicides (Underwood et al. 1999), a finding we replicated in a second cohort using PH-8 (TPH) immunoautoradiography (Boldrini et al. 2005). More TPH indicates an up-regulatory homeostatic response to impaired 5-HT release or less serotonergic autoreceptor activation. Alternatively, the 5-HT impairment in suicide may be due to hypofunctional 5-HT-synthesizing enzyme. While the change in PH-8 immunoautoradiography was not replicated by another laboratory using depressed suicides (Bonkale et al. 2004), greater PH-8 expression was found in the dorsal subnucleus of the DRN of depressed suicides who were also alcohol dependent (Bonkale et al. 2006). This inconsistency in immunoautoradiography may be reconciled by variation in methodologies between the two laboratories. Bonkale et al. (2004) sampled five representative rostral-most sections of the DRN, while our group sampled the anteroposterior length of the DRN at 1 mm intervals, thereby increasing the resolution in which changes in expression can be found.

With the discovery of the neuronal form of TPH in 2003 came the ability to detect and measure transcript levels of the 5-HT-producing enzyme in postmortem human brain tissue. We developed a riboprobe specific for TPH2 mRNA and found robust expression specific to the large multipolar neurons within the DRN and MRN. Consistent with our protein (immunocytochemical) findings, we discovered that TPH2 mRNA expression was elevated in depressed suicides compared to matched controls and further that the 5-HT-synthesizing neurons in suicide had higher transcriptional capacity (Bach-Mizrachi et al. 2006, 2008). More TPH-immunoreactive neurons (Underwood et al. 1999), when taken together with more TPH protein (Boldrini et al. 2005; Underwood et al. 1999), and more TPH2 mRNA (Bach-Mizrachi et al. 2006) would favor higher 5-HT levels and yet most studies report lower 5-HT levels in suicides. Lower brain stem 5-HT levels in depressed suicides and lower CSF 5-HIAA suggest a compensatory mechanism in which the enzyme level may be increased but catalytic activity or release of 5-HT is impaired. To date, there is only one functional yet rare single nucleotide polymorphism (SNP) in the human *TPH2* gene that affects TPH2 catalytic activity (Zhang et al. 2005). While this SNP was shown to be associated with MDD in one population, the association was not found in any of the several other studies that attempted to replicate this finding (Glatt et al. 2005; Van Den et al. 2005; Zhou et al. 2005). While an undiscovered functional SNP in the *TPH2* gene may still remain a reasonable explanation for deficits in TPH2 catalytic activity, it should be noted that TPH2 function is dependent on a number of factors including phosphorylation state (Winge et al. 2006), regulation by signal transduction pathways (Beaulieu et al. 2008), presence of cofactors and adapters (Winge et al. 2008), and potentially other regulatory TPH isoforms (Haghighi et al. 2008). These mechanisms likely regulate TPH2 catalytic activity by gauging brain 5-HT levels and therefore in suicides where 5-HT is low, the TPH2 enzyme produced in affected individuals may be of lower catalytic activity.

Previous studies have shown the presence of low levels of TPH2 transcript in the terminal fields of serotonergic neurons including the cortex, hippocampus, and amygdala (De Luca et al. 2005; Zill et al. 2007) using quantitative Reverse Transcriptase-Polymerase Chain Reaction (RT-PCR). However, abundant levels of TPH protein have been found in the PFC by western blot (Ono et al. 2002). The presence of TPH in serotonergic terminals suggests a potential mechanism for the regulation of 5-HT synthesis locally, at the synapse in terminal regions far from the site of origin in the raphe nuclei. Therefore, it is possible that inefficient axoplasmic flow of TPH to cortical terminal fields could lead to decreases in TPH protein at terminals and increased levels at cell bodies. This notion is proposed in one postmortem study of Alzheimer's patients in which TPH activity and the serotonergic metabolite 5-HIAA were found to be increased in the DRN but decreased in terminal fields in the amygdala. These authors propose that high levels of toxic serotonergic metabolites in the raphe contribute to degeneration of serotonergic neurons seen in Alzheimer's disease (Burke et al. 1990).

2.5 5-HT RECEPTORS AND THE 5-HT TRANSPORTER IN SUICIDE

The 5-HT_{1A} receptor is a G protein-coupled receptor that is expressed at pre- and postsynaptic sites and plays a role in multiple physiological functions including mood regulation, neuroendocrine functions, thermoregulation, sexual behavior, and food intake (for review, see Raymond et al. 2001). Dysfunction of the 5-HT_{1A} receptor has been implicated in the pathology of various psychiatric disorders including major depression, anxiety, and suicide (Blier et al. 1993; Mann 2003). In the DRN, the 5-HT_{1A} receptor functions as a somatodendritic inhibitory autoreceptor on 5-HT neurons (Middlemiss and Fozard 1983; Verge et al. 1985). Locally released 5-HT acts on the autoreceptor to inhibit further release (Wang and Aghajanian 1977). Greater autoinhibition of the 5-HT_{1A} receptor in the brain stem raphe nuclei may be a mechanism that contributes to reduced serotonergic neurotransmission in the PFC in the context of suicide and depression (Stockmeier et al. 1998).

In postmortem studies, 5-HT_{1A} autoreceptor levels have been shown to be elevated in the midbrain of suicides by some (Stockmeier et al. 1998), but not all (Arango et al. 2001), investigators. The apparently discrepant findings were reconciled by our recent work (Boldrini et al. 2008) that shows an increase in 5-HT_{1A} receptors in the rostral part of the dorsal raphe (DRN) in suicides and a decrease in the remaining 15 mm (~75% of the DRN), for a net decrease in binding throughout the DRN (see 5-HT_{1A} receptor binding profile in the DRN in Figure 2.3). Stockmeier et al. (1998) examined the most rostral 5 mm of the DRN and, like in our study (Boldrini et al. 2008), found an increase in binding. Another determining factor may be found in the sex composition of the patient cohorts from which the various postmortem

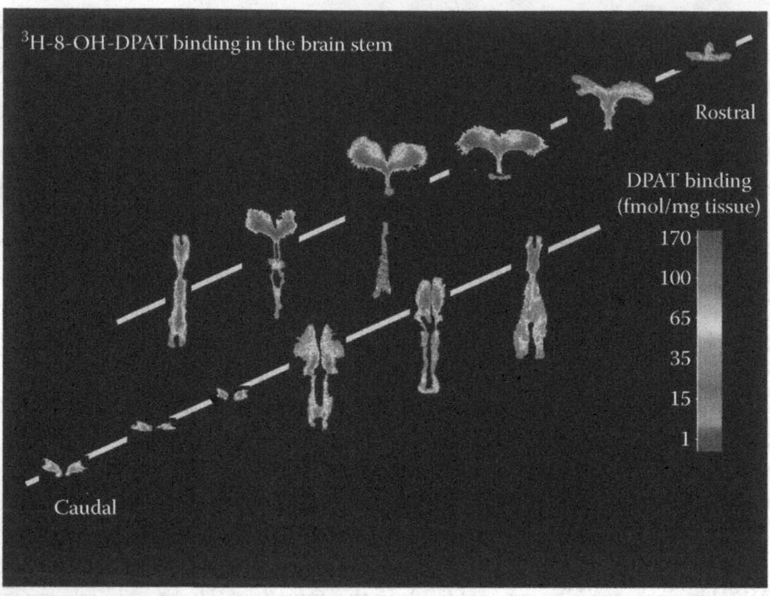

FIGURE 2.3 (See color insert.) Pseudocolor receptor autoradiograms representing ^{3}H-8-OH-DPAT binding, a 5-HT agonist, within the DRN and MRN along the rostrocaudal axis of the brain stem.

samples were obtained. For example, we find that females have significantly higher 5-HT$_{1A}$ receptor binding than males (Arango et al. 2001), a finding replicated by our group *in vivo* with positron emission tomography (PET) (Parsey et al. 2002). Goswami et al. (2010) reported that mRNA concentrations of the 5-HT$_{1D}$ receptor as well as transcription regulators were significantly increased in DRN laser-captured neurons of female MDD subjects compared to female control subjects, while no differences were reported in male subjects from both groups.

Some postmortem studies showed greater 5-HT$_{1A}$ receptor binding in the ventrolateral PFC (Arango et al. 1995; Lowther et al. 1997), while other studies did not find any changes (Stockmeier et al. 1997). We also reported an increase in 5-HT$_{1A}$ receptor binding across the ventral, lateral, orbital, and medial aspects of the hemisphere compared to the dorsal gyri and sulci in the PFC of depressed suicides compared to controls (Arango et al. 2002). We considered increased cortical 5-HT$_{1A}$ receptor binding to be a compensatory mechanism in response to decreased 5-HT levels partly because binding to the 5-HT transporter and receptor is inversely correlated in the PFC.

The 5-HT$_{1A}$ autoreceptor is speculated to be involved in the mechanisms underlying the delayed onset of therapeutic effects of selective serotonin reuptake inhibitors (SSRIs). SSRIs increase 5-HT levels present in the extracellular space and therefore increase activation of postsynaptic 5-HT receptors. However, the desired increase in presynaptic 5-HT release is not achieved until the somatodendritic 5-HT$_{1A}$ autoreceptors in the raphe are desensitized, thereby releasing the inhibition on serotonergic firing. 5-HT$_{1A}$ receptor-mediated serotonergic neurotransmission is a passive process dependent on intrasynaptic 5-HT availability and therefore the time it takes to desensitize the receptor contributes to the delayed onset of the therapeutic effects of SSRIs (reviewed in Hensler 2006). One interpretation of the mechanism of SSRI action is that release of inhibition on serotonergic neurons by desensitization of the 5-HT_{1A} autoreceptor does result not only in increased serotonergic firing and subsequent increased release of 5-HT into the synapse but also in *de novo* synthesis of 5-HT by serotonergic neurons (Kim et al. 2002; Meller et al. 1990). This was demonstrated by one study in which long-term treatment of rats with the potent SSRI sertraline resulted in marked up-regulation of TPH transcript and protein expression. When tested in cell culture, this group showed that TPH activity and total 5-HT were increased in the presence of sertraline (Kim et al. 2002). The idea that sertraline stimulates *de novo* synthesis of 5-HT is intriguing and has a potential therapeutic value.

In the rat, serotonergic neurons can be distinguished from non-serotonergic neurons by their electrophysiological properties (Kirby et al. 2003). In addition, double-label immunocytochemical experiments in rat with antibodies against 5-HT and 5-HT_{1A} receptor show that the modulatory 5-HT_{1A} receptor protein does not totally colocalize with all 5-HT-containing neurons in the raphe. In fact, in this species at least, a significant non-serotonergic population of cells exist that *also* express 5-HT_{1A} receptors (Kirby et al. 2003). Stimulation of the 5-HT_{1A} autoreceptor by 5-HT or an agonist results in the feedback inhibition of the serotonergic neuron, with subsequent suppression of 5-HT synthesis and turnover (Sibille and Hen 2001). However, the role of 5-HT_{1A} receptors on non-serotonergic cells is unclear. There is evidence that serotonergic and non-serotonergic neurons in the DRN may be affected by 5-HT_{1A} and 5-HT_{2A} receptors (Craven et al. 2001). Studies by the Beck laboratory (Beck et al. 2004; Kirby et al. 2003) have determined that both 5-HT and non-5-HT DRN neurons respond to 5-HT_{1A} receptor agonists. Furthermore, they were able to determine that DRN neurons, whether or not serotonergic, have similar characteristics traditionally associated with 5-HT neurons. In contrast, 5-HT and non-5-HT MRN neurons have very different characteristics (Beck et al. 2004). Presumably, stimulation of receptors on non-serotonergic neurons will act indirectly to modulate local 5-HT levels in the raphe nuclei.

The presumption that non-serotonergic neurons in the DRN and MRN have a regulatory role in controlling local 5-HT levels is further supported by the finding that the majority of neurons in this population, at least in rodents, are GABAergic (Stamp and Semba 1995). A GABAergic coexpression with 5-HT_{1A} receptors in humans would suggest a postsynaptic regulation of 5-HT in the raphe by GABA neurons. This may help explain the time delay in patient response to antidepressant drugs that block 5-HT_{1A} autoreceptors. However, it is not yet known whether there are GABAergic neurons in the human DRN that also express 5-HT_{1A} receptors.

Most, but not all, studies of postsynaptic serotonergic receptor binding studies in suicides report an increase in 5-HT_{2A} receptor binding in the PFC (reviewed in Mann et al. 1996). Using [^{3}H]ketanserin to measure 5-HT_{2A} receptor density in frontal cortex, some groups reported no difference between suicides and control subjects (Arranz et al. 1994; Cheetham et al. 1988; Crow et al. 1984; Lowther et al. 1994; Owen et al. 1983; Rosel et al. 2000) while others found an increase in the number of 5-HT_{2A} receptors in the PFC of suicide subjects (Gross-Isseroff et al. 1990;

Hrdina et al. 1993; Turecki et al. 1999). Using other ligands to study 5-HT_{2A} binding sites, other groups have reported higher 5-HT_{2A} cortical receptor binding in suicide subjects (Arango et al. 1990; Arora and Meltzer 1989; Pandey et al. 2002; Stanley and Mann 1983). Differences in methodology and age and sex of cases studied may explain some of the discrepancies in the studies, but, nevertheless, most studies find an increase in 5-HT_{2A} postsynaptic receptor binding in the PFC. We recently reported a positive correlation of prefrontal cortical 5-HT_{2A} receptor binding sites and lifetime aggression scores in suicides (Oquendo et al. 2006). In another postmortem study, cases with depression had an increase in 5-HT_{2A} receptor abundance in Brodmann area 10, which was correlated with decreased Protein Kinase A (PKA) activity (Shelton et al. 2009), a finding consistent with previous studies demonstrating decreased PKA expression and activity in the PFC of teenage suicides compared to controls (Dwivedi et al. 2004). The postsynaptic 5-HT_{2c} receptor is also reported to be specifically elevated in the PFC of suicides compared to controls (Pandey et al. 2006) Taken together, these studies demonstrate the vulnerability of the serotonergic system in cortical targets that regulate mood and behavior in suicide.

Serotonergic neurotransmission between the DRN/MRN and the PFC is regulated by the presynaptic 5-HT transporter (SERT) and by pre- and postsynaptic serotonergic receptors. SERT regulates intrasynaptic 5-HT levels via reuptake into the presynaptic neuron. Using ^{3}H-cyanoimipramine and receptor autoradiography, we found reduced SERT binding in the ventral PFC in depressed suicides (Arango et al. 1995, 2002), a finding replicated by immunocytochemical studies that showed a deficit in the length and density of 5-HT transporter-immunoreactive neurons (Austin et al. 2002). Lower SERT binding in the ventromedial PFC was related specifically to suicide independent of diagnosis, a finding consistent with the role of this region of the PFC (PFC) in regulating behavioral inhibition (Shallice and Burgess 1996). Lower SERT binding throughout the PFC was related to MDD, which was confirmed by *in vivo* PET studies showing lower SERT binding that was widespread throughout the PFC in patients with MDD (Parsey et al. 2006). In the brain stem, *in situ* hybridization was used to measure transporter mRNA expression (McLaughlin et al. 1996) and no difference was found in total SERT mRNA between suicides and controls (Arango et al. 2001; Little et al. 1997). However, MDD suicides had fewer SERT expressing neurons and these neurons had higher SERT transcriptional capacity (Arango et al. 2001). Taken together, less SERT binding in the PFC and fewer SERT expressing neurons in the brain stem support a compensatory homeostatic mechanism for deficits in 5-HT neurotransmission.

2.6 CONCLUSIONS

Postmortem human brain studies have made significant contributions to unraveling the neuroanatomical and biochemical profile of suicide. We and others have used the postmortem brain of suicides to understand the alterations present in the serotonergic system that underlie suicidal behaviors. In reviewing the current data from postmortem studies of suicides, it is clear that there are anatomically specific alterations in the serotonergic system that are specific to suicide and consistent with a homeostatic brain response both in source 5-HT synthesizing neurons in the raphe nuclei and in postsynaptic target neurons in the cortex, to deficits in serotonergic neurotransmission. While the data from postmortem studies are compelling and define the molecular profile of the brain in suicide, further studies are necessary to pinpoint whether these changes define causality for 5-HT deficits or are alternatively a normal brain response to a preexisting hyposerotonergic environment.

ACKNOWLEDGMENTS

We would like to thank the NIH, the American Foundation for Suicide Prevention, and the Diane Goldberg Foundation for supporting the research described in this chapter. We would also like to thank Suham A. Kassir, Mihran J. Bakalian, and Virginia Johnson for their help with the postmortem human brain studies.

REFERENCES

Arango, V., Ernsberger, P., Marzuk, P. M., Chen, J. S., Tierney, H., Stanley, M., Reis, D. J., and Mann, J. J. 1990. Autoradiographic demonstration of increased serotonin 5-HT2 and beta-adrenergic receptor binding sites in the brain of suicide victims. *Archives of General Psychiatry* 47:1038–1047.

Arango, V., Underwood, M. D., Boldrini, M., Tamir, H., Kassir, S. A., Hsiung, S., Chen, J. J. X., and Mann, J. J. 2001. Serotonin 1A receptors, serotonin transporter binding and serotonin transporter mRNA expression in the brainstem of depressed suicide victims. *Neuropsychopharmacology* 25:892–903.

Arango, V., Underwood, M. D., Gubbi, A. V., and Mann, J. J. 1995. Localized alterations in pre- and postsynaptic serotonin binding sites in the ventrolateral prefrontal cortex of suicide victims. *Brain Research* 688:121–133.

Arango, V., Underwood, M. D., and Mann, J. J. 2002. Serotonin brain circuits involved in major depression and suicide. *Progress in Brain Research* 136:443–453.

Arora, R. C. and Meltzer, H. Y. 1989. Serotonergic measures in the brains of suicide victims: 5-HT2 binding sites in the frontal cortex of suicide victims and control subjects. *American Journal of Psychiatry* 146:730–736.

Arranz, B., Eriksson, A., Mellerup, E., Plenge, P., and Marcusson, J. 1994. Brain 5-HT1A, 5-HT1D, and 5-HT2 receptors in suicide victims. *Biological Psychiatry* 35:457–463.

Åsberg, M. 1976. Treatment of depression with tricyclic drugs—Pharmacokinetic and pharmacodynamic aspects. *Pharmakopsychiatrie Neuropsychopharmakologie* 9:18–26.

Åsberg, M., Träskman, L., and Thorén, P. 1976. 5-HIAA in the cerebrospinal fluid. A biochemical suicide predictor? *Archives of General Psychiatry* 33:1193–1197.

Austin, M. C. and O'Donnell, S. M. 1999. Regional distribution and cellular expression of tryptophan hydroxylase messenger RNA in postmortem human brainstem and pineal gland. *Journal of Neurochemistry* 72:2065–2073.

Austin, M. C., Whitehead, R. E., Edgar, C. L., Janosky, J. E., and Lewis, D. A. 2002. Localized decrease in serotonin transporter-immunoreactive axons in the prefrontal cortex of depressed subjects committing suicide. *Neuroscience* 114:807–815.

Azmitia, E. C. and Gannon, P. J. 1986. The primate serotonergic system: A review of human and animal studies and a report on *Macaca fascicularis*. *Advances in Neurology* 43:407–468.

Bach-Mizrachi, H., Underwood, M. D., Kassir, S. A., Bakalian, M. J., Sibille, E., Tamir, H., Mann, J. J., and Arango, V. 2006. Neuronal tryptophan hydroxylase mRNA expression in the human dorsal and median raphe nuclei: Major depression and suicide. *Neuropsychopharmacology* 31:814–824.

Bach-Mizrachi, H., Underwood, M. D., Tin, A., Ellis, S. P., Mann, J. J., and Arango, V. 2008. Elevated expression of tryptophan hydroxylase-2 mRNA at the neuronal level in the dorsal and median raphe nuclei of depressed suicides. *Molecular Psychiatry* 13:507–513.

Baker, K. G., Halliday, G. M., and Tork, I. 1990. Cytoarchitecture of the human dorsal raphe nucleus. *Journal of Comparative Neurology* 301:147–161.

Banki, C. M., Arato, M., Papp, Z., and Kurcz, M. 1984. Biochemical markers in suicidal patients. Investigations with cerebrospinal fluid amine metabolites and neuroendocrine tests. *Journal of Affective Disorders* 6:341–350.

Beaulieu, J. M., Zhang, X., Rodriguiz, R. M., Sotnikova, T. D., Cools, M. J., Wetsel, W. C., Gainetdinov, R. R., and Caron, M. G. 2008. Role of GSK3beta in behavioral abnormalities induced by serotonin deficiency. *Proceedings of the National Academy of Sciences of the United States of America* 105:1333–1338.

Beck, S. G., Pan, Y. Z., Akanwa, A. C., and Kirby, L. G. 2004. Median and dorsal raphe neurons are not electrophysiologically identical. *Journal of Neurophysiology* 91:994–1005.

Blier, P., Lista, A., and De Montigny, C. 1993. Differential properties of pre- and postsynaptic 5-hydroxytryptamine$_{1A}$ receptors in the dorsal raphe and hippocampus: I. Effect of spiperone. *Journal of Pharmacology and Experimental Therapeutics* 265:7–15.

Boldrini, M., Underwood, M. D., Mann, J. J., and Arango, V. 2005. More tryptophan hydroxylase in the brainstem dorsal raphe nucleus in depressed suicides. *Brain Research* 1041:19–28.

Boldrini, M., Underwood, M. D., Mann, J. J., and Arango, V. 2008. Serotonin-1A autoreceptor binding in the dorsal raphe nucleus of depressed suicides. *Journal of Psychiatric Research* 42:433–442.

Bonkale, W. L., Murdock, S., Janosky, J. E., and Austin, M. C. 2004. Normal levels of tryptophan hydroxylase immunoreactivity in the dorsal raphe of depressed suicide victims. *Journal of Neurochemistry* 88:958–964.

Bonkale, W. L., Turecki, G., and Austin, M. C. 2006. Increased tryptophan hydroxylase immunoreactivity in the dorsal raphe nucleus of alcohol-dependent, depressed suicide subjects is restricted to the dorsal subnucleus. *Synapse* 60:81–85.

Brodal, A., Taber, E., and Walberg, F. 1960. The raphe nuclei of the brain stem of the cat: Efferent projections. *Journal of Comparative Neurology* 114:239–259.

Burke, W. J., Park, D. H., Chung, H. D., Marshall, G. L., Haring, J. H., and Joh, T. H. 1990. Evidence for decreased transport of tryptophan hydroxylase in Alzheimer's disease. *Brain Research* 537:83–87.

Carlsson, A., Svennerholm, L., and Winblad, B. 1980. Seasonal and circadian monoamine variations in human brains examined post mortem. *Acta Psychiatrica Scandinavica* 280:75–85.

Cheetham, S. C., Crompton, M. R., Katona, C. L. E., and Horton, R. W. 1988. Brain 5-HT$_2$ receptor binding sites in depressed suicide victims. *Brain Research* 443:272–280.

Clark, M. S. and Russo, A. F. 1997. Tissue-specific glucocorticoid regulation of tryptophan hydroxylase mRNA levels. *Brain Research Molecular Brain Research* 48:346–354.

Commons, K. G., Connolley, K. R., and Valentino, R. J. 2003. A neurochemically distinct dorsal raphe-limbic circuit with a potential role in affective disorders. *Neuropsychopharmacology* 28:206–215.

Cotton, R. G. H., McAdam, W., Jennings, I., and Morgan, F. J. 1988. A monoclonal antibody to aromatic amino acid hydroxylases. Identification of the epitope. *Biochemical Journal* 255:193–196.

Craven, R. M., Grahame-Smith, D. G., and Newberry, N. R. 2001. 5-HT$_{1A}$ and 5-HT$_2$ receptors differentially regulate the excitability of 5-HT-containing neurones of the guinea pig dorsal raphe nucleus *in vitro*. *Brain Research* 899:159–168.

Crawford, L. K., Craige, C. P., and Beck, S. G. 2010. Increased intrinsic excitability of lateral wing serotonin neurons of the dorsal raphe: A mechanism for selective activation in stress circuits. *Journal of Neurophysiology* 103:2652–2663.

Crow, T. J., Cross, A. J., Cooper, S. J., Deakin, J. F. W., Ferrier, I. N., Johnson, J. A., Joseph, M. H., Owen, F., Poulter, M., Lofthouse, R., Corsellis, J. A. N., Chambers, D. R., Blessed, G., Perry, E. K., Perry, R. H., and Tomlinson, B. E. 1984. Neurotransmitter receptors and monoamine metabolites in the brains of patients with Alzheimer-type dementia and depression, and suicides. *Neuropharmacology* 23:1561–1569.

De Luca, V., Likhodi, O., Van Tol, H. H., Kennedy, J. L., and Wong, A. H. 2005. Tryptophan hydroxylase 2 gene expression and promoter polymorphisms in bipolar disorder and schizophrenia. *Psychopharmacology (Berlin)* 183:378–382.

Dulchin, M. C., Oquendo, M. A., Malone, K. M., Ellis, S. P., Li, S., and Mann, J. J. 2001. Prolactin response to DL-fenfluramine challenge before and after treatment with paroxetine. *Neuropsychopharmacology* 25:395–401.

Duval, F., Mokrani, M. C., Correa, H., Bailey, P., Valdebenito, M., Monreal, J., Crocq, M. A., and Macher, J. P. 2001. Lack of effect of HPA axis hyperactivity on hormonal responses to D-fenfluramine in major depressed patients: Implications for pathogenesis of suicidal behaviour. *Psychoneuroendocrinology* 26:521–537.

Dwivedi, Y., Rizavi, H. S., Shukla, P. K., Lyons, J., Faludi, G., Palkovits, M., Sarosi, A., Conley, R. R., Roberts, R. C., Tamminga, C. A., and Pandey, G. N. 2004. Protein kinase A in postmortem brain of depressed suicide victims: Altered expression of specific regulatory and catalytic subunits. *Biological Psychiatry* 55:234–243.

Glatt, C. E., Carlson, E., Taylor, T. R., Risch, N., Reus, V. I., and Schaefer, C. A. 2005. Response to Zhang et al. (2005): Loss-of-function mutation in tryptophan hydroxylase-2 identified in unipolar major depression. *Neuron* 45:11–16; *Neuron* 48:704–705.

Goswami, D. B., May, W. L., Stockmeier, C. A., and Austin, M. C. 2010. Transcriptional expression of serotonergic regulators in laser-captured microdissected dorsal raphe neurons of subjects with major depressive disorder: Sex-specific differences. *Journal of Neurochemistry* 112:397–409.

Gross-Isseroff, R., Salama, D., Israeli, M., and Biegon, A. 1990. Autoradiographic analysis of age-dependent changes in serotonin 5-HT_2 receptors of the human brain postmortem. *Brain Research* 519:223–227.

Gundersen, H. J. G., Bagger, P., Bendtsen, T. F., Evans, S. M., Korbo, L., Marcussen, N., Moller, A., Nielsen, K., Nyengaard, J. R., Pakkenberg, B., Sorensen, F. B., Vesterby, A., and West, M. J. 1988. The new stereological tools: Disector, fractionator, nucleator and point sampled intercepts and their use in pathological research and diagnosis. *Acta Pathologica Microbiologica et Immunologica Scandinavica* 96:857–881.

Haan, E. A., Jennings, I. G., Cuello, A. C., Nakata, H., Fujisawa, H., Chow, C. W., Kushinsky, R., Brittingham, J., and Cotton, R. G. H. 1987. Identification of serotonergic neurons in human brain by a monoclonal antibody binding to all three aromatic amino acid hydroxylases. *Brain Research* 426:19–27.

Haghighi, F., Bach-Mizrachi, H., Huang, Y. Y., Arango, V., Shi, S., Dwork, A. J., Rosoklija, G., Sheng, H. T., Morozova, I., Ju, J., Russo, J. J., and Mann, J. J. 2008. Genetic architecture of the human tryptophan hydroxylase 2 gene: Existence of neural isoforms and relevance for major depression. *Molecular Psychiatry* 13:813–820.

Halliday, G. M. and Törk, I. 1986. Comparative anatomy of the ventromedial mesencephalic tegmentum in the rat, cat, monkey and human. *Journal of Comparative Neurology* 252:423–445.

Hensler, J. G. 2006. Serotonergic modulation of the limbic system. *Neuroscience and Behavioral Reviews* 30:203–214.

Hornung, J. P. 2003. The human raphe nuclei and the serotonergic system. *Journal of Chemical Neuroanatomy* 26:331–343.

Hornung, J. P. and Fritschy, J. M. 1988. Serotoninergic system in the brainstem of the marmoset: A combined immunocytochemical and three-dimensional reconstruction study. *Journal of Comparative Neurology* 270:471–487.

Hrdina, P. D., Demeter, E., Vu, T. B., Sótónyi, P., and Palkovits, M. 1993. 5-HT uptake sites and 5-HT_2 receptors in brain of antidepressant-free suicide victims/depressives: Increase in 5-HT_2 sites in cortex and amygdala. *Brain Research* 614:37–44.

Kim, S. W., Park, S. Y., and Hwang, O. 2002. Up-regulation of tryptophan hydroxylase expression and serotonin synthesis by sertraline. *Molecular Pharmacology* 61:778–785.

Kirby, L. G., Pernar, L., Valentino, R. J., and Beck, S. G. 2003. Distinguishing characteristics of serotonin and non-serotonin-containing cells in the dorsal raphe nucleus: Electrophysiological and immunohistochemical studies. *Neuroscience* 116:669–683.

Kosofsky, B. E. and Molliver, M. E. 1987. The serotoninergic innervation of cerebral cortex: Different classes of axon terminals arise from dorsal and median raphe nuclei. *Synapse* 1:153–168.

Laruelle, M., Abi-Dargham, A., Casanova, M. F., Toti, R., Weinberger, D. R., and Kleinman, J. E. 1993. Selective abnormalities of prefrontal serotonergic receptors in schizophrenia: A postmortem study. *Archives of General Psychiatry* 50:810–818.

Little, K. Y., McLauglin, D. P., Ranc, J., Gilmore, J., Lopez, J. F., Watson, S. J., Carroll, F. I., and Butts, J. D. 1997. Serotonin transporter binding sites and mRNA levels in depressed persons committing suicide. *Biological Psychiatry* 41:1156–1164.

Lowther, S., De Paermentier, F., Cheetham, S. C., Crompton, M. R., Katona, C. L., and Horton, R. W. 1997. 5-HT1A receptor binding sites in post-mortem brain samples from depressed suicides and controls. *Journal of Affective Disorders* 42:199–207.

Lowther, S., De Paermentier, F., Crompton, M. R., Katona, C. L. E., and Horton, R. W. 1994. Brain 5-HT_2 receptors in suicide victims: Violence of death, depression and effects of antidepressant treatment. *Brain Research* 642:281–289.

Malone, K. M., Corbitt, E. M., Li, S., and Mann, J. J. 1996. Prolactin response to fenfluramine and suicide attempt lethality in major depression. *British Journal of Psychiatry* 168:324–329.

Mamounas, L. A., Mullen, C. A., O'Hearn, E., and Molliver, M. E. 1991. Dual serotoninergic projections to forebrain in the rat: Morphologically distinct 5-HT axon terminals exhibit differential vulnerability to neurotoxic amphetamine derivatives. *Journal of Comparative Neurology* 314:558–586.

Mann, J. J. 2003. Neurobiology of suicidal behaviour. *Nature Reviews Neuroscience* 4:819–828.

Mann, J. J., Huang, Y. Y., Underwood, M. D., Kassir, S. A., Oppenheim, S., Kelly, T. M., Dwork, A. J., and Arango, V. 2000. A serotonin transporter gene promoter polymorphism (5-HTTLPR) and prefrontal cortical binding in major depression and suicide. *Archives of General Psychiatry* 57:729–738.

Mann, J. J. and Malone, K. M. 1997. Cerebrospinal fluid amines and higher-lethality suicide attempts in depressed inpatients. *Biological Psychiatry* 41:162–171.

Mann, J. J., Malone, K. M., and Arango, V. 1996. The neurobiology of suicidal behavior. *Primary Psychiatry* 3:45–48.

Mann, J. J., McBride, P. A., Malone, K. M., DeMeo, M. D., and Keilp, J. G. 1995. Blunted serotonergic responsivity in depressed patients. *Neuropsychopharmacology* 13:53–64.

McLaughlin, D. P., Little, K. Y., Lopez, J. F., and Watson, S. J. 1996. Expression of serotonin transporter mRNA in human brainstem raphe nuclei. *Neuropsychopharmacology* 15:523–529.

Meller, E., Goldstein, M., and Bohmaker, K. 1990. Receptor reserve for 5-hydroxytryptamine1A-mediated inhibition of serotonin synthesis: Possible relationship to anxiolytic properties of 5-hydroxytryptamine1A agonists. *Molecular Pharmacology* 37:231–237.

Middlemiss, D. N. and Fozard, J. R. 1983. 8-Hydroxy-2-(di-*n*-propylamino)-tetralin discriminates between subtypes of the 5-HT_1 recognition site. *European Journal of Pharmacology* 90:151–153.

Mockus, S. M. and Vrana, K. E. 1998. Advances in the molecular characterization of tryptophan hydroxylase. *Journal of Molecular Neuroscience* 10:163–179.

Olszewski, J. and Baxter, D. 1954. In *Cytoarchitecture of the Human Brain Stem*, S. Karger, Basel, Switzerland, pp. 186.

Ono, H., Shirakawa, O., Kitamura, N., Hashimoto, T., Nishiguchi, N., Nishimura, A., Nushida, H., Ueno, Y., and Maeda, K. 2002. Tryptophan hydroxylase immunoreactivity is altered by the genetic variation in postmortem brain samples of both suicide victims and controls. *Molecular Psychiatry* 7:1127–1132.

Oquendo, M. A., Russo, S. A., Underwood, M. D., Kassir, S. A., Ellis, S. P., Mann, J. J., and Arango, V. 2006. Higher postmortem prefrontal 5-HT2A receptor binding correlates with lifetime aggression in suicide. *Biological Psychiatry* 59:235–243.

Owen, F., Cross, A. J., Crow, T. J., Deakin, J. F. W., Ferrier, I. N., Lofthouse, R., and Poulter, M. 1983. Brain 5-HT_2 receptors and suicide. *Lancet* 2:1256.

Pandey, G. N. 1997. Altered serotonin function in suicide. Evidence from platelet and neuroendocrine studies. *Annals of the New York Academy of Sciences* 836:182–200.

Pandey, G. N., Dwivedi, Y., Ren, X., Rizavi, H. S., Faludi, G., Sarosi, A., and Palkovits, M. 2006. Regional distribution and relative abundance of serotonin(2c) receptors in human brain: Effect of suicide. *Neurochemical Research* 31:167–176.

Pandey, G. N., Dwivedi, Y., Rizavi, H. S., Ren, X., Pandey, S. C., Pesold, C., Roberts, R. C., Conley, R. R., and Tamminga, C. A. 2002. Higher expression of serotonin 5-HT(2A) receptors in the postmortem brains of teenage suicide victims. *American Journal of Psychiatry* 159:419–429.

Parsey, R. V., Hastings, R. S., Oquendo, M. A., Huang, Y. Y., Simpson, N., Arcement, J., Huang, Y., Ogden, R. T., Van Heertum, R. L., Arango, V., and Mann, J. J. 2006. Lower serotonin transporter binding potential in the human brain during major depressive episodes. *American Journal of Psychiatry* 163:52–58.

Parsey, R. V., Oquendo, M. A., Simpson, N. R., Ogden, R. T., Van Heertum, R., Arango, V., and Mann, J. J. 2002. Effects of sex, age, and aggressive traits in man on brain serotonin 5-HT(1A) receptor binding potential measured by PET using [C-11]WAY-100635. *Brain Research* 954:173–182.

Patel, P. D., Pontrello, C., and Burke, S. 2004. Robust and tissue-specific expression of TPH2 versus TPH1 in rat raphe and pineal gland. *Biological Psychiatry* 55:428–433.

Peyron, C., Petit, J. M., Rampon, C., Jouvet, M., and Luppi, P. H. 1998. Forebrain afferents to the rat dorsal raphe nucleus demonstrated by retrograde and anterograde tracing methods. *Neuroscience* 82:443–468.

Placidi, G. P., Oquendo, M. A., Malone, K. M., Huang, Y. Y., Ellis, S. P., and Mann, J. J. 2001. Aggressivity, suicide attempts, and depression: Relationship to cerebrospinal fluid monoamine metabolite levels. *Biological Psychiatry* 50:783–791.

Raymond, J. R., Mukhin, Y. V., Gelasco, A., Turner, J., Collinsworth, G., Gettys, T. W., Grewal, J. S., and Garnovskaya, M. N. 2001. Multiplicity of mechanisms of serotonin receptor signal transduction. *Pharmacology and Therapeutics* 92:179–212.

Rosel, P., Arranz, B., San, L., Vallejo, J., Crespo, J. M., Urretavizcaya, M., and Navarro, M. A. 2000. Altered 5-HT_{2A} binding sites and second messenger inositol trisphosphate (IP_3) levels in hippocampus but not in frontal cortex from depressed suicide victims. *Psychiatry Research: Neuroimaging* 99:173–181.

Roy, A., Ågren, H., Pickar, D., Linnoila, M., Doran, A., Cutler, N., and Paul, S. M. 1986. Reduced CSF concentrations of homovanillic acid and homovanillic acid to 5-hydroxyindoleacetic acid ratios in depressed patients: Relationship to suicidal behavior and dexamethasone nonsuppression. *American Journal of Psychiatry* 143:1539–1545.

Shallice, T. and Burgess, P. 1996. The domain of supervisory processes and temporal organization of behaviour. *Philosophical Transactions of the Royal Society of London* 351:1405–1412.

Shelton, R. C., Sanders-Bush, E., Manier, D. H., and Lewis, D. A. 2009. Elevated 5-HT 2A receptors in postmortem prefrontal cortex in major depression is associated with reduced activity of protein kinase A. *Neuroscience* 158:1406–1415.

Sibille, E. and Hen, R. 2001. Serotonin(1A) receptors in mood disorders: A combined genetic and genomic approach. *Behavioural Pharmacology* 12:429–438.

Stamp, J. A. and Semba, K. 1995. Extent of colocalization of serotonin and GABA in the neurons of the rat raphe nuclei. *Brain Research* 677:39–49.

Stanley, M. and Mann, J. J. 1983. Increased serotonin-2 binding sites in frontal cortex of suicide victims. *Lancet* 1:214–216.

Stockmeier, C. A., Dilley, G. E., Shapiro, L. A., Overholser, J. C., Thompson, P. A., and Meltzer, H. Y. 1997. Serotonin receptors in suicide victims with major depression. *Neuropsychopharmacology* 16:162–73.

Stockmeier, C. A., Shapiro, L. A., Dilley, G. E., Kolli, T. N., Friedman, L., and Rajkowska, G. 1998. Increase in serotonin-1A autoreceptors in the midbrain of suicide victims with major depression-postmortem evidence for decreased serotonin activity. *Journal of Neuroscience* 18:7394–7401.

Törk, I. 1990. Anatomy of the serotonergic system. *Annals of the New York Academy of Sciences* 600:9–35.

Törk, I. and Hornung, J.-P. 1990. Raphe nuclei and the serotonergic system. In *The Human Nervous System*, G. Paxinos, ed. Academic Press, San Diego, CA, pp. 1001–1022.

Träskman, L., Åsberg, M., Bertilsson, L., and Sjöstrand, L. 1981. Monoamine metabolites in CSF and suicidal behavior. *Archives of General Psychiatry* 38:631–636.

Turecki, G., Briere, R., Dewar, K., Antonetti, T., Lesage, A. D., Seguin, M., Chawky, N., Vanier, C., Alda, M., Joober, R., Benkelfat, C., and Rouleau, G. A. 1999. Prediction of level of serotonin 2A receptor binding by serotonin receptor 2A genetic variation in postmortem brain samples from subjects who did or did not commit suicide. *American Journal of Psychiatry* 156:1456–1458.

Underwood, M. D., Khaibulina, A. A., Ellis, S. P., Moran, A., Rice, P. M., Mann, J. J., and Arango, V. 1999. Morphometry of the dorsal raphe nucleus serotonergic neurons in suicide victims. *Biological Psychiatry* 46:473–483.

Van Bockstaele, E. J., Biswas, A., and Pickel, V. M. 1993. Topography of serotonin neurons in the dorsal raphe nucleus that send axon collaterals to the rat prefrontal cortex and nucleus accumbens. *Brain Research* 624:188–198.

Van Den, B. A., De, Z. S., Heyrman, L., Mendlewicz, J., Adolfsson, R., Van Broeckhoven, C., and Del-Favero, J. 2005. Response to Zhang et al. (2005): Loss-of-function mutation in tryptophan hydroxylase-2 identified in unipolar major depression. *Neuron* 45, 11–16. *Neuron* 48:704–706.

Verge, D., Daval, G., Patey, A., Gozlan, H., El Mestikawy, S., and Hamon, M. 1985. Presynaptic 5-HT autoreceptors on serotonergic cell bodies and/or dendrites but not terminals are of the 5-HT1A subtype. *European Journal of Pharmacology* 113:463–464.

Walther, D. J., Peter, J. U., Bashammakh, S., Hortnagl, H., Voits, M., Fink, H., and Bader, M. 2003. Synthesis of serotonin by a second tryptophan hydroxylase isoform. *Science* 299:76.

Wang, R. Y. and Aghajanian, G. K. 1977. Antidromically identified serotonergic neurons in the rat midbrain raphe: Evidence for collateral inhibition. *Brain Research* 132:186–193.

Weiss, D. and Coccaro, E. F. 1997. Neuroendocrine challenge studies of suicidal behavior. *Psychiatric Clinics of North America* 20:563–579.

West, M. J. 1993. New stereological methods for counting neurons. *Neurobiology of Aging* 14:275–285.

Wilson, M. A. and Molliver, M. E. 1991a. The organization of serotonergic projections to cerebral cortex in primates: Regional distribution of axon terminals. *Neuroscience* 44:537–553.

Wilson, M. A. and Molliver, M. E. 1991b. The organization of serotonergic projections to cerebral cortex in primates: Retrograde transport studies. *Neuroscience* 44:555–570.

Wilson, M. A., Ricaurte, G. A., and Molliver, M. E. 1989. Distinct morphologic classes of serotonergic axons in primates exhibit differential vulnerability to the psychotropic drug 3,4-methylenedioxymethamphetamine. *Neuroscience* 28:121–137.

Winge, I., McKinney, J. A., Knappskog, P. M., and Haavik, J. 2006. Characterization of wild-type and mutant forms of human tryptophan hydroxylase 2. *Journal of Neurochemistry* 100:1648–1657.

Winge, I., McKinney, J. A., Ying, M., D'Santos, C. S., Kleppe, R., Knappskog, P. M., and Haavik, J. 2008. Activation and stabilization of human tryptophan hydroxylase 2 by phosphorylation and 14-3-3 binding. *Biochemical Journal* 410:195–204.

Zhang, X., Beaulieu, J. M., Sotnikova, T. D., Gainetdinov, R. R., and Caron, M. G. 2004. Tryptophan hydroxylase-2 controls brain serotonin synthesis. *Science* 305:217.

Zhang, X., Gainetdinov, R. R., Beaulieu, J. M., Sotnikova, T. D., Burch, L. H., Williams, R. B., Schwartz, D. A., Krishnan, K. R., and Caron, M. G. 2005. Loss-of-function mutation in tryptophan hydroxylase-2 identified in unipolar major depression. *Neuron* 45:11–16.

Zhou, Z., Peters, E. J., Hamilton, S. P., McMahon, F., Thomas, C., McGrath, P. J., Rush, J., Trivedi, M. H., Charney, D. S., Roy, A., Wisniewski, S., Lipsky, R., and Goldman, D. 2005. Response to Zhang et al. (2005): Loss-of-function mutation in tryptophan hydroxylase-2 identified in unipolar major depression. *Neuron* 45:11–16; *Neuron* 48:702–703.

Zill, P., Buttner, A., Eisenmenger, W., Moller, H. J., Ackenheil, M., and Bondy, B. 2007. Analysis of tryptophan hydroxylase I and II mRNA expression in the human brain: A post-mortem study. *Journal of Psychiatric Research* 41:168–173.

3 Noradrenergic Dysfunction in Depression and Suicide

Michelle J. Chandley and Gregory A. Ordway

CONTENTS

Norepinephrine (NE) is one of three catecholamine neurotransmitters in the brain and has been studied extensively in relation to the biology of suicide as well as psychiatric disorders that significantly increase the risk of suicide. NE became a candidate for the pathology of depression in the 1950s, but not because of a discovery of altered concentrations of NE in depressed patients or suicide victims. Instead, NE was one of the neurotransmitters along with dopamine and serotonin that was directly affected by newly discovered antidepressant drugs. Since that time, NE has been one of the most studied neurotransmitters with regard to depression biology and suicide, second only to serotonin. However, interest in the role of NE in suicide and depression has dwindled considerably over the past 10 years. In fact, interest in monoamines appears to be waning overall, possibly driven by a push by the National

Institutes of Health for paradigm shifts in understanding psychiatric disease biology. The move away from interest in the monoamines is also being driven by high-throughput technologies such as microarrays, which divert investigators from traditional disease candidates to novel proteins and pathways. Despite the current trends, evidence that points to dysfunction in the central noradrenergic system in depression and suicide is very strong and it remains quite possible that deficits in NE signaling may lie at the very root of psychiatric disorders that contribute to suicide. This chapter reviews the neurobiology and functional output of the brain noradrenergic system in relation to the potential involvement of NE in depression and suicide.

3.1 BRIEF REVIEW OF THE BRAIN NORADRENERGIC SYSTEM

3.1.1 ANATOMY

NE is produced primarily by neurons in the locus coeruleus (LC), which is located at the pontomesencephalic junction at the floor of the fourth ventricle in the pontine region of the brain.[1] Other nuclei with noradrenergic neurons include the lateral tegmentum and the nucleus of the solitary tract. Retrograde and ultrastructural examinations reveal that the core of the LC and its surrounding regions receive afferent projections from the insular cortex, the central amygdala, preoptic area, bed nucleus of the stria terminalis, hypothalamus, nucleus paragigantocellularis, prepositus hypoglossi, paraventricular nucleus, nucleus of solitary tract, lateral habenula, and cerebral cortex.[2–5] These afferent projections provide multiple neurochemical inputs to the LC with changes in LC neuronal firing being a highly coordinated event. Neurochemical interactions within the LC have been documented for numerous neurotransmitters including glutamate, gamma-aminobutyric acid (GABA), glycine, serotonin, dopamine, orexin, corticotropin-releasing factor (CRF), enkephalin, acetylcholine, and substance P.[6–15] Noradrenergic neurons of the LC express multiple receptor types for these various neurochemical inputs to the LC.[16,17] Some of the neurochemical inputs to the LC have received a great deal of attention in the recent literature with regard to depression and these are considered in greater detail later.

NE containing fibers from the LC innervate nearly the entire brain. These fibers leave the LC largely by five distinct efferent tracts. The central tegmental tract, central gray dorsal longitudinal fasciculus, and the ventral tegmental medial forebrain tract leave the LC to innervate the hypothalamus, limbic regions, thalamus, cortex, and olfactory bulb. A fourth pathway travels through the cerebellar peduncle to terminate in the cerebellum. A fifth tract descends into the spinal cord and mesencephalon.[18] NE release can occur at individual synaptic boutons in terminal fields or through *en passant* connections along the noradrenergic axons, thus allowing NE to be released throughout innervated regions of the brain.

3.1.2 NEUROCHEMISTRY

NE is synthesized from the precursor amino acid, tyrosine. An important step in the synthesis of NE, or other catecholamines, is an initial rate-limiting hydroxylation of tyrosine by tyrosine hydroxylase. NE biosynthesis is regulated by changes in phosphorylation-dependent tyrosine hydroxylase activity or by changes in the amount of available tyrosine hydroxylase protein. The regulation of tyrosine hydroxylase is cell specific, and possibly catecholamine specific, wherein conditions that stimulate tyrosine hydroxylase gene expression in noradrenergic neurons of the LC do not increase expression in midbrain dopaminergic neurons.[18] Once NE is synthesized in the brain noradrenergic neuron, it is packaged into vesicles via the vesicular monoamine transporters (VMAT2) and released into the synaptic cleft via calcium-dependent exocytosis. NE release is regulated by adrenoceptors located on noradrenergic neurons. Activation of presynaptic α_2-adrenoceptors (autoreceptors) inhibits NE release, while activation of presynaptic β_2-adrenoceptors facilitates NE release. The action of NE at the synaptic cleft is terminated largely through reuptake of the neurotransmitter by the NE transporter. Once in the neuron, NE is repackaged into vesicles for release or degraded by monoamine oxidase (MAO).

In the brain, NE signals through either $\alpha_{1(A,B,D)}$-, $\alpha_{2(A,B,C)}$-, β_1-, or β_2-adrenoceptors. Signaling of NE through α-adrenoceptors increases phospholipase (A, C, D) activity and decreases adenylyl cyclase activity, while activation of β-adrenoceptors increases adenylyl cyclase activity. The function of each receptor type in each of the projection areas, occurring on different cell types, is not fully characterized. The complexity with which NE actions are mediated is exemplified in the cerebral cortex, where different α- and β-adrenoceptor subtype receptor densities are displayed across the neocortical layers, as well as in multiple neuronal and glia cell types.[19–23]

3.2 NOREPINEPHRINE AND BEHAVIOR

NE, acting through its receptors, participates in the modulation of numerous behaviors, including the stress response, attention, memory, the sleep–wake cycle, decision making, and regulation of sympathetic states. An increase in NE activity can result in insomnia, anxiety, irritability, and hyperactivity, while reduced NE activity can result in lethargy and loss of alertness and focus. Various forms of stress and pain activate the LC[24] and sustained stress increases the demand for NE. Pharmacological reduction of the action of NE can be sedating, while pharmacologically increasing NE action modulates attention and focus, and can elevate mood in a depressed individual. In mood disorders and in attention disorders, NE reuptake inhibitors (NRIs) have been successfully used as antidepressants and attention enhancers, respectively, for many years. Although NE signaling participates in promoting specific behaviors such as arousal, attentiveness, memory formation, and the stress response, much remains to be discovered regarding its role in normal and abnormal behavior.

3.2.1 Norepinephrine and the Biology of Stress

A wealth of information exists demonstrating a very strong relationship between stress and the development of depression, suicidal behavior, and suicide.[25,26] It follows that biological systems that mediate stress may be particularly vulnerable to damage produced by sustained stress, or alternatively, deficiencies in these systems may be pathognomonic and contribute to the underlying vulnerability of a human to develop stress-related disorders such as depression or suicidal behaviors. Hence, a discussion of the role of the noradrenergic system in suicide and depression requires an understanding of how stress and NE biology are inextricable.

The central noradrenergic system, along with the CRF-median eminence system, is a principal component of the body's neurobiological response to stress. Exposure to stress activates the LC resulting in an increased demand for NE, driving an increase in synthetic capacity through phosphorylation of tyrosine hydroxylase and elevated levels of tyrosine hydroxylase protein. Stress-induced increases in LC activity and elevations in tyrosine hydroxylase are mediated in part by increased hypothalamic–pituitary–adrenal (HPA) axis activity, as well as through activation of stress-sensitive neuronal circuits that directly influence LC activity.[24,27–31] Although the precise mechanism of stress-induced activation of the LC has yet to be completely characterized, efferents from the CRF system are key mediators. The LC receives direct CRF synaptic input from the nucleus paragigantocellularis and Barrington's nucleus, while receiving pericoerulear CRF input from the amygdala, stria terminalis, and hypothalamus.[32,33] The LC also receives stress-sensitive glutamatergic input that is excitatory as noted later. Interestingly, ~30% of CRF terminals in the LC colocalize with glutamate,[15] thereby demonstrating dual regulation of LC activity by these two excitatory influences. CRF increases LC firing rate and increases levels of NE in projection areas during stress, facilitating a shift from focused attention to scanning attention[34,35] likely a mechanism that evolved to enhance survival.

Animal stressors including uncontrollable foot shock, hypotensive challenge, and even increased handling activate the LC, which can be blocked by CRF antagonists.[36–42] Exposure to repeated stress ultimately results in depletion of NE from noradrenergic neurons, as demand outweighs supply, and this depletion occurs in projection areas as well as in the LC itself.[43,44] Interestingly, repeated antidepressant treatments block the elevation of tyrosine hydroxylase in rat LC induced by repeated stress or chemical depletion of NE,[45,46] and decreases the activity of LC neurons, including decreasing sensory-induced stimulation of the LC.[47,48]

CRF regulates LC activity, but NE also regulates CRF release. Neurons from the LC innervate the paraventricular nucleus (PVN) of the hypothalamus, although a more substantial direct noradrenergic influence may come from non-LC neurons contributing to the ventral noradrenergic bundle.[2,49] It is possible that LC activity affects the PVN indirectly through contacts in noradrenergic projection areas such as the prefrontal cortex, amygdala, or the bed nucleus stria terminalis.[50] Direct innervation of the PVN comes from noradrenergic neurons in the nucleus of the solitary tract.[51] Noradrenergic receptors are found in the PVN[52,53] and activation of α_1-adrenoceptors stimulates adrenocorticotropic hormone (ACTH) secretion.[54] Infusion of NE into the PVN stimulates elevation of plasma corticosterone levels[55] and depletion of NE inhibits the corticosterone response to restraint stress.[56]

Regardless of the order of activation during stress with regard to CRF and NE neurons, the importance of LC activation in exacting a biological response to stress cannot be overstated. Interactions between NE and CRF effectively create a loop that produces both appropriate physiological and behavioral responses to stress. Interestingly, low doses of CRF reduce LC response to swim stress in rats indicating that noradrenergic neurons could desensitize to CRF.[57] It seems reasonable to suggest that a compromise or maladaptation in this loop could manifest as an inappropriate or faulty response to stress and ultimately result in a psychiatric disorder, such as depression. Evidence that maladaptation of the NE–CRF circuit occurs and can have enduring biological and behavioral consequences comes from studies demonstrating the effects of maternal separation of rodent pups from dams early in development. Maternal separation is used in rodents to model early life adversity, a factor known to contribute to psychiatric disease and suicide. Early maternal separation of rat pups results in elevated tonic activity of the LC, blunted activation in response to CRF, and reduced LC dendritic branching in adolescent rats.[58] Hence, stress insults during development can disrupt CRF–NE circuitry and this disruption may provide the initial neurobiological foundation for susceptibility to stress-induced psychiatric illness.

3.2.2 Norepinephrine and Cognition

In humans, the prefrontal cortex is responsible for executive processing, including functions known to be modulated by NE such as memory formation and retrieval, decision making, and selective attention.[59] NE release in the prefrontal cortex facilitates cognitive behaviors as demonstrated by numerous pharmacological manipulations involving adrenergic receptor signaling.[60] Treatment with the α_2-adrenoceptor agonists, clonidine, guanfacine, and meditomidine, improves memory and attention,[61–63] whereas administration of α_2-adrenoceptor antagonists suppresses working memory.[64] Conversely, high levels of NE acting at the α_1-adrenoceptor impair working memory as shown by cortical infusions of α_1-agonists in both rats and monkeys.[64,65] At the level of the LC, monitoring of task performance and selective attention behaviors in primates demonstrates that both of these higher cognitive functions are closely related to the phasic or tonic firing pattern of NE neurons. Higher levels of tonic firing of LC neurons result in poorer task performance, while cue elicited phasic firing of LC neurons facilitates good task performance during behavior assessment.[66]

Attenuation of prefrontal cortical activity is associated with depression (as reviewed by Savits and Drevets[67]), probably contributes to deficiencies in executive function as measured in MDD subjects,[68–70] and may result at least in part from disrupted noradrenergic signaling. The monoamine-depleting agent reserpine can induce depression-like behaviors in nonhuman primates[71] and reduces success on tests of cognition that are dependent on prefrontal cortical activity, the latter effect of which is reversed with the α_2-agonist, clonidine.[72] Functional imaging studies in humans performing selective attention tasks have shown that patients with MDD have increased blood flow in the prefrontal cortex after infusions of clonidine as compared to reduced blood flow in psychiatrically normal controls who also received clonidine infusions, directly implicating altered noradrenergic signaling specifically associated with depression.[73]

Cognitive impairments, that is, poor concentration, inappropriate choices, and impaired memory, are found in psychiatric illnesses other than depression as well, and recent findings demonstrate that cognitive deficits are linked to suicide. Subjects who had previous suicide attempts demonstrate deficits in problem solving, decision making,[74–76] and verbal fluency.[77] While some of the early studies[74,77] could not assess the confounds of depression on suicide attempts, one study found no significant differences in these cognitive functions when comparing depressed suicide attempters to depressed nonattempters using an older patient population.[78] However, subjects with bipolar disorder consistently exhibit deficiencies in attention, memory, and decision making, and those with suicide attempts score significantly worse on measures of decision making.[79] Depressed high-lethality suicide victims or suicide attempters perform significantly worse than psychiatrically normal controls on several measures of cognition, including general intelligence, attention, memory, and executive function.[68] Also, reduced orbitofrontal activation is associated with impaired decision-making behavior that is significantly reduced in MDD subjects who have previous suicide attempts when compared to MDD subjects without attempts.[80] A recent longitudinal single-photon emission computed tomography (SPECT) analysis of subjects prior to suicide completion demonstrated reduced premortem perfusion in the prefrontal cortex of depressed subjects with or without suicide completion as compared to psychiatric normal controls.[81] In this study, perfusion in the depressed suicide group during periods of concentration was significantly lower when compared to the depressed non-suicide group. These findings indicate that cognitive deficits associate with affective disorders such as MDD and are worse in those with suicide attempts, suggesting that substantially poor executive functioning may predispose suicidal behavior. Given the influence of NE in modulating cognition, it seems reasonable to assume that NE alterations could be a factor in the cognitive component of depression and suicide.

3.2.3 Norepinephrine, Sleep, and Depression

Sleep disturbances are considered a prominent feature in unipolar and bipolar depression and are listed as a diagnostic criterion of chronic depression (DSMIV-TR). Multiple studies of MDD have reported altered patterns of activity during all stages of sleep when compared to controls.[82,83] Altered sleep activity has been considered indicative of depression or identified as a comorbid behavior for depression and suicidal ideation.[84–89] Also, the converse must be considered that continually disrupted sleep or insomnia is causal of clinically defined depression.[90–92] Patients with sleep disturbances with and without diagnosed depression have demonstrated increases in suicide behavior.[93–96] This reciprocal relationship between sleep and depression indicates that common elements of the sleep–wake cycle and the neurobiology of depression may be intertwined. Activation of the LC is highly correlated with the state of arousal with higher neuronal firing during the awake state, complete LC neuronal inhibition during rapid eye movement (REM) sleep, and with lower firing rates during non-REM sleep.[97,98] Readers are referred elsewhere for excellent reviews of the role of the LC in sleep.[99,100] Interestingly, sleep deprivation in healthy controls increases plasma NE,[101,102] and sleep deprivation has antidepressant effects.[103] Also, rats kept in constant darkness for 6 weeks demonstrate robust reductions in cortical noradrenergic innervation, apoptosis of LC neurons, disruptions of the sleep–wake rhythm, and a depressive behavioral phenotype, some effects of which are reversed by the NE uptake inhibitor desipramine.[104] Given the physiological role of NE in the sleep–wake cycle, it is tempting to speculate that pathophysiological deficiencies of the noradrenergic system facilitate sleep disturbances that ultimately contribute to increased risk of suicidal behavior.

3.3 NORADRENERGIC NEUROBIOLOGY IN SUICIDE AND MOOD DISORDERS

3.3.1 PHARMACOLOGY

In the middle of the twentieth century, it was noted that some hypertensive patients treated with reserpine, a monoamine-depleting drug, developed a depressive state. In contrast, dysphoric tuberculosis patients treated with the antituberculosis agent iproniazid, which is also an inhibitor of monoamine catabolism, experienced mood elevation. These seminal observations led to the hypothesis that depressed mood may result from pathological deficiencies in the brain monoamines, NE, dopamine, and serotonin, which can be corrected by reducing monoamine catabolism.[105–107] Animal studies have since shown that drugs that alleviate depression in humans generally increase the availability of synaptic monoamines, and drugs that can induce depressive symptoms deplete monoamines. Today, multiple lines of pharmacological evidence indicate that central noradrenergic and serotonergic systems are involved in the therapeutic actions of antidepressants in mood disorders. Whether antidepressants specifically reduce suicide risk in general is debated, particularly when used in children. However, management of depressive symptoms with antidepressants is thought to lower the risk of suicide.[108]

With respect to alleviating depression, the pharmacological effects of antidepressants on central noradrenergic neurotransmission contribute to therapeutic actions. This conclusion is supported by several lines of evidence, from both laboratory animal and human studies. Animal experimentation has shown that lesions to the noradrenergic LC interfere with the antidepressant effects of the tricyclic antidepressant desipramine in animals.[109] Relative to other brain regions, the LC in humans and rats possesses high densities of binding sites for the key proteins to which antidepressant drugs bind, that is, NE transporter, MAO-A, and serotonin transporters,[110–113] suggesting that the LC is a major anatomical target for antidepressants that are MAO inhibitors, NRIs, or selective serotonin reuptake inhibitors (SSRIs). Independent research groups have shown that repeated treatment of rodents with antidepressants of different pharmacological classes, including SSRIs, results in a reduction in the activity of the LC,[48,114] contrasting the increase in LC activity produced by stress.[115–117] In humans, rapid catecholamine depletion in patients taking antidepressants with prominent noradrenergic actions causes a rapid relapse of depression, demonstrating a direct relationship between NE and therapeutic effect. Collectively, these studies demonstrate that NE plays a prominent role in antidepressant drug action. Whether noradrenergic antidepressants are more or less anti-suicide, as compared to serotonergic antidepressants, is a matter of debate.

3.3.2 CATECHOLAMINE DEPLETION STUDIES

The pathology of the noradrenergic system in depression and suicide has been studied using a variety of methods, including the use of blood, urine, and CSF samples from depressed or suicidal patients, catecholamine depletion in living patients, and postmortem brain tissues from depressed subjects and suicide victims. Of these studies, depletion studies provide some of the most convincing evidence to date that a deficiency of catecholamines is directly linked to depressive symptoms. Two independent studies have shown that rapid α-methyl-*p*-tyrosine (AMPT)-induced depletion of catecholamines in antidepressant-free euthymic subjects with a history of MDD results in a rapid relapse into depression,[118,119] whereas catecholamine depletion in psychiatrically normal subjects does not elicit depressive symptoms.[120] Treatment with AMPT failed to significantly worsen the depressive state of untreated, depressed patients,[121,122] nor did AMPT illicit depressive symptoms in subjects with other psychiatric disorders.[123,124] But administration of AMPT was sufficient to induce depressive responses from patients who were in remission using concurrent treatment with NRIs, but not SSRIs,[125,126] and induce changes in isolated behaviors, such as reward processing and attention.[127,128] Several studies indicate catecholamine depletion is associated with a depressed state, but current data is insufficient to determine if this relationship is causal or simply a trait of the disorder.[129–131] It should be noted that low CSF 3-methoxy-4-hydroxyphenylglycol (MHPG), the major metabolite of NE, predicts the risk of suicide attempts and the greater the MHPG deficiency the more lethal the suicide attempts.[25] Collectively, these findings suggest that patients with depression have a vulnerability to catecholamine depletion-induced depressive symptoms, and that low NE-induced depressive symptoms contribute to suicidal behavior.

A weakness of catecholamine depletion studies is that they do not distinguish between effects of the depleting agent on NE and dopamine, since synthesis of both transmitters is inhibited by the AMPT used to induce catecholamine depletion.[132,133] Some researchers have used findings of depletion studies to argue the weaknesses of the monoamine deficiency theory of depression, since normal humans do not become depressed when catecholamines are depleted. Even in depressed humans, treatment with AMPT fails to significantly worsen the depressive state of untreated, depressed patients.[121,122] Hence, it has been argued that monoamines are indirect in the illness, working through some primary neurotransmitter to cause mood dysfunction. However, the time course of the depletion studies is very short and is not likely to recapitulate the behavioral sequelae of sustained dysfunction. It remains reasonable to conjecture that a primary insult to a monoamine system, each of which have diffuse brain targets, could result in secondary disruptions of other transmitter systems that over some period of time results in depression. The conjecture that the reverse is true, that is, a primary disruption of another system may ultimately result in disruption of monoamines, is neither more nor less likely, given our current understanding of the biology of depression. Certainly, it seems unlikely that there is a single failing transmitter system that produces any complex psychiatric illness such as depression. Studies examining monoamine depletion in depressed subjects are also confounded by a potential ceiling effect of monoamine depletion, as well as a placebo effect. In fact, some depressed patients' mood improved after monoamine depletion.[121] Overall, arguments that monoamines, including NE, are not likely to be involved in the pathophysiology of the illness seem weak in the face of the collective data supporting the role of these transmitters in depression.

3.3.3 Postmortem Brain Pathology

Several laboratories have utilized postmortem brain tissues to study the putative pathology of noradrenergic neurons in suicide and depression. A key confounding issue for studies of the noradrenergic system in postmortem brain tissues is that it is well known from animal studies that antidepressant drug treatments regulate the expression of a variety of noradrenergic receptors and their signaling proteins in several brain regions including the LC. Hence, it is critical that postmortem toxicology rule out the presence of these drugs. The earliest of these postmortem studies typically utilized tissues collected from suicide victims and natural death control subjects, for whom psychiatric and medical histories were not available. Over the past 15–20 years, however, most postmortem brain studies of suicide and depression have used tissues from subjects for whom psychiatric diagnosis was known, or was determined posthumously through psychiatric autopsy, using standardized questionnaires administered to next of kin. For many of these studies, it has been difficult to separate suicide from the behaviors associated with depression because most suicide victims suffer premortem depressive symptoms,[134,135] although they may not have had sufficient symptoms to be diagnosed with an Axis I MDD. Another approach to the issue of depression vs. suicide is to study natural or accidental death subjects who die with an active diagnosis of MDD along with matched MDD suicide victims to determine the differences in the comparison of these two groups to psychiatrically normal control subjects. However, there have been only a few studies that have used this approach because tissues from non-suicide MDD subjects that do not have antidepressants in their blood are not nearly as available as tissues from MDD suicide victims without positive antidepressant toxicology.

Using postmortem brain tissue from psychiatrically characterized subjects, researchers have focused study on noradrenergic signaling molecules, for example, noradrenergic receptors and linked second messenger systems, in noradrenergic projection areas and also on a variety of biochemical and cellular indices in the cell body region of noradrenergic neurons. Readers are referred to other review articles for a detailed summary of this research.[136–139] For many years, noradrenergic receptors, particularly α_2- and β-adrenoceptors, received considerable attention in studies of postmortem brain pathology in depression and suicide. Elevated or unchanged densities of agonist binding to α_2-adrenoceptors have been reported for various projection regions in depression and suicide, as compared to matched control subjects.[140–144] Increased levels of mRNA for the α_2-adrenoceptors were demonstrated in frontal cortex from suicide postmortem tissue, where 9 of 12 suicide subjects had a previous diagnosis of MDD.[145] In addition, significant elevations in α_2-adrenoceptor immunoreactivity have been described, as well as increases in signaling-related G-proteins and kinases in prefrontal cortex in depressed suicide subjects.[146] α_2-Adrenoceptor-linked Gαi signaling was increased in depressed victims of suicide, but serotonin, muscarinic, and μ-opioid signaling was unchanged.[147,148] Collectively, these studies demonstrate elevated α_2-adrenoceptor binding and signaling in depression and suicide. α_2-Adrenoceptors occur presynaptically on noradrenergic and serotonergic neuronal terminals (as well as on other types of neurons) and inhibit neurotransmitter release upon stimulation. One interpretation of these findings is that supersensitive presynaptic α_2-adrenoceptors could lead to reduced NE and serotonin release. An alternative explanation derives from rodent studies that demonstrate that pharmacological depletion of NE results in an adaptive upregulation of α_2-adrenoceptors and associated Gi proteins.[149–151] Hence, elevations of α_2-adrenoceptors in depressed suicide may reflect an adaptation to reduced NE. Regardless of the interpretation, it can be argued that the abnormalities in α_2-adrenoceptors in brains from depressed suicide victims suggest that depression and/or suicide is associated with reduced brain NE.

There is little support in the literature to implicate β-adrenoceptors in the pathology of depression or suicide. Early studies of the β-adrenoceptor in psychiatrically uncharacterized brain tissue of suicide victims indicated elevated receptor binding in the prefrontal cortex.[110,152] Subsequently, studies using postmortem tissue from antidepressant-free suicide subjects have reported either decreased or unchanged β-adrenoceptor densities.[153–155] Using tissue from depressed or other psychiatric disorders has also failed to conclusively support alterations in β-adrenoceptors in suicide.[144,156]

Numerous researchers have investigated putative noradrenergic pathology in suicide and depression by studying the cellular source of brain NE, the LC. Using quantitative Western blotting or quantitative immunoblotting methods, elevated levels of the rate-limiting enzyme in NE, tyrosine hydroxylase, were found in the LC of psychiatrically uncharacterized subjects who died by suicide compared to controls[157] and in MDD depressive subjects (most of which died by suicide) compared to psychiatrically normal control subjects.[158] No changes in the levels of mRNA for tyrosine hydroxylase in the LC of suicide victims with MDD have been reported.[159] Our laboratory has also consistently found a lack of change in the levels of tyrosine hydroxylase mRNA in the LC of suicide victims with MDD (unpublished observations). Given repeated demonstrations of elevated tyrosine hydroxylase protein in the LC of suicide victims, we conclude that elevated LC tyrosine hydroxylase protein is not a result of an elevation of tyrosine hydroxylase gene expression, but occurs through other nontranscriptional mechanisms as known to be in control of LC tyrosine hydroxylase in the rodent brain.[30]

Investigators have also estimated the number of LC neurons in suicide, and these findings are mixed. Low numbers of LC neurons have been reported in psychiatrically uncharacterized suicide victims compared to control subjects, depending on age.[160] In contrast, normal LC neuron numbers have been observed in depressed suicide victims[111,158,161] and in depressed elderly non-suicide subjects,[162] albeit these latter studies did not estimate total cell number but numbers of cells per tissue section along the axis of the LC.

Other measures of LC biochemistry reveal a complex pathology in depression and suicide. Radioligand binding to α_2-adrenoceptors in the LC was found elevated in uncharacterized suicide victims[157] and in suicide victims with MDD[163] relative to normal control subjects, as was generally observed for studies of this receptor in the frontal cortex (see above). In addition, reduced levels of NE transporter were found in the LC of suicide subjects with MDD.[111]

Interpreting the biochemical basis of findings of noradrenergic pathology in human postmortem tissue can be facilitated by examining animal studies wherein noradrenergic neural activity is manipulated pharmacologically or behaviorally. Depletion of NE by administration of reserpine or 6-hydroxydopamine, both of which increase LC activity, results in upregulation of tyrosine hydroxylase and α_2-adrenoceptor levels,[28,150] paralleling elevations in LC tyrosine hydroxylase and α_2-adrenoceptors observed in MDD and suicide. Exposure to chronic stressors in rodents activates the LC,[24] depletes NE,[24,43,44] and causes an upregulation of tyrosine hydroxylase[28,30,31] in the LC, an effect similar to what is observed in the LC of MDD and suicide subjects. Likewise, depleting catecholamines with reserpine and 6-hydroxydopamine significantly reduces NE transporter levels,[164,165] paralleling reduced NE transporter levels in depression noted earlier. In contrast, repeated treatment of rodents with antidepressants reduces LC activity[48,114,166,167] and results in reductions in the expression of α_2-adrenoceptors and tyrosine hydroxylase,[168–170] changes opposite in direction to what is observed in depression and suicide. Interestingly, a NE transporter knockout mouse model displays increased mRNA and protein levels of α_2-adrenoceptors in the brain stem and hippocampus,[171] providing a mechanistic link between reduced NE transporter and elevated α_2-adrenoceptors. Overall, the similarity of findings from catecholamine depletion and chronic stress studies with rodent and the human postmortem findings from depressed and/or suicide subjects strongly imply elevated premortem activity of the central noradrenergic system, and possibly a synaptic deficiency in NE. One might conclude from this literature that therapies that reduce the activity of the LC, particularly activity that is a result of stress,[172] will have utility in the treatment of depression. In this regard, NE uptake inhibitors have the dual ability to reduce the noradrenergic activity through uptake inhibition in the LC, while simultaneously maintaining or even elevating synaptic noradrenergic activity through local uptake inhibition in projection areas. As noted earlier, repeated treatment of rats with SSRIs also reduces LC activity. Figure 3.1 provides a diagram hypothesizing how changes in LC activity, or reactivity to stress, may play a modulatory role in the development of susceptibility to depression and suicide.

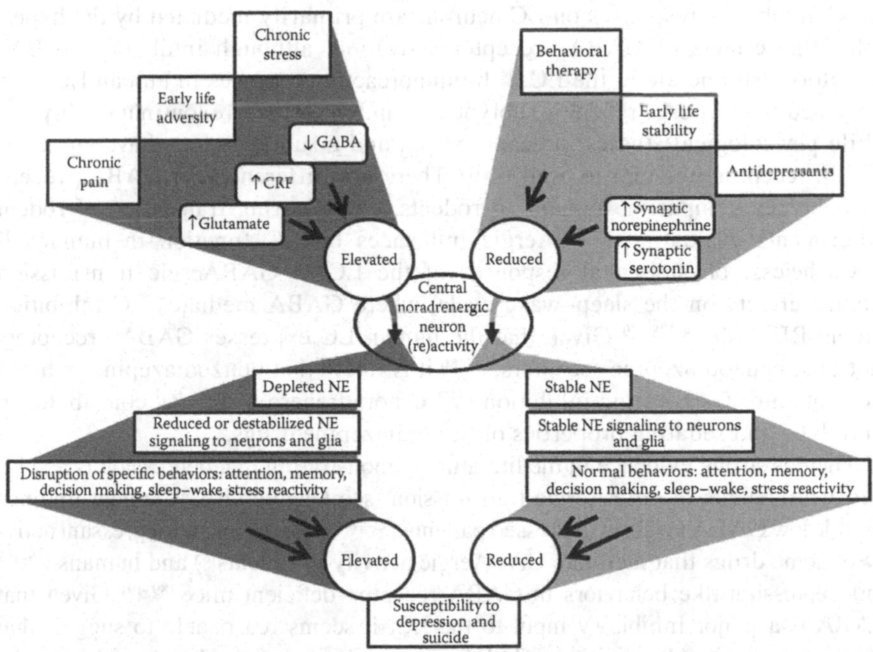

FIGURE 3.1 Role of central noradrenergic neurons in mediating the biological impact of factors that increase or decrease susceptibility to depression and suicide. It is assumed that early life stability and behavior therapy would counteract the biological impact of stress and adversity on the stress reactivity of noradrenergic neurons and subsequently reduce susceptibility to depression and suicide, although no specific evidence of this is presented.

3.4 NORADRENERGIC CIRCUITRY: CURRENT AND FUTURE TARGETS

The noradrenergic LC receives a variety of neurotransmitter and paracrine inputs that modulate its activity. Among these, several are particularly sensitive to stress or modulate stress activation of the LC. Here, we focus on three systems that influence central noradrenergic activity and/or are influenced by NE, emphasizing systems that have received a great deal of interest recently in the field of suicide and depression biology either because of a growing wealth of information implicating them or because of the development of new promising antidepressant drugs that act through those systems.

3.4.1 GABA and Norepinephrine

Unlike many brain regions, the LC has no GABAergic interneurons of which we are aware. Nevertheless, this noradrenergic nucleus is under inhibitory GABAergic control that originates in the nucleus prepositus hypoglossi (PrH) in the medulla.[3,9] Synaptic junctions within the LC have immunoreactivity for GABA decarboxylase (GAD-67).[173] Stimulation of the PrH inhibits firing of LC neurons, which can be substantially reduced with either systemic administration of picrotoxin or localized LC microinfusion of bicuculline, both GABA antagonists.[9] Inhibitory responses on LC neurons are primarily mediated by the hyperpolarizing effects of $GABA_A$ receptor activation, although inhibitory $GABA_B$ receptors also operate in the LC.[174] Immunoreactivity studies in human LC have identified the $GABA_A$ receptor subtypes α_3 and γ_2 as the predominant subtypes, while physiological studies indicate the $\beta_{1/3}$ and ε subtypes may have an active role in receptor function as well.[175–177] There are differences in $GABA_A$ receptor subtypes comparing humans to rodents, complicating translation of rodent experiments evaluating GABAergic influences on LC function in humans.[176] Nevertheless, physiological responses of the LC to GABAergic transmission include effects on the sleep–wake cycle, where GABA mediates LC inhibition during REM sleep.[178,179] Given that the human LC expresses $GABA_A$ receptors that bind benzodiazepine receptors,[176,180] it is likely that benzodiazepine-induced facilitation of GABAergic inhibition of LC noradrenergic activity contributes to anxiolytic and sedating properties of benzodiazepine drugs.

There is strong evidence in the literature demonstrating that depression is associated with deficits in GABAergic transmission (see Chapter 10).[181] Salient findings include low GABA levels in depressed patients,[182–184] evidence of antidepressant activity of some drugs that facilitate GABAergic activity in rodents[185] and humans,[186–188] and depression-like behaviors in GABA receptor–deficient mice.[189–191] Given that GABA is a major inhibitory input to the LC, it seems reasonable to suggest that drugs that specifically facilitate GABAergic inhibition of the LC would have the ability to alleviate stress-induced disease, such as depression. Given that the LC expresses specific subtypes of $GABA_A$ receptors, agonists specific for these receptors would be expected to enhance antidepressant action by facilitating the calming of stress-induced increases in LC activity.

3.4.2 Glutamate and Norepinephrine

Glutamate has robust effects on the activity of noradrenergic neurons in the LC. The primary glutamatergic input to the LC is the nucleus paragigantocellularis located in the brain stem,[2] although there are glutamate neurons in the prefrontal cortex that also project to the LC.[11] There is also excitatory input (presumably glutamatergic) from the habenula to the LC.[5,192,193] Glutamatergic terminals form synapses with tyrosine hydroxylase–containing dendrites in the LC.[32,194] Gene expression of several different glutamate receptors, including N-methyl-D-aspartate (NMDA) subtypes, kainate subtypes, AMPA subtypes, and metabotropic subtypes, occurs in the rodent[195] and primate LC.[196] At least in brain slice preparations, glutamate activates LC neurons through NMDA, AMPA, kainate, and presumably metabotropic receptors.[197,198] The origin of excitatory glutamatergic input to the LC may be aligned with specific glutamate receptors expressed by LC neurons because PGi activation (and noxious stimuli-induced activation) of the LC is blocked by broad spectrum ionotropic glutamate receptor antagonists, but not by selective NMDA receptor antagonists.[199] In contrast, stimulation of the medial prefrontal cortex glutamatergic input to the LC can be inhibited by NMDA and non-NMDA glutamate receptor antagonists.[11] The metabotropic glutamate receptors, mGluR2/3, appear to play a role in opiate withdrawal-induced activation of the LC[200] and blockade of mGluR5 receptors results in a significant reduction of NE in the frontal cortex of rodents,[198] although whether these actions are mediated by metabotropic receptors occurring on LC neurons and their processes is not known.

In the recent past, there has been growing interest in the putative role of glutamate in depressive disorders and suicide. The detailed evidence for this is discussed elsewhere in this book. Briefly, multiple lines of evidence from both animal and human studies indicate that glutamate signaling is altered in mood disorders. MRI studies have shown regionally specific abnormalities in glutamate levels in the brain in subjects suffering from depression.[184,201–203] Altered expression of several glutamate-related genes has been reported in depression using postmortem tissue from temporal cortex.[204,205] Glutamatergic inputs to the LC are activated by a variety of stressors, visceral stimulation, and pain pathways.[206,207] Ketamine, a NMDA receptor antagonist, has been shown in randomized clinical trials to have uniquely rapid antidepressant effects within hours of postinfusion, with effects persisting for 2 weeks.[208] Most recently, a single infusion of ketamine had significant beneficial effects for patients experiencing suicide ideation.[209,210] Interestingly, animal models indicate that ketamine's blockade of NMDA receptors drives glutamate selection for AMPA-mediated signaling, because antidepressant-like effects of ketamine were lost when mice were pretreated with the AMPA receptor antagonist, NBQX.[211] Clinical treatment with the stimulator of glutamate uptake, riluzole, also has beneficial effects in treatment-resistant depression.[212,213] In addition to antidepressant effects of these glutamate drugs, preclinical studies with animals using antagonists for mGluR1, mGluR3, and mGluR5 improve behaviors associated with mood disorders (as reviewed by Witkin et al.[214]). Collectively, these findings demonstrate that glutamatergic activity is elevated in depression and that antagonism of glutamatergic activity through NMDA and metabotropic glutamate receptors has antidepressant effects.

The robust modulation of noradrenergic LC activity by glutamate and evidence of glutamate and NE pathologies in depression raises the possibility that dysfunction of glutamate–NE interactions may be an important component of depression and suicide pathology. Recent studies support this contention. Measurement of receptor immunoreactivities in the postmortem LC revealed elevated levels of the NR2C subunit of the NMDA receptor in subjects suffering from MDD (mostly suicides) as compared to psychiatrically normal control subjects.[215] In addition, reduced levels of neuronal nitric oxide synthase (nNOS) protein, a key mediator in glutamate receptor signaling, in the LC were associated with MDD.[216] Recently, Bernard et al.[204] demonstrated several abnormalities in glutamate-related gene expressions in the LC of MDD suicides using gene expression microarrays, including reduced glutamate transporters and increased glutamate receptors. Elevated glutamate receptor expression in depression may be secondary to elevated stress-sensitive glutamate input to the LC since exposure of rats to chronic stress results in upregulated gene expression of ionotropic glutamate receptors including NMDA receptors.[217–220] Reduced glutamate transporter expression, presumably resulting in sustained synaptic glutamate, could simultaneously intensify elevated excitatory input to the LC during stress. Regardless of the interpretation, thesc findings demonstrate that glutamatergic input to the LC is abnormal in MDD and suicide.

To summarize, glutamate is a major excitatory input to the LC and can be activated by stress. Elevated glutamate has been repeatedly reported in MDD subjects. Postmortem studies demonstrate altered glutamate input to the LC in MDD suicide victims. Hence, it seems reasonable to suggest that elevated glutamatergic activation of the LC, particularly during exposure to stress, may be an important aspect of the pathology of depression and suicide. While other theories are proposed to account for the antidepressant actions of ketamine,[221] blockade of excitatory glutamatergic input to the LC could contribute to the overall antidepressant properties of ketamine, particularly given that most, if not all, current antidepressant drugs reduce LC activity following repeated treatment. Ketamine, which has rapid antidepressant effects, would be expected to rapidly antagonize glutamate activation of the LC, whereas traditionally antidepressants (which produce slow antidepressant response) silence the LC only after repeated administration (especially SSRIs) over a period of 2 weeks. It is likely that ketamine produces antidepressant responses in humans through a variety of mechanisms, but inhibition of glutamatergic activation of the LC is likely to contribute to the therapeutic effect of this drug.

3.4.3 Glia and Norepinephrine

NE is a neurotransmitter, implying transmission from one neuron to another. However, another major function of NE is communication with glia, including astrocytes, oligodendrocytes, and microglia. In fact, many noradrenergic terminals and boutons do not form synaptic contacts with neurons. In some areas of the brain, noradrenergic contacts with astrocytes are more common than contacts with neurons. For example, noradrenergic neurons terminate in the ventral tegmental area but these terminals contact astrocyte processes rather than tegmental neurons.[222] Direct contacts between noradrenergic terminals and glia have also been demonstrated in other regions of the brain, such as the cerebral cortex, where nearly approximately one-third of contacts are with glia rather than neurons.[223] Noradrenergic neurons appear to communicate with glial cells through noradrenergic receptors that are expressed by glia.[224–228] In fact, astrocytes and oligodendrocytes, as well as microglia, express several different noradrenergic receptors, the activation of which stimulates intracellular signaling cascades, and ultimately regulates a variety of cellular processes including glycogenolysis,[229] glycolysis,[230,231] glutamate uptake,[232,233] metabolism,[234] and calcium mobilization.[223]

Until the late 1990s, most investigations of the pathology of depression were "neurocentric," focusing on neurons and their synaptic interactions. However, a growing body of literature shows that depression and probably suicide also involve glial pathology. Cell-counting studies using human postmortem tissue have demonstrated a loss of glial cells in the anterior cingulate cortex, dorsolateral prefrontal cortex, and the orbitofrontal cortex in depression, many cases of which were suicides.[235,236] Reductions in glia have also been reported for the amygdala in MDD, many cases of which were suicides.[237,238] In contrast to the cortex and amygdala, elevated glial cell density has been reported in the hippocampus of MDD subjects, most of which died by suicide.[239] Hence, morphometric analyses indicate glial cell pathology in MDD and suicide is brain-region specific. Other indices of dysfunctional glia in depression and suicide include glial fibrillary acidic protein (GFAP) loss in the frontal cortex of subjects with MDD and bipolar disorder, some of whom died by suicide,[240,241] and microgliosis in postmortem cortical and thalamic regions from suicide victims with schizophrenia and mood disorders.[242] Readers are referred to a recent review by Hercher et al.[243] regarding detailed evidence of glia abnormalities in depression and suicide.

Given the functionally dynamic relationship between NE and glia, particularly astrocytes, it is possible that NE–glial interactions are dysfunctional in depression and suicide. Studies that support or address this possibility are presently few, but intriguing nonetheless. Bernard et al.[204] probed homogenized LC tissue for gene expression differences between MDD subjects (half of which were suicides) and matched control subjects and discovered a number of gene expression changes, including GFAP, specifically implicating glia pathology in the LC. As noted earlier, Bernard et al.[204] also reported a reduction in the gene expression of glutamate transporters in the LC, transporters known to be expressed predominately by glia. This reduction may partly explain why riluzole, a drug that increases Glt-1 glutamate transporter function[244] and gene expression,[245,246] demonstrates antidepressant activity in rodents and humans.[247] Whether NE plays a role in the reduced excitatory amino acid transporter (EAAT) expression in the LC is not known, but it is noted that NE regulates glutamate uptake by astrocytes via α_1- and β-adrenoceptors, as shown *in vivo*[248] and by using *in vitro* cultured astrocytes. At low concentrations, NE reduces glutamate uptake through β-adrenoceptors, while higher concentrations of NE increase glutamate uptake through α_{1b}-adrenoceptors.[232,233,249] Hence, deficient NE concentrations in MDD could facilitate reduced glutamate uptake by astrocytes in the LC region, resulting in prolonged stress-sensitive excitatory activation of the LC and further NE depletion.

Recently, our group used laser capture microdissection to examine gene expression in neurons, astrocytes, and oligodendrocytes from the LC of MDD suicide subjects and psychiatrically normal control subjects. We focused on a family of developmental/trophic factors, that is, bone morphogenetic proteins (BMPs), and found significantly lower BMP7 expression specifically in astrocytes from MDD subjects relative to controls. *BMP7* has unique developmental and trophic actions on catecholamine neurons suggesting that reduced astrocyte support for pontine LC neurons may contribute to pathology of brain noradrenergic neurons in MDD.[250] Interestingly, chronic social defeat exposure of rats results in reduced BMP7 expression in the LC,[250] although preliminary evidence shows that antidepressant treatment of rats does not upregulate *BMP7* (G.A. Ordway, unpublished findings).

In conclusion, there is overwhelming evidence of dysfunctional glia in MDD, and to some extent in suicide. Recent studies point specifically to glial deficits in the noradrenergic LC, including evidence of reduced glutamate transporter expression and reduced neurotrophic support for LC neurons. Given the intimate functional relationship between NE and glia, particularly astrocytes, it seems reasonable to hypothesize that glia deficits may be secondary to a deficiency of NE. On the other hand, it is equally possible that glia deficits arise first and the outcome of these is dysfunctional noradrenergic (and serotonergic) neurons. This quandary will only be unraveled through additional clinical and basic research.

3.5 SUMMARY AND CONCLUSIONS

NE is a chemical transmitter synthesized and released by a relatively small number of neurons in the brain that communicate information to many different types of neurons and glia, and to vascular elements. Since noradrenergic neurons project to nearly all brain regions, NE modulates a vast array of cellular processes, modifies networks of neurons and glia, and ultimately influences a large number of behaviors. Despite the diversity of noradrenergic projections, NE's actions in the brain appear to have profound effects on a smaller set of behaviors, including those that are commonly disrupted in depression. Recent postmortem findings demonstrate pathology of NE neurons and their surrounding glia in brains from depressed victims of suicide, strongly implicating a role of NE in the pathogenesis/etiology of the depression. Since depression is associated with 80% of suicides as assessed retrospectively,[134,135] it is likely that NE dysfunction also contributes to suicide, although this effect may be indirect through negative effects of NE depletion on mood. Stress, a common precipitator of depression and suicide, activates the noradrenergic LC and chronic stress depletes NE in noradrenergic projection areas as well as in the LC. Depletion of NE precipitates depression in individuals with a history of depression, but not in individuals who have no history of psychiatric illness. Hence, depression occurs in individuals that have a particular susceptibility to NE depletion-induced mood changes. The perplexing unknown is the constitution of the "susceptibility," which is presumably biological and likely to involve stress-activated transmitters such as glutamate and GABA, and stress hormones such as CRF and cortisol, systems of which have profound effects on central noradrenergic neuronal activity. Also highly likely is a fundamental role of glia dysfunction in susceptibility to depression, and glial dysfunction may be intertwined neurobiologically with monoamine deficits. Given the central role of NE in mediating biological response to stress and the critical role of stress in the development of depression, it seems likely that manipulation of noradrenergic transmission will continue to be a direct or indirect therapeutic endpoint in the development of new antidepressant drugs. In fact, drugs that combine actions at traditional noradrenergic targets (NE transporters, α_2-adrenoceptors) with other stress-sensitive transmitter targets (glutamate transporters and receptors, GABA receptors) may provide a broader spectrum of activity as well as an accelerated onset of action, and would be expected to contribute to a reduction in suicide.

REFERENCES

1. Dahlström A, Fuxe K. Localization of monoamines in the lower brain stem. *Experientia*. 1964;20(7):398–399. Available at: http://www.ncbi.nlm.nih.gov/pubmed/5856530
2. Aston-Jones G, Ennis M, Pieribone VA, Nickell WT, Shipley MT. The brain nucleus locus coeruleus: Restricted afferent control of a broad efferent network. *Science (New York, N.Y.)*. 1986;234(4777):734–737. Available at: http://www.ncbi.nlm.nih.gov/pubmed/3775363
3. Cedarbaum JM, Aghajanian GK. Afferent projections to the rat locus coeruleus as determined by a retrograde tracing technique. *The Journal of Comparative Neurology*. 1978;178(1):1–16. Available at: http://www.ncbi.nlm.nih.gov/pubmed/632368
4. Mantyh PW, Hunt SP, Maggio JE. Substance P receptors: Localization by light microscopic autoradiography in rat brain using [^{3}H]SP as the radioligand. *Brain Research*. 1984;307(1–2):147–165. Available at: http://www.ncbi.nlm.nih.gov/pubmed/6087984
5. Sutherland RJ. The dorsal diencephalic conduction system: A review of the anatomy and functions of the habenular complex. *Neuroscience and Biobehavioral Reviews*. 1982;6(1):1–13. Available at: http://www.ncbi.nlm.nih.gov/pubmed/7041014
6. Curtis AL, Lechner SM, Pavcovich LA, Valentino RJ. Activation of the locus coeruleus noradrenergic system by intracoerulear microinfusion of corticotropin-releasing factor: Effects on discharge rate, cortical norepinephrine levels and cortical electroencephalographic activity. *The Journal of Pharmacology and Experimental Therapeutics*. 1997;281(1):163–172. Available at: http://www.ncbi.nlm.nih.gov/pubmed/9103494
7. Date Y, Ueta Y, Yamashita H et al. Orexins, orexigenic hypothalamic peptides, interact with autonomic, neuroendocrine and neuroregulatory systems. *Proceedings of the National Academy of Sciences of the United States of America*. 1999;96(2):748–753. Available at: http://www.pubmedcentral.nih.gov/articlerender.fcgi?artid=15208&tool=pmcentrez&rendertype=abstract
8. Ennis M, Aston-Jones G. A potent excitatory input to the nucleus locus coeruleus from the ventrolateral medulla. *Neuroscience Letters*. 1986;71(3):299–305. Available at: http://www.ncbi.nlm.nih.gov/pubmed/3025783
9. Ennis M, Aston-Jones G. GABA-mediated inhibition of locus coeruleus from the dorsomedial rostral medulla. *The Journal of Neuroscience: The Official Journal of the Society for Neuroscience*. 1989;9(8):2973–2981. Available at: http://www.ncbi.nlm.nih.gov/pubmed/2769374
10. Fodor M, Görcs TJ, Palkovits M. Immunohistochemical study on the distribution of neuropeptides within the pontine tegmentum—Particularly the parabrachial nuclei and the locus coeruleus of the human brain. *Neuroscience*. 1992;46(4):891–908. Available at: http://www.ncbi.nlm.nih.gov/pubmed/1542421

11. Jodo E, Aston-Jones G. Activation of locus coeruleus by prefrontal cortex is mediated by excitatory amino acid inputs. *Brain Research*. 1997;768(1–2):327–332. Available at: http://www.ncbi.nlm.nih.gov/pubmed/9369332
12. Luppi PH, Charlety PJ, Fort P et al. Anatomical and electrophysiological evidence for a glycinergic inhibitory innervation of the rat locus coeruleus. *Neuroscience Letters*. 1991;128(1):33–36. Available at: http://www.ncbi.nlm.nih.gov/pubmed/1717896
13. Pickel VM, Joh TH, Reis DJ. A serotonergic innervation of noradrenergic neurons in nucleus locus coeruleus: Demonstration by immunocytochemical localization of the transmitter specific enzymes tyrosine and tryptophan hydroxylase. *Brain Research*. 1977;131(2):197–214. Available at: http://www.ncbi.nlm.nih.gov/pubmed/19125
14. Sakai K. Physiological properties and afferent connections of the locus coeruleus and adjacent tegmental neurons involved in the generation of paradoxical sleep in the cat. *Progress in Brain Research*. 1991;88:31–45. Available at: http://www.ncbi.nlm.nih.gov/pubmed/1687620
15. Valentino RJ, Rudoy C, Saunders A, Liu XB, Van Bockstaele EJ. Corticotropin-releasing factor is preferentially colocalized with excitatory rather than inhibitory amino acids in axon terminals in the peri-locus coeruleus region. *Neuroscience*. 2001;106(2):375–384. Available at: http://www.ncbi.nlm.nih.gov/pubmed/11566507
16. Aston-Jones G, Shipley MT, Chouvet G et al. Afferent regulation of locus coeruleus neurons: Anatomy, physiology and pharmacology. *Progress in Brain Research*. 1991;88:47–75. Available at: http://www.ncbi.nlm.nih.gov/pubmed/1687622
17. Van Bockstaele EJ, Reyes BAS, Valentino RJ. The locus coeruleus: A key nucleus where stress and opioids intersect to mediate vulnerability to opiate abuse. *Brain Research*. 2010;1314:162–174. Available at: http://www.ncbi.nlm.nih.gov/pubmed/19765557
18. Cooper JR, Bloom FE, Roth RH. *The Biochemical Basis of Neuropharmacology*. Oxford, U.K.: Oxford University Press, 2003, pp. 181–224.
19. Goldman-Rakic PS, Lidow MS, Gallager DW. Overlap of dopaminergic, adrenergic, and serotoninergic receptors and complementarity of their subtypes in primate prefrontal cortex. *The Journal of Neuroscience: The Official Journal of the Society for Neuroscience*. 1990;10(7):2125–2138. Available at: http://www.ncbi.nlm.nih.gov/pubmed/2165520
20. Rainbow TC, Biegon A. Quantitative autoradiography of [^{3}H]prazosin binding sites in rat forebrain. *Neuroscience Letters*. 1983;40(3):221–226. Available at: http://www.ncbi.nlm.nih.gov/pubmed/6316208
21. Rajkowska G. Histopathology of the prefrontal cortex in major depression: What does it tell us about dysfunctional monoaminergic circuits? *Progress in Brain Research*. 2000;126:397–412. Available at: http://www.ncbi.nlm.nih.gov/pubmed/11105659
22. Stone EA, John SM. Further evidence for a glial localization of rat cortical beta-adrenoceptors: Studies of *in vivo* cyclic AMP responses to catecholamines. *Brain Research*. 1991;549(1):78–82. Available at: http://www.ncbi.nlm.nih.gov/pubmed/1654173
23. Young WS, Kuhar MJ. Noradrenergic alpha 1 and alpha 2 receptors: Light microscopic autoradiographic localization. *Proceedings of the National Academy of Sciences of the United States of America*. 1980;77(3):1696–1700. Available at: http://www.pubmedcentral.nih.gov/articlerender.fcgi?artid=348564&tool=pmcentrez&rendertype=abstract
24. Pavcovich LA, Cancela LM, Volosin M, Molina VA, Ramirez OA. Chronic stress-induced changes in locus coeruleus neuronal activity. *Brain Research Bulletin*. 1990;24(2):293–296. Available at: http://www.ncbi.nlm.nih.gov/pubmed/2157529
25. Mann JJ, Currier DM. Stress, genetics and epigenetic effects on the neurobiology of suicidal behavior and depression. *European Psychiatry: The Journal of the Association of European Psychiatrists*. 2010;25(5):268–271. Available at: http://www.pubmedcentral.nih.gov/articlerender.fcgi?artid=2896004&tool=pmcentrez&rendertype=abstract

26. Mundt C, Reck C, Backenstrass M, Kronmüller K, Fiedler P. Reconfirming the role of life events for the timing of depressive episodes. A two-year prospective follow-up study. *Journal of Affective Disorders*. 2000;59(1):23–30. Available at: http://www.ncbi.nlm.nih.gov/pubmed/10814767
27. Makino S, Smith MA, Gold PW. Regulatory role of glucocorticoids and glucocorticoid receptor mRNA levels on tyrosine hydroxylase gene expression in the locus coeruleus during repeated immobilization stress. *Brain Research*. 2002;943(2):216–223. Available at: http://www.ncbi.nlm.nih.gov/pubmed/12101044
28. Melia KR, Rasmussen K, Terwilliger RZ et al. Coordinate regulation of the cyclic AMP system with firing rate and expression of tyrosine hydroxylase in the rat locus coeruleus: Effects of chronic stress and drug treatments. *Journal of Neurochemistry*. 1992;58(2):494–502. Available at: http://www.ncbi.nlm.nih.gov/pubmed/1345939
29. Ordway G, Klimek V, Mann JJ. Neurocircuitry of mood disorders. In: Davis KL, Charney D, Coyle JT and Nemeroff C, eds. *Neuropsychopharmacology: The Fifth Generation of Progress*. Philadelphia, PA: Lippincott Williams & Wilkins, 2002, pp. 1051–1064.
30. Osterhout CA, Sterling CR, Chikaraishi DM, Tank AW. Induction of tyrosine hydroxylase in the locus coeruleus of transgenic mice in response to stress or nicotine treatment: Lack of activation of tyrosine hydroxylase promoter activity. *Journal of Neurochemistry*. 2005;94(3):731–741. Available at: http://www.ncbi.nlm.nih.gov/pubmed/16033421
31. Wang P, Kitayama I, Nomura J. Tyrosine hydroxylase gene expression in the locus coeruleus of depression-model rats and rats exposed to short-and long-term forced walking stress. *Life Sciences*. 1998;62(23):2083–2092. Available at: http://www.ncbi.nlm.nih.gov/pubmed/9627087
32. Van Bockstaele EJ, Colago EE, Valentino RJ. Corticotropin-releasing factor-containing axon terminals synapse onto catecholamine dendrites and may presynaptically modulate other afferents in the rostral pole of the nucleus locus coeruleus in the rat brain. *The Journal of Comparative Neurology*. 1996;364(3):523–534. Available at: http://www.ncbi.nlm.nih.gov/pubmed/8820881
33. Van Bockstaele EJ, Colago EE, Valentino RJ. Amygdaloid corticotropin-releasing factor targets locus coeruleus dendrites: Substrate for the co-ordination of emotional and cognitive limbs of the stress response. *Journal of Neuroendocrinology*. 1998;10(10):743–757. Available at: http://www.ncbi.nlm.nih.gov/pubmed/9792326
34. Valentino RJ, Foote SL, Aston-Jones G. Corticotropin-releasing factor activates noradrenergic neurons of the locus coeruleus. *Brain Research*. 1983;270(2):363–367. Available at: http://www.ncbi.nlm.nih.gov/pubmed/6603889
35. Valentino RJ, Van Bockstaele E. Convergent regulation of locus coeruleus activity as an adaptive response to stress. *European Journal of Pharmacology*. 2008;583(2–3):194–203. Available at: http://www.pubmedcentral.nih.gov/articlerender.fcgi?artid=2349983&tool=pmcentrez&rendertype=abstract
36. Curtis AL, Grigoriadis DE, Page ME, Rivier J, Valentino RJ. Pharmacological comparison of two corticotropin-releasing factor antagonists: *In vivo* and *in vitro* studies. *The Journal of Pharmacology and Experimental Therapeutics*. 1994;268(1):359–365. Available at: http://www.ncbi.nlm.nih.gov/pubmed/8301577
37. Curtis AL, Bello NT, Connolly KR, Valentino RJ. Corticotropin-releasing factor neurones of the central nucleus of the amygdala mediate locus coeruleus activation by cardiovascular stress. *Journal of Neuroendocrinology*. 2002;14(8):667–682. Available at: http://www.ncbi.nlm.nih.gov/pubmed/12153469
38. Heinsbroek RP, Haaren F van, Feenstra MG, Boon P, Poll NE van de. Controllable and uncontrollable footshock and monoaminergic activity in the frontal cortex of male and female rats. *Brain Research*. 1991;551(1–2):247–255. Available at: http://www.ncbi.nlm.nih.gov/pubmed/1913155

39. Jedema HP, Grace AA. Corticotropin-releasing hormone directly activates noradrenergic neurons of the locus ceruleus recorded *in vitro*. *The Journal of Neuroscience: The Official Journal of the Society for Neuroscience*. 2004;24(43):9703–9713. Available at: http://www.ncbi.nlm.nih.gov/pubmed/15509759
40. Kollack-Walker S, Watson SJ, Akil H. Social stress in hamsters: Defeat activates specific neurocircuits within the brain. *The Journal of Neuroscience: The Official Journal of the Society for Neuroscience*. 1997;17(22):8842–8855. Available at: http://www.ncbi.nlm.nih.gov/pubmed/9348352
41. Melia KR, Duman RS. Involvement of corticotropin-releasing factor in chronic stress regulation of the brain noradrenergic system. *Proceedings of the National Academy of Sciences of the United States of America*. 1991;88(19):8382–8386. Available at: http://www.pubmedcentral.nih.gov/articlerender.fcgi?artid=52512&tool=pmcentrez&rendertype=abstract
42. Valentino RJ, Page ME, Curtis AL. Activation of noradrenergic locus coeruleus neurons by hemodynamic stress is due to local release of corticotropin-releasing factor. *Brain Research*. 1991;555(1):25–34. Available at: http://www.ncbi.nlm.nih.gov/pubmed/1933327
43. Weiss JM, Glazer HI, Pohorecky LA, Brick J, Miller NE. Effects of chronic exposure to stressors on avoidance-escape behavior and on brain norepinephrine. *Psychosomatic Medicine*. 1975;37(6):522–534. Available at: http://www.ncbi.nlm.nih.gov/pubmed/711
44. Weiss JM, Bailey WH, Pohorecky LA, Korzeniowski D, Grillione G. Stress-induced depression of motor activity correlates with regional changes in brain norepinephrine but not in dopamine. *Neurochemical Research*. 1980;5(1):9–22. Available at: http://www.ncbi.nlm.nih.gov/pubmed/7366796
45. Melia KR, Nestler EJ, Duman RS. Chronic imipramine treatment normalizes levels of tyrosine hydroxylase in the locus coeruleus of chronically stressed rats. *Psychopharmacology*. 1992;108(1–2):23–26. Available at: http://www.ncbi.nlm.nih.gov/pubmed/1357707
46. Schultzberg M, Austin MC, Crawley JN, Paul SM. Repeated administration of desmethylimipramine blocks the reserpine-induced increase in tyrosine hydroxylase mRNA in locus coeruleus neurons of the rat. *Brain Research. Molecular Brain Research*. 1991;10(4):307–314. Available at: http://www.ncbi.nlm.nih.gov/pubmed/1717808
47. Szabo ST, Blier P. Effect of the selective noradrenergic reuptake inhibitor reboxetine on the firing activity of noradrenaline and serotonin neurons. *The European Journal of Neuroscience*. 2001;13(11):2077–2087. Available at: http://www.ncbi.nlm.nih.gov/pubmed/11422448
48. Grant MM, Weiss JM. Effects of chronic antidepressant drug administration and electroconvulsive shock on locus coeruleus electrophysiologic activity. *Biological Psychiatry*. 2001;49(2):117–129. Available at: http://www.ncbi.nlm.nih.gov/pubmed/11164758
49. Sawchenko PE, Swanson LW. Central noradrenergic pathways for the integration of hypothalamic neuroendocrine and autonomic responses. *Science (New York, N.Y.)*. 1981;214(4521):685–687. Available at: http://www.ncbi.nlm.nih.gov/pubmed/7292008
50. Herman JP, Cullinan WE. Neurocircuitry of stress: Central control of the hypothalamo–pituitary–adrenocortical axis. *Trends in Neurosciences*. 1997;20(2):78–84. Available at: http://www.ncbi.nlm.nih.gov/pubmed/9023876
51. Cunningham ET, Sawchenko PE. Anatomical specificity of noradrenergic inputs to the paraventricular and supraoptic nuclei of the rat hypothalamus. *The Journal of Comparative Neurology*. 1988;274(1):60–76. Available at: http://www.ncbi.nlm.nih.gov/pubmed/2458397

52. Boundy VA, Cincotta AH. Hypothalamic adrenergic receptor changes in the metabolic syndrome of genetically obese (ob/ob) mice. *American Journal of Physiology. Regulatory, Integrative and Comparative Physiology*. 2000;279(2):R505–R514. Available at: http://www.ncbi.nlm.nih.gov/pubmed/10938239
53. Capuano CA, Leibowitz SF, Barr GA. The pharmaco-ontogeny of the paraventricular alpha 2-noradrenergic receptor system mediating norepinephrine-induced feeding in the rat. *Brain Research. Developmental Brain Research*. 1992;68(1):67–74. Available at: http://www.ncbi.nlm.nih.gov/pubmed/1325877
54. Szafarczyk A, Malaval F, Laurent A, Gibaud R, Assenmacher I. Further evidence for a central stimulatory action of catecholamines on adrenocorticotropin release in the rat. *Endocrinology*. 1987;121(3):883–892. Available at: http://www.ncbi.nlm.nih.gov/pubmed/3040380
55. Cole R, Sawchenko P. Neurotransmitter regulation of cellular activation and neuropeptide gene expression in the paraventricular nucleus of the hypothalamus. *Journal of Neuroscience*. 2002;22(3):959–969.
56. Gibson A, Hart SL, Patel S. Effects of 6-hydroxydopamine-induced lesions of the paraventricular nucleus, and of prazosin, on the corticosterone response to restraint in rats. *Neuropharmacology*. 1986;25(3):257–260. Available at: http://www.ncbi.nlm.nih.gov/pubmed/2871514
57. Curtis AL, Pavcovich LA, Valentino RJ. Long-term regulation of locus ceruleus sensitivity to corticotropin-releasing factor by swim stress. *The Journal of Pharmacology and Experimental Therapeutics*. 1999;289(3):1211–1219. Available at: http://www.ncbi.nlm.nih.gov/pubmed/10336508
58. Swinny JD, O'Farrell E, Bingham BC et al. Neonatal rearing conditions distinctly shape locus coeruleus neuronal activity, dendritic arborization, and sensitivity to corticotrophin-releasing factor. *The International Journal of Neuropsychopharmacology/Official Scientific Journal of the Collegium Internationale Neuropsychopharmacologicum (CINP)*. 2010;13(4):515–525. Available at: http://www.pubmedcentral.nih.gov/articlerender.fcgi?artid=2857591&tool=pmcentrez&rendertype=abstract
59. Knight RT, Staines WR, Swick D, Chao LL. Prefrontal cortex regulates inhibition and excitation in distributed neural networks. *Acta Psychologica*. 1999;101(2–3):159–178. Available at: http://www.ncbi.nlm.nih.gov/pubmed/10344184
60. Ramos BP, Arnsten AFT. Adrenergic pharmacology and cognition: Focus on the prefrontal cortex. *Pharmacology and Therapeutics*. 2007;113(3):523–536. Available at: http://www.pubmedcentral.nih.gov/articlerender.fcgi?artid=2151919&tool=pmcentrez&rendertype=abstract
61. Arnsten AF, Cai JX, Goldman-Rakic PS. The alpha-2 adrenergic agonist guanfacine improves memory in aged monkeys without sedative or hypotensive side effects: Evidence for alpha-2 receptor subtypes. *The Journal of Neuroscience: The Official Journal of the Society for Neuroscience*. 1988;8(11):4287–4298. Available at: http://www.ncbi.nlm.nih.gov/pubmed/2903226
62. Rämä P, Linnankoski I, Tanila H, Pertovaara A, Carlson S. Medetomidine, atipamezole, and guanfacine in delayed response performance of aged monkeys. *Pharmacology, Biochemistry, and Behavior*. 1996;55(3):415–422. Available at: http://www.ncbi.nlm.nih.gov/pubmed/8951983
63. Tanila H, Rämä P, Carlson S. The effects of prefrontal intracortical microinjections of an alpha-2 agonist, alpha-2 antagonist and lidocaine on the delayed alternation performance of aged rats. *Brain Research Bulletin*. 1996;40(2):117–119. Available at: http://www.ncbi.nlm.nih.gov/pubmed/8724429
64. Li BM, Mao ZM, Wang M, Mei ZT. Alpha-2 adrenergic modulation of prefrontal cortical neuronal activity related to spatial working memory in monkeys. *Neuropsychopharmacology: Official Publication of the American College of Neuropsychopharmacology*. 1999;21(5):601–610. Available at: http://www.ncbi.nlm.nih.gov/pubmed/10516956

65. Mao ZM, Arnsten AF, Li BM. Local infusion of an alpha-1 adrenergic agonist into the prefrontal cortex impairs spatial working memory performance in monkeys. *Biological Psychiatry*. 1999;46(9):1259–1265. Available at: http://www.ncbi.nlm.nih.gov/pubmed/10560031
66. Usher M, Cohen JD, Servan-Schreiber D, Rajkowski J, Aston-Jones G. The role of locus coeruleus in the regulation of cognitive performance. *Science (New York, N.Y.)*. 1999;283(5401):549–554. Available at: http://www.ncbi.nlm.nih.gov/pubmed/9915705
67. Savitz J, Drevets WC. Bipolar and major depressive disorder: Neuroimaging the developmental-degenerative divide. *Neuroscience and Biobehavioral Reviews*. 2009;33(5):699–771. Available at: http://www.pubmedcentral.nih.gov/articlerender.fcgi?artid=2858318&tool=pmcentrez&rendertype=abstract
68. Keilp JG, Sackeim HA, Brodsky BS et al. Neuropsychological dysfunction in depressed suicide attempters. *The American Journal of Psychiatry*. 2001;158(5):735–741. Available at: http://www.ncbi.nlm.nih.gov/pubmed/11329395
69. Paelecke-Habermann Y, Pohl J, Leplow B. Attention and executive functions in remitted major depression patients. *Journal of Affective Disorders*. 2005;89(1–3):125–135. Available at: http://www.ncbi.nlm.nih.gov/pubmed/16324752
70. Veiel HO. A preliminary profile of neuropsychological deficits associated with major depression. *Journal of Clinical and Experimental Neuropsychology*. 1997;19(4):587–603. Available at: http://www.ncbi.nlm.nih.gov/pubmed/9342691
71. McKinney WT, Eising RG, Moran EC, Suomi SJ, Harlow HF. Effects of reserpine on the social behavior of rhesus monkeys. *Diseases of the Nervous System*. 1971;32(11):735–741. Available at: http://www.ncbi.nlm.nih.gov/pubmed/5002259
72. Cai JX, Ma YY, Xu L, Hu XT. Reserpine impairs spatial working memory performance in monkeys: Reversal by the alpha 2-adrenergic agonist clonidine. *Brain Research*. 1993;614(1–2):191–196. Available at: http://www.ncbi.nlm.nih.gov/pubmed/8102313
73. Fu CH, Reed LJ, Meyer JH et al. Noradrenergic dysfunction in the prefrontal cortex in depression: An $[^{15}O]H_2O$ PET study of the neuromodulatory effects of clonidine. *Biological Psychiatry*. 2001;49(4):317–325. Available at: http://www.ncbi.nlm.nih.gov/pubmed/11239902
74. Patsiokas AT, Clum GA, Luscomb RL. Cognitive characteristics of suicide attempters. *Journal of Consulting and Clinical Psychology*. 1979;47(3):478–484. Available at: http://www.ncbi.nlm.nih.gov/pubmed/528716
75. Pollock LR, Williams JMG. Problem-solving in suicide attempters. *Psychological Medicine*. 2004;34(1):163–167. Available at: http://www.ncbi.nlm.nih.gov/pubmed/14971637
76. Roskar S, Zorko M, Bucik V, Marusic A. Problem solving for depressed suicide attempters and depressed individuals without suicide attempt. *Psychiatria Danubina*. 2007;19(4):296–302. Available at: http://www.ncbi.nlm.nih.gov/pubmed/18000480
77. Bartfai A, Winborg IM, Nordström P, Asberg M. Suicidal behavior and cognitive flexibility: Design and verbal fluency after attempted suicide. *Suicide and Life-Threatening Behavior*. 1990;20(3):254–266. Available at: http://www.ncbi.nlm.nih.gov/pubmed/2238017
78. King DA, Conwell Y, Cox C et al. A neuropsychological comparison of depressed suicide attempters and nonattempters. *The Journal of Neuropsychiatry and Clinical Neurosciences*. 2000;12(1):64–70. Available at: http://www.ncbi.nlm.nih.gov/pubmed/10678515
79. Malloy-Diniz LF, Neves FS, Abrantes SSC, Fuentes D, Corrêa H. Suicide behavior and neuropsychological assessment of type I bipolar patients. *Journal of Affective Disorders*. 2009;112(1–3):231–236. Available at: http://www.ncbi.nlm.nih.gov/pubmed/18485487
80. Jollant F, Lawrence NS, Olie E et al. Decreased activation of lateral orbitofrontal cortex during risky choices under uncertainty is associated with disadvantageous decision-making and suicidal behavior. *NeuroImage*. 2010;51(3):1275–1281. Available at: http://www.ncbi.nlm.nih.gov/pubmed/20302946

81. Amen DG, Prunella JR, Fallon JH, Amen B, Hanks C. A comparative analysis of completed suicide using high resolution brain SPECT imaging. *The Journal of Neuropsychiatry and Clinical Neurosciences*. 2009;21(4):430–439. Available at: http://www.ncbi.nlm.nih.gov/pubmed/19996252
82. Armitage R. Microarchitectural findings in sleep EEG in depression: Diagnostic implications. *Biological Psychiatry*. 1995;37(2):72–84. Available at: http://www.ncbi.nlm.nih.gov/pubmed/7718683
83. Benca RM, Obermeyer WH, Thisted RA, Gillin JC. Sleep and psychiatric disorders. A meta-analysis. *Archives of General Psychiatry*. 1992;49(8):651–668; discussion 669–670. Available at: http://www.ncbi.nlm.nih.gov/pubmed/1386215
84. Chang PP, Ford DE, Mead LA, Cooper-Patrick L, Klag MJ. Insomnia in young men and subsequent depression. The Johns Hopkins Precursors Study. *American Journal of Epidemiology*. 1997;146(2):105–114. Available at: http://www.ncbi.nlm.nih.gov/pubmed/9230772
85. Ford DE, Kamerow DB. Epidemiologic study of sleep disturbances and psychiatric disorders. An opportunity for prevention? *JAMA: The Journal of the American Medical Association*. 1989;262(11):1479–1484. Available at: http://www.ncbi.nlm.nih.gov/pubmed/2769898
86. Gregory AM, Rijsdijk FV, Lau JYF, Dahl RE, Eley TC. The direction of longitudinal associations between sleep problems and depression symptoms: A study of twins aged 8 and 10 years. *Sleep*. 2009;32(2):189–199. Available at: http://www.pubmedcentral.nih.gov/articlerender.fcgi?artid=2635583&tool=pmcentrez&rendertype=abstract
87. Holsboer-Trachsler E, Seifritz E. Sleep in depression and sleep deprivation: A brief conceptual review. *The World Journal of Biological Psychiatry: The Official Journal of the World Federation of Societies of Biological Psychiatry*. 2000;1(4):180–186. Available at: http://www.ncbi.nlm.nih.gov/pubmed/12607213
88. Ohayon MM. Prevalence of DSM-IV diagnostic criteria of insomnia: Distinguishing insomnia related to mental disorders from sleep disorders. *Journal of Psychiatric Research*. 1997;31(3):333–346. Available at: http://www.ncbi.nlm.nih.gov/pubmed/9306291
89. Ohayon MM, Roth T. Place of chronic insomnia in the course of depressive and anxiety disorders. *Journal of Psychiatric Research*. 2003;37(1):9–15. Available at: http://www.ncbi.nlm.nih.gov/pubmed/12482465
90. Breslau N, Roth T, Rosenthal L, Andreski P. Sleep disturbance and psychiatric disorders: A longitudinal epidemiological study of young adults. *Biological Psychiatry*. 1996;39(6):411–418. Available at: http://www.ncbi.nlm.nih.gov/pubmed/8679786
91. Johnson EO, Roth T, Breslau N. The association of insomnia with anxiety disorders and depression: Exploration of the direction of risk. *Journal of Psychiatric Research*. 2006;40(8):700–708. Available at: http://www.ncbi.nlm.nih.gov/pubmed/16978649
92. Pigeon WR, Hegel M, Unützer J et al. Is insomnia a perpetuating factor for late-life depression in the IMPACT cohort? *Sleep*. 2008;31(4):481–488. Available at: http://www.pubmedcentral.nih.gov/articlerender.fcgi?artid=2279755&tool=pmcentrez&rendertype=abstract
93. Ağargün MY, Kara H, Solmaz M. Sleep disturbances and suicidal behavior in patients with major depression. *The Journal of Clinical Psychiatry*. 1997;58(6):249–251. Available at: http://www.ncbi.nlm.nih.gov/pubmed/9228889
94. Fawcett J, Scheftner WA, Fogg L et al. Time-related predictors of suicide in major affective disorder. *The American Journal of Psychiatry*. 1990;147(9):1189–1194. Available at: http://www.ncbi.nlm.nih.gov/pubmed/2104515
95. Krakow B, Ribeiro JD, Ulibarri VA, Krakow J, Joiner TE. Sleep disturbances and suicidal ideation in sleep medical center patients. *Journal of Affective Disorders*. 2011. Available at: http://www.ncbi.nlm.nih.gov/pubmed/21211850

96. Li SX, Lam SP, Yu MWM, Zhang J, Wing YK. Nocturnal sleep disturbances as a predictor of suicide attempts among psychiatric outpatients: A clinical, epidemiologic, prospective study. *The Journal of Clinical Psychiatry*. 2010;71(11):1440–1446. Available at: http://www.ncbi.nlm.nih.gov/pubmed/21114949
97. Aston-Jones G, Bloom FE. Activity of norepinephrine-containing locus coeruleus neurons in behaving rats anticipates fluctuations in the sleep-waking cycle. *The Journal of Neuroscience: The Official Journal of the Society for Neuroscience*. 1981;1(8):876–886. Available at: http://www.ncbi.nlm.nih.gov/pubmed/7346592
98. Hobson JA, McCarley RW, Freedman R, Pivik RT. Time course of discharge rate changes by cat pontine brain stem neurons during sleep cycle. *Journal of Neurophysiology*. 1974;37(6):1297–1309. Available at: http://www.ncbi.nlm.nih.gov/pubmed/4436702
99. Aston-Jone G, Gonzalez M, Doran S. Role of the locus coeruleus–norepinephrine system in arousal and circadian regulation of the sleep–wake cycle. In: Ordway G, Schwartz M, Frazer A, eds. *Brain Norepinephrine*. Cambridge, U.K.: Cambridge University Press, 2007, pp. 157–195.
100. Gottesmann C. Noradrenaline involvement in basic and higher integrated REM sleep processes. *Progress in Neurobiology*. 2008;85(3):237–272. Available at: http://www.ncbi.nlm.nih.gov/pubmed/18514380
101. Müller HU, Riemann D, Berger M, Müller WE. The influence of total sleep deprivation on urinary excretion of catecholamine metabolites in major depression. *Acta Psychiatrica Scandinavica*. 1993;88(1):16–20. Available at: http://www.ncbi.nlm.nih.gov/pubmed/8396844
102. Schreiber W, Opper C, Dickhaus B et al. Alterations of blood platelet MAO-B activity and LSD-binding in humans after sleep deprivation and recovery sleep. *Journal of Psychiatric Research*. 1997;31(3):323–331. Available at: http://www.ncbi.nlm.nih.gov/pubmed/9306290
103. Gillin JC, Buchsbaum M, Wu J, Clark C, Bunney W. Sleep deprivation as a model experimental antidepressant treatment: Findings from functional brain imaging. *Depression and Anxiety*. 2001;14(1):37–49. Available at: http://www.ncbi.nlm.nih.gov/pubmed/11568981
104. Gonzalez MMC, Aston-Jones G. Light deprivation damages monoamine neurons and produces a depressive behavioral phenotype in rats. *Proceedings of the National Academy of Sciences of the United States of America*. 2008;105(12):4898–4903. Available at: http://www.pubmedcentral.nih.gov/articlerender.fcgi?artid=2290795&tool=pmcentrez&rendertype=abstract
105. Bunney WE, Davis JM. Norepinephrine in depressive reactions. A review. *Archives of General Psychiatry*. 1965;13(6):483–494. Available at: http://www.ncbi.nlm.nih.gov/pubmed/5320621
106. Prange AJ. The pharmacology and biochemistry of depression. *Diseases of the Nervous System*. 1964;25:217–221. Available at: http://www.ncbi.nlm.nih.gov/pubmed/14140032
107. Schildkraut JJ. The catecholamine hypothesis of affective disorders: A review of supporting evidence. *The American Journal of Psychiatry*. 1965;122(5):509–522. Available at: http://www.ncbi.nlm.nih.gov/pubmed/5319766
108. Gibbons RD, Mann JJ. Strategies for quantifying the relationship between medications and suicidal behaviour: What has been learned? *Drug Safety: An International Journal of Medical Toxicology and Drug Experience*. 2011;34(5):375–395. Available at: http://www.ncbi.nlm.nih.gov/pubmed/21513361
109. Danysz W, Kostowski W, Hauptmann M. Evidence for the locus coeruleus involvement in desipramine action in animal models of depression. *Polish Journal of Pharmacology and Pharmacy*. 1985;37(6):855–864. Available at: http://www.ncbi.nlm.nih.gov/pubmed/3938536

110. Biegon A, Israeli M. Regionally selective increases in beta-adrenergic receptor density in the brains of suicide victims. *Brain Research*. 1988;442(1):199–203. Available at: http://www.ncbi.nlm.nih.gov/pubmed/2834015
111. Klimek V, Stockmeier C, Overholser J et al. Reduced levels of norepinephrine transporters in the locus coeruleus in major depression. *The Journal of Neuroscience: The Official Journal of the Society for Neuroscience*. 1997;17(21):8451–8458. Available at: http://www.ncbi.nlm.nih.gov/pubmed/9334417
112. Klimek V, Roberson G, Stockmeier CA, Ordway GA. Serotonin transporter and MAO-B levels in monoamine nuclei of the human brainstem are normal in major depression. *Journal of Psychiatric Research*. 2003;37(5):387–397. Available at: http://www.ncbi.nlm.nih.gov/pubmed/12849931
113. Richards JG, Saura J, Ulrich J, Da Prada M. Molecular neuroanatomy of monoamine oxidases in human brainstem. *Psychopharmacology*. 1992;106(Suppl):S21–S23. Available at: http://www.ncbi.nlm.nih.gov/pubmed/1546134
114. West CHK, Ritchie JC, Boss-Williams KA, Weiss JM. Antidepressant drugs with differing pharmacological actions decrease activity of locus coeruleus neurons. *The International Journal of Neuropsychopharmacology/Official Scientific Journal of the Collegium Internationale Neuropsychopharmacologicum (CINP)*. 2009;12(5):627–641. Available at: http://www.pubmedcentral.nih.gov/articlerender.fcgi?artid=2700044&tool=pmcentrez&rendertype=abstract
115. Morilak DA, Fornal CA, Jacobs BL. Effects of physiological manipulations on locus coeruleus neuronal activity in freely moving cats. I. Thermoregulatory challenge. *Brain Research*. 1987;422(1):17–23. Available at: http://www.ncbi.nlm.nih.gov/pubmed/3676779
116. Pacak K, Palkovits M, Kopin IJ, Goldstein DS. Stress-induced norepinephrine release in the hypothalamic paraventricular nucleus and pituitary-adrenocortical and sympathoadrenal activity: *In vivo* microdialysis studies. *Frontiers in Neuroendocrinology*. 1995;16(2):89–150. Available at: http://www.ncbi.nlm.nih.gov/pubmed/7621982
117. Page ME, Akaoka H, Aston-Jones G, Valentino RJ. Bladder distention activates noradrenergic locus coeruleus neurons by an excitatory amino acid mechanism. *Neuroscience*. 1992;51(3):555–563. Available at: http://www.ncbi.nlm.nih.gov/pubmed/1336819
118. Berman RM, Narasimhan M, Miller HL et al. Transient depressive relapse induced by catecholamine depletion: Potential phenotypic vulnerability marker? *Archives of General Psychiatry*. 1999;56(5):395–403. Available at: http://www.ncbi.nlm.nih.gov/pubmed/10232292
119. Hasler G, Fromm S, Carlson PJ et al. Neural response to catecholamine depletion in unmedicated subjects with major depressive disorder in remission and healthy subjects. *Archives of General Psychiatry*. 2008;65(5):521–531. Available at: http://www.pubmedcentral.nih.gov/articlerender.fcgi?artid=2676777&tool=pmcentrez&rendertype=abstract
120. Krahn LE, Lin SC, Klee GG et al. The effect of presynaptic catecholamine depletion on 6-hydroxymelatonin sulfate: A double blind study of alpha-methyl-*para*-tyrosine. *European Neuropsychopharmacology: The Journal of the European College of Neuropsychopharmacology*. 1999;9(1–2):61–66. Available at: http://www.ncbi.nlm.nih.gov/pubmed/10082229
121. Berman RM, Sanacora G, Anand A et al. Monoamine depletion in unmedicated depressed subjects. *Biological Psychiatry*. 2002;51(6):469–473. Available at: http://www.ncbi.nlm.nih.gov/pubmed/11922881
122. Miller HL, Delgado PL, Salomon RM, Heninger GR, Charney DS. Effects of alpha-methyl-*para*-tyrosine (AMPT) in drug-free depressed patients. *Neuropsychopharmacology: Official Publication of the American College of Neuropsychopharmacology*. 1996;14(3):151–157. Available at: http://www.ncbi.nlm.nih.gov/pubmed/8866698

123. Anand A, Darnell A, Miller HL et al. Effect of catecholamine depletion on lithium-induced long-term remission of bipolar disorder. *Biological Psychiatry*. 1999;45(8):972–978. Available at: http://www.ncbi.nlm.nih.gov/pubmed/10386179
124. Longhurst JG, Carpenter LL, Epperson CN, Price LH, McDougle CJ. Effects of catecholamine depletion with AMPT (alpha-methyl-*para*-tyrosine) in obsessive–compulsive disorder. *Biological Psychiatry*. 1999;46(4):573–576. Available at: http://www.ncbi.nlm.nih.gov/pubmed/10459409
125. Bremner JD, Vythilingam M, Ng CK et al. Regional brain metabolic correlates of alpha-methyl-*para*-tyrosine-induced depressive symptoms: Implications for the neural circuitry of depression. *JAMA: The Journal of the American Medical Association*. 2003;289(23):3125–3134. Available at: http://www.ncbi.nlm.nih.gov/pubmed/12813118
126. Miller HL, Delgado PL, Salomon RM et al. Clinical and biochemical effects of catecholamine depletion on antidepressant-induced remission of depression. *Archives of General Psychiatry*. 1996;53(2):117–128. Available at: http://www.ncbi.nlm.nih.gov/pubmed/8629887
127. Hasler G, Luckenbaugh DA, Snow J et al. Reward processing after catecholamine depletion in unmedicated, remitted subjects with major depressive disorder. *Biological Psychiatry*. 2009;66(3):201–205. Available at: http://www.pubmedcentral.nih.gov/articlerender.fcgi?artid=3073352&tool=pmcentrez&rendertype=abstract
128. Hasler G, Mondillo K, Drevets WC, Blair JR. Impairments of probabilistic response reversal and passive avoidance following catecholamine depletion. *Neuropsychopharmacology: Official Publication of the American College of Neuropsychopharmacology*. 2009;34(13):2691–2698. Available at: http://www.pubmedcentral.nih.gov/articlerender.fcgi?artid=2783713&tool=pmcentrez&rendertype=abstract
129. Booij L, Van der Does AJW, Riedel WJ. Monoamine depletion in psychiatric and healthy populations: Review. *Molecular Psychiatry*. 2003;8(12):951–973. Available at: http://www.ncbi.nlm.nih.gov/pubmed/14647394
130. Delgado PL. Depression: The case for a monoamine deficiency. *The Journal of Clinical Psychiatry*. 2000;61(Suppl 6):7–11. Available at: http://www.ncbi.nlm.nih.gov/pubmed/10775018
131. Ruhé HG, Mason NS, Schene AH. Mood is indirectly related to serotonin, norepinephrine and dopamine levels in humans: A meta-analysis of monoamine depletion studies. *Molecular Psychiatry*. 2007;12(4):331–359. Available at: http://www.ncbi.nlm.nih.gov/pubmed/17389902
132. Engelman K, Horwitz D, Jéquier E, Sjoerdsma A. Biochemical and pharmacologic effects of alpha-methyltyrosine in man. *The Journal of Clinical Investigation*. 1968;47(3):577–594. Available at: http://www.pubmedcentral.nih.gov/articlerender.fcgi?artid=297204&tool=pmcentrez&rendertype=abstract
133. Widerlöv E, Lewander T. Inhibition of the *in vivo* biosynthesis and changes of catecholamine levels in rat brain after alpha-methyl-*p*-tyrosine; time- and dose-response relationships. *Naunyn-Schmiedeberg's Archives of Pharmacology*. 1978;304(2):111–123. Available at: http://www.ncbi.nlm.nih.gov/pubmed/703854
134. Beskow J. Suicide in mental disorder in Swedish men. *Acta Psychiatrica Scandinavica. Supplementum*. 1979;(277):1–138. Available at: http://www.ncbi.nlm.nih.gov/pubmed/286500
135. Rich CL, Young D, Fowler RC. San Diego suicide study. I. Young vs old subjects. *Archives of General Psychiatry*. 1986;43(6):577–582. Available at: http://www.ncbi.nlm.nih.gov/pubmed/3707290
136. Ordway GA. Pathophysiology of the locus coeruleus in suicide. *Annals of the New York Academy of Sciences*. 1997;836:233–252. Available at: http://www.ncbi.nlm.nih.gov/pubmed/9616802

137. Ordway GA, Klimek V. Noradrenergic pathology in psychiatric disorders: Postmortem studies. *CNS Spectrums*. 2001;6(8):697–703. Available at: http://www.ncbi.nlm.nih.gov/pubmed/15520616
138. Ordway GA. Neuropathology of central norepinephrine in psychiatric disorders: Postmortem research. In: Ordway GA, Schwartz M, Frazer A, eds. *Brain Norepinephrine*, Cambridge, U.K.: Cambridge University Press, 2007, pp. 341–362.
139. Pandey GN, Dwivedi Y. Noradrenergic function in suicide. *Archives of Suicide Research: Official Journal of the International Academy for Suicide Research*. 2007;11(3):235–246. Available at: http://www.ncbi.nlm.nih.gov/pubmed/17558608
140. Arango V, Ernsberger P, Sved AF, Mann JJ. Quantitative autoradiography of alpha 1- and alpha 2-adrenergic receptors in the cerebral cortex of controls and suicide victims. *Brain Research*. 1993;630(1–2):271–282. Available at: http://www.ncbi.nlm.nih.gov/pubmed/8118693
141. Callado LF, Meana JJ, Grijalba B et al. Selective increase of alpha2A-adrenoceptor agonist binding sites in brains of depressed suicide victims. *Journal of Neurochemistry*. 1998;70(3):1114–1123. Available at: http://www.ncbi.nlm.nih.gov/pubmed/9489732
142. De Paermentier F, Mauger JM, Lowther S et al. Brain alpha-adrenoceptors in depressed suicides. *Brain Research*. 1997;757(1):60–68. Available at: http://www.ncbi.nlm.nih.gov/pubmed/9200499
143. Gross-Isseroff R, Weizman A, Fieldust SJ, Israeli M, Biegon A. Unaltered alpha(2)-noradrenergic/imidazoline receptors in suicide victims: A postmortem brain autoradiographic analysis. *European Neuropsychopharmacology: The Journal of the European College of Neuropsychopharmacology*. 2000;10(4):265–271. Available at: http://www.ncbi.nlm.nih.gov/pubmed/10871708
144. Klimek V, Rajkowska G, Luker SN et al. Brain noradrenergic receptors in major depression and schizophrenia. *Neuropsychopharmacology: Official Publication of the American College of Neuropsychopharmacology*. 1999;21(1):69–81. Available at: http://www.ncbi.nlm.nih.gov/pubmed/10379521
145. Escribá PV, Ozaita A, García-Sevilla JA. Increased mRNA expression of alpha2A-adrenoceptors, serotonin receptors and mu-opioid receptors in the brains of suicide victims. *Neuropsychopharmacology: Official Publication of the American College of Neuropsychopharmacology*. 2004;29(8):1512–1521. Available at: http://www.ncbi.nlm.nih.gov/pubmed/15199368
146. García-Sevilla JA, Escribá PV, Ozaita A et al. Up-regulation of immunolabeled alpha2A-adrenoceptors, Gi coupling proteins, and regulatory receptor kinases in the prefrontal cortex of depressed suicides. *Journal of Neurochemistry*. 1999;72(1):282–291. Available at: http://www.ncbi.nlm.nih.gov/pubmed/9886080
147. González-Maeso J, Rodríguez-Puertas R, Meana JJ, García-Sevilla JA, Guimón J. Neurotransmitter receptor-mediated activation of G-proteins in brains of suicide victims with mood disorders: Selective supersensitivity of alpha(2A)-adrenoceptors. *Molecular Psychiatry*. 2002;7(7):755–767. Available at: http://www.ncbi.nlm.nih.gov/pubmed/12192620
148. Valdizán EM, Díez-Alarcia R, González-Maeso J et al. α_2-Adrenoceptor functionality in postmortem frontal cortex of depressed suicide victims. *Biological Psychiatry*. 2010;68(9):869–872. Available at: http://www.ncbi.nlm.nih.gov/pubmed/20864091
149. Ordway GA. Effect of noradrenergic lesions on subtypes of alpha 2-adrenoceptors in rat brain. *Journal of Neurochemistry*. 1995;64(3):1118–1126. Available at: http://www.ncbi.nlm.nih.gov/pubmed/7861142
150. U'Prichard DC, Bechtel WD, Rouot BM, Snyder SH. Multiple apparent alpha-noradrenergic receptor binding sites in rat brain: Effect of 6-hydroxydopamine. *Molecular Pharmacology*. 1979;16(1):47–60. Available at: http://www.ncbi.nlm.nih.gov/pubmed/39248

151. Ribas C, Miralles A, Busquets X, García-Sevilla JA. Brain alpha(2)-adrenoceptors in monoamine-depleted rats: Increased receptor density, G coupling proteins, receptor turnover and receptor mRNA. *British Journal of Pharmacology*. 2001;132(7):1467–1476. Available at: http://www.pubmedcentral.nih.gov/articlerender.fcgi?artid=1572698&tool=pmcentrez&rendertype=abstract
152. Mann JJ, Stanley M, McBride PA, McEwen BS. Increased serotonin2 and beta-adrenergic receptor binding in the frontal cortices of suicide victims. *Archives of General Psychiatry*. 1986;43(10):954–959. Available at: http://www.ncbi.nlm.nih.gov/pubmed/3019268
153. De Paermentier F, Cheetham SC, Crompton MR, Katona CL, Horton RW. Brain beta-adrenoceptor binding sites in antidepressant-free depressed suicide victims. *Brain Research*. 1990;525(1):71–77. Available at: http://www.ncbi.nlm.nih.gov/pubmed/2173963
154. Little KY, Clark TB, Ranc J, Duncan GE. Beta-adrenergic receptor binding in frontal cortex from suicide victims. *Biological Psychiatry*. 1993;34(9):596–605. Available at: http://www.ncbi.nlm.nih.gov/pubmed/8292688
155. Stockmeier CA, Meltzer HY. Beta-adrenergic receptor binding in frontal cortex of suicide victims. *Biological Psychiatry*. 1991;29(2):183–191. Available at: http://www.ncbi.nlm.nih.gov/pubmed/1847309
156. Crow TJ, Cross AJ, Cooper SJ et al. Neurotransmitter receptors and monoamine metabolites in the brains of patients with Alzheimer-type dementia and depression, and suicides. *Neuropharmacology*. 1984;23(12B):1561–1569. Available at: http://www.ncbi.nlm.nih.gov/pubmed/6084823
157. Ordway GA, Widdowson PS, Smith KS, Halaris A. Agonist binding to alpha 2-adrenoceptors is elevated in the locus coeruleus from victims of suicide. *Journal of Neurochemistry*. 1994;63(2):617–624. Available at: http://www.ncbi.nlm.nih.gov/pubmed/8035185
158. Zhu MY, Klimek V, Dilley GE et al. Elevated levels of tyrosine hydroxylase in the locus coeruleus in major depression. *Biological Psychiatry*. 1999;46(9):1275–1286. Available at: http://www.ncbi.nlm.nih.gov/pubmed/10560033
159. Sanchez-Bahillo A, Bautista-Hernandez V, Barcia Gonzalez C et al. Increased mRNA expression of cytochrome oxidase in dorsal raphe nucleus of depressive suicide victims. *Neuropsychiatric Disease and Treatment*. 2008;4(2):413–416. Available at: http://www.pubmedcentral.nih.gov/articlerender.fcgi?artid=2518385&tool=pmcentrez&rendertype=abstract
160. Arango V, Underwood MD, Mann JJ. Fewer pigmented locus coeruleus neurons in suicide victims: Preliminary results. *Biological Psychiatry*. 1996;39(2):112–120. Available at: http://www.ncbi.nlm.nih.gov/pubmed/8717609
161. Baumann B, Danos P, Diekmann S et al. Tyrosine hydroxylase immunoreactivity in the locus coeruleus is reduced in depressed non-suicidal patients but normal in depressed suicide patients. *European Archives of Psychiatry and Clinical Neuroscience*. 1999;249(4):212–219. Available at: http://www.ncbi.nlm.nih.gov/pubmed/10449597
162. Syed A, Chatfield M, Matthews F et al. Depression in the elderly: Pathological study of raphe and locus coeruleus. *Neuropathology and Applied Neurobiology*. 2005;31(4):405–413. Available at: http://www.ncbi.nlm.nih.gov/pubmed/16008824
163. Ordway GA, Schenk J, Stockmeier CA, May W, Klimek V. Elevated agonist binding to alpha2-adrenoceptors in the locus coeruleus in major depression. *Biological Psychiatry*. 2003;53(4):315–323. Available at: http://www.ncbi.nlm.nih.gov/pubmed/12586450
164. Lee CM, Javitch JA, Snyder SH. Recognition sites for norepinephrine uptake: Regulation by neurotransmitter. *Science (New York, N.Y.)*. 1983;220(4597):626–629. Available at: http://www.ncbi.nlm.nih.gov/pubmed/6301013

165. Weinshenker D, White SS, Javors MA, Palmiter RD, Szot P. Regulation of norepinephrine transporter abundance by catecholamines and desipramine *in vivo*. *Brain Research*. 2002;946(2):239–246. Available at: http://www.ncbi.nlm.nih.gov/pubmed/12137927
166. Nybäck HV, Walters JR, Aghajanian GK, Roth RH. Tricyclic antidepressants: Effects on the firing rate of brain noradrenergic neurons. *European Journal of Pharmacology*. 1975;32(02):302–312. Available at: http://www.ncbi.nlm.nih.gov/pubmed/1149813
167. Szabo ST, Montigny C de, Blier P. Progressive attenuation of the firing activity of locus coeruleus noradrenergic neurons by sustained administration of selective serotonin reuptake inhibitors. *The International Journal of Neuropsychopharmacology/Official Scientific Journal of the Collegium Internationale Neuropsychopharmacologicum (CINP)*. 2000;3(1):1–11. Available at: http://www.ncbi.nlm.nih.gov/pubmed/11343573
168. Giaroni C, Canciani L, Zanetti E et al. Effects of chronic desipramine treatment on alpha2-adrenoceptors and mu-opioid receptors in the guinea pig cortex and hippocampus. *European Journal of Pharmacology*. 2008;579(1–3):116–125. Available at: http://www.ncbi.nlm.nih.gov/pubmed/18028907
169. Nestler EJ, McMahon A, Sabban EL, Tallman JF, Duman RS. Chronic antidepressant administration decreases the expression of tyrosine hydroxylase in the rat locus coeruleus. *Proceedings of the National Academy of Sciences of the United States of America*. 1990;87(19):7522–7526. Available at: http://www.pubmedcentral.nih.gov/articlerender.fcgi?artid=54779&tool=pmcentrez&rendertype=abstract
170. Subhash MN, Nagaraja MR, Sharada S, Vinod KY. Cortical alpha-adrenoceptor downregulation by tricyclic antidepressants in the rat brain. *Neurochemistry International*. 2003;43(7):603–609. Available at: http://www.ncbi.nlm.nih.gov/pubmed/12892647
171. Gilsbach R, Faron-Górecka A, Rogóz Z et al. Norepinephrine transporter knockout-induced up-regulation of brain alpha2A/C-adrenergic receptors. *Journal of Neurochemistry*. 2006;96(4):1111–1120. Available at: http://www.ncbi.nlm.nih.gov/pubmed/16417582
172. Dazzi L, Seu E, Cherchi G, Biggio G. Antagonism of the stress-induced increase in cortical norepinephrine output by the selective norepinephrine reuptake inhibitor reboxetine. *European Journal of Pharmacology*. 2003;476(1–2):55–61. Available at: http://www.ncbi.nlm.nih.gov/pubmed/12969749
173. Berod A, Chat M, Paut L, Tappaz M. Catecholaminergic and GABAergic anatomical relationship in the rat substantia nigra, locus coeruleus, and hypothalamic median eminence: Immunocytochemical visualization of biosynthetic enzymes on serial semithin plastic-embedded sections. *The Journal of Histochemistry and Cytochemistry: Official Journal of the Histochemistry Society*. 1984;32(12):1331–1338. Available at: http://www.ncbi.nlm.nih.gov/pubmed/6150057
174. Shefner SA, Osmanović SS. GABAA and GABAB receptors and the ionic mechanisms mediating their effects on locus coeruleus neurons. *Progress in Brain Research*. 1991;88:187–195. Available at: http://www.ncbi.nlm.nih.gov/pubmed/1667544
175. Belujon P, Baufreton J, Grandoso L et al. Inhibitory transmission in locus coeruleus neurons expressing GABAA receptor epsilon subunit has a number of unique properties. *Journal of Neurophysiology*. 2009;102(4):2312–2325. Available at: http://www.ncbi.nlm.nih.gov/pubmed/19625540
176. Hellsten KS, Sinkkonen ST, Hyde TM et al. Human locus coeruleus neurons express the GABA(A) receptor gamma2 subunit gene and produce benzodiazepine binding. *Neuroscience Letters*. 2010;477(2):77–81. Available at: http://www.ncbi.nlm.nih.gov/pubmed/20417252
177. Waldvogel HJ, Baer K, Eady E et al. Differential localization of gamma-aminobutyric acid type A and glycine receptor subunits and gephyrin in the human pons, medulla oblongata and uppermost cervical segment of the spinal cord: An immunohistochemical study. *The Journal of Comparative Neurology*. 2010;518(3):305–328. Available at: http://www.ncbi.nlm.nih.gov/pubmed/19950251

178. Gervasoni D, Darracq L, Fort P et al. Electrophysiological evidence that noradrenergic neurons of the rat locus coeruleus are tonically inhibited by GABA during sleep. *The European Journal of Neuroscience*. 1998;10(3):964–970. Available at: http://www.ncbi.nlm.nih.gov/pubmed/9753163
179. Kaur S, Saxena RN, Mallick BN. GABAergic neurons in prepositus hypoglossi regulate REM sleep by its action on locus coeruleus in freely moving rats. *Synapse (New York, N.Y.)*. 2001;42(3):141–150. Available at: http://www.ncbi.nlm.nih.gov/pubmed/11746711
180. Zhu H, Karolewicz B, Nail E et al. Normal [^{3}H]flunitrazepam binding to GABAA receptors in the locus coeruleus in major depression and suicide. *Brain Research*. 2006;1125(1):138–146. Available at: http://www.pubmedcentral.nih.gov/articlerender.fcgi?artid=1783976&tool=pmcentrez&rendertype=abstract
181. Luscher B, Shen Q, Sahir N. The GABAergic deficit hypothesis of major depressive disorder. *Molecular Psychiatry*. 2011;16(4):383–406. Available at: http://www.ncbi.nlm.nih.gov/pubmed/21079608
182. Petty F, Kramer GL, Gullion CM, Rush AJ. Low plasma gamma-aminobutyric acid levels in male patients with depression. *Biological Psychiatry*. 1992;32(4):354–363. Available at: http://www.ncbi.nlm.nih.gov/pubmed/1420649
183. Sanacora G, Mason GF, Rothman DL et al. Reduced cortical gamma-aminobutyric acid levels in depressed patients determined by proton magnetic resonance spectroscopy. *Archives of General Psychiatry*. 1999;56(11):1043–1047. Available at: http://www.ncbi.nlm.nih.gov/pubmed/10565505
184. Sanacora G, Gueorguieva R, Epperson CN et al. Subtype-specific alterations of gamma-aminobutyric acid and glutamate in patients with major depression. *Archives of General Psychiatry*. 2004;61(7):705–713. Available at: http://www.ncbi.nlm.nih.gov/pubmed/15237082
185. Ye Z-Y, Zhou K-Q, Xu T-L, Zhou J-N. Fluoxetine potentiates GABAergic IPSCs in rat hippocampal neurons. *Neuroscience Letters*. 2008;442(1):24–29. Available at: http://www.ncbi.nlm.nih.gov/pubmed/18606211
186. Bhagwagar Z, Wylezinska M, Taylor M et al. Increased brain GABA concentrations following acute administration of a selective serotonin reuptake inhibitor. *The American Journal of Psychiatry*. 2004;161(2):368–370. Available at: http://www.ncbi.nlm.nih.gov/pubmed/14754790
187. Robinson RT, Drafts BC, Fisher JL. Fluoxetine increases GABA(A) receptor activity through a novel modulatory site. *The Journal of Pharmacology and Experimental Therapeutics*. 2003;304(3):978–984. Available at: http://www.ncbi.nlm.nih.gov/pubmed/12604672
188. Sanacora G, Mason GF, Rothman DL, Krystal JH. Increased occipital cortex GABA concentrations in depressed patients after therapy with selective serotonin reuptake inhibitors. *The American Journal of Psychiatry*. 2002;159(4):663–665. Available at: http://www.ncbi.nlm.nih.gov/pubmed/11925309
189. Crestani F, Lorez M, Baer K et al. Decreased GABAA-receptor clustering results in enhanced anxiety and a bias for threat cues. *Nature Neuroscience*. 1999;2(9):833–839. Available at: http://www.ncbi.nlm.nih.gov/pubmed/10461223
190. Earnheart JC, Schweizer C, Crestani F et al. GABAergic control of adult hippocampal neurogenesis in relation to behavior indicative of trait anxiety and depression states. *The Journal of Neuroscience: The Official Journal of the Society for Neuroscience*. 2007;27(14):3845–3854. Available at: http://www.pubmedcentral.nih.gov/articlerender.fcgi?artid=2441879&tool=pmcentrez&rendertype=abstract
191. Shen Q, Lal R, Luellen BA et al. Gamma-aminobutyric acid-type A receptor deficits cause hypothalamic–pituitary–adrenal axis hyperactivity and antidepressant drug sensitivity reminiscent of melancholic forms of depression. *Biological Psychiatry*. 2010;68(6):512–520. Available at: http://www.pubmedcentral.nih.gov/articlerender.fcgi?artid=2930197&tool=pmcentrez&rendertype=abstract

192. Herkenham M, Nauta WJ. Efferent connections of the habenular nuclei in the rat. *The Journal of Comparative Neurology.* 1979;187(1):19–47. Available at: http://www.ncbi.nlm.nih.gov/pubmed/226566
193. Sartorius A, Henn FA. Deep brain stimulation of the lateral habenula in treatment resistant major depression. *Medical Hypotheses.* 2007;69(6):1305–1308. Available at: http://www.ncbi.nlm.nih.gov/pubmed/17498883
194. Somogyi J, Llewellyn-Smith IJ. Patterns of colocalization of GABA, glutamate and glycine immunoreactivities in terminals that synapse on dendrites of noradrenergic neurons in rat locus coeruleus. *The European Journal of Neuroscience.* 2001;14(2):219–228. Available at: http://www.ncbi.nlm.nih.gov/pubmed/11553275
195. Anon. Allen Brain Atlas Resources [Internet]. *Allen Institute for Brain Science.* 2009. Available at: http://www.brain-map.org, ed.
196. Noriega NC, Garyfallou VT, Kohama SG, Urbanski HF. Glutamate receptor subunit expression in the rhesus macaque locus coeruleus. *Brain Research.* 2007;1173:53–65. Available at: http://www.pubmedcentral.nih.gov/articlerender.fcgi?artid=2067256&tool=pmcentrez&rendertype=abstract
197. Olpe HR, Steinmann MW, Brugger F, Pozza MF. Excitatory amino acid receptors in rat locus coeruleus. An extracellular *in vitro* study. *Naunyn-Schmiedeberg's Archives of Pharmacology.* 1989;339(3):312–314. Available at: http://www.ncbi.nlm.nih.gov/pubmed/2566932
198. Page ME, Szeliga P, Gasparini F, Cryan JF. Blockade of the mGlu5 receptor decreases basal and stress-induced cortical norepinephrine in rodents. *Psychopharmacology.* 2005;179(1):240–246. Available at: http://www.ncbi.nlm.nih.gov/pubmed/15717212
199. Ennis M, Aston-Jones G, Shiekhattar R. Activation of locus coeruleus neurons by nucleus paragigantocellularis or noxious sensory stimulation is mediated by intracoerulear excitatory amino acid neurotransmission. *Brain Research.* 1992;598(1–2):185–195. Available at: http://www.ncbi.nlm.nih.gov/pubmed/1336704
200. Rasmussen K, Hsu M-A, Vandergriff J. The selective mGlu2/3 receptor antagonist LY341495 exacerbates behavioral signs of morphine withdrawal and morphine-withdrawal-induced activation of locus coeruleus neurons. *Neuropharmacology.* 2004;46(5):620–628. Available at: http://www.ncbi.nlm.nih.gov/pubmed/14996539
201. Auer DP, Pütz B, Kraft E et al. Reduced glutamate in the anterior cingulate cortex in depression: An *in vivo* proton magnetic resonance spectroscopy study. *Biological Psychiatry.* 2000;47(4):305–313. Available at: http://www.ncbi.nlm.nih.gov/pubmed/10686265
202. Hashimoto K, Sawa A, Iyo M. Increased levels of glutamate in brains from patients with mood disorders. *Biological Psychiatry.* 2007;62(11):1310–1316. Available at: http://www.ncbi.nlm.nih.gov/pubmed/17574216
203. Michael N, Erfurth A, Ohrmann P et al. Metabolic changes within the left dorsolateral prefrontal cortex occurring with electroconvulsive therapy in patients with treatment resistant unipolar depression. *Psychological Medicine.* 2003;33(7):1277–1284. Available at: http://www.ncbi.nlm.nih.gov/pubmed/14580081
204. Bernard R, Kerman IA, Thompson RC et al. Altered expression of glutamate signaling, growth factor, and glia genes in the locus coeruleus of patients with major depression. *Molecular Psychiatry.* 2011;16:634–646. Available at: http://www.pubmedcentral.nih.gov/articlerender.fcgi?artid=2927798&tool=pmcentrez&rendertype=abstract
205. Choudary PV, Molnar M, Evans SJ et al. Altered cortical glutamatergic and GABAergic signal transmission with glial involvement in depression. *Proceedings of the National Academy of Sciences of the United States of America.* 2005;102(43):15653–15658. Available at: http://www.pubmedcentral.nih.gov/articlerender.fcgi?artid=1257393&tool=pmcentrez&rendertype=abstract

206. Singewald N, Zhou GY, Schneider C. Release of excitatory and inhibitory amino acids from the locus coeruleus of conscious rats by cardiovascular stimuli and various forms of acute stress. *Brain Research.* 1995;704(1):42–50. Available at: http://www.ncbi.nlm.nih.gov/pubmed/8750960
207. Timmerman W, Cisci G, Nap A, de Vries JB, Westerink BH. Effects of handling on extracellular levels of glutamate and other amino acids in various areas of the brain measured by microdialysis. *Brain Research.* 1999;833(2):150–160. Available at: http://www.ncbi.nlm.nih.gov/pubmed/10375690
208. Zarate CA, Singh JB, Carlson PJ et al. A randomized trial of an *N*-methyl-D-aspartate antagonist in treatment-resistant major depression. *Archives of General Psychiatry.* 2006;63(8):856–864. Available at: http://www.ncbi.nlm.nih.gov/pubmed/16894061
209. Diazgranados N, Ibrahim LA, Brutsche NE et al. Rapid resolution of suicidal ideation after a single infusion of an *N*-methyl-D-aspartate antagonist in patients with treatment-resistant major depressive disorder. *The Journal of Clinical Psychiatry.* 2010;71(12):1605–1611. Available at: http://www.pubmedcentral.nih.gov/articlerender.fcgi?artid=3012738&tool=pmcentrez&rendertype=abstract
210. Price RB, Nock MK, Charney DS, Mathew SJ. Effects of intravenous ketamine on explicit and implicit measures of suicidality in treatment-resistant depression. *Biological Psychiatry.* 2009;66(5):522–526. Available at: http://www.pubmedcentral.nih.gov/articlerender.fcgi?artid=2935847&tool=pmcentrez&rendertype=abstract
211. Maeng S, Zarate CA, Du J et al. Cellular mechanisms underlying the antidepressant effects of ketamine: Role of alpha-amino-3-hydroxy-5-methylisoxazole-4-propionic acid receptors. *Biological Psychiatry.* 2008;63(4):349–352. Available at: http://www.ncbi.nlm.nih.gov/pubmed/17643398
212. Sanacora G, Kendell SF, Levin Y et al. Preliminary evidence of riluzole efficacy in antidepressant-treated patients with residual depressive symptoms. *Biological Psychiatry.* 2007;61(6):822–825. Available at: http://www.pubmedcentral.nih.gov/articlerender.fcgi?artid=2754299&tool=pmcentrez&rendertype=abstract
213. Zarate CA, Payne JL, Quiroz J et al. An open-label trial of riluzole in patients with treatment-resistant major depression. *The American Journal of Psychiatry.* 2004;161(1):171–174. Available at: http://www.ncbi.nlm.nih.gov/pubmed/14702270
214. Witkin JM, Marek GJ, Johnson BG, Schoepp DD. Metabotropic glutamate receptors in the control of mood disorders. *CNS and Neurological Disorders Drug Targets.* 2007;6(2):87–100. Available at: http://www.ncbi.nlm.nih.gov/pubmed/17430147
215. Karolewicz B, Stockmeier CA, Ordway GA. Elevated levels of the NR2C subunit of the NMDA receptor in the locus coeruleus in depression. *Neuropsychopharmacology: Official Publication of the American College of Neuropsychopharmacology.* 2005;30(8):1557–1567. Available at: http://www.pubmedcentral.nih.gov/articlerender.fcgi?artid=2921564&tool=pmcentrez&rendertype=abstract
216. Karolewicz B, Szebeni K, Stockmeier CA et al. Low nNOS protein in the locus coeruleus in major depression. *Journal of Neurochemistry.* 2004;91(5):1057–1066. Available at: http://www.pubmedcentral.nih.gov/articlerender.fcgi?artid=2923201&tool=pmcentrez&rendertype=abstract
217. Bartanusz V, Aubry JM, Pagliusi S et al. Stress-induced changes in messenger RNA levels of *N*-methyl-D-aspartate and AMPA receptor subunits in selected regions of the rat hippocampus and hypothalamus. *Neuroscience.* 1995;66(2):247–252. Available at: http://www.ncbi.nlm.nih.gov/pubmed/7477869
218. Fitzgerald LW, Ortiz J, Hamedani AG, Nestler EJ. Drugs of abuse and stress increase the expression of GluR1 and NMDAR1 glutamate receptor subunits in the rat ventral tegmental area: Common adaptations among cross-sensitizing agents. *The Journal of Neuroscience: The Official Journal of the Society for Neuroscience.* 1996;16(1):274–282. Available at: http://www.ncbi.nlm.nih.gov/pubmed/8613793

219. Schwendt M, Jezová D. Gene expression of two glutamate receptor subunits in response to repeated stress exposure in rat hippocampus. *Cellular and Molecular Neurobiology*. 2000;20(3):319–329. Available at: http://www.ncbi.nlm.nih.gov/pubmed/10789831
220. Watanabe Y, Weiland NG, McEwen BS. Effects of adrenal steroid manipulations and repeated restraint stress on dynorphin mRNA levels and excitatory amino acid receptor binding in hippocampus. *Brain Research*. 1995;680(1–2):217–225. Available at: http://www.ncbi.nlm.nih.gov/pubmed/7663979
221. Autry AE, Adachi M, Nosyreva E et al. NMDA receptor blockade at rest triggers rapid behavioural antidepressant responses. *Nature*. 2011;475(7354):91–95. Available at: http://www.ncbi.nlm.nih.gov/pubmed/21677641
222. Liprando LA, Miner LH, Blakely RD, Lewis DA, Sesack SR. Ultrastructural interactions between terminals expressing the norepinephrine transporter and dopamine neurons in the rat and monkey ventral tegmental area. *Synapse (New York, N.Y.)*. 2004;52(4):233–244. Available at: http://www.ncbi.nlm.nih.gov/pubmed/15103690
223. Bekar LK, He W, Nedergaard M. Locus coeruleus alpha-adrenergic-mediated activation of cortical astrocytes *in vivo*. *Cerebral Cortex (New York, N.Y.: 1991)*. 2008;18(12):2789–2795. Available at: http://www.pubmedcentral.nih.gov/articlerender.fcgi?artid=2583159&tool=pmcentrez&rendertype=abstract
224. Hösli L, Hösli E, Zehntner C, Lehmann R, Lutz TW. Evidence for the existence of alpha- and beta-adrenoceptors on cultured glial cells—An electrophysiological study. *Neuroscience*. 1982;7(11):2867–2872. Available at: http://www.ncbi.nlm.nih.gov/pubmed/6296723
225. Mantyh PW, Rogers SD, Allen CJ et al. Beta 2-adrenergic receptors are expressed by glia *in vivo* in the normal and injured central nervous system in the rat, rabbit, and human. *The Journal of Neuroscience: The Official Journal of the Society for Neuroscience*. 1995;15(1 Pt 1):152–164. Available at: http://www.ncbi.nlm.nih.gov/pubmed/7823126
226. Papay R, Gaivin R, Jha A et al. Localization of the mouse alpha1A-adrenergic receptor (AR) in the brain: Alpha1AAR is expressed in neurons, GABAergic interneurons, and NG2 oligodendrocyte progenitors. *The Journal of Comparative Neurology*. 2006;497(2):209–222. Available at: http://www.ncbi.nlm.nih.gov/pubmed/16705673
227. Peng L, Li B, Du T et al. Astrocytic transactivation by alpha2A-adrenergic and 5-HT2B serotonergic signaling. *Neurochemistry International*. 2010;57(4):421–431. Available at: http://www.ncbi.nlm.nih.gov/pubmed/20450946
228. Tanaka KF, Kashima H, Suzuki H, Ono K, Sawada M. Existence of functional beta1- and beta2-adrenergic receptors on microglia. *Journal of Neuroscience Research*. 2002;70(2):232–237. Available at: http://www.ncbi.nlm.nih.gov/pubmed/12271472
229. Cambray-Deakin M, Pearce B, Morrow C, Murphy S. Effects of neurotransmitters on astrocyte glycogen stores *in vitro*. *Journal of Neurochemistry*. 1988;51(6):1852–1857. Available at: http://www.ncbi.nlm.nih.gov/pubmed/2903222
230. Hertz L, Chen Y, Gibbs ME, Zang P, Peng L. Astrocytic adrenoceptors: A major drug target in neurological and psychiatric disorders? *Current Drug Targets. CNS and Neurological Disorders*. 2004;3(3):239–267. Available at: http://www.ncbi.nlm.nih.gov/pubmed/15180484
231. Porter JT, McCarthy KD. Astrocytic neurotransmitter receptors *in situ* and *in vivo*. *Progress in Neurobiology*. 1997;51(4):439–455. Available at: http://www.ncbi.nlm.nih.gov/pubmed/9106901
232. Fahrig T. Receptor subtype involved and mechanism of norepinephrine-induced stimulation of glutamate uptake into primary cultures of rat brain astrocytes. *Glia*. 1993;7(3):212–218. Available at: http://www.ncbi.nlm.nih.gov/pubmed/8095921
233. Hansson E, Rönnbäck L. Regulation of glutamate and GABA transport by adrenoceptors in primary astroglial cell cultures. *Life Sciences*. 1989;44(1):27–34. Available at: http://www.ncbi.nlm.nih.gov/pubmed/2563301

234. Bezzi P, Gundersen V, Galbete JL et al. Astrocytes contain a vesicular compartment that is competent for regulated exocytosis of glutamate. *Nature Neuroscience*. 2004;7(6):613–620. Available at: http://www.ncbi.nlm.nih.gov/pubmed/15156145
235. Ongür D, Drevets WC, Price JL. Glial reduction in the subgenual prefrontal cortex in mood disorders. *Proceedings of the National Academy of Sciences of the United States of America*. 1998;95(22):13290–13295. Available at: http://www.pubmedcentral.nih.gov/articlerender.fcgi?artid=23786&tool=pmcentrez&rendertype=abstract
236. Rajkowska G, Miguel-Hidalgo JJ, Wei J et al. Morphometric evidence for neuronal and glial prefrontal cell pathology in major depression. *Biological Psychiatry*. 1999;45(9):1085–1098. Available at: http://www.ncbi.nlm.nih.gov/pubmed/10331101
237. Altshuler LL, Abulseoud OA, Foland-Ross L et al. Amygdala astrocyte reduction in subjects with major depressive disorder but not bipolar disorder. *Bipolar Disorders*. 2010;12(5):541–549. Available at: http://www.ncbi.nlm.nih.gov/pubmed/20712756
238. Bowley MP, Drevets WC, Ongür D, Price JL. Low glial numbers in the amygdala in major depressive disorder. *Biological Psychiatry*. 2002;52(5):404–412. Available at: http://www.ncbi.nlm.nih.gov/pubmed/12242056
239. Stockmeier CA, Mahajan GJ, Konick LC et al. Cellular changes in the postmortem hippocampus in major depression. *Biological Psychiatry*. 2004;56(9):640–650. Available at: http://www.pubmedcentral.nih.gov//articlerender.fcgi?artid=2929806&tool=pmcentrez&rendertype=abstract
240. Johnston-Wilson NL, Sims CD, Hofmann JP et al. Disease-specific alterations in frontal cortex brain proteins in schizophrenia, bipolar disorder, and major depressive disorder. The Stanley Neuropathology Consortium. *Molecular Psychiatry*. 2000;5(2):142–149. Available at: http://www.ncbi.nlm.nih.gov/pubmed/10822341
241. Si X, Miguel-Hidalgo JJ, O'Dwyer G, Stockmeier CA, Rajkowska G. Age-dependent reductions in the level of glial fibrillary acidic protein in the prefrontal cortex in major depression. *Neuropsychopharmacology: Official Publication of the American College of Neuropsychopharmacology*. 2004;29(11):2088–2096. Available at: http://www.ncbi.nlm.nih.gov/pubmed/15238995
242. Steiner J, Bielau H, Brisch R et al. Immunological aspects in the neurobiology of suicide: Elevated microglial density in schizophrenia and depression is associated with suicide. *Journal of Psychiatric Research*. 2008;42(2):151–157. Available at: http://www.ncbi.nlm.nih.gov/pubmed/17174336
243. Hercher C, Turecki G, Mechawar N. Through the looking glass: Examining neuroanatomical evidence for cellular alterations in major depression. *Journal of Psychiatric Research*. 2009;43(11):947–961. Available at: http://www.ncbi.nlm.nih.gov/pubmed/19233384
244. Samuel D, Blin O, Dusticier N, Nieoullon A. Effects of riluzole (2-amino-6-trifluoromethoxy benzothiazole) on striatal neurochemical markers in the rat, with special reference to the dopamine, choline, GABA and glutamate synaptosomal high affinity uptake systems. *Fundamental & Clinical Pharmacology*. 1992;6(4–5):177–184. Available at: http://www.ncbi.nlm.nih.gov/pubmed/1385285
245. Azbill RD, Mu X, Springer JE. Riluzole increases high-affinity glutamate uptake in rat spinal cord synaptosomes. *Brain Research*. 2000;871(2):175–180. Available at: http://www.ncbi.nlm.nih.gov/pubmed/10899284
246. Gourley SL, Espitia JW, Sanacora G, Taylor JR. Antidepressant-like properties of oral riluzole and utility of incentive disengagement models of depression in mice. *Psychopharmacology*. 2012;219(3):805–14. Available at: http://www.ncbi.nlm.nih.gov/pubmed/21779782

247. Valentine GW, Sanacora G. Targeting glial physiology and glutamate cycling in the treatment of depression. *Biochemical Pharmacology*. 2009;78(5):431–439. Available at: http://www.pubmedcentral.nih.gov/articlerender.fcgi?artid=2801154&tool=pmcentrez&rendertype=abstract
248. Alexander GM, Grothusen JR, Gordon SW, Schwartzman RJ. Intracerebral microdialysis study of glutamate reuptake in awake, behaving rats. *Brain Research*. 1997;766(1–2):1–10. Available at: http://www.ncbi.nlm.nih.gov/pubmed/9359581
249. Hansson E, Rönnbäck L. Adrenergic receptor regulation of amino acid neurotransmitter uptake in astrocytes. *Brain Research Bulletin*. 1992;29(3–4):297–301. Available at: http://www.ncbi.nlm.nih.gov/pubmed/1356597
250. Ordway GA, Szebeni A, Chandley MJ et al. Reduced gene expression of bone morphogenetic protein 7 in brain astrocytes in major depression. *Neuropsychopharmacology*. 2011 September 6:1–14 [epub ahead of print] PMID:21896235.

4 Gamma-Aminobutyric Acid Involvement in Depressive Illness

Interactions with Corticotropin-Releasing Hormone and Serotonin

Hymie Anisman, Zul Merali, and Michael O. Poulter

CONTENTS

There is little doubt that genetic and experiential factors contribute to the neurochemical processes responsible for the development of major depressive disorder (MDD) (Caspi et al., 2003; Kendler et al., 2005; Millan, 2006). In this regard, MDD is a biochemically heterogeneous disorder, and any of several neurochemical and/or receptor alterations provoked by stressful experiences might contribute to the development of depressive symptoms. Moreover, the effectiveness of antidepressants in attenuating MDD symptoms might be tied to the particular neurochemical alterations elicited by stressors in any given individual, and multitargeting as a strategy for the treatment of depression has received increased attention (Millan, 2006, 2009).

Although considerable evidence had pointed to a role for serotonergic processes in subserving MDD (Pineyro and Blier, 1999), it is clear that attributing MDD *uniquely* to serotonin (5-HT) is not a sustainable perspective. Considerable evidence has indicated that the nature of the 5-HT changes associated with depression (e.g., in postmortem analyses of depressed individuals that died by suicide) are highly variable (Anisman, 2009; Stockmeier, 2003). Moreover, drug treatments that affect 5-HT processes are effective in only a portion of patients, not all symptoms resolve with treatment, and recurrence rates are exceedingly high (Millan, 2006). Although not dismissing a role for 5-HT in the evolution or maintenance of MDD, it has been maintained that other processes might contribute in this regard. These have included several growth factors and cytokines, such as brain-derived neurotrophic factor (BDNF) (Duman and Monteggia, 2006) and various interleukins (Anisman et al., 2008; Dantzer et al., 2008), corticotropin-releasing hormone (CRH), and other peptides, such as neuromedin B and somatostatin (Merali et al., 2004, 2006; Nemeroff, 1996; Reul and Holsboer, 2002). There has also been a rejuvenation of the view that γ-aminobutyric acid A ($GABA_A$) functioning might contribute to depressive illness (Rupprecht et al., 2006; Sanacora and Saricicek, 2007; Sequeira and Turecki, 2006; Tunnicliff and Malatynska, 2003), possibly by moderating the interplay between CRH and 5-HT (Hayley et al., 2005).

4.1 γ-AMINOBUTYRIC ACID FUNCTIONING WITHIN THE BRAIN

GABA and $GABA_A$ receptors are ubiquitous within the central nervous systems, playing a fundamental role in controlling neural inhibition and timing of neural networks (i.e., gating pyramidal cell synchrony) (Traub et al., 1999; Whitington and Traub, 2003; Whittington et al., 1995). $GABA_A$ receptors are pentameric protein complexes constructed from protein subunits named α, β, γ, δ, ε, and π that are derived from a repertoire (cassette) of 21 different proteins/genes (Olsen and Sieghart, 2008). Often these subunits exist as subtypes: α has six, β has three, and γ has three. The functionality of $GABA_A$ receptors is determined by specific subunit configurations. The stoichiometry of $GABA_A$ receptors is thought to be 2α;2β and one of the other subunits (γ, δ, ε, π) is "chosen" to complete the pentamer. Although there are many possible subunit combinations that could potentially make up a $GABA_A$ receptor, there are two primary classes that have been broadly defined. One class comprises receptors that have a γ subunit and the other class is one that has a δ subunit. For insertion into a synaptic site, a $GABA_A$ receptor must contain a γ subunit (which most often is the γ_2 subtype in the adult CNS).

Synaptic receptors mediate what is usually termed "phasic inhibition," but those having the δ subunit are not usually located in postsynaptic densities and are important for providing "tonic (extrasynaptic) inhibition." Importantly, the variability of $GABA_A$ receptor structure as defined by other subunits (α and β) seems to control the timing of phasic inhibition. The timing is primarily manifested by the variability of the rate of decay of the synaptic currents. It seems that once GABA is released from the synaptic vesicle it saturates the synaptic receptors, activating them within a millisecond (although there are exceptions to this generalization). Although the GABA concentration declines very quickly (within 2 or 3 ms), the synaptic current does not, as GABA tends to unbind relatively slowly from its receptors (Hutcheon et al., 2000; Maconochie et al., 1994). The rate at which it unbinds is controlled by structure (subunit expression) (Burgard et al., 1999; Hutcheon et al., 2000; Verdoorn, 1994). So, depending on the subunits comprising a given receptor, synaptic currents can last from 5 to 10 ms to hundreds of milliseconds. This variability in duration seems to be important in determining neuronal network synchronization. For example, fast phasic inhibition tends to create fast brain rhythms (gamma oscillations in hippocampus and cortex; Klausberger et al., 2002, 2004), whereas slow inhibition produces slow brain rhythms (delta in renticular nucleus; Bentivoglio et al., 1990; Zhang et al., 1997). Finally, to some extent, extrasynaptic/tonic inhibition (which is essentially active continuously) also seems important for controlling brain rhythms. In this regard, a recent report implicated tonic inhibition in controlling hippocampal gamma rhythms (Mann and Mody, 2010). Given the heterogeneous expression of $GABA_A$ receptor subunits regionally and even at the subcellular level, the understanding of the potential complexity of how all these factors control normal and abnormal brain function is daunting. Nevertheless, the implications are relatively easy to appreciate; perturbations in the structure of $GABA_A$ receptors have the potential to alter neural network activity and thus alter behavior.

4.2 GABA INVOLVEMENT IN ANXIETY AND DEPRESSION

The data supporting a role for GABAergic processes in mediating anxiety-related disorders have come from several lines of research. Among other things, it has been reported that (a) GABA levels in plasma and in CSF were increased in stress situations, (b) stressors influenced $GABA_A$ receptor functioning, (c) treatments that increase vulnerability to elevated anxiety and depression-like behaviors, such as early life stressors, also influence $GABA_A$ subunit expression, and (d) drugs that affect $GABA_A$ activity are effective in attenuating anxiety (reviewed in Anisman et al., 2008). Beyond the GABA–anxiety relationship, there is reason to believe that these processes may also contribute to MDD and may be important for the comorbidity that often occurs between anxiety and depression. In particular, it was reported that depression was accompanied by lower levels of GABA in cerebrospinal fluid (Sanacora and Saricicek, 2007) and based on neuroimaging analyses, GABA was reported to be reduced in the dorsolateral prefrontal and occipital cortex of depressed patients (Bhagwagar et al., 2007; Hasler et al., 2007; Price et al., 2009; Sanacora et al., 1999, 2004) and a reduction in the density and size of calbindin-immunoreactive (CB-IR) GABAergic neurons was reported to be evident in the prefrontal and occipital cortex (Maciag et al., 2010; Rajkowska et al., 2007). Furthermore, it was reported that GABA levels within the PFC were inversely related to severity of depression (Honig et al., 1988), and $GABA_A$ receptor subunit expression was likely altered in depressed suicides given that $GABA_A$/BDZ (benzodiazepine)-binding sites were elevated (Cheetham et al., 1988). In fact, mood disorders in female patients were reported to be associated with $GABA_A$ α_1 and α_6 polymorphisms (Yamada et al., 2003).

In addition to these GABA variations, it was reported that the relative density of a primary enzyme for GABA synthesis, glutamic acid decarboxylase (GAD) neuropil, was elevated in the hippocampal region of depressed suicidal patients relative to controls, but such an outcome was not apparent in several cortical regions, such as the orbitofrontal, anterior cingulate, dorsolateral prefrontal, and the entorhinal cortex (Gos et al., 2009), although in other studies variations of GAD were detected in diverse prefrontal cortical regions (Bielau et al., 2007; Fatemi et al., 2005; Gos et al., 2009; Karolewicz et al., 2010). Furthermore, it was reported that drugs that act as antidepressants influence interneuron functioning (Akinci and Johnston, 1993;

Brambilla et al., 2003; Krystal et al., 2002; Shiah and Yatham, 1998; Tunnicliff and Malatynska, 2003), raising the possibility that the actions of these agents stem from their GABAergic effects.

Despite these positive findings, there have also been reports indicating that neither GABA levels (Korpi et al., 1988), GABA-related enzymes (Cheetham et al., 1988; Sherif et al., 1991), nor the GABA transporter (Sundman-Eriksson and Allard, 2002) differed between drug-free depressed suicide victims and controls (Arranz et al., 1992; Cross et al., 1988; Sundman et al., 1997). Not surprisingly, there were many differences between the depressed populations assessed across studies, as well as the procedures used to assess GABA functioning. Thus, as in many other analyses of relations between pathology and biological substrates, it is difficult to define what factors might have been fundamental in accounting for the between-study differences that have been reported. These inconsistencies notwithstanding, there does seem to be appreciable support for the view that MDD was accompanied by disturbances in key metabolic enzymes involved in the synthesis of glutamate and GABA as well as proteins involved in membrane expression of GABA and the uptake of glutamate by glial cells.

Beyond the variations of GABA functioning, it appears that the mRNA expression of $GABA_A$ subunits may be either up- or down-regulated in association with MDD, depending on the brain region assessed (Choudary et al., 2005; Merali et al., 2004; Poulter et al., 2010b; Rupprecht et al., 2006; Sequeira and Turecki, 2006). Consistent with these findings, a broad gene expression analysis, supported by semiquantitative reverse transcription polymerase chain reaction (RT-PCR) analyses, revealed that glutamatergic (GLU) and GABAergic-related genes were altered across numerous cortical and subcortical brain regions (being particularly notable within the prefrontal cortex and hippocampus) of suicides with and without major depression and controls (Sequeira et al., 2009). Most of the GLU-related probe sets corresponded to ionotropic N methyl D aspartic acid (NMDA) receptor subunits (GRINA, GRIN2A, and GRINL1A) and 2-amino-3-(5-methyl-3-oxo-1,2-oxazol-4-yl)propanoic acid (AMPA) receptors (GRIA3, GRIA4, GRIA1, and GRIA2). Generally, AMPA receptors were up-regulated among the suicides with major depression relative to the control samples and those from suicides without a history of major depression. Conversely, in portions of the prefrontal cortex, Brodmans Area (BA) 46 and BA47, as well as aspects of the parietal cortex (BA38 and BA20), the glutamate metabotropic 3 receptor (GRM3) was down-regulated among the suicides with and without major depression. Thus, the former effects are likely tied to depression, whereas the latter might be related to suicide itself, rather than the depression associated with it.

Our own research has been consistent with the view that $GABA_A$ subunit expression was altered in MDD. As seen in Figure 4.1, in the frontopolar cortex (FPC) of depressed suicides the mRNA expression of the α_1, α_3, α_4, and δ subunits was lower in depressed suicides than in controls as was the expression of CRH Type 1 receptors; the latter was coupled with elevated levels of the peptide itself and thus may have reflected a compensatory down-regulation (Merali et al., 2004).

Beyond the frank changes of mRNA expression of $GABA_A$ subunits, using RT-qPCR analysis, we also showed relatively subtle changes in their expression patterns (Merali et al., 2004; Poulter et al., 2010a,b). It appeared that the expressional organization of the $GABA_A$ gene cassette may be altered among depressed

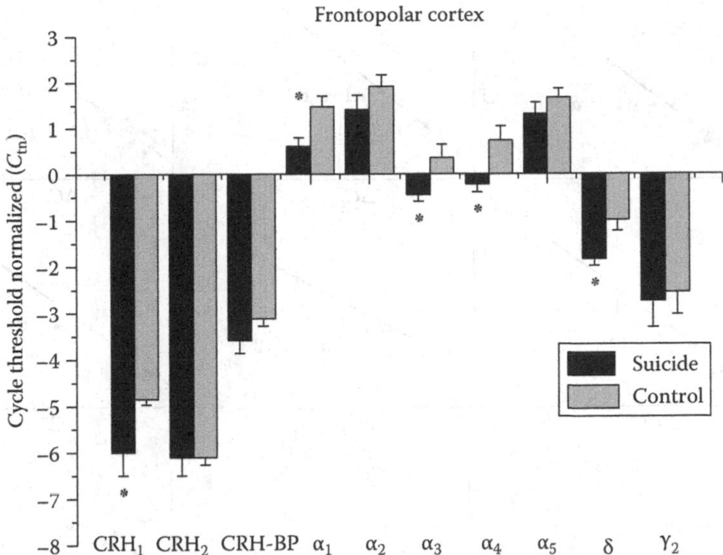

FIGURE 4.1 Mean (±SEM) expression of mRNA CRH_1, CRH_2, and CRH binding protein (CRH-BP), as well as mRNA expression of $GABA_A$ subunits in the frontopolar cortex of depressed individuals that died by suicide and that of controls (nondepressed individuals that died of causes other than suicide). Data are presented as normalized cycle thresholds (C_{tn}) wherein the expression of each species was normalized by subtracting its cycle threshold (C_t) from the housekeeping C_t. Thus, a negative C_{tn} indicates that the mRNA species was less abundant than that of housekeeping. A change of 1 C_{tn} is equivalent to a twofold difference in the abundance of that species. (From Merali, Z. et al., *J. Neurosci.*, 24, 1478, 2004. Reprinted with permission.)

individuals that died by suicide. Specifically, in several brain regions of individuals that died suddenly of causes other than suicide, there was appreciable "coordination" between several $GABA_A$ subunit mRNA expressions. Essentially, we found that the variations of subunit mRNA expression from individual to individual were often matched by expression levels of another subunit, so that numerous high positive interrelations existed between subunit mRNA expressions. For example, Figure 4.2A shows the relationships between the α_1 subunit and several other subunits in the FPC, and very similar patterns were evident with respect to other subunits as well. In fact, of 21 possible correlations involving the α_1, α_2, α_3, α_4, α_5, δ, and γ subunits, 18 were statistically significant. In contrast, as shown in Figure 4.2B, the comparable subunit interrelations were not apparent among depressed individuals that died by suicide, as only 3 of the 21 possible correlations were statistically significant. The RIN values and the pH in these samples were acceptable and were comparable in the two conditions, and, in this study, as in our other reports, samples were collected from individuals with brief agonal periods. Moreover, these brains were obtained within a few hours of death, as in Hungary, where the brains were collected (by our collaborators Drs. Miklos Palkovits and Gabor Faludi), the law requires that brains be taken soon after death, and permission from family members to use the tissue for experimental purposes is obtained thereafter. Thus, the findings

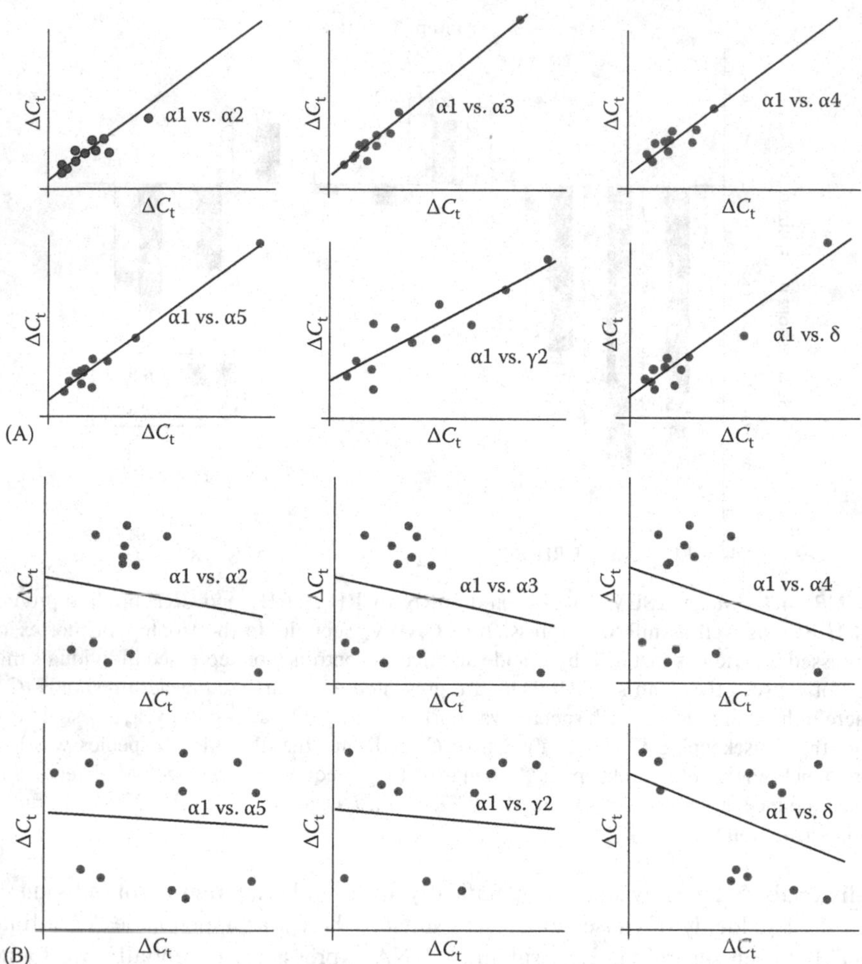

FIGURE 4.2 Coordinated expression was diminished in the frontopolar cortex of depressed individuals that died by suicide. In (A), regression graphs show examples of several significant correlations between α_1 and other subunits. (B) Regressions between these same subunits among depressed individuals that died by suicide. Note that the coordination between these subunits is markedly disturbed in the latter condition.

that we reported cannot be readily attributable to nonspecific effects related to tissue harvesting or contamination.

When the reduced subunit interrelations in the FPC of depressed suicides were first observed, we considered this to reflect a dysfunctional profile. That is, we assumed that the norm ought to be one of high interrelations reflecting coordination within this system, and indeed, these high interrelations were not only evident in humans, but were observed in mice as well (Poulter et al., 2010a). Since then we have also observed a similar coordination of subunit mRNA expression in the hippocampus and amygdala of humans; however, in other regions such as the orbital frontal cortex and the paraventricular nucleus of the hypothalamus, the significant interrelations between subunits were low (2–3 significant) among individuals that died of causes other than suicide, whereas in these same regions of depressed suicides, these interrelations were more frequent (Poulter et al., 2010b). Thus, we moved from our original perspective that the reduced interrelations reflect a dysfunctional outcome, and instead we simply view the altered interrelationships as potentially reflecting a reorganization of subunit expression.

This leaves us with several very fundamental questions. First, are these interrelations an unimportant epiphenomenon or an artifact of the measurement approach? Second, if these are genuine relationships, then what is their functional significance? Third, why are the relationships between subunits altered with suicide/depression, and why do these relationships decline in some regions of the depressed/suicide brain, but increase in others?

The initial thought concerning the interrelations might be that the genes for subunits might be close to one another on a chromosome, and as a result they might vary together (linkage). Indeed, although the gene for the δ subunit appears on its own on chromosome 1, genes for GABA subunit genes appear in clusters (McKernan and Whiting, 1996). For instance, in humans, genes encoding the α_1, α_6, β_2, and γ_2 subunits are clustered on chromosome 5, whereas genes for α_2, α_4, β_1, and γ_1 subunits appear on chromosome 4, the genes for the α_5, β_3, and γ_3 subunit appear on chromosome 15, and α_3 and ε genes are found on the X-chromosome (McKernan and Whiting, 1996; Steiger and Russek, 2004). However, we found that intercorrelations appear between subunits even if they are not on the same chromosomes, and, further, the intercorrelations vary appreciably across brain regions irrespective of the chromosomes on which they appear (Poulter et al., 2010b). Thus, the correlations cannot be attributed to factors related to certain genes being inherited as clusters.

This brings us to whether high interrelations between $GABA_A$ subunits are the norm, and whether the interrelations are disturbed in pathologies other than depression/suicide? There have, however, been few studies that assessed $GABA_A$ subunit expression in relation to pathology in humans or in animal models, and fewer still that assessed intercorrelations in relation to pathology. Yet, in studies of epilepsy in humans and in animal models, interrelations between GABA receptor subunits were reported (Brooks-Kayal et al., 1999), indicating that our findings were not anomalous and that coordinated expression of genes for the various subunits likely is the norm.

Accepting this view raises the obvious question concerning the functional significance of the subunit interrelations. It is widely accepted that $GABA_A$ receptor subunits control the timing of synaptic inhibition. It is also clear that synaptic inhibition is fundamental for generating nearly all types of brain rhythms including gamma, theta, and fast ripple oscillations. No region of the brain generates only one type of rhythm, and nearly every region has homogeneous expression of $GABA_A$ receptors. How the homogenous expression relates to differing brain rhythms and behavior is not completely understood. For example, in the hippocampus, one class of interneurons expresses a high abundance of α_1 subunits, whereas high expression of α_2 subunits appears in pyramidal cells (Klausberger and Somogyi, 2008). Thus, the timing of inhibition in pyramidal cells is different than in interneurons. Adding to this complexity are observations that synaptic inhibition is heterogeneous even on the same cell. In effect, it seems that the brain has evolved a highly complex mosaic of synaptic timing patterns that permit one region to generate preferred brain rhythms and perhaps restrict the generation of others.

Coordinated expression has been reported for other ion channels that are heteromeric, such as gap junctions and potassium channels, so this patterning is not unique to GABA receptor genes. The question is basically one of defining the biological significance for the presence of coordinated regulation of receptors and ion channels. One possibility is that transcriptionally regulating mRNA abundance of subunits would be a potential mechanism to ensure a proportional abundance of protein, as it would be energetically inefficient to produce excess of one protein over another if the stoichiometry of their assembly is 1:1. In this regard, for heteromeric proteins with variable functionality, such as $GABA_A$ receptors, having the correct balance of receptors involved in phasic vs. tonic inhibition may be particularly important. In addition, maintaining a balance of synaptic timing as some receptors (e.g., α_1) give rise to fast synaptic currents, whereas others are associated with relatively slow synaptic currents (α_5). In effect, maintaining balance between fast and slow synaptic currents might be facilitated by having the expression of one subunit with a particular physiological characteristic being balanced by the concurrent expression of another that acts in opposition to the first.

This raises the question as to how these timing patterns interact to produce a rhythm? $GABA_A$ receptor activity is not the only one that regulates neural network functioning. Ordinarily, the activity of neural networks or firing patterns involves several cellular characteristics (channel densities, calcium buffering, and cell morphology) and network parameters (distribution of neurons, as well as the abundance and location of synaptic contacts). Computational models that varied these as well as several other parameters revealed considerable "resiliency" as the timing of neural networks could function within a wide range of circuit parameters (Prinz et al., 2004). Nevertheless, critical interrelationships appeared to exist between certain parameters (e.g., synaptic strength), in that being outside of a particular range of values gave rise to the generation of aberrant rhythms (Prinz et al., 2004). As such, it might be expected that variations in the relative amounts of the subunits (or their coordination) that influence the stoichiometric ratios of subunits that make up $GABA_A$ receptors would lead to variations of the duration of inhibitory currents. Whether this reorganization creates a situation that puts the circuit outside some critical range is not known. So, at this point, the answer to the question regarding the impact of this organization (vs. disorganization) is not clear. Nevertheless, it seems that there are processes that control the relative strength of one timing pattern over another. These processes may ensure an appropriate "mix" of synaptic timing patterns that ultimately controls rhythms within a brain area. Importantly, our observations and those of others (Brooks-Kayal et al., 1999) suggest that the coordinated expression of the $GABA_A$ receptor gene cassette is plastic. Indeed, we have shown that these patterns were perturbed in mice that had been acutely or chronically stressed and that these responses varied in stressor-sensitive and stressor-resilient strains of mice (Poulter et al., 2010a).

Supposing our suggestion that coordinated subunits influence electrical rhythmicity, it would be expected that a stimulus that normally generates a gamma rhythm, for example, would not do so as readily in response to sufficiently intense challenges.

This could mean that the rhythm would not be generated at all or the strength of the oscillation (higher or lower) would be inadequate or inappropriate. Altered brain synchrony has been documented for many neurological disorders including epilepsy and sleep disorders, and altered EEG rhythms have been documented in those with MDD (Howland and Thase, 1991; Koyama and Yamashita, 1992; Pollock and Schneider, 1990). Coordinated $GABA_A$ receptor subunit expression may be a mechanism that ensures that normal brain electrical activity occurs and disturbances in these patterns may alter brain activity in such a way that abnormal cognitive function is created.

From this perspective, a certain degree of coordination between subunits would be advantageous for appropriate neural activity being maintained. Deviations from this coordinated pattern, in the form of elevated or reduced subunit interrelations across brain regions, might be viewed as engendering dysregulated neuronal firing patterns. In this regard, following our initial report of disturbed coordination of subunit expression in the FPC of individuals who died by suicide (Merali et al., 2004), we observed that this pattern was also apparent in the hippocampus and amygdala, but in some brain regions (orbital frontal cortex and paraventricular nucleus of the hypothalamus) coordinated expression of the same subunits was appreciably increased in depressed/suicides relative to nondepressed controls (Poulter et al., 2010b).

Studies of neuronal changes in postmortem tissue in humans do not permit analyses of the processes "causally" related to these outcomes, as experimental manipulations obviously cannot be undertaken. Thus, we assessed $GABA_A$ subunit gene expression and coordination in brain regions of mice that were highly stressor reactive (BALB/cByJ mice) relative to that evident in the stress-resilient C57BL/6ByJ strain (Poulter et al., 2010a). We observed that in most brain regions examined an acute stressor moderately increased the expression of these subunits, and these effects were still more pronounced following a chronic stressor, especially within the hippocampus. This was very different from the profile evident in the cortical regions of the depressed suicides where the subunit expression was diminished relative to controls, although in the mouse strains a high degree of coordinated expression of the subunits was apparent, just as they were in humans.

Interestingly, different profiles of $GABA_A$ subunit coordination were apparent in the highly stressor-reactive BALB/cByJ mice relative to that evident in the stress-resilient C57BL/6ByJ strain (the profiles were highly similar in other brain regions). Importantly, in these strains acute and chronic stressors differentially influenced subunit coordination. Unfortunately, the observed outcomes, while very pronounced, were exceptionally complex. In some regions the stressors reduced subunit organization, but in other regions organization appeared to be increased. Moreover, whereas some effects of acute stressors were exaggerated following chronic stressors, depending on the strain and the brain region examined, an apparent normalization of the coordinated gene expression occurred following a chronic stressor regimen (Poulter et al., 2010a). At this point, it is difficult to make definitive conclusions concerning the relationship between the stressor-provoked subunit organization and particular behavioral changes. But it might turn out to be meaningful that this organization is tied to stressor reactivity and is altered (increased or decreased) by acute and chronic stressor experiences. Nevertheless, we suggest that analyses that involve coordinated expression of factors that influence neuronal firing, in addition to analyses of the extent of subunit gene expression *per se*, may provide fundamental information regarding brain–behavior relations.

4.3 EPIGENETIC REGULATION

Epigenetics broadly defines two processes that alter or control DNA structure and ultimately the degree that DNA is transcribed. The first process is the chemical modification of DNA by the methylation of cytosines that are paired with guanines (CpGs) (D'Alessio and Szyf, 2006). This occurs through the activity of a family of enzymes called DNA methyl transferases (DNMT). There are three kinds of DNMTs that methylate DNA, namely DNMT-1, -3A, and -3B (type 2 was improperly named based on sequence homology and seems to methylate RNA; Schaefer and Lyko, 2010; Szyf and Detich, 2001). The methylation of DNA attracts proteins that recognize methylated CpGs (meCpGs) that, in turn, attract other proteins (histones) that tend to wind and condense the DNA. Condensed DNA is unable to be transcribed and hence the gene expression is prevented. The other process occurs through "histone rearrangements," which is controlled by their covalent modification. Histones, as already mentioned, are proteins that bind and wrap DNA. Depending on how they are positioned on DNA they can either enhance or repress transcription (Miller et al., 2008; Roth and Sweatt, 2009; Roth et al., 2009). This occurs through their acetylation and/or methylation by specific enzymes. These covalent modifications are reversible and so there is a dynamic regulation of the degree to which DNA is exposed to transcription factors and RNA polymerase by histone positioning. Importantly, DNA methylation and histone rearrangements also act in concert with one another. It should also be mentioned that chromatin structure is under the control of noncoding RNA, which are small untranslated RNA molecules that bind to chromatin and control its interaction with DNA (Malecova and Morris, 2010). Although epigenetic mechanisms often involve covalent modifications (methylation and acetylation), they are reversible and all these mechanisms are surprisingly "well tuned" to environmental and experiential stimuli. Thus, the view developed that the genome is under constant epigenetic surveillance, permitting cells to respond to their environment as the need arises. It is particularly exciting that this surveillance occurs in brain (D'Alessio and Szyf, 2006; Lubin et al., 2008; McGowan et al., 2009; Miller et al., 2008; Mullins-Sweatt et al., 2009; Poulter et al., 2008; Roth and Sweatt, 2009; Roth et al., 2009; Sweatt, 2009; Szyf et al., 2008; Weaver et al., 2004, 2005) and may contribute to the processes associated with mental illness.

Despite the allure of this relatively new line of research in studying brain–behavior relations, which is admittedly still in its infancy, several fundamental questions need to be addressed, some of which are not unlike those that have plagued research that involve microarrays used to assess the relationships between discrete behaviors and variations associated with thousands of genes concurrently. Specifically, there is good reason to believe that a very large number of genes encounter epigenetic changes, and thus one must wonder how meaningful this is when any one of these changes is tied to a particular phenotype. Studies that involve analyses of phenotypes in relation to genome wide scans have dealt with this through the use of a very large number of participants in an effort to deal with the possibility of spurious findings (i.e., alpha errors).

Analyses that involve epigenetic changes, such as our own research involving $GABA_A$ subunit promotors (Poulter et al., 2008), or those that involve glucocorticoid receptor promoters (McGowan et al., 2008) might thus be victims of this very problem. This said, the findings that have been reported might turn out to provide valuable information concerning the processes associated with stress and depression.

Based on studies in animals, it was reported that early life negative experiences and chronic stressors can induce DNA methylation and histone acetylation (Weaver et al., 2004, 2006). As such experiences can precipitate MDD, it prompted us to ask whether epigenetic mechanisms may be at play in the brain samples obtained from MDD/suicide completers. Indeed, we found that DNMT-3B expression was increased in both males and females and that this was associated with the hypermethylation of the $GABA_A$ receptor α_1 subunit promoter. Furthermore, the expression of DNMT-3B mRNA was negatively correlated to the expression of α_1 subunit mRNA (Poulter et al., 2008). These data suggest that the DNA methylation of the promoter may be responsible for altered expression of the α_1 subunit that we had previously reported to be related to depression/suicide. Similar data have also been reported in an analogous cohort of brain samples where the RNA polymerase promoter was shown to be hypermethylated (McGowan et al., 2008). Unlike our data where only a few sites were specifically methylated on the α_1 promoter, this study indicated that the hypermethylation showed no specificity as all sites assessed seemed to be more or less similarly affected. The exact mechanism by which the α_1 subunit is down-regulated by these processes is not clear. We suggested several possibilities in this regard. One was that the specific methylation sites are within transcription factor–binding regions and when methylated cannot be occupied by the transcription factor. The other possibility is that the methylation sites block the binding/activation of adjacent transcription factor–binding sites. Indeed, one hypermethylated site was next to a putative responsi element binding (CREB) regulatory site and so we speculated that mRNA expression might be reduced if a methyl-binding domain protein occluded CREB binding. This interaction has been found to occur on the *BDNF* gene (Levenson et al., 2006), but there is no reason to believe that this would not occur on *GABA* genes as well.

Hypermethylation might not be the only factor that accounts for the down-regulation of $GABA_A$ receptor subunits in the suicide MDD brains, as we found that the α_3 subunit was decreased in expression, although it has few or no CpGs in its proximal promoter region. Alternatively, global alterations of chromatin structure may occur that create "broad" control signals that lead to the exposure of numerous promoter sequences to transcription factors/RNA polymerase. This would "globally" up-regulate gene expression, which may be an efficient way to control coordinated expression of $GABA_A$ subunits discussed earlier. By contrast, methylation events may be more dependent on the absence or presence of CpGs within the promoter regions. Thus, methylation may affect only a few genes, shutting down their activity, thereby reducing the coordinated expression patterns. As indicated earlier, conclusions derived from epigenetic studies in postmortem human brain might be subject to difficulties related to the potentially large number of other concurrent changes that might have occurred. Nonetheless, it does seem that brain plasticity in both animal and humans might be influenced by epigenetic processes and the possibility that these processes contribute to pathology warrants further investigation.

4.4 INTERRELATIONS BETWEEN GABA, CRH, AND 5-HT

We have attempted to make a case for $GABA_A$ involvement in depressive illness, but this should not be misconstrued as other potential processes being excluded. To the contrary, we have argued elsewhere that GABA functioning in the context of depression and anxiety ought to be considered in relation to both CRH and 5-HT functioning at hypothalamic and at extrahypothalamic sites (Anisman et al., 2008).

A detailed review of CRH and 5-HT involvement in depression/anxiety is not within the scope of the present chapter (but see reviews in Anisman et al., 2008; Holsboer, 2003; Nemeroff and Vale, 2005). Suffice it that there is considerable biochemical and pharmacological evidence supporting CRH involvement in depressive illness, including our own findings that CRH levels were elevated and CRH_1 receptor mRNA expression was reduced within several aspects of the frontal cortex of depressed individuals that died by suicide (Merali et al., 2004). Postmortem analyses of depressed suicides have also revealed altered 5-HT receptor binding and variations of the expression of 5-HT receptor subtypes within the frontal cortex and/or hippocampus (Arango et al., 2003; Bhagwagar et al., 2006; Mintun et al., 2004; Stockmeier, 2003). As well, controls and depressed suicides differed with respect to the expression of several 5-HT receptor mRNAs, including $5\text{-}HT_{1A}$, $5\text{-}HT_{1B}$, and p11 (a protein involved in 5-HT receptor membrane expression; Anisman et al., 2008; Svenningsson and Greengard, 2007; Svenningsson et al., 2006). To be sure, the data supporting 5-HT involvement in MDD have not been unequivocal, as there have been reports indicating that depression/suicide was not associated with particular 5-HT receptor variations (Lowther et al., 1997; Rosel et al., 1997) and DNA microarray analyses indicated few molecular genetic differences within the dorsolateral and ventral prefrontal cortex of suicides and controls (Sibille et al., 2004). Perhaps, given that (a) the wide array of symptoms that characterize depressive illness vary across individuals, (b) depressive subtypes exist, (c) depression may be treatment-responsive or treatment-nonresponsive, and (d) depressed suicides are not necessarily reflective of depression *per se*, it should not be surprising that such studies have yielded inconsistent results. As described earlier, MDD is likely a biochemically heterogeneous illness, and as such it might be productive to think of the illness in terms of CRH, 5-HT, and GABA acting together, in some fashion, to influence depressive disorders. Thus, although it is highly likely that 5-HT plays some role in MDD, it seems that other factors/processes contribute in this regard, possibly by interacting with factors such as CRH. It does appear, after all, that 5-HT acting agents may be effective in managing MDD (Pineyro and Blier, 1999) especially when combined with treatments that target other processes (Millan, 2009).

Of the candidates that might be operating in conjunction with 5-HT, there is considerable evidence pointing to interactions with CRH. For instance, it was reported that stressor-provoked CRH release may be fundamental in promoting hippocampal 5-HT changes (Linthorst et al., 2002). As well, CRH appears to regulate a subpopulation of raphe neurons that promote 5-HT release at terminal regions within the PFC (Kirby et al., 2000; Valentino et al., 2001). In fact, CRH administered to the DRN influenced forebrain 5-HT release (Price and Lucki, 2001), and chronic treatment with a CRH_1 antagonist, NBI 30775 (which has antidepressant actions), altered hippocampal 5-HT functioning (Oshima et al., 2003). Moreover, in genetically engineered mice with altered CRH or CRH receptors, the activity of 5-HT was increased as were signs of anxiety (Penalva et al., 2002; van Gaalen et al., 2002a,b).

An additional process by which CRH and 5-HT might interact was recently proposed to involve a multistep mechanism in which activation of CRH_1 receptors increased $5\text{-}HT_2$ signaling by increasing the number of $5\text{-}HT_2$ receptors on the cell surface. Specifically, after activation, CRH_1 receptors are internalized to endosomes where they dimerize with $5\text{-}HT_{2c}$ receptors. This facilitates the recycling of $5\text{-}HT_{2c}$ receptor from endosomes to the cell surface when the dimer is recycled. Thus, the availability of the $5\text{-}HT_{2c}$ receptor at the cell surface is increased, which has the effect of altering anxiety-related behaviors elicited by selective $5\text{-}HT_{2c}$ acting drugs. There is no reason to dismiss the possibility that CRH might, through this same process, affect symptoms of MDD (Magalhaes et al., 2010).

Beyond the interrelations between CRH and 5-HT, reciprocal innervation also appears to occur between $GABA_A$ and 5-HT activity within the PFC and hippocampus, and might contribute to MDD (Brambilla et al., 2003). It was suggested, in this regard, that $5\text{-}HT_{1A}$ receptors influence $GABA_A$ receptor expression (Sibille et al., 2000), hence regulating GABAergic inhibitory transmission. Moreover, it was proposed that chronic SSRI treatments may actually have their therapeutic effects through actions on GABA processes within limbic brain regions (Zhong and Yan, 2004), which could come about through several different mechanisms. This said, it was reported that the activity of 5-HT neurons within the dorsal raphe nucleus seemed to be regulated by $GABA_A$ and $5\text{-}HT_{1A}$ inhibitory receptors (Boothman et al., 2006; Cremers et al., 2007; Judge et al., 2006), which would affect forebrain 5-HT release. As well, other 5-HT receptor subtypes may interact with $GABA_A$ functioning, and it was reported that $5\text{-}HT_{2c}$ antagonists augmented the acute effect of SSRIs on hippocampal 5-HT release, an outcome that was modifiable by GABA manipulations (Cremers et al., 2007).

Just as communication relevant to depression occurs between 5-HT and GABA, it appears that CRH and $GABA_A$ functioning within the hypothalamus (and in limbic neural circuits) may be related. For example, it was shown that the basal expression of transcripts encoding several subunits of the $GABA_A$ receptor was present within CRH neurons fundamental in the stress response (Cullinan, 2000; Cullinan and Wolfe, 2000). Moreover, *in vivo*, *CRH* and arginine vasopressin (AVP) gene expression was increased by bicuculline methiodide, a $GABA_A$ antagonist (Cole and Sawchenko, 2002), although *in vitro* application of the $GABA_A$ antagonist altered AVP mRNA, without affecting CRH expression (Bali and Kovacs, 2003).

In addition to these hypothalamic changes, it was reported that CRH was uniquely expressed in GAD-positive interneurons in rat cortex and that $GABA_A$ receptor expression was altered upon chronic stressor exposure (Cullinan and Wolfe, 2000; Yan et al., 1998). Importantly, transcripts encoding several $GABA_A$ receptor subunits were altered within CRH neurons that are ordinarily responsive to stressors (Cullinan, 2000; Cullinan and Wolfe, 2000), and pharmacological treatments that ordinarily influence GABA functioning affected CRH mRNA expression within limbic sites (Cullinan and Wolfe, 2000; Gilmor et al., 2003; Skelton et al., 2000; Stout et al., 2001).

It seems that connections exist between CRH and 5-HT, CRH and GABA, and between 5-HT and GABA functioning. Thus, it should not be surprising that three-way interrelationships exist between CRH, 5-HT, and GABA neuronal functioning. In fact, studies using dual immunoelectron microscopy to assess synaptic contacts indicated that CRH has both direct and indirect effects on dorsal raphe 5-HT neurons, with GABA serving as a mediator in this regard (Waselus et al., 2005). This directional process is, to be sure, not the only one that might exist, as CRH variations may instigate 5-HT receptor changes, which then influence frontal cortical $GABA_A$ functioning (Tan et al., 2004).

In addition to these connections, it seems that the $GABA_A$-mediated inhibition of dorsal raphe 5-HT neurons was potentiated by the progesterone metabolite, allopregnanolone (Kaura et al., 2007). Further to this point, $GABA_A$ δ subunit expression may be influenced by progesterone, and the stressor-sensitive neurosteroid, $3\alpha,5\alpha$-tetrahydrodeoxycorticosterone (THDOC) (Reddy, 2003), might mediate this effect (Maguire and Mody, 2007). It should also be mentioned that among male mice long-term social isolation reduced responsiveness to GABA mimetic agents, an effect that was attributed to down-regulated biosynthesis of neurosteroids and decreased α_1/α_2 and γ_2 subunits coupled with an increase of α_4 and α_5 subunits. These effects were reversed by selective serotonin reuptake inhibitors (SSRI; fluoxetine and norfluoxetine) when administered systemically at nmol/kg doses (Girdler and Klatzkin, 2007; Matsumoto et al., 2007). These data raise the possibility that stressor effects on depressive symptoms, and the sex differences that exist, may involve interactions between $GABA_A$ subunits and THDOC and/or allopregnanolone (Birzniece et al., 2006), and as such support the view that treatments targeting neuroactive steroids and the $GABA_A$ receptor may be fruitful in the treatment of depression.

4.5 CONCLUDING COMMENTS

As indicated earlier, GABA is the most ubiquitous neurotransmitter in the CNS and GABA-containing interneurons, acting as an inhibitory neurotransmitter, are essential in regulating hyperexcitability as well as the synchronization and shaping of cortical neuronal activity (Gelman and Marín, 2010). As such, it should come as no surprise that GABA functioning might be involved in some fashion. In the present report, we offer three basic take-home messages: (a) Following almost a decade of $GABA_A$ functioning being largely overlooked in the analysis of MDD, there has, for good reason, been a resurgence of interest in the possible role of this neurotransmitter in depressive illness. (b) GABA likely works in collaboration with other neurotransmitters, notably 5-HT and CRH, in affecting depressive illness, and these actions are moderated by neurosteroids. (c) The coordination in the appearance of the subunits that comprise $GABA_A$ receptors may be fundamental in the timing and synchronization of neuronal activity and might thus influence depressive illness. These suggestions are clearly not independent of one another, and it is likely that the interactions between the systems, as well as the within system regulation that occurs, are modifiable by stressors that promote depressive illness.

ACKNOWLEDGMENT

The research from the authors was supported by grants from the Canadian Institutes of Health Research. H.A. is a Canada Research Chair in Neuroscience.

REFERENCES

Akinci, M.K., Johnston, G.A. 1993. Sex differences in acute swim stress-induced changes in the binding of MK-801 to the NMDA subclass of glutamate receptors in mouse forebrain. *J Neurochem* 61:2290–2293.

Anisman, H. 2009. Cascading effects of stressors and inflammatory immune system activation: Implications for major depressive disorder. *J Psychiatry Neurosci* 34:4–20.

Anisman, H., Du, L., Palkovits, M., Faludi, G., Kovacs, G.G., Szontagh-Kishazi, P., Merali, Z., Poulter, M.O. 2008. Serotonin receptor subtype and p11 mRNA expression in stress-relevant brain regions of suicide and control subjects. *J Psychiatry Neurosci* 33:131–141.

Arango, V., Huang, Y.Y., Underwood, M.D., Mann, J.J. 2003. Genetics of the serotonergic system in suicidal behavior. *J Psychiatry Res* 37:375–386.

Arranz, B., Cowburn, R., Eriksson, A., Vestling, M., Marcusson, J. 1992. Gamma-aminobutyric acid-B (GABAB) binding sites in postmortem suicide brains. *Neuropsychobiology* 26:33–36.

Bali, B., Kovacs, K.G. 2003. GABAergic control of neuropeptide gene expression in parvocellular neurons of the hypothalamic paraventricular nucleus. *Eur J Neurosci* 18:1518–1526.

Bentivoglio, M., Spreafico, R., Alvarez-Bolado, G., Sanchez, M.P., Fairen, A. 1990. Differential expression of the $GABA_A$ receptor complex in the dorsal thalamus and reticular nucleus: An immunohistochemical study in the adult and developing rat. *Eur J Neurosci* 3:118–125.

Bhagwagar, Z., Hinz, R., Taylor, M., Fancy, S., Cowen, P., Grasby, P. 2006. Increased 5-HT(2A) receptor binding in euthymic, medication-free patients recovered from depression: A positron emission study with [(11)C]MDL 100,907. *Am J Psychiatry* 163:1580–1587.

Bhagwagar, Z., Wylezinska, M., Jezzard, P., Evans, J., Ashworth, F., Sule, A., Cohen, P.J. 2007. Reduction in occipital cortex gamma-aminobutyric acid concentrations in medication-free recovered unipolar depressed and bipolar subjects. *Biol Psychiatry* 61:806–812.

Bielau, H., Steiner, J., Mawrin, C., Trübner, K., Brisch, R., Meyer-Lotz, G., Brodhun, M., Dobrowolny, H., Baumann, B., Gos, T., Bernstein, H.G., Bogerts, B. 2007. Dysregulation of GABAergic neurotransmission in mood disorders: A postmortem study. *Ann N Y Acad Sci* 1096:157–169.

Birzniece, V., Backstrom, T., Johansson, I.M., Lindblad, C., Lundgren, P., Lofgren, M., Olsson, T., Ragagnin, G., Taube, M., Turkmen, S., Wahlstrom, G., Wang, M.D., Wihlback, C., Zhu, D. 2006. Neuroactive steroid effects on cognitive functions with a focus on the serotonin and GABA systems. *Brain Res Rev* 51:212–239.

Boothman, L., Raley, J., Denk, F., Hirani, E., Sharp, T. 2006. In vivo evidence that 5-HT(2C) receptors inhibit 5-HT neuronal activity via a GABAergic mechanism. *Br J Pharmacol* 149:861–869.

Brambilla, P., Perez, J., Barale, F., Schettini, G., Soares, J.C. 2003. GABAergic dysfunction in mood disorders. *Mol Psychiatry* 8:721–737.

Brooks-Kayal, A.R., Shumate, M.D., Jin, H., Lin, D.D., Rikhter, T.Y., Holloway, K.L., Coulter, D.A. 1999. Human neuronal gamma-aminobutyric acid(A) receptors: Coordinated subunit mRNA expression and functional correlates in individual dentate granule cells. *J Neurosci* 19:8312–8318.

Burgard, E.C., Haas, K.F., Macdonald, R.L. 1999. Channel properties determine the transient activation kinetics of recombinant GABA(A) receptors. *Mol Brain Res* 73:28–36.

Caspi, A., Sugden, K., Moffitt, T.E., Taylor, A., Craig, I.W., Harrington, H., McClay, J., Mill, J., Martin, J., Braithwait, A., Poulton, R. 2003. Influence of life stress on depression: Moderation by a polymorphism in the *5-HTT* gene. *Science* 301:386–389.

Cheetham, S.C., Crompton, M.R., Katona, C.L., Parker, S.J., Horton, R.W. 1988. Brain $GABA_A$/benzodiazepine binding sites and glutamic acid decarboxylase activity in depressed suicide victims. *Brain Res* 460:114–123.

Choudary, P.V., Molnar, M., Evans, S.J., Tomita, H., Li, J.Z., Vawter, M.P., Myers, R.M., Bunney, W.E. Jr., Akil, H., Watson, S.J., Jones, E.G. 2005. Altered cortical glutamatergic and GABAergic signal transmission with glial involvement in depression. *Proc Natl Acad Sci USA* 102:15653–15658.

Cole, R.L., Sawchenko, P.E. 2002. Neurotransmitter regulation of cellular activation and neuropeptide gene expression in the paraventricular nucleus of the hypothalamus. *J Neurosci* 22:959–969.

Cotter, D., Mackay, D., Chana, G., Beasley, C., Landau, S., Everall, I.P. 2002. Reduced neuronal size and glial cell density in area 9 of the dorsolateral prefrontal cortex in subjects with major depressive disorder. *Cereb Cortex* 12(4):386–394.

Cremers, T.I., Rea, K., Bosker, F.J., Wikstrom, H.V., Hogg, S., Mork, A., Westerink, B.H. 2007. Augmentation of SSRI effects on serotonin by $5\text{-}HT_{2C}$ antagonists: Mechanistic studies. *Neuropsychopharmacology* 32:1550–1557.

Cross, J.A., Cheetham, S.C., Crompton, M.R., Katona, C.L., Horton, R.W. 1988. Brain GABA(B) binding sites in depressed suicide victims. *Psychiatry Res* 26:119–129.

Cullinan, W.E. 2000. GABA(A) receptor subunit expression within hypophysiotropic CRH neurons: A dual hybridization histochemical study. *J Comp Neurol* 419:344–351.

Cullinan, W.E., Wolfe, T.J. 2000. Chronic stress regulates levels of mRNA transcripts encoding beta subunits of the GABA(A) receptor in the rat stress axis. *Brain Res* 887:118–124.

D'Alessio, A.C., Szyf, M. 2006. Epigenetic tete-a-tete: The bilateral relationship between chromatin modifications and DNA methylation. *Biochem Cell Biol* 84:463–476.

Dantzer, R., O'Connor, J.C., Freund, G.G., Johnson, R.W., Kelley, K.W. 2008. From inflammation to sickness and depression: When the immune system subjugates the brain. *Nat Rev Neurosci* 9:46–56.

Duman, R.S., Monteggia, M. 2006. A neurotrophic model for stress-related mood disorders. *Biol Psychiatry* 59:1116–1127.

Fatemi, S.H., Stary, J.M., Earle, J.A., Araghi-Niknam, M., Eagan, E. 2005. GABAergic dysfunction in schizophrenia and mood disorders as reflected by decreased levels of glutamic acid decarboxylase 65 and 67 kDa and Reelin proteins in cerebellum. *Schizophr Res* 72:109–122.

Gelman, D.M., Marín, O. 2010. Generation of interneuron diversity in the mouse cerebral cortex. *Eur J Neurosci* 31:2136–2141.

Gilmor, M.L., Skelton, K.H., Nemeroff, C.B., Owens, M.J. 2003. The effects of chronic treatment with the mood stabilizers valproic acid and lithium on corticotropin-releasing factor neuronal systems. *J Pharmacol Exp Ther* 305:434–439.

Girdler, S.S., Klatzkin, R. 2007. Neurosteroids in the context of stress: Implications for depressive disorders. *Pharmacol Ther* 116:125–139.

Gos, T., Günther, K., Bielau, H., Dobrowolny, H., Mawrin, C., Trübner, K., Brisch, R., Steiner, J., Bernstein, H.G., Jankowski, Z., Bogerts, B. 2009. Suicide and depression in the quantitative analysis of glutamic acid decarboxylase-immunoreactive neuropil. *J Affect Disord* 113:45–55.

Hamidi, M., Drevets, W.C., Price, J.L. 2004. Glial reduction in amygdala in major depressive disorder is due to oligodendrocytes. *Biol Psychiatry* 55(6):563–569.

Hasler, G., van der Veen, J.W., Tumonis, T., Meyers, N., Shen, J., Drevets, W.C. 2007. Reduced prefrontal glutamate/glutamine and gamma-aminobutyric acid levels in major depression determined using proton magnetic resonance spectroscopy. *Arch Gen Psychiatry* 64:193–200.

Hayley, S., Poulter, M., Merali, Z., Anisman, H. 2005. The pathogenesis of clinical depression: Stressor- and cytokine-induced alterations of neuroplasticity. *Neuroscience* 135:659–678.

Holsboer, F. 2003. Corticotropin-releasing hormone modulators and depression. *Curr Opin Investig Drugs* 4(1):46–50.

Honig, A., Bartlett, J.R., Bouras, N., Bridges, P.K. 1988. Amino acid levels in depression: A preliminary investigation. *J Psychiatr Res* 22:159–164.

Howland, R.H., Thase, M.E. 1991. Biological studies of dysthymia. *Biol Psychiatry* 30:283–304.

Hutcheon, B., Morley, P., Poulter, M.O. 2000. Developmental change in $GABA_A$ receptor desensitization kinetics and its role in synapse function in rat cortical neurons. *J Physiol* 522(Pt 1):3–17.

Judge, S.J., Young, L., Gartside, S.E. 2006. GABA(A) receptor modulation of 5-HT neuronal firing in the median raphe nucleus: Implications for the action of anxiolytics. *Eur Neuropsychopharmacol* 16:612–619.

Karolewicz, B., Maciag, D., O'Dwyer, G., Stockmeier, C.A., Rajkowska, G. 2010. Reduced level of glutamic acid decarboxylase-67kDa in the prefrontal cortex in major depression. *Int J Neuropsychopharmacol* 13(4):411–420.

Kaura, V., Ingram, C.D., Gartside, S.E., Young, A.H., Judge, S.J. 2007. The progesterone metabolite allopregnanolone potentiates GABA(A) receptor-mediated inhibition of 5-HT neuronal activity. *Eur. Neuropsychopharmacol.* 17:108–115.

Kendler, K.S., Kuhn, J.W., Vittum, J., Prescott, C.A., Riley, B. 2005. The interaction of stressful life events and a serotonin transporter polymorphism in the prediction of episodes of major depression: A replication. *Arch Gen Psychiatry* 62:529–635.

Kirby, L.G., Rice, K.C., Valentino, R.J. 2000. Effects of corticotropin-releasing factor on neuronal activity in the serotonergic dorsal raphe nucleus. *Neuropsychopharmacology* 22:148–162.

Klausberger, T., Marton, L.F., Baude, A., Roberts, J.D., Magill, P.J., Somogyi, P. 2004. Spike timing of dendrite-targeting bistratified cells during hippocampal network oscillations in vivo. *Nat Neurosci* 7:41–47.

Klausberger, T., Roberts, J.D.B., Somogyi, P. 2002. Cell type and input specific differences in the number and subtypes of synaptic $GABA_A$ receptors in the hippocampus. *J Neurosci* 22:2513–2521.

Korpi, E.R., Kleinman, J.E., Wyatt, R.J. 1988. GABA concentrations in forebrain areas of suicide victims. *Biol Psychiatry* 23:109–114.

Koyama, T., Yamashita, I. 1992. Biological markers of depression: WHO multi-center studies and future perspective. *Prog Neuropsychopharmacol Biol Psychiatry* 16:791–796.

Krystal, J.H., Sanacora, G., Blumberg, H., Anand, A., Charney, D.S., Marek, G., Epperson, C.N., Goddard, A., Mason, G.F. 2002. Glutamate and GABA systems as targets for novel antidepressant and mood-stabilizing treatments. *Mol Psychiatry* 7(1):S71–S80.

Levenson, J.M., Roth, T.L., Lubin, F.D., Miller, C.A., Huang, I.C., Desai, P., Malone, L.M., Sweatt, J.D. 2006. Evidence that DNA (cytosine-5) methyltransferase regulates synaptic plasticity in the hippocampus. *J Biol Chem* 281:15763–15773.

Linthorst, A.C., Penalva, R.G., Flachskamm, C., Holsboer, F., Reul, J.M. 2002. Forced swim stress activates rat hippocampal serotonergic neurotransmission involving a corticotropin-releasing hormone receptor-dependent mechanism. *Eur J Neurosci* 16:2441–2452.

Lowther, S., De Paermentier, F., Cheetham, S.C., Crompton, M.R., Katona, C.L., Horton, R.W. 1997. $5\text{-}HT_{1A}$ receptor binding sites in post-mortem brain samples from depressed suicides and controls. *J Affect Disord* 42:199–207.

Lubin, F.D., Roth, T.L., Sweatt, J.D. 2008. Epigenetic regulation of *BDNF* gene transcription in the consolidation of fear memory. *J Neurosci* 28:10576–10586.

Maciag, D., Hughes, J., O'Dwyer, G., Pride, Y., Stockmeier, C.A., Sanacora, G., Rajkowska, G. 2010. Reduced density of calbindin immunoreactive GABAergic neurons in the occipital cortex in major depression: Relevance to neuroimaging studies. *Biol Psychiatry* 67:465–470.

Maconochie, D.J., Zempel, J.M., Steinbach, J.H. 1994. How quickly can $GABA_A$ receptors open? *Neuron* 12:61–71.

Magalhaes, A.C., Holmes, K.D., Dale, L.B., Comps-Agrar, L., Lee, D., Yadav, P.N., Drysdale, L., Poulter, M.O., Roth, B.L., Pin, J.P., Anisman, H., Ferguson, S.S. 2010. CRF receptor 1 regulates anxiety behavior via sensitization of 5-HT_2 receptor signaling. *Nat Neurosci* 13:622–629.

Maguire, J., Mody, I. 2007. Neurosteroid synthesis-mediated regulation of GABA(A) receptors: Relevance to the ovarian cycle and stress. *J Neurosci* 27(9):2155–2162.

Malecova, B., Morris, K.V. 2010. Transcriptional gene silencing through epigenetic changes mediated by non-coding RNAs. *Curr Opin Mol Ther* 12:214–222.

Mann, E.O., Mody, I. 2010. Control of hippocampal gamma oscillation frequency by tonic inhibition and excitation of interneurons. *Nat Neurosci* 13:205–212.

Matsumoto, K., Puia, G., Dong, E., Pinna, G. 2007. GABA(A) receptor neurotransmission dysfunction in a mouse model of social isolation-induced stress: Possible insights into a non-serotonergic mechanism of action of SSRIs in mood and anxiety disorders. *Stress* 10:3–12.

McGowan, P.O., Sasaki, A., D'Alessio, A.C., Dymov, S., Labonté, B., Szyf, M., Turecki, G., Meaney, M.J. 2009. Epigenetic regulation of the glucocorticoid receptor in human brain associates with childhood abuse. *Nat Neurosci* 12:342–348.

McGowan, P.O., Sasaki, A., Huang, T.C., Unterberger, A., Suderman, M., Ernst, C., Meaney, M.J., Turecki, G., Szyf, M. 2008. Promoter-wide hypermethylation of the ribosomal RNA gene promoter in the suicide brain. *PLoS One* 3(5):e2085.

McKernan, R.M., Whiting, P.J. 1996. Which $GABA_A$-receptor subtypes really occur in the brain? *Trends Neurosci* 19:139–143.

Merali, Z., Du, L., Hrdina, P., Palkovits, M., Faludi, G., Poulter, M.O., Anisman, H. 2004. Dysregulation in the suicide brain: mRNA expression of corticotropin releasing hormone receptors and $GABA_A$ receptor subunits in frontal cortical brain region. *J Neurosci* 24:1478–1485.

Merali, Z., Kent, P., Du, L., Hrdina, P., Palkovits, M., Faludi, G., Poulter, M.O., Bedard, T., Anisman, H. 2006. Corticotropin-releasing hormone, arginine vasopressin, gastrin-releasing peptide, and neuromedin B alterations in stress-relevant brain regions of suicides and control subjects. *Biol Psychiatry* 59:594–602.

Millan, M.J. 2009. Dual- and triple-acting agents for treating core and co-morbid symptoms of major depression: Novel concepts, new drugs. *Neurotherapeutics* 6(1):53–77.

Millan, M.J., 2006. Multi-target strategies for the improved treatment of depressive states: Conceptual foundations and neuronal substrates, drug discovery and therapeutic application. *Pharmacol Ther.* 110:135–370.

Miller, C.A., Campbell, S.L., Sweatt, J.D. 2008. DNA methylation and histone acetylation work in concert to regulate memory formation and synaptic plasticity. *Neurobiol Learn Mem* 89:599–603.

Mintun, M.A., Sheline, Y.I., Moerlein, S.M., Vlassenko, A.G., Huang, Y., Snyder, A.Z. 2004. Decreased hippocampal 5-HT_{2A} receptor binding in major depressive disorder: In vivo measurement with [^{18}F]altanserin positron emission tomography. *Biol Psychiatry* 55:217–224.

Mullins-Sweatt, S.N., Smit, V., Verheul, R., Oldham, J., Widiger, T.A. 2009. Dimensions of personality: Clinicians' perspectives. *Can J Psychiatry* 54:247–259.

Nemeroff, C.B. 1996. The corticotropin-releasing factor (CRF) hypothesis of depression: New findings and new directions. *Mol Psychiatry* 1:336–342.

Nemeroff, C.B., Vale, W. 2005. The neurobiology of depression: Inroads to treatment and new drug discovery. *J Clin Psychiatry* 66(7):5–13.

Olsen, R.W., Sieghart, W. 2008. International Union of Pharmacology. LXX. Subtypes of gamma-aminobutyric acid(A) receptors: Classification on the basis of subunit composition, pharmacology, and function. Update. *Pharmacol Rev* 60:243–260.

Oshima, A., Flachskamm, C., Reul, J.M., Holsboer, F., Linthorst, A.C. 2003. Altered serotonergic neurotransmission but normal hypothalamic–pituitary–adrenocortical axis activity in mice chronically treated with the corticotropin-releasing hormone receptor type 1 antagonist NBI 30775. *Neuropsychopharmacology* 28:2148–2159.

Penalva, R.G., Flachskamm, C., Zimmermann, S., Wurst, W., Holsboer, F., Reul, J.M., Linthorst, C. 2002. Corticotropin-releasing hormone receptor type 1-deficiency enhances hippocampal serotonergic neurotransmission: An in vivo microdialysis study in mutant mice. *Neuroscience* 109:253–266.

Pineyro, G., Blier, P. 1999. Autoregulation of serotonin neurons: Role in antidepressant drug action. *Pharmacol Rev* 51:533–591.

Pollock, V.E., Schneider, L.S. 1990. Quantitative, waking EEG research on depression. *Biol Psychiatry* 27:757–780.

Poulter, M.O., Du, L., Weaver, I.C., Palkovits, M., Faludi, G., Merali, Z., Szyf, M., Anisman, H. 2008. $GABA_A$ receptor promoter hypermethylation in suicide brain: Implications for the involvement of epigenetic processes. *Biol Psychiatry* 64:645–652.

Poulter, M.O., Du, L., Zhurov, V., Merali, Z., Anisman, H. 2010a. Plasticity of the GABA(A) receptor subunit cassette in response to stressors in reactive versus resilient mice. *Neuroscience* 165:1039–1051.

Poulter, M.O., Du, L., Zhurov, V., Palkovits, M., Faludi, G., Merali, Z., Anisman, H. 2010b. Altered organization of $GABA_A$ receptor mRNA expression in the depressed suicide brain. *Front Neurosci* 3:3–11.

Price, M.L., Lucki, I. 2001. Regulation of serotonin release in the lateral septum and striatum by corticotropin-releasing factor. *J Neurosci* 21:2833–2841.

Price, R.B., Shungu, D.C., Mao, X., Nestadt, P., Kelly, C., Collins, K.A., Murrough, J.W., Charney, D.S., Matthew, S.J. 2009. Amino acid neurotransmitters assessed by proton magnetic resonance spectroscopy: Relationship to treatment resistance in major depressive disorder. *Biol Psychiatry* 65:792–800.

Prinz, A.A., Bucher, D., Marder, E. 2004. Similar network activity from disparate circuit parameters. *Nat Neurosci.* 7:1345–1352.

Rajkowska, G., O'Dwyer, G., Teleki, Z., Stockmeier, C.A., Miguel-Hidalgo, J.J. 2007. GABAergic neurons immunoreactive for calcium binding proteins are reduced in the prefrontal cortex in major depression. *Neuropsychopharmacology* 32:471–482.

Reddy, D.S. 2003. Is there a physiological role for the neurosteroid THDOC in stress-sensitive conditions? *Trends Pharmacol Sci* 24:103–106.

Reul, J.M., Holsboer, F. 2002. On the role of corticotropin-releasing harmone receptors in anxiety and depression. *Dial Clin Neurosci.* 4:31–46.

Rosel, P., Arranz, B., Vallejo, J., Oros, M., Menchon, J.M., Alvarez, P., Navarro, M.A. 1997. High affinity [^{3}H]imipramine and [^{3}H]paroxetine binding sites in suicide brains. *J Neural Transm* 104:921–929.

Roth, T.L., Lubin, F.D., Funk, A.J., Sweatt, J.D. 2009. Lasting epigenetic influence of early-life adversity on the *BDNF* gene. *Biol Psychiatry* 65:760–769.

Roth, T.L., Sweatt, J.D. 2009. Regulation of chromatin structure in memory formation. *Curr Opin Neurobiol* 19:336–342.

Rupprecht, R., Eser, D., Zwanzger, P., Möller, H.J. 2006. $GABA_A$ receptors as targets for novel anxiolytic drugs. *World J Biol Psychiatry* 7:231–237.

Sanacora, G., Gueorguieva, R., Epperson, C.N., Wu, Y.T., Appel, M., Rothman, D.L., Krystal, J.H. 2004. Subtype-specific alterations of GABA and glutamate in major depression. *Arch Gen Psychiatry* 61:705–713.

Sanacora, G., Mason, G.F., Rothman, D.L., Behar, K.L., Hyder, F., Petroff, O.A., Berman, R.M., Charney, D.S., Krystal, J.J. 1999. Reduced cortical gamma-aminobutyric acid levels in depressed patients determined by proton magnetic resonance spectroscopy. *Arch Gen Psychiatry* 56:1043–1047.

Sanacora, G., Saricicek, A. 2007. GABAergic contributions to the pathophysiology of depression and the mechanism of antidepressant action. *CNS Neurol Disord Drug Targets* 6:127–140.

Schaefer, M., Lyko, F. 2010. Solving the Dnmt2 enigma. *Chromosoma* 119:35–40.

Sequeira, A., Mamdani, F., Ernst, C., Vawter, M.P., Bunney, W.E., Lebel, V., Rehal, S., Klempan, T., Gratton, A., Benkelfat, C., Rouleau, G.A., Mechawar, N., Turecki, G. 2009. Global brain gene expression analysis links glutamatergic and GABAergic alterations to suicide and major depression. *PLoS One* 4(8):e6585.

Sequeira, A., Turecki, G. 2006. Genome wide gene expression studies in mood disorders. *OMICS* 10:444–454.

Sherif, F., Marcusson, J., Oreland, L. 1991. Brain gamma-aminobutyrate transaminase and monoamine oxidase activities in suicide victims. *Eur Arch Psychiatry Clin Neurosci* 241:139–144.

Shiah, I.S., Yatham, L.N. 1998. GABA function in mood disorders: An update and critical review. *Life Sci* 63:1289–1303.

Sibille, E., Arango, V., Galfalvy, H.C., Pavlidis, P., Erraji-Benchekroun, L., Ellis, S.P., Mann, J. 2004. Gene expression profiling of depression and suicide in human prefrontal cortex. *Neuropsychopharmacology* 29:351–361.

Sibille, E., Pavlides, C., Benke, D., Toth, M. 2000. Genetic inactivation of the serotonin(1A) receptor in mice results in downregulation of major GABA(A) receptor alpha subunits, reduction of GABA(A) receptor binding, and benzodiazepine-resistant anxiety. *J Neurosci* 20:2758–2765.

Skelton, K.H., Nemeroff, C.B., Knight, D.L., Owens, M.J. 2000. Chronic administration of the triazolobenzodiazepine alprazolam produces opposite effects on corticotropin-releasing factor and urocortin neuronal systems. *J Neurosci* 20:1240–1248.

Steiger, J.L., Russek, S.J. 2004. GABAA receptors: building the bridge between subunit mRNAs, their promoters, and cognate transcription factors. *Pharmacol Ther.* 101:259–281.

Stockmeier, C.A. 2003. Involvement of serotonin in depression: Evidence from postmortem and imaging studies of serotonin receptors and the serotonin transporter. *J Psychiatr Res* 37:357–373.

Stout, S.C., Owens, M.J., Lindsey, K.P., Knight, D.L., Nemeroff, C.B. 2001. Effects of sodium valproate on corticotropin-releasing factor systems in rat brain. *Neuropsychopharmacology* 24:624–631.

Sundman, I., Allard, P., Eriksson, A., Marcusson, J. 1997. GABA uptake sites in frontal cortex from suicide victims and in aging. *Neuropsychobiology* 35(1):11–15.

Sundman-Eriksson, I., Allard, P. 2002. [(3)H]Tiagabine binding to GABA transporter-1 (GAT-1) in suicidal depression. *J Affect Disord* 71:29–33.

Svenningsson, P., Chergui, K., Rachleff, I., Flajolet, M., Zhang, X., El Yacoubi, M., Vaugeois, J.M., Nomikos, G.G., Greengard, P. 2006. Alterations in 5-HT_{1B} receptor function by p11 in depression-like states. *Science* 311:77–80.

Svenningsson, P., Greengard, P. 2007. p11 (S100A10)—An inducible adaptor protein that modulates neuronal functions. *Curr Opin Pharmacol* 7(1):27–32.

Sweatt, J.D. 2009. Experience-dependent epigenetic modifications in the central nervous system. *Biol Psychiatry* 65:191–197.

Szyf, M., Detich, N. 2001. Regulation of the DNA methylation machinery and its role in cellular transformation. *Prog Nucleic Acid Res Mol Biol* 69:47–79.

Szyf, M., McGowan, P., Meaney, M.J. 2008. The social environment and the epigenome. *Environ Mol Mutagen* 49:46–60.

Tan, H., Zhong, P., Yan, Z. 2004. Corticotropin-releasing factor and acute stress prolongs serotonergic regulation of GABA transmission in prefrontal cortical pyramidal neurons. *J Neurosci* 24:5000–5008.

Traub, R.D., Jefferys, J.G.R., Whittington, M.A. 1999. *Fast Oscillations in Cortical Circuits*. MIT Press, Cambridge, MA.

Tunnicliff, G., Malatynska, E. 2003. Central GABAergic systems and depressive illness. *Neurochem Res* 28:965–976.

Valentino, R.J., Liouterman, L., Van Bockstaele, E.J. 2001. Evidence for regional heterogeneity in corticotropin-releasing factor interactions in the dorsal raphe nucleus. *J Comp Neurol* 435:450–463.

van Gaalen, M.M., Reul, J.M., Gesing, A., Stenzel-Poore, M.P., Holsboer, F., Steckler, T. 2002a. Mice overexpressing CRH show reduced responsiveness in plasma corticosterone after a 5-HT_{1A} receptor challenge. *Genes Brain Behav* 1:174–177.

van Gaalen, M.M., Stenzel-Poore, M.P., Holsboer, F., Steckler, T. 2002b. Effects of transgenic overproduction of CRH on anxiety-like behaviour. *Eur J Neurosci* 15:2007–2015.

Verdoorn, T.A. 1994. Formation of heteromeric g-aminobutyric acid type A receptors containing two different alpha subunits. *Mol Pharmacol* 45:475–480.

Waselus, M., Valentino, R.J., Van Bockstaele, E.J. 2005. Ultrastructural evidence for a role of gamma-aminobutyric acid in mediating the effects of corticotropin-releasing factor on the rat dorsal raphe serotonin system. *J Comp Neurol* 482:155–165.

Weaver, I.C., Cervoni, N., Champagne, F.A., D'Alessio, A.C., Sharma, S., Seckl, J.R., Dymov, S., Szyf, M., Meaney, M.J. 2004. Epigenetic programming by maternal behavior. *Nat Neurosci* 7:847–884.

Weaver, I.C., Champagne, F.A., Brown, S.E., Dymov, S., Sharma, S., Meaney, M.J., Szyf, M. 2005. Reversal of maternal programming of stress responses in adult offspring through methyl supplementation: Altering epigenetic marking later in life. *J Neurosci* 25:11045–11054.

Weaver, I.C., Meaney, M.J., Szyf, M. 2006. Maternal care effects on the hippocampal transcriptome and anxiety-mediated behaviors in the offspring that are reversible in adulthood. *Proc Natl Acad Sci USA*. 103:3480–3485.

Whittington, M.A., Traub, R.D. 2003. Interneuron diversity series: Inhibitory interneurons and network oscillations in vitro. *Trends Neurosci* 26:676–682.

Whittington, M.A., Traub, R.D., Jefferys, J.G. 1995. Synchronized oscillations in interneuron networks driven by metabotropic glutamate receptor activation. *Nature* 373:612–615.

Yamada, K., Watanabe, A., Iwayama-Shigeno, Y., Yoshikawa, T. 2003. Evidence of association between gamma-aminobutyric acid type A receptor genes located on 5q34 and female patients with mood disorders. *Neurosci Lett* 349:9–12.

Yan, X.X., Baram, T.Z., Gerth, A., Schultz, L., Ribak, C.E. 1998. Co-localization of corticotropin-releasing hormone with glutamate decarboxylase and calcium-binding proteins in infant rat neocortical interneurons. *Exp Brain Res* 123:334–340.

Zhang, S.J., Huguenard, J.R., Prince, D.A. 1997. $GABA_A$ receptor-mediated Cl-currents in rat thalamic reticular and relay neurons. *J Neurophysiol* 78:2280–2286.

Zhong, P., Yan, Z. 2004. Chronic antidepressant treatment alters serotonergic regulation of GABA transmission in prefrontal cortical pyramidal neurons. *Neuroscience* 129:65–73.

5 Role of the Endocannabinoid System in the Neurobiology of Suicide

K. Yaragudri Vinod

CONTENTS

5.1 INTRODUCTION

In the past decade, remarkable advances have been made in cannabinoid (CB) research. The brain endocannabinoid (eCB) system modulates several neurobiological processes and its dysfunction is suggested to be involved in the pathophysiology of mood and drug use disorders. The CB1 receptor–mediated signaling, in particular, has been shown to play a critical role in the neural circuitry that mediates mood, motivation, and emotional behaviors. This chapter presents the data pertaining to the involvement of the eCB system in depression, suicide, and alcohol addiction.

5.2 NEURONAL CIRCUITRY THAT MEDIATES MOOD, COGNITION, AND REWARD

Several brain regions are involved in the regulation of mood and are targets of stress and stress hormones (Manji et al. 2001; McEwen 2005; Heimer and Van Hoesen 2006). Prefrontal cortex, in particular, is believed to be an important cortical area participating in the brain circuitry that regulates mood. It is involved in working memory, extinction of learning, and executive functions, and it might become dysfunctional in mood disorders (Manji et al. 2001). The stress-induced impairment in the executive function might also contribute to suicide vulnerability. The postmortem and neuroimaging studies have revealed physiological abnormalities in multiple areas of prefrontal cortex and its linked brain regions in patients with major depression (Manji et al. 2001; Sibille et al. 2004). Alterations in glucose metabolism, and reduced activity and volume of prefrontal cortex, have been shown in depressed patients (Drevets 2000; Manji et al. 2001). The treatment of depression appears to reverse some of these abnormalities (Drevets 2000; Drevets et al. 2002). In addition, the development of depression and/or impulsivity is often associated with injuries to prefrontal cortex and could also lead to behavioral inhibition, altered decision making, and emotional disturbance (Mann 2003; Bechara and Van Der Linden 2005), which have been impaired in patients with depression and suicidal behavior. Among other cortical regions, dorsolateral prefrontal cortex (DLPFC) is critically involved in decision making and in spatial working memory (Krawczyk 2002; Huettel et al. 2006). The spatial working memory is important for maintaining decision goals, considering options, and integrating the two processes to predict future outcomes and probabilities of meeting goals (Krawczyk 2002). An impairment in decision making, due to altered mood, might be a neuropsychological risk factor for suicidal behavior. In addition to prefrontal cortex, ventral striatum is an important component of the reward circuitry. Nucleus accumbens, a region of ventral striatum, contributes to the motivational salience of stimuli and reward-dependent behaviors through reciprocal cortical and subcortical connections (Berridge and Robinson 2003). The dysfunction of ventral striatum could contribute to drug addiction, possibly by affecting impulsive decision making (Kalivas and Volkow 2005; Eisch et al. 2003). It might also lead to *anhedonia* (an inability to experience pleasure from previously pleasurable activities), a major symptom of depression. Notably, the stimulation of nucleus accumbens is shown to alleviate anhedonia in treatment-resistant human depression (Schlaepfer et al. 2008). Taken together, the dysfunction of these brain regions and associated structures could predispose an individual to drug addiction, depression, and suicidality.

5.3 MONOAMINE HYPOTHESIS OF DEPRESSION AND SUICIDE

Although there have been considerable advances in our understanding of the neurobiology of depression, this disorder remains a major cause of suicide in the present society. Previous biological studies on depression and suicide have most frequently focused on the monoamine neurotransmitter pathways in prefrontal cortex, especially serotonin (5-HT) and norepinephrine (NE). Most of the studies support the hypothesis of a deficiency in 5-HT neurotransmission in the pathophysiology of major depressive and suicidal behavior (Arango et al. 2002; Caspi et al. 2003; Mann 2003). Also, alterations in the 5-HT system are likely to be independently associated with depression and suicide (Arango et al. 2002).

Among therapeutic agents, antidepressants are the most widely used drugs for the treatment of depression-related disorders. They exert their therapeutic action through elevation of the synaptic content of monoamine neurotransmitters, mainly 5-HT and NE. However, the mood-elevating effects of antidepressants can occur after a prolonged administration. This suggests that enhancement of serotonergic or noradrenergic neurotransmission *per se* is not sufficient for the clinical benefits. While currently available treatments are inadequate in many patients, the search for an additional biological substrate(s) that could be a therapeutic target(s) for depressive behavior is continuing. Indeed, there is accumulating evidence to suggest that the eCB system is involved in the regulation of mood, motivation, and emotional behavior.

5.4 eCB SYSTEM IN THE CENTRAL NERVOUS SYSTEM

The eCB system consists of endogenous CB receptor agonists (i.e., eCBs), CB receptors, and proteins that are involved in the metabolism and regulation of eCBs. eCBs are a class of lipid mediators, including amides, esters, and ethers of long-chain polyunsaturated fatty acids. The first eCB was isolated from porcine brain in 1992 and was characterized to be arachidonoyl ethanolamide (AEA) (Devane et al. 1992). This compound was later named *anandamide*, derived from the Indian Sanskrit word *ananda*, which means inner bliss. Subsequently, the second eCB, 2-arachidonoylglycerol (2-AG), was discovered in 1994 (Mechoulam et al. 1995; Sugiura et al. 1995). 2-AG is present at higher level in the mammalian central nervous system (CNS) compared with AEA (Blankman et al. 2007). 2-AG and AEA act as full and partial agonists at CB1 receptor, respectively. Several other eCBs have been identified lately, and their physiological roles are yet to be understood. eCBs are abundantly present in cerebral cortex, basal ganglia, and limbic structures and exert their effects mainly through CB receptors (Matias et al. 2006). Some of these eCBs, especially AEA, also act through the vanilloid receptors (Di Marzo 1998) (Figure 5.1).

There are currently two known CB receptor subtypes: CB1 and CB2. CB1 receptors are primarily localized in the CNS, whereas CB2 receptors are expressed in peripheral tissue and are mainly associated with the immune system (Howlett 2002). However, recent studies have indicated the existence of CB2 receptors in the CNS (Van Sickle et al. 2005). The tissue-specific distribution of these receptors may suggest that CB1 and CB2 play different roles in mediating exocannabinoid (exoCB)- and eCB-induced effects. The human CB1 and CB2 receptors contain 472 and 360 amino acid residues, respectively, and both are members of the G-protein-coupled receptors. These receptors possess some striking difference in biochemical and pharmacological properties, despite their high homology (44% sequence identity overall and 68% in the transmembrane regions) in their amino acid sequence (Cabral and Griffin-Thomas 2009). CB1 receptors are thought to be among the most abundant neuromodulatory G-protein-coupled receptors in the mammalian brain. They are highly expressed in cerebral cortex, hippocampus, cerebellum, and basal ganglia (Howlett 2002). These receptors are negatively coupled to adenylyl cyclase and N- and P/Q type Ca^{2+}

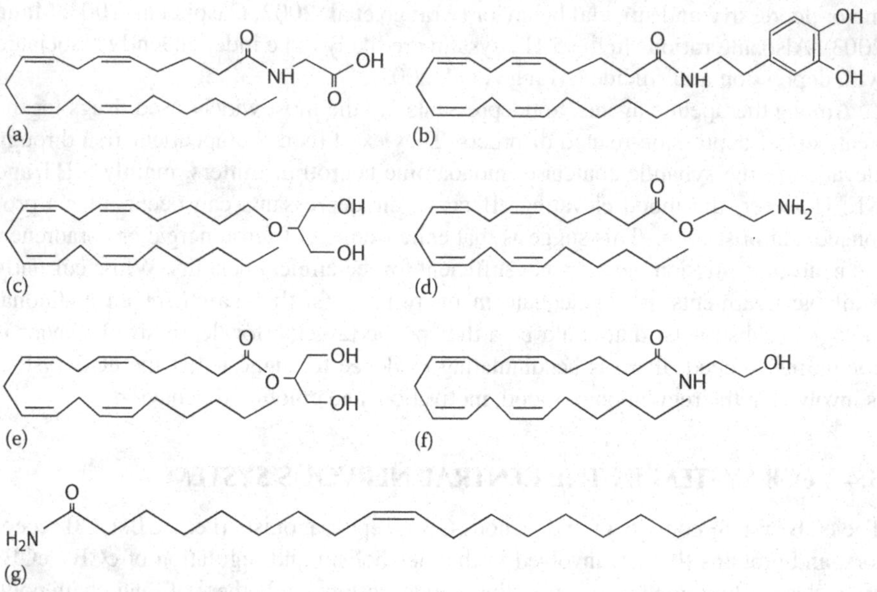

FIGURE 5.1 Structure of eCBs: (a) *N*-arachidonoylglycine, (b) *N*-arachidonoyldopamine, (c) 2-arachidonoylglyceryl ether, (d) *N*-arachidonoylglycerol, (e) 2-arachidonoylglycine, (f) *N*-arachidonoylethanolamide, and (g) 9-octadecenoamide. (Adapted from Vinod, K.Y. and Hungund, B.L., *Life Sci.*, 77, 1569, 2005.)

channels and positively to A-type and inwardly rectifying K^+ channels and mitogen-activated protein kinases through $G_{i/o}$ proteins (Howlett 2002).

Presently, the regulatory mechanism of the eCB system is not clearly understood. Nevertheless, the neuroanatomical and electrophysiological studies of the mammalian CNS have provided evidence for the presynaptic localization of CB1 receptor (Howlett 2002; Wilson and Nicoll 2002). The biosynthesis of eCBs appears to occur through several pathways. For instance, the synthesis of AEA could occur via the condensation of arachidonic acid and ethanolamine, which subsequently could be released from membrane phospholipids through the activation of phospholipases. Alternatively, AEA might be synthesized by phospholipase D–mediated hydrolysis of *N*-arachidonoylphosphatidylethanolamine in a calcium-dependent manner (Di Marzo et al. 1994). The formation of 2-AG is calcium dependent and is mediated by the enzymes phospholipase C and diacylglycerol lipase (Blankman et al. 2007). 2-AG is translocated to the presynaptic cell, where it acts at CB1 receptor. It is later inactivated by being resorbed into the cell and mainly metabolized by monoacylglycerol lipase. However, the mechanism of eCB transport across the membrane is yet to be conclusively elucidated.

Unlike classic neurotransmitters, eCBs are not stored in intracellular compartments (i.e., synaptic vesicles). They are synthesized in the postsynaptic neurons and released into the synaptic cleft on demand by stimulus-dependent cleavage of membrane phospholipids, which then act as retrograde messengers (Wilson and Nicoll 2002). Because eCBs are lipophilic, they could also diffuse through the membrane if their levels in the synaptic cleft are higher than inside the cells. The removal of eCBs from the extracellular compartment appears to be facilitated by esterification into membrane phospholipids (Di Marzo et al. 1999). AEA can also be rapidly removed from the extracellular space through an uptake mechanism by a membrane transporter protein known as the anandamide membrane transporter. An intracellular membrane-bound enzyme, fatty acid amide hydrolase (FAAH), which is located in the somatodendritic compartments of neurons, is involved in the inactivation of AEA and related lipids (Di Marzo et al. 1994; Day et al. 2001; Deutsch et al. 2001). In the CNS, eCBs activate CB1 receptors and regulate synaptic transmission of excitatory and inhibitory circuits by modulating the release of monoamine neurotransmitters (Howlett 2002; Wilson and Nicoll 2002). This action is likely to be dependent on the localization of CB1 receptors within the excitatory or inhibitory neural circuits (Figure 5.2).

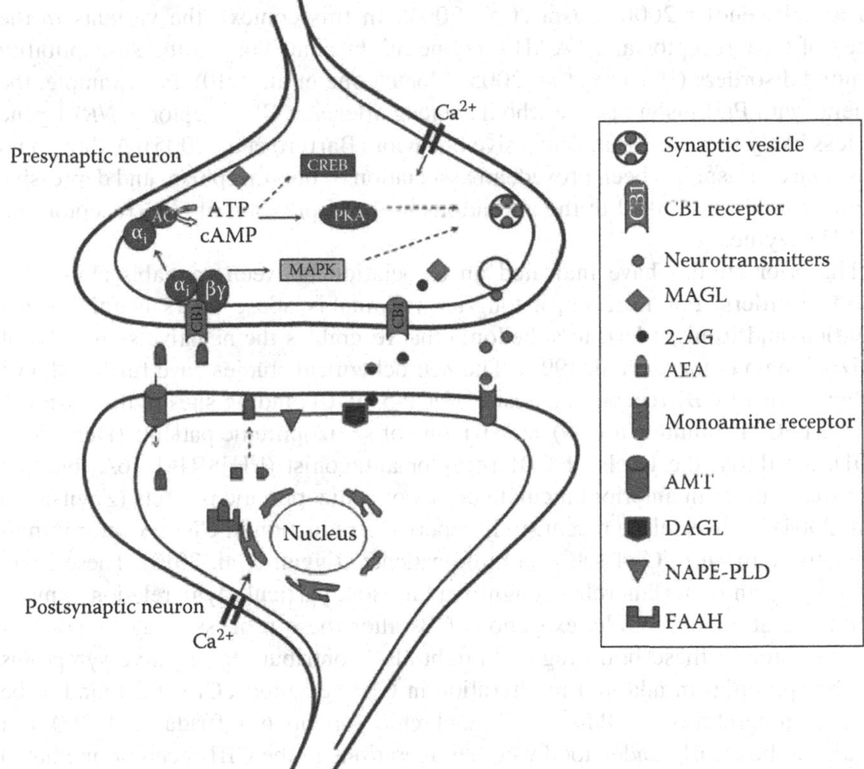

FIGURE 5.2 **(See color insert.)** Schematic illustration of the eCB system in the brain. AEA and 2-AG are synthesized in the postsynaptic membrane and then act as retrograde signaling molecules to stimulate the presynaptic CB1 receptor. This leads to the activation of various effectors including adenylyl cyclase (AC), mitogen-activated protein kinase (MAPK), K^+ and Ca^{2+} channels, etc., through $G\alpha_{i/o}$ proteins. Inhibition of AC activity and subsequent decrease in cAMP content could lead to a reduction in the activity of protein kinases (PKA), resulting in the modulation of ion channels and neurotransmitter release. The activities of AEA and 2-AG are limited by uptake mechanism (transporter) following hydrolysis by fatty acid amide hydrolase (FAAH) and monoacylglycerol lipase (MAGL), respectively. (Adapted from Vinod, K.Y. and Hungund, B.L., *Trends in pharmacol. Sci.*, 27, 539, 2006.)

5.5 ROLE OF THE eCB SYSTEM IN DEPRESSION AND SUICIDALITY

5.5.1 Human Studies

To date, a paucity of literature exists on the role of eCB system in the pathophysiology of major depression and suicide. A potential involvement of the eCB system in the neurobiology of depressive behavior has been revealed by the postmortem study of depressed suicide victims (Hungund et al. 2004). This study showed an elevation in CB1 receptor and CB1 receptor–stimulated G-protein activation in DLPFC of depressed suicide victims compared to non-psychiatric controls (Figure 5.3). Several studies have also indicated the genetic contributions to depressive disorder through family studies and molecular genetics. Furthermore, stressful life events could precipitate depressive illness in individuals with genetic predispositions (Fava and Kendler 2000; Caspi et al. 2003). In this context, the variants in the genes of CB1 receptor and FAAH enzyme might contribute to the susceptibility to mood disorders (Barrero et al. 2005; Monteleone et al. 2010). For example, the patients with Parkinson disease who have long alleles in CB receptor (*CNR*) 1 gene are less likely to suffer from depressive behavior (Barrero et al. 2005). Although no causal mechanism has been proved, an association of polymorphism and depressive behavior might be linked to the alterations in the expression of CB1 receptor and FAAH enzyme.

The prior studies have indicated an association between cannabis abuse and mood disorders. For instance, a long-term cannabis abuse alters cognition and attention, and it might lead to anhedonia that resembles the negative symptoms of schizophrenia (Emrich et al. 1997). The neurochemical studies have further shown higher levels of CB1 receptor agonist ([^{3}H]CP-55,940) binding sites in the postmortem DLPFC (Brodmann area 9) and striatum of schizophrenic patients (Dean et al. 2001). Similarly, the levels of CB1 receptor antagonist ([^{3}H]SR141716A) binding sites were higher in anterior cingulate cortex of schizophrenic patients (Zavitsanou et al. 2004). Conversely, a recent study reported a reduction in CB1 receptor immunoreactivity in DLPFC of schizophrenic patients (Eggan et al. 2010). These brain regions play an important role in cognitive function, particularly in relation to motivation and attention. While, exogenous CBs alter these processes, dysfunction of CB1 receptor in these brain regions might also contribute to negative symptoms of schizophrenia. In addition to alteration in CB1 receptor, eCBs are found to be elevated in cerebrospinal fluid of schizophrenic patients (Giuffrida et al. 2004). It remains to be clearly understood whether alterations in the CB1 receptor–mediated signaling in selective brain regions are linked to the symptoms of psychotic and/or mood disorders.

The neurochemical abnormalities in prefrontal cortex of depressed suicide victims, however, may not delineate whether they contribute to the pathophysiology of depression or suicide *per se*. In this regard, elevations in the levels of CB1 receptor and CB1 receptor–stimulated G-protein activation are shown in the postmortem prefrontal cortex of alcoholic suicide victims compared to alcoholic non-suicide subjects (Vinod et al. 2005). Consistent with the previous observation in depressed suicide victims (Hungund et al. 2004),

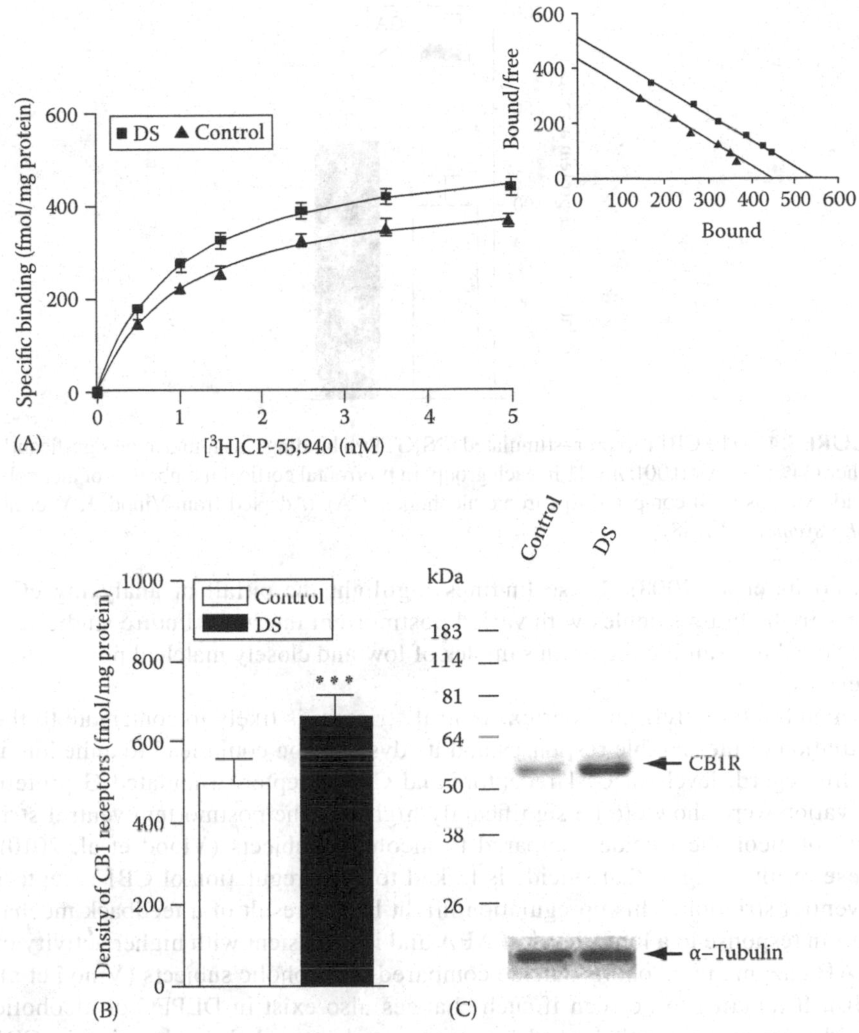

FIGURE 5.3 The saturation binding of CB1 receptor agonist, [^{3}H]CP-55,940, to the synaptic membrane revealed a higher level of CB1 receptor density in dorsolateral prefrontal cortex of depressed suicide victims (DS) compared to normal control subjects (24%; ***$p < 0.0001$; $n = 10$ in each group; (A) and (B)). Western blot analysis also indicated a higher level of CB1 receptor immunoreactivity in prefrontal cortex of DS compared to normal controls (C). (Adapted from Hungund, B.L. et al., *Mol. Psychiatry*, 9, 184, 2004.)

this study further provided an association of upregulation of frontal cortical CB1 receptor with suicide (Figure 5.4). In addition, it found elevated levels of eCBs (AEA and 2-AG) in DLPFC of alcoholic suicide victims. However, it is important to note that eCBs are labile to the postmortem delay. The brain AEA levels increase approximately sevenfold by 6 h of postmortem delay. While, 2-AG levels rapidly decline within the first hour and remain relatively stable thereafter

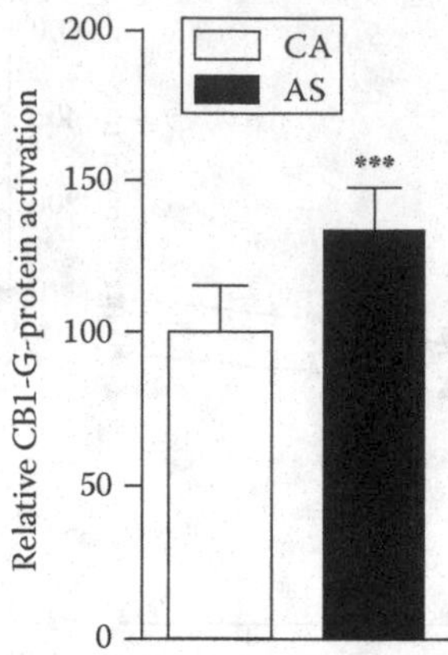

FIGURE 5.4 The CB1 receptor–stimulated [^{35}S]GTPγS binding was found to be significantly higher (34%; *** $p < 0.001$; $n = 11$ in each group) in prefrontal cortical membranes of alcoholic suicide victims (AS) compared to chronic alcoholics (CA). (Adapted from Vinod, K.Y. et al., *Biol. Psychiatry*, 57, 480, 2005.)

(Palkovits et al. 2008). These findings highlight the pitfall of analyzing eCB levels in the brain samples with varied postmortem intervals. Future studies are warranted to examine the brain samples of low and closely matched postmortem intervals.

In addition to prefrontal cortex, ventral striatum is likely to contribute to the mediation of pleasurable responses and its dysfunction could lead to anhedonia. In this regard, levels of CB1 receptors and CB1 receptor–stimulated G-protein activation were shown to be significantly higher in the postmortem ventral striatum of alcoholic suicide compared to alcoholic subjects (Vinod et al. 2010). These results suggest that suicide is linked to an upregulation of CB1 receptors in ventral striatum. This upregulation might be the result of a feedback mechanism in response to a lower level of AEA and is consistent with higher activity of FAAH enzyme in alcoholic suicide compared to alcoholic subjects (Vinod et al. 2010). It remains to be seen if such changes also exist in DLPFC of alcoholic suicide victims. Nevertheless, the previous study revealed an elevation in CB1 receptors and G-protein activation (Figure 5.4) in DLPFC but not in occipital cortex of alcoholic suicide victims (Vinod et al. 2005), suggesting a region-specific dysfunction of eCB signaling. Further studies are required to examine the eCB system in the brain of depressed suicide victims compared to non-suicide depressed subjects. This would determine whether the pathophysiological features of depression and suicide have overlapping dysfunction in the eCB system. It is also important to examine whether dysfunction of the eCB system in other brain regions is associated with the pathophysiology of depression and/or suicide. The comorbidity of other psychiatric disorders and drugs of abuse with suicidal behavior (Suominen et al. 1996; Rich et al. 1998; Kessler et al. 1999; Potash et al. 2000) are potential confounding factors. In addition, changes in mood (e.g., stress and anxiety) at the time of suicidal act could also affect the brain eCB system because the previous studies have shown the involvement of the eCB system in stress and anxiety-related disorders (Patel et al. 2004; Vinod and Hungund 2006; Mangieri and Piomelli 2007; Patel and Hillard 2008; Steiner et al. 2008b; Kamprath et al. 2009).

Some of the previous studies have suggested the utility of CB1 receptor antagonist/inverse agonist for the treatment of depressive behavior. Although rimonabant (SR141716) and other CB1 receptor antagonists had therapeutic potentials in treating obesity and other pathological conditions, recent clinical studies have reported adverse effects of rimonabant. Rimonabant at a dose of 20 mg/day is shown to increase depressive behavior (Christensen et al. 2007). Recently, the U.S. Food and Drug Administration (FDA) has also reported an increase in risk of depression and suicidal ideations during treatment with rimonabant (FDA briefing document, 2007).

5.5.2 Animal Studies

Pharmacological studies in rodents have revealed a critical role of the eCB system in depressive-like behavior. For instance, an anandamide uptake inhibitor, AM404, and CB1 receptor agonist, HU-210, have been shown to exert antidepressant-like responses in the rat forced swim test (Hill and Gorzalka 2005). In addition, an increase in AEA level through inhibition of FAAH enzyme had a similar effect (Gobbi et al. 2005). A long-term exposure (20 days) of adolescent rats to CB1 receptor agonist, WIN 55,212-2 (0.2 and 1 mg/kg), is also shown to exert depressive-like behavior (Bambico et al. 2010b) and cognitive dysfunction (O'Shea et al. 2004, 2006). Moreover, an overexpression of CB2 receptor has recently been linked to a reduction in depressive-like behavior in mice (Garcia-Gutierrez et al. 2010), suggesting an antidepressant-like effect via enhancement of the CB receptor signaling pathway. The underlying mechanism by which CBs exert their antidepressant properties is not clearly understood. Nevertheless, the monoamine systems of the midbrain serve an important adaptive function in response to stress, and long-term alterations in their activity might contribute to the development of depression (Manji et al. 2001; Arango et al. 2002). A key component in the action of clinically effective antidepressants is their ability to increase the levels of central monoamine neurotransmitters. In this regard, exoCBs, and an inhibitor of FAAH (URB597), have been shown to increase the firing activity of serotonergic and noradrenergic neurons (Muntoni et al. 2006). The CB1 receptor agonist, WIN 55,212-2, also elevates NE levels in frontal cortex of rats (Oropeza et al. 2005). Furthermore, CB1 receptor partial agonist, Δ^9-THC, is found to increase dopamine (DA) and glutamate levels in prefrontal cortex (Pistis 2002). In agreement with previously described findings, a long-term treatment with rimonabant is shown to elicit depression-like phenotype and decrease 5-HT levels in frontal cortex (Beyer et al. 2010).

An important insight into the role of eCB system in regulation of mood is derived from the studies revealing the effect of antidepressants on the components of eCB system. A long-term treatment with tranylcypromine (a monoamine oxidase inhibitor) and fluoxetine (a selective 5-HT reuptake inhibitor) has been shown to significantly increase CB1 receptor density in the prefrontal cortex (Hill et al. 2008). A chronic treatment with fluoxetine (10 mg/kg/day) also enhanced the CB1 receptor–mediated inhibition of adenylyl cyclase in prefrontal cortex (Mato et al. 2010). Interestingly, fluoxetine treatment completely reverses the increase in the CB1 receptor signaling in prefrontal cortex, following olfactory bulbectomy in rat (Rodriguez-Gaztelumendi et al. 2009). In addition, downregulation of eCB signaling in nucleus accumbens, which occurs after chronic unpredicted stress, could be reversed following fluoxetine treatment (Wang et al. 2010). These findings suggest that monoaminergic neurotransmission could regulate the eCB system and is indicative of a role of the cortical and accumbal eCB system in mood disorder and its treatment.

The CB1 receptor antagonists and/or inverse agonists (i.e., rimonabant and AM251) have also been shown to exert antidepressant-like effects (Shearman et al. 2003; Tzavara et al. 2003; Witkin et al. 2005) similar to those of fluoxetine in various animal models of depressive behavior (Griebel et al. 2005). Such an effect was also found to be absent in CB1 receptor knockout mice treated with AM251 (Shearman et al. 2003). Furthermore, the lack of CB1 receptor is shown to induce a facilitation of serotonergic activity in dorsal raphe nucleus by increasing 5-HT extracellular levels in prefrontal cortex in mice (Aso et al. 2009). The treatment with rimonabant is shown to increase 5-HT, NE, and DA levels in prefrontal cortex (Tzavara et al. 2003; Need et al. 2006). It is interesting to note that a long-term treatment with fluoxetine decreases the expression of CB1 receptor in frontal, cingulate, and piriform cortices, without significant alterations in parietal, temporal, and occipital regions of rodents (Oliva et al. 2005; Zarate et al. 2008). An enhancement in the 5-HT neurotransmission appears to be associated with this effect. The discrepancy pertaining to the antidepressant activity and the CB1 receptor–stimulated neurotransmitter release is yet to be clearly determined. However, the dose and duration of treatment, and the brain region under investigation (e.g., prefrontal vs. frontal cortex), might be some contributing factors.

Recent studies have highlighted the modulation of the central eCB system due to stress. For instance, downregulation of the eCB system in rat hippocampus is shown following chronic unpredictable stress (Hill et al. 2005). Acute stress, however, is found to induce elevation in prefrontal cortical AEA (Fride and Sanudo-Pena 2002) and midbrain AEA and 2-AG (Hohmann et al. 2005). Although the occurrence of major depression has been linked to an increased vulnerability to stress, the impact of stress on the eCB system and how it modulates the function of other brain regions, particularly prefrontal cortex, and how it affects mood and decision making, is not clearly understood.

Immobility tests in rodents have been extensively used as measures of depressive-like symptoms. The *FAAH* gene deleted ($FAAH^{-/-}$) mice exhibit a reduction in immobility in the forced swim and tail suspension tests, predicting antidepressant-like activity. Electrophysiological studies further revealed an increase in dorsal raphe 5-HT neural firing. These two parameters were shown to be attenuated by rimonabant (Bambico et al. 2010a,b). Behavioral studies using CB1 receptor knockout mice, however, yielded mixed results, such as both a decrease (Zimmer et al. 1999; Martin et al. 2002) and an increase (Ledent et al. 1999) in spontaneous locomotor activity. The reasons for these contrasting results are not clear.

Nevertheless, use of different genetic backgrounds and dosage of the drug might be among some contributing factors. Although the effect of *CNR*1 gene deletion may not be behavior-specific, neuroadaptation to the receptor deletion may also play an important role. Presently, there is an inadequate understanding of the mechanisms of CBs in the regulation of mood. Whether eCBs exert their mood-altering effects through CB1-like or non-CB receptors (e.g. vanilloid) needs to be further investigated.

5.6 MODULATION OF HPA FUNCTION BY THE eCB SYSTEM

The hypothalamic–pituitary–adrenal (HPA) axis, a neuroendocrine system, plays an important role in the regulation of mood, and its dysregulation is believed to be involved in increased susceptibility to depression and suicidal behavior and in the development of alcohol addiction (Manji et al. 2001; Mann 2003; Sher 2007; Richardson et al. 2008; Steiner et al. 2008a). It is a major regulator of circulating levels of glucocorticoid hormones (cortisol in humans and corticosterone in rodents), which are found to be elevated in major depression and in response to stress. Importantly, animal studies have shown the modulation of the HPA axis by the eCB system (Patel et al. 2004; Vinod and Hungund 2006; Steiner et al. 2008a; Kamprath et al. 2009). Notably, studies have indicated the activation of the HPA axis through CB1 receptors (Di Marzo 1998; Wenger et al. 2003; Manzanares et al. 2004) by stimulating the neurons containing corticotropin-releasing factor (Rodriguez de Fonseca et al. 1997). The stimulation of CB1 receptor leads to an elevation in the levels of corticotropin and corticosterone. This increase is shown to be attenuated by rimonabant (Murphy et al. 1998; Manzanares et al. 1999). The level of adrenocorticotropin was also found to be lower in CB1 receptor knockout mice compared to wild-type mice (Uriguen et al. 2004). Conversely, the eCB signaling has also been shown to inhibit stress-induced corticosterone release via CB1 receptor in rodents (Barna et al. 2004; Patel et al. 2004). The basal and stress-induced plasma levels of adrenocorticotropin and corticosterone are also reported to be higher in CB1 receptor knockout mice, suggesting a dysfunctional HPA axis (Haller et al. 2004). Although the discrepancy between these findings is yet to be determined, they suggest an interaction between the eCB and the neuroendocrine systems. Considering a critical role of the HPA axis in the pathophysiology of depression and suicidal behavior (Pfennig et al. 2005), it appears that the eCB system might have a critical role in the regulation of mood and emotional responses that are impaired in patients with depression and suicidal behavior.

5.7 POSSIBLE RELEVANCE OF A DYSFUNCTIONAL eCB SYSTEM TO DEPRESSION AND SUICIDE

The underlying mechanism of elevation in the levels of CB1 receptor in DLPFC of depressed suicide victims is not currently known. The upregulation of CB1 receptors because of a feedback response to low levels of eCBs in depression *per se* might be a possibility. This assumption is based on the findings obtained using rodent models.

However, the observed sensitization of CB1 receptor and its G-protein activation, despite higher eCB levels in DLPFC of alcoholic suicide victims (Vinod et al. 2005), is of particular interest. Such a trend in the brain of depressed suicide victims could not be excluded. Although the underlying mechanism remains to be established, the changes in metabolism and uptake of eCBs appear to be responsible for altered levels of eCBs. Elevated levels of eCBs and CB1 receptors in DLPFC of suicide victims raise the following questions: what might be the mechanism that causes these changes and what are the functional consequences? Investigations into whether alterations in CB1 receptors reflect a primary pathological condition or a compensatory homeostatic adaptation in response to dysfunction in other neuronal systems remain to be elucidated.

It is important to note that the monoaminergic systems, which are involved in the regulation of mood, anxiety, reward, and impulsive behaviors, functionally interact with the eCB system (Patel et al. 2003; Vinod and Hungund 2006; Mangieri and Piomelli 2007; Kamprath et al. 2009). The cAMP response element–binding protein (CREB) pathway is also a target for several monoamine and neuromodulatory systems and is shown to play a pivotal role in neuronal plasticity associated with stress, drug addiction, and suicidal behavior (Self and Nestler 1998; Reiach et al. 1999; Dwivedi et al. 2002). Because CB1 receptors are among the most abundant neuromodulatory G-protein-coupled receptors, the alteration in their levels in prefrontal cortex and ventral striatum is likely to have a greater impact on the cAMP pathway. This appears to account for the dysfunctional cAMP-dependent protein kinase A–CREB pathway that might play a critical role in the pathophysiology of depression and suicide. Furthermore, exogenous CB (e.g., Δ^9-THC) that exerts its effect mainly through CB1 receptor appears to modulate impulsive behavior (McDonald et al. 2003; Pattij and Vanderschuren 2008). In addition, AEA, is shown to elicit Δ^9-THC-like discriminative and neurochemical effects (Solinas et al. 2007). Hence, the impulsive behavior, which is one of the contributing factors for suicidal behavior (Mann et al. 1999; Koller et al. 2002), might be associated with the dysfunction in the central eCB system. Importantly, ventral striatum is shown to mediate anhedonia and impulsive behavior (Eisch et al. 2003; Tremblay et al. 2005; Juckel et al. 2006; Kumar et al. 2008). Considering the reported abnormalities in CB1 receptor function in prefrontal cortex and ventral striatum of suicide victims (Hungund et al. 2004; Vinod et al. 2005), the dysfunction of eCB system in the frontocorticostriatal circuitry is likely to produce behavioral deficits associated with suicidality.

5.8 COMORBIDITY OF DRUGS OF ABUSE WITH DEPRESSION AND SUICIDAL BEHAVIOR

Psychosocial problems might contribute to suicide to some degree. However, most suicides occur in context with psychiatric illness. Mood and substance abuse disorders, in particular, are major risk factors for suicide, although clinical studies of cannabis abuse in patients with mood disorders have provided contrasting results. Recent studies have suggested the beneficial effects of CB-based drugs for the treatment of depressive behavior, while there is also a negative impact of a long-term cannabis abuse. For example, cannabis dependence is associated with increased rates of psychotic depressive symptoms and even suicidal behavior (Degenhardt et al. 2003; Friedman et al. 2004; Lynskey et al. 2004; Raphael et al. 2005). Clinical studies have also demonstrated that a short-term CB intoxication produces deficits in mood and cognition (i.e., social withdrawal with affective flattening, poor motivation, and apathy) in patients with schizophrenia (Emrich et al. 1997). Although Δ^9-THC is beneficial in certain disease conditions (Manzanares et al. 2004), this psychoactive ingredient and abuse of cannabis and alcohol induce certain forms of impulsive behavior in humans (Askenazy et al. 2003; McDonald et al. 2003; Simons et al. 2005; Ramaekers et al. 2006) that might also be associated with suicidal behavior and suicide (Mann 2003; Price et al. 2009).

5.8.1 Role of eCB System in Alcohol Addiction

Besides cannabis abuse, a strong association of alcohol use disorder with depression and suicide has been shown (Friedman et al. 2004; Sher 2006). Interestingly, there is overwhelming evidence to suggest involvement of the eCB system in the regulation of alcohol-related behavior. The pharmacological manipulation of CB1 receptor function is shown to regulate alcohol-drinking behavior in rodents. For instance, CB1 receptor agonists (CP-55,940 and WIN 55,212-2) enhance alcohol-drinking behavior (Gallate et al. 1999; Colombo et al. 2004a,b; Vinod et al. 2008a). In line with these findings, a reduction in motivation to drink alcohol and relapse-like drinking behavior could be achieved by genetic deletion and antagonism of CB1 receptor function in rodents (Colombo et al. 1998; Hungund et al. 2003; Wang et al. 2003; Vinod and Hungund 2005; Malinen and Hyytia 2008; Vinod et al., 2008a, 2012; Maccioni et al. 2010). Notably, the blockade of CB1 receptor–mediated signaling specifically in nucleus accumbens is shown to suppress alcohol-drinking behavior in rats (Malinen and Hyytia 2008; Alvarez-James et al. 2009).

Preliminary clinical studies, however, have reported the inability of rimonabant to reduce alcohol drinking in alcohol-dependent subjects (Soyka et al. 2008; George et al. 2010). It remains to be examined if different dosage or other CB1 receptor antagonists have beneficial effects in alcohol dependence. The activation of CB1 receptor through an increase in endogenous AEA is also reported to enhance alcohol-drinking behavior. In this regard, the genetic deletion (Basavarajappa et al. 2006; Blednov et al. 2007; Vinod et al. 2008b) and the pharmacological inhibition of FAAH (Blednov et al. 2007; Vinod et al. 2008b) have been shown to enhance motivation to drink alcohol in mice. Prefrontal cortex is also likely to play a critical role in motivation to drink more alcohol. For example, an increase in alcohol self-administration is evident when rats are given an injection of FAAH inhibitor, URB597, into prefrontal cortex (Hansson et al. 2007). The biochemical studies have further suggested that a difference in CB1 receptor function is likely a contributing factor for variation in alcohol-seeking behavior. In this context, higher levels of CB1 receptors are correlated with greater alcohol-drinking behavior (Vinod et al. 2008a, 2012). Lower expression and activity of FAAH in prefrontal cortex in alcohol-preferring rats is also likely to be linked with increased alcohol consumption (Hansson et al. 2007).

An increased vulnerability to alcohol abuse in humans is suggested to be the result of polymorphisms in the genes of CB1 receptor (Schmidt et al. 2002; Zuo et al. 2007) and FAAH, and reduced expression and activity of FAAH (Sipe et al. 2002; Chiang et al. 2004). In addition, a relationship between *CNR*1 gene and attention-deficit hyperactivity disorder is also reported in alcoholic patients (Ponce et al. 2003). Some of these patients also exhibit suicidal behavior (Johann et al. 2005; Kenemans et al. 2005) that appears to be associated with impulsive behavior and impairment in the decision-making process. It remains to be established whether this polymorphism is related to the alterations in CB1 receptor levels. Nevertheless, these studies clearly indicate a role of the eCB system in alcohol addiction and underscore the importance of corticostriatal dysfunction of the eCB system in motivation/impulsive behavior.

The aggressive and impulsive behaviors are likely to be linked to alcohol addiction and suicide (Rich et al. 1998; Potash et al. 2000; Roy 2000; Koller et al. 2002; Preuss et al. 2002; Makhija and Sher 2007). Notably, the addiction to alcohol in humans is associated with the dysfunction in the brain eCB system. For instance, CB1 receptor levels were found to be significantly lower in postmortem ventral striatum of chronic alcoholic individuals compared to the healthy control group (Vinod et al. 2010), suggesting that alcohol dependence is associated with downregulation of CB1 receptors (Figure 5.5). The lower levels of CB1 receptors in chronic alcoholics

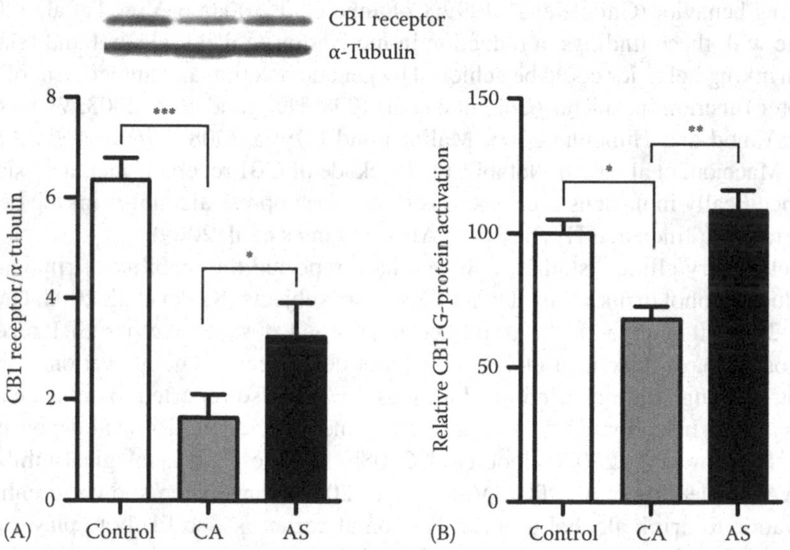

FIGURE 5.5 CB1 receptor levels were found to be lower in the membranes isolated from ventral striatum of chronic alcoholics (CA, 74%, ***$p < 0.0001$; (A)) and alcoholic suicide victims (AS, 48%, **$p < 0.001$; (A)) compared to normal controls ($n = 9$ in each group). However, a marked higher level of CB1 receptors was evident in the CA (98%, *$p < 0.05$; (A)) compared to CA subjects. A representative immunoblot of CB1 receptor is shown in the upper panel. The CB1 receptor–mediated G-protein activation was also significantly lower in ventral striatum of CA (35%, *$p < 0.01$; (B)) compared to normal controls; it was found to be higher in AS (32%, **$p < 0.001$; (B)) compared to the CA group. (Adapted from Vinod, K.Y. et al., *J. Psychiatr. Res.*, 44, 591, 2010.)

is consistent with the studies in rodents showing downregulation of CB1 receptors by a long-term alcohol exposure (Basavarajappa et al. 1998; Ortiz et al. 2004; Vinod and Hungund 2006; Mitrirattanakul et al. 2007; Vinod et al. 2012). The FAAH activity was also found to be reduced in ventral striatum of chronic alcoholics compared to healthy controls (Vinod et al. 2010). While FAAH is a key degrading enzyme of AEA, a decrease in its activity might eventually elevate AEA level in chronic alcoholics and vice versa. Indeed, previous studies in rodents have shown an increase in AEA content in striatum and "limbic" brain by chronic alcohol exposure (Gonzalez et al. 2004; Vinod 2006) through reduction in the activity of FAAH (Vinod and Hungund 2006). Thus, repeated alcohol consumption could desensitize the CB1 receptor function in ventral striatum of alcohol-dependent patients as a consequence of a compensatory adaptation to increased AEA. It is inferred from these studies that the impaired eCB function might confer a phenotype of high voluntary alcohol intake. An association of the eCB system with alcohol addiction, and the existence of a high incidence rate of suicide in cannabis and alcohol abusers, indicate that dysfunction of the eCB system might be one of the major contributing factors for suicidal behavior. Although the available literature points to the existence of a strong association among suicide with many neuropsychiatric and substance use disorders, the nature of such a relation is complex and might vary depending on the disorder in question and the substance used. A causative relationship among drug abuse, depression, and suicidality is yet to be clearly established.

5.8.2 Neurobiology of Alcohol Addiction Involving the eCB System

The dopaminergic neurotransmitter system in prefrontal cortex and ventral striatum has long been implicated in the drug-reinforcing mechanism. The mesolimbic dopaminergic system mainly consists of dopaminergic neurons whose cell bodies are located in ventral tegmental area and project terminals into nucleus accumbens, frontal cortex, amygdala, and septal area. This neuronal circuitry is likely to be involved in the process that mediates the reward mechanisms of various drugs of abuse, including alcohol (Koob 1992). Alcohol is shown to elevate extracellular DA levels in nucleus accumbens, which may increase the hedonic experience. Thus, accumbal DA may play a role in the development of addiction to alcohol. The eCB system also seems to contribute to this effect. For example, pharmacological blockade and deletion of *CB1* receptor gene reduce the acute alcohol-induced release of DA in nucleus accumbens in mice (Hungund et al. 2003). In addition, alcohol consumption is shown to increase AEA content in limbic forebrain (Gonzalez et al. 2004), which appears to activate mesolimbic dopaminergic transmission by increasing DA release in nucleus accumbens. In this regard, an intravenous administration of AEA and methanandamide (a stable derivative of AEA) and pharmacological inhibition of FAAH with URB597 (which increases the brain levels of AEA) have been shown to increase DA in the shell region of nucleus accumbens (Solinas et al. 2006). On the contrary, antagonism of CB1 receptor function reduces DA release in nucleus accumbens (Tanda et al. 1999), indicating a critical role for the AEA–CB1 receptor signaling in mediating a reward effect of alcohol (Hungund et al., 2003; Vinod et al. 2008b) through the stimulation of the mesolimbic dopaminergic system.

The persistent drug use might be linked to repeated activation of the mesolimbic DA system, which could enhance the incentive value of the drug of abuse. The alcohol-induced changes in the function of the accumbal DA and eCB systems might lead to the progression from reward to addiction. This might provide one of the mechanistic explanations for the pathophysiology of alcohol addiction.

5.9 CONCLUSION

One of the important tasks in psychiatry is to protect patients from their suicidal behavior. Preventive strategies could be improved by increasing our knowledge of the neurochemical abnormalities underlying this disorder. An increase in CB1 receptor levels is associated with suicide, whereas alcohol dependence is linked to the downregulation of these receptors. The changes in eCB levels might, in turn, explain differences in the CB1 receptor function in alcohol dependence, depression, and suicide. The dysfunction in the CB1 receptor–mediated signaling in frontocorticostriatal circuitry might be one of the etiological factors involved in the pathophysiology of suicide, in addition to depression and alcohol addiction. Based on preclinical studies, the antagonists of CB1 receptor appear to have therapeutic potential in the treatment of alcohol addiction. While majority of studies suggest hypoactivity of the CB1 receptor–mediated signaling in the pathophysiology of depressive behavior. Because depression is one of the major risk factors for suicide, a deficiency of this signaling might contribute to suicidal behavior. An abnormal interaction of the eCB system with the HPA axis and monoamine neurotransmitter systems might also constitute, at least in part, the underlying pathophysiology of depression and suicidal behavior. While many drugs are currently available for the treatment of these neuropsychiatric disorders, they only partially ameliorate the symptoms and elicit significant adverse effects. The data provided in this chapter support the notion that the eCB system might be an additional target for the development of a drug against alcohol use, depression, and suicidal behavior.

Although the focus of this chapter has been on the eCB system, it is important to note that its role might constitute just one facet of a very complex mental illness. A further study on the functional interactions between the eCB system and other monoamine neurotransmitter systems will be essential in understanding a given mental disorder. Numerous questions and contradictory findings exist in the field of CB research. An important question to consider is how a single CB1 receptor accounts for several of these behavioral manifestations. It is quite possible that activation of multiple signaling pathways by the CB1 receptor and/or existence of subtypes of CB or CB1 receptors in the CNS might account for the heterogeneity. Whether the dysfunction of the eCB system is directly associated with the pathophysiology of depression and suicide or if they are part of neuroadaptative changes in response to alteration in some other neuronal substrates still remains to be examined. Future studies should also focus on the effects (beneficial and adverse) of different dosage and duration of treatment of a given drug. Finally, alcohol addiction and stress-related disorders are shown to be risk factors for suicide attempts and suicide; treatment of these disorders could eventually reduce the rate of suicide.

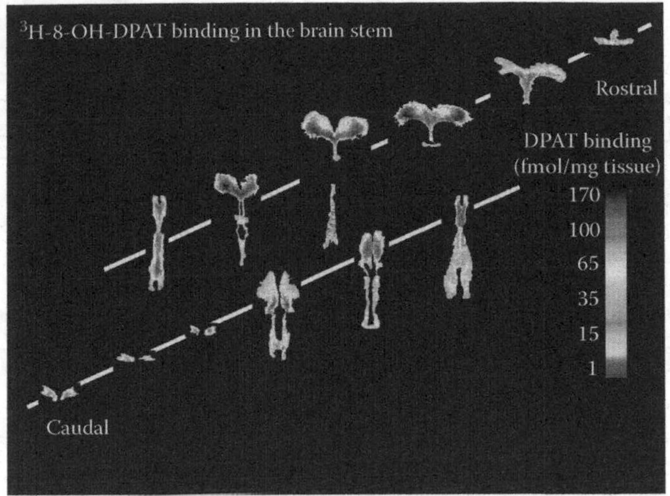

FIGURE 2.3 Pseudocolor receptor autoradiograms representing ^{3}H-8-OH-DPAT binding, a 5-HT agonist, within the DRN and MRN along the rostrocaudal axis of the brain stem.

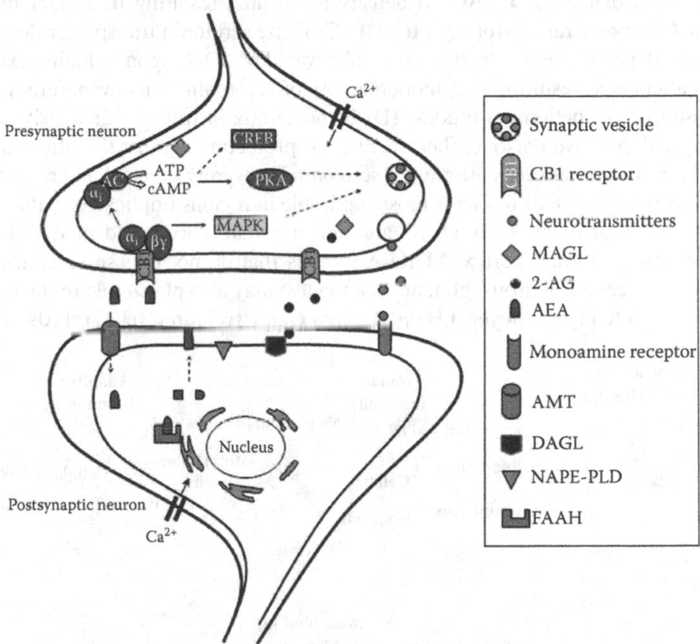

FIGURE 5.2 Schematic illustration of the eCB system in the brain. AEA and 2-AG are synthesized in the postsynaptic membrane and then act as retrograde signaling molecules to stimulate the presynaptic CB1 receptor. This leads to the activation of various effectors including adenylyl cyclase (AC), mitogen-activated protein kinase (MAPK), K^+ and Ca^{2+} channels, etc., through $G\alpha_{i/o}$ proteins. Inhibition of AC activity and subsequent decrease in cAMP content could lead to a reduction in the activity of protein kinases (PKA), resulting in the modulation of ion channels and neurotransmitter release. The activities of AEA and 2-AG are limited by uptake mechanism (transporter) following hydrolysis by the fatty acid amide hydrolase (FAAH) and monoacylglycerol lipase (MAGL), respectively. (Adapted from Vinod, K.Y. and Hungund, B.L., *Trends in pharmacol. Sci.*, 27, 539, 2006.)

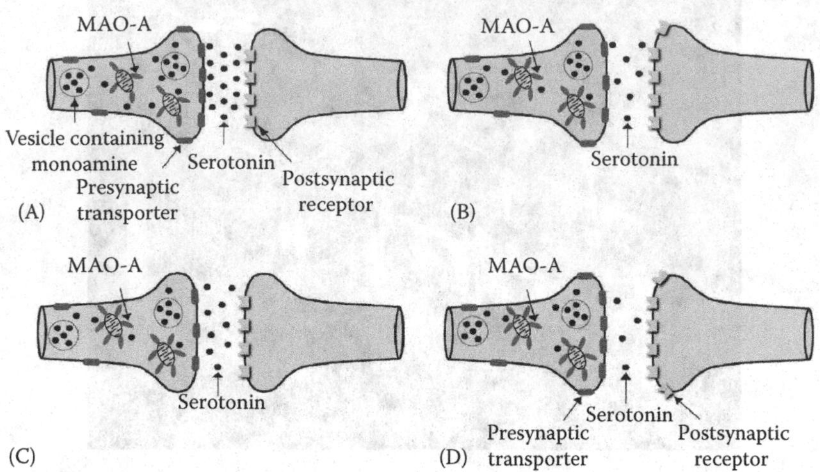

FIGURE 9.6 Modern model of excessive extracellular serotonin loss during major depressive disorder (A) serotonin release in a synapse in health. (B) During a major depressive episode, monoamine oxidase A (MAO-A) density is elevated resulting in greater metabolism of serotonin. Outcomes range from (C) to (D). (C) If the serotonin transporter density is low during a major depressive episode, the effect of elevated MAO-A upon reducing extracellular serotonin is attenuated resulting in a moderate loss of serotonin. This eventually results in a moderate rise in dysfunctional attitudes. (D) If the serotonin transporter density is not low during a major depressive episode, then there is no protection against the effect of elevated MAO-A. The extracellular concentration of serotonin is severely reduced and rise in dysfunctional attitudes is severe. This model is most applicable in regions implicated in the generation of optimism and pessimism such as the anterior cingulate cortex and medial/dorsolateral subregions of the prefrontal cortex. MAO-A in cells that do not release serotonin (such as norepinephrine-releasing neurons, glia, and astrocytes) may also play a role in the metabolism of serotonin. (Adapted from Meyer, J.H. et al., *Arch Gen Psychiatry*, 63(11), 1209, 2006.)

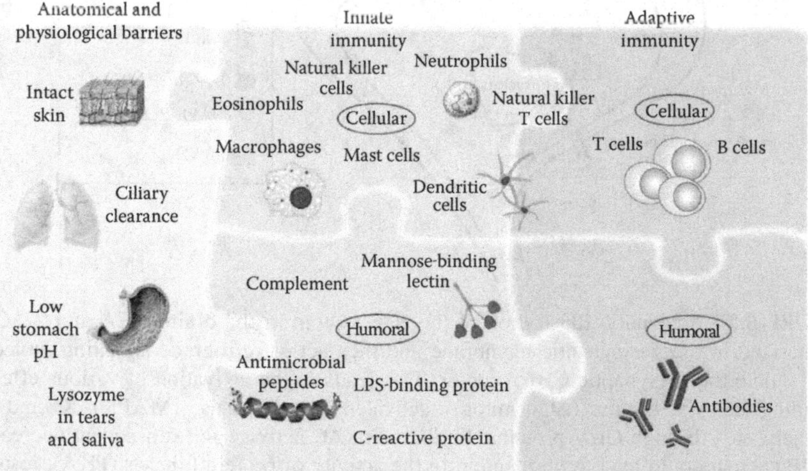

FIGURE 19.1 Various subdivisions of the human immune system. (From Turvey, S.E. and Broide, D.H., *J. Allergy*, 25, S24, 2010. With permission.)

CHEMICAL NAMES

CP-55,940: 5-(1,1-Dimethylheptyl)-2-[(1R,2R,5R)-5-hydroxy-2-(3-hydroxypropyl) cyclohexyl]-phenol

SR141716: *N*-(Piperidin-1-yl)-5-(4-chlorophenyl)-1-(2,4-dichlorophenyl)-4-methyl-1*H*-pyrazole-3-carboxamide

AM251: 1-(2,4-Dichlorophenyl-5-(4-iodophenyl)-4-methyl-*N*-1-piperidinyl-1*H*-pyrozole-3-carboxamide

AM404: *N*-(4-Hydroxyphenyl)arachidonoylamide

HU-210: 3-(1,1-Dimethylheptyl)-tetrahydro-1-hydroxy-6,6-dimethyl-6*H*-dibenzopyran-9-methanol

Δ^9-THC: Delta-9-tetrahydrocannibinol

URB597: 3′-Carbamoyl-biphenyl-3-yl-cyclohexylcarbamate

WIN 55,212-2: 2,3-Dihydro-5-methyl-3-(4-morpholinyl-methyl-pyrrolo[1,2,3-*de*]-1,4-benzoxazin-6yl]-1-naphthalenylmethanone mesylate

ACKNOWLEDGMENTS

The research studies cited in this chapter were partly supported by the National Institutes of Health, the National Alliance for Research on Schizophrenia and Depression, and the American Foundation for Suicide Prevention. The author is thankful to Dr. Raymond Suckow for editing this book chapter.

REFERENCES

Alvarez-Jaimes, L., Polis, I., Parsons, L.H. 2009. Regional influence of cannabinoid CB1 receptors in the regulation of ethanol self-administration by Wistar rats. *Neuropsychopharmacol. J.* 2:77–85.

Arango, V., Underwood, M.D., Mann, J.J. 2002. Serotonin brain circuits involved in major depression and suicide. *Prog. Brain Res.* 136:443–453.

Askenazy, F.L., Sorci, K., Benoit, M., Lestideau, K., Myquel, M., Lecrubier, Y. 2003. Anxiety and impulsivity levels identify relevant subtypes in adolescents with at-risk behavior. *J. Affect. Disord.* 74:219–227.

Aso, E., Renoir, T., Mengod, G. et al. 2009. Lack of CB1 receptor activity impairs serotonergic negative feedback. *J. Neurochem.* 109:935–944.

Bambico, F.R., Cassano, T., Dominguez-Lopez, S. et al. 2010a. Genetic deletion of fatty acid amide hydrolase alters emotional behavior and serotonergic transmission in the dorsal raphe, prefrontal cortex, and hippocampus. *Neuropsychopharmacology* 35:2083–2100.

Bambico, F.R., Nguyen, N.T., Katz, N., Gobbi, G. 2010b. Chronic exposure to cannabinoids during adolescence but not during adulthood impairs emotional behaviour and monoaminergic neurotransmission. *Neurobiol. Dis.* 37:641–655.

Barna, I., Zelena, D., Arszovszki, A.C., Ledent, C. 2004. The role of endogenous cannabinoids in the hypothalamo–pituitary–adrenal axis regulation: In vivo and in vitro studies in CB1 receptor knockout mice. *Life Sci.* 75:2959–2970.

Barrero, F.J., Ampuero, I., Morales, B. et al. 2005. Depression in Parkinson's disease is related to a genetic polymorphism of the cannabinoid receptor gene (CNR1). *Pharmacogenomics J.* 5:135–141.

Basavarajappa, B.S., Cooper, T.B., Hungund, B.L. 1998. Chronic ethanol administration down-regulates cannabinoid receptors in mouse brain synaptic plasma membrane. *Brain Res.* 79:212–218.

Basavarajappa, B.S., Yalamanchili, R., Cravatt, B.F., Cooper, T.B., Hungund, B.L. 2006. Increased ethanol consumption and preference and decreased ethanol sensitivity in female FAAH knockout mice. *Neuropharmacology* 50:834–844.

Bechara, A., Van Der Linden, M. 2005. Decision-making and impulse control after frontal lobe injuries. *Curr. Opin. Neurol.* 18:734–739.

Berridge, K.C., Robinson, T.E. 2003. Parsing reward. *Trends Neurosci.* 26:507–513.

Beyer, C.E., Dwyer, J.M., Piesla, M.J. et al. 2010. Depression-like phenotype following chronic CB1 receptor antagonism. *Neurobiol. Dis.* 39:148–155.

Blankman, J.L., Simon, G.M., Cravatt, B.F. 2007. A comprehensive profile of brain enzymes that hydrolyze the endocannabinoid 2-arachidonoylglycerol. *Chem. Biol.* 14:1347–1356.

Blednov, Y., Cravatt, B.F., Boehm II, S.L., Walker, D., Harris, R.A. 2007. Role of endocannabinoids in alcohol consumption and intoxication: Studies of mice lacking fatty acid amide hydrolase. *Neuropsychopharmacology* 32:1570–1582.

Cabral, G.A., Griffin-Thomas, L. 2009. Emerging role of the cannabinoid receptor CB2 in immune regulation: Therapeutic prospects for neuroinflammation. *Exp. Rev. Mol. Med.* 11:e3.

Caspi, A., Sugden, K., Moffitt, T.E. et al. 2003. Influence of life stress on depression: Moderation by a polymorphism in the *5-HTT* gene. *Science* 301:386–389.

Chiang, K.P., Gerber, A.L., Sipe, J.C., Cravatt, B.F. 2004. Reduced cellular expression and activity of the P129T mutant of human fatty acid amide hydrolase: Evidence for a link between defects in the endocannabinoid system and problem drug use. *Hum. Mol. Genet.* 13:2113–2119.

Christensen, R., Kristensen, P.K., Bartels, E.M., Bliddal, H., Astrup, A. 2007. Efficacy and safety of the weight-loss drug rimonabant: A meta-analysis of randomised trials. *Lancet* 370:1706–1713.

Colombo, G., Agabio, R., Fa, M. et al. 1998. Reduction of voluntary ethanol intake in ethanol-preferring sP rats by the cannabinoid antagonist SR-141716. *Alcohol Alcoholism* 33:126–130.

Colombo, G., Serra, S., Vacca, G., Gessa, G.L., Carai, M.A. 2004a. Suppression by baclofen of the stimulation of alcohol intake induced by morphine and WIN 55,212-2 in alcohol-preferring rats. *Eur. J. Pharmacol.* 492:189–193.

Colombo, G., Vacca, G., Serra, S., Carai, M.A.M., Gessa, G.L. 2004b. Suppressing effect of the cannabinoid CB_1 receptor antagonist, SR 141716, on alcohol's motivational properties in alcohol-preferring rats. *Eur. J. Pharmacol.* 498:119–123.

Day, T.A., Rakhshan, F., Deutsch, D.G., Barker, E.L. 2001. Role of fatty acid amide hydrolase in the transport of the endogenous cannabinoid anandamide. *Mol. Pharmacol.* 59:1369–1375.

Dean, B., Sundram, S., Bradbury, R., Scarr, E., Copolov, D. 2001. Studies on [^{3}H]CP-55940 binding in the human central nervous system: Regional specific changes in density of CB-1 receptors associated with schizophrenia and cannabis use. *Neuroscience* 103:9–15.

Degenhardt, L., Hall, W., Lynskey, M. 2003. Exploring the association between cannabis use and depression. *Addiction* 98:1493–1404.

Deutsch, D.G., Glaser, S.T., Howell, J.M. et al. 2001. The cellular uptake of anandamide is coupled to its breakdown by fatty-acid amide hydrolase. *J. Biol. Chem.* 276:6967–6973.

Devane, W.A., Hanus, L., Breuer, A. et al. 1992. Isolation and structure of a brain constituent that binds to the CB receptor. *Science* 258:1946–1949.

Di Marzo, V. 1998. Endocannabinoids: Endogenous cannabinoid receptor ligands with neuromodulatory action. *Trends Neurosci.* 21:521–518.

Di Marzo, V., De Petrocellis, L., Bisogno, T., Melck, D. 1999. Metabolism of anandamide and 2-arachidonoylglycerol: An historical overview and some recent developments. *Lipids* 34:S319–S325.

Di Marzo, V., Fontana, A., Cadas, H. et al. 1994. Formation and inactivation of endogenous cannabinoid anandamide in central neurons. *Nature* 372:686–691.

Drevets, W.C. 2000. Functional anatomical abnormalities in limbic and prefrontal cortical structures in major depression. *Prog. Brain Res.* 126:413–431.

Drevets, W.C., Bogers, W., Raichle, M.E. 2002. Functional anatomical correlates of antidepressant drug treatment assessed using PET measures of regional glucose metabolism. *Eur. Neuropsychopharmacol.* 12:27–44.

Dwivedi, Y., Conley, R.R., Roberts, R.C., Tamminga, C.A., Pandey, G.N. 2002. [^{3}H]cAMP binding sites and protein kinase a activity in the prefrontal cortex of suicide victims. *Am. J. Psychiatry* 159:66–73.

Eggan, S.M., Stoyak, S.R., Verrico, C.D., Lewis, D.A. 2010. Cannabinoid CB1 receptor immunoreactivity in the prefrontal cortex: Comparison of schizophrenia and major depressive disorder. *Neuropsychopharmacology* 35:2060–2071.

Eisch, A.J., Bolanos, C.A., de Wit, J. et al. 2003. Brain-derived neurotrophic factor in the ventral midbrain-nucleus accumbens pathway: A role in depression. *Biol. Psychiatry* 54:994–905.

Emrich, H.M., Leweke, F.M., Schneider, U. 1997. Towards a cannabinoid hypothesis of schizophrenia: Cognitive impairments due to dysregulation of the endogenous cannabinoid system. *Pharmacol. Biochem. Behav.* 56:803–807.

Fava, M., Kendler, K.S. 2000. Major depressive disorder. *Neuron* 8:335–341.

FDA briefing document. 2007. Briefing Information for FDA Advisory Committee Meeting on ZIMULTI® (Rimonabant), May 10, 2007. http://www.fda.gov/ohrms/dockets/ac/07/briefing/2007-4306b1-01-sponsor-backgrounder.pdf

Fride, E., Sanudo-Pena, C. 2002. Cannabinoids and endocannabinoids: Behavioral and developmental aspects. In *The Biology of Marijuana*. E. Onaivi, Ed., Harwood Academic, Reading, PA.

Friedman, A.S., Terras, A., Zhu, W., McCallum, J. 2004. Depression, negative self-image, and suicidal attempts as effects of substance use and substance dependence. *J. Addict. Dis.* 23:55–71.

Gallate, J.E., Saharov, T., Mallet, P.E., McGregor, I.S. 1999. Increased motivation for beer in rats following administration of a cannabinoid CB1 receptor agonist. *Eur. J. Pharmacol.* 370:233–240.

Garcia-Gutierrez, M.S., Perez-Ortiz, J.M., Gutierrez-Adan, A., Manzanares, J. 2010. Depression-resistant endophenotype in mice overexpressing cannabinoid CB(2) receptors. *Br. J. Pharmacol.* 160:1773–1784.

George, D.T., Herion, D.W., Jones, C.L. et al. 2010. Rimonabant (SR141716) has no effect on alcohol self-administration or endocrine measures in nontreatment-seeking heavy alcohol drinkers. *Psychopharmacology* 208:37–44.

Giuffrida, A., Leweke, F.M., Gerth, C.W. et al. 2004. Cerebrospinal anandamide levels are elevated in acute schizophrenia and are inversely correlated with psychotic symptoms. *Neuropsychopharmacology* 29:2108–2114.

Gobbi, G., Bambico, F.R., Mangieri, R. et al. 2005. Antidepressant-like activity and modulation of brain monoaminergic transmission by blockade of anandamide hydrolysis. *Proc. Natl. Acad. Sci. USA* 102:18620–18625.

Gonzalez, S., Valenti, M., de Miguel, R. et al. 2004. Changes in endocannabinoid contents in reward-related brain regions of alcohol-exposed rats, and their possible relevance to alcohol relapse. *Br. J. Pharmacol.* 143:455–464.

Griebel, G., Stemmelin, J., Scatton, B. 2005. Effects of the cannabinoid CB1 receptor antagonist rimonabant in models of emotional reactivity in rodents. *Biol. Psychiatry* 57:261–267.

Haller, J., Varga, B., Ledent, C., Barna, I., Freund, T.F. 2004. Context-dependent effects of *CB1* cannabinoid gene disruption on anxiety-like and social behaviour in mice. *Eur. J. Neurosci.* 19:1906–1912.

Hansson, A.C., Bermudez-Silva, F.J., Malinen, H. et al 2007. Genetic impairment of frontocortical endocannabinoid degradation and high alcohol preference. *Neuropsychopharmacology* 32:117–126.

Heimer, L., Van Hoesen, G.W. 2006. The limbic lobe and its output channels: Implications for emotional functions and adaptive behavior. *Neurosci. Biobehav. Rev.* 30:126–147.

Hill, M.N., Gorzalka, B.B. 2005. Pharmacological enhancement of cannabinoid CB1 receptor activity elicits an antidepressant-like response in the rat forced swim test. *Eur. Neuropsychopharmacol.* 15:593–599.

Hill, M.N., Ho, W.S., Hillard, C.J., Gorzalka, B.B. 2008. Differential effects of the antidepressants tranylcypromine and fluoxetine on limbic cannabinoid receptor binding and endocannabinoid contents. *J. Neural. Trans.* 115:1673–1679.

Hill, M.N., Patel, S., Carrier, E.J. et al. 2005. Downregulation of endocannabinoid signaling in the hippocampus following chronic unpredictable stress. *Neuropsychopharmacology* 30:508–515.

Hohmann, A.G., Suplita, R.L., Bolton, N.M. et al. 2005. An endocannabinoid mechanism for stress-induced analgesia. *Nature* 435:1108–1112.

Howlett, A.C. 2002. The cannabinoid receptors. *Prost. Lipid Mediat.* 68–69:619–631.

Huettel, S.A., Stowe, C.J., Gordon, E.M., Warner, B.T., Platt, M.L. 2006. Neural signatures of economic preferences for risk and ambiguity. *Neuron* 49:765–775.

Hungund, B.L., Szakall, I., Adam, A. et al. 2003. Cannabinoid CB1 receptor-knockout mice exhibit markedly reduced voluntary alcohol consumption and lack alcohol-induced dopamine release in the nucleus accumbens. *J. Neurochem.* 84:698–604.

Hungund, B.L., Vinod, K.Y., Kassir, S.A. et al. 2004. Upregulation of CB_1 receptors and agonist-stimulated [^{35}S]GTPγS binding in the prefrontal cortex of depressed suicide victims. *Mol. Psychiatry* 9:184–190.

Johann, M., Putzhammer, A., Eichhammer, P. et al. 2005. Association of the −141C Del variant of the dopamine D2 receptor (DRD2) with positive family history and suicidality in German alcoholics. *Am. J. Med. Genet. B. Neuropsychiatry Genet.* 132:46–49.

Juckel, G., Schlagenhauf, F., Koslowski, M. et al. 2006. Dysfunction of ventral striatal reward prediction in schizophrenic patients treated with typical, not atypical, neuroleptics. *Psychopharmacology* 187:222–228.

Kalivas, P.W., Volkow, N.D. 2005. The neural basis of addiction: A pathology of motivation and choice. *Am. J. Psychiatry* 162:1403–1413.

Kamprath, K., Plendl, W., Marsicano, G. et al. 2009. Endocannabinoids mediate acute fear adaptation via glutamatergic neurons independently of corticotropin-releasing hormone signaling. *Genes Brain Behav.* 8:203–211.

Kenemans, J.L., Bekker, E.M., Lijffijt, M., Overtoom, C.C., Jonkman, L.M., Verbaten, M.N. 2005. Attention deficit and impulsivity: Selecting, shifting, and stopping. *Int. J. Psychophysiol.* 58:59–70.

Kessler, R.C., Borges, G., Walters, E.E. 1999. Prevalence of risk factors for lifetime suicide attempts in the National Comorbidity Survey. *Arch. Gen. Psychiatry* 56:617–626.

Koller, G., Preuss, U.W., Bottlender, M., Wenzel, K., Soyka, M. 2002. Impulsivity and aggression as predictors of suicide attempts in alcoholics. *Arch. Psychiatry Clin. Neurosci.* 252:155–160.

Koob, G.F. 1992. Drugs of abuse: Anatomy, pharmacology and function of reward pathways. *Trends Pharmacol. Sci.* 13:177–184.

Krawczyk, D.C. 2002. Contributions of the prefrontal cortex to the neural basis of human decision making. *Neurosci. Biobehav. Rev.* 26:631–664.

Kumar, P., Waiter, G., Ahearn, T., Milders, M., Reid, I., Steele, J.D. 2008. Abnormal temporal difference reward-learning signals in major depression. *Brain* 131:2084–2093.

Ledent, C., Valverde, O., Cossu, G. et al. 1999. Unresponsiveness to cannabinoids and reduced addictive effects of opiates in CB_1 receptor knockout mice. *Science* 283:401–404.

Lynskey, M.T., Glowinski, A.L., Todorov, A.A. 2004. Major depressive disorder, suicidal ideation, and suicide attempt in twins discordant for cannabis dependence and early-onset cannabis use. *Arch. Gen. Psychiatry* 61:1026–1032.

Maccioni, P., Colombo, G., Carai, M.A.M. 2010. Blockade of the cannabinoid CB_1 receptor and alcohol dependence: Preclinical evidence and preliminary clinical data. *CNS Neurol. Disord. Drug Targets* 9:55–59.

Makhija, N.J., Sher, L. 2007. Preventing suicide in adolescents with alcohol use disorders. *Int. J. Adolesc. Med. Health* 19:53–59.

Malinen, H., Hyytia, P. 2008. Ethanol self-administration is regulated by CB1 receptors in the nucleus accumbens and ventral tegmental area in alcohol-preferring AA rats. *Alcohol Clin. Exp. Res.* 32:1976–1983.

Mangieri, R.A., Piomelli, D. 2007. Enhancement of endocannabinoid signaling and the pharmacotherapy of depression. *Pharmacol. Res.* 56:360–366.

Manji, H.K., Drevets, W.C., Charney, D.S. 2001. The cellular neurobiology of depression. *Nat. Med.* 7:541–547.

Mann, J.J. 2003. Neurobiology of suicidal behaviour. *Nat. Rev. Neurosci.* 4:819–828.

Mann, J.J., Waternaux, C., Haas, G.L., Malone, K.M. 1999. Toward a clinical model of suicidal behavior in psychiatric patients. *Am. J. Psychiatry* 156:181–189.

Manzanares, J., Corchero, J., Fuentes, J.A. 1999. Opioid and cannabinoid receptor-mediated regulation of the increase in adrenocorticotropin hormone and corticosterone plasma concentrations induced by central administration of delta(9)-tetrahydrocannabinol in rats. *Brain Res.* 839:173–179.

Manzanares, J., Uriguen, L., Rubio, G., Palomo, T. 2004. Role of endocannabinoid system in mental diseases. *Neurotoxicol. Res.* 6:213–224.

Martin, M., Ledent, C., Parmentier, M., Maldonado, R., Valverde, O. 2002. Involvement of CB1 cannabinoid receptors in emotional behaviour. *Psychopharmacology* 159:379–387.

Matias, I., Bisogno, T., Di Marzo, V. 2006. Endogenous cannabinoids in the brain and peripheral tissues: Regulation of their levels and control of food intake. *Int. J. Obes. (Lond.)* 30:S7–S12.

Mato, S., Vidal, R., Castro, E., Diaz, A., Pazos, A., Valdizan, E.M. 2010. Long-term fluoxetine treatment modulates cannabinoid type 1 receptor-mediated inhibition of adenylyl cyclase in the rat prefrontal cortex through 5-hydroxytryptamine 1A receptor-dependent mechanisms. *Mol. Pharmacol.* 77:424–434.

McDonald, J., Schleifer, L., Richards, J.B., de Wit, H. 2003. Effects of THC on behavioral measures of impulsivity in humans. *Neuropsychopharmacology* 28:1356–1365.

McEwen, B.S. 2005. Glucocorticoids, depression, and mood disorders: Structural remodeling in the brain. *Metabolism* 54:20–23.

Mechoulam, R., Ben-Shabat, S., Hanus, L. et al. 1995. Identification of an endogenous 2-monoglyceride, present in canine gut, that binds to cannabinoid receptors. *Biochem. Pharmacol.* 50:83–90.

Mitrirattanakul, S., López-Valdés, H.E., Liang, J. et al. 2007. Bidirectional alterations of hippocampal cannabinoid 1 receptors and their endogenous ligands in a rat model of alcohol withdrawal and dependence. *Alcohol Clin. Exp. Res.* 31:855–867.

Monteleone, P., Bifulco, M., Maina, G. et al. 2010. Investigation of *CNR*1 and *FAAH* endocannabinoid gene polymorphisms in bipolar disorder and major depression. *Pharmacol. Res*. 61:400–404.

Muntoni, A.L., Pillolla, G., Melis, M., Perra, S., Gessa, G.L., Pistis, M. 2006. Cannabinoids modulate spontaneous neuronal activity and evoked inhibition of locus coeruleus noradrenergic neurons. *Eur. J. Neurosci*. 23:2385–2394.

Murphy, L.L., Munoz, R.M., Adrian, B.A., Villanua, M.A. 1998. Function of cannabinoid receptors in the neuroendocrine regulation of hormone secretion. *Neurobiol. Dis*. 5:432–446.

Need, A.B., Davis, R.J., Alexander-Chacko, J.T. et al. 2006. The relationship of in vivo central CB1 receptor occupancy to changes in cortical monoamine release and feeding elicited by CB1 receptor antagonists in rats. *Psychopharmacology* 184:26–35.

Oliva, J.M., Uriguen, L., Perez-Rial, S., Manzanares, J. 2005. Time course of opioid and cannabinoid gene transcription alterations induced by repeated administration with fluoxetine in rat brain. *Neuropharmacology* 49:618–626.

Oropeza, V.C., Page, M.E., Van Bockstaele, E.J. 2005. Systemic administration of WIN 55,212-2 increases norepinephrine release in the rat frontal cortex. *Brain Res*. 1046:45–54.

Ortiz, S., Oliva, J.M., Perez-Rial, S., Palomo, T., Manzanares, J. 2004. Chronic ethanol consumption regulates cannabinoid *CB1* receptor gene expression in selected regions of rat brain. *Alcohol Alcoholism* 39:88–92.

O'Shea, M., McGregor, I.S., Mallet, P.E. 2006. Repeated cannabinoid exposure during perinatal, adolescent or early adult ages produces similar long-lasting deficits in object recognition and reduced social interaction in rats. *J. Psychopharmacol*. 20:611–621.

O'Shea, M., Singh, M.E., McGregor, I.S., Mallet, P.E. 2004. Chronic cannabinoid exposure produces lasting memory impairment and increased anxiety in adolescent but not adult rats. *J. Psychopharmacol*. 18:502–508.

Palkovits, M., Harvey-White, J., Liu, J. et al. 2008. Regional distribution and effects of postmortal delay on endocannabinoid content of the human brain. *Neuroscience* 152:1032–1039.

Patel, S., Hillard, C.J. 2008. Adaptations in endocannabinoid signaling in response to repeated homotypic stress: A novel mechanism for stress habituation. *Eur. J. Neurosci*. 27:2821–2829.

Patel, S., Rademacher, D.J., Hillard, C.J. 2003. Differential regulation of the endocannabinoids anandamide and 2-arachidonoylglycerol within the limbic forebrain by dopamine receptor activity. *J. Pharmacol. Exp. Ther*. 306:880–888.

Patel, S., Roelke, C.T., Rademacher, D.J., Cullinan, W.E., Hillard, C.J. 2004. Endocannabinoid signaling negatively modulates stress-induced activation of the hypothalamic–pituitary–adrenal axis. *Endocrinology* 145:5431–5438.

Pattij, T., Vanderschuren, L.J. 2008. The neuropharmacology of impulsive behaviour. *Trends Pharmacol. Sci*. 29:192–199.

Pfennig, A., Kunzel, H.E., Kern, N. et al. 2005. Hypothalamus–pituitary–adrenal system regulation and suicidal behavior in depression. *Biol. Psychiatry* 57:336–342.

Pistis, M. 2002. Delta(9)-tetrahydrocannabinol decreases extracellular GABA and increases extracellular glutamate and dopamine levels in the rat prefrontal cortex: An in vivo microdialysis study. *Brain Res*. 948:155–158.

Ponce, G., Hoenicka, J., Rubio, G. et al. 2003. Association between cannabinoid receptor gene (*CNR*1) and childhood attention deficit/hyperactivity disorder in Spanish male alcoholic patients. *Mol. Psychiatry* 8:466–467.

Potash, J.B., Kane, H.S., Chiu, Y.F. et al. 2000. Attempted suicide and alcoholism in bipolar disorder: Clinical and familial relationships. *Am. J. Psychiatry* 157:2048–2050.

Preuss, U.W., Schuckit, M.A., Smith, T.L. et al. 2002. A comparison of alcohol-induced and independent depression in alcoholics with histories of suicide attempts. *J. Stud. Alcohol* 63:498–502.

Price, C., Hemmingsson, T., Lewis, G., Zammit, S., Allebeck, P. 2009. Cannabis and suicide: Longitudinal study. *Br. J. Psychiatry* 195:492–497.

Ramaekers, J.G., Kauert, G., van Ruitenbeek, P., Theunissen, E.L., Schneider, E., Moeller, M.R. 2006. High-potency marijuana impairs executive function and inhibitory motor control. *Neuropsychopharmacology* 31:2296–2303.

Raphael, B., Wooding, S., Stevens, G., Connor, J. 2005. Comorbidity: Cannabis and complexity. *J. Psychiatr. Pract.* 11:161–176.

Reiach, J.S., Li, P.P., Warsh, J.J., Kish, S.J., Young, L.T. 1999. Reduced adenylyl cyclase immunolabeling and activity in postmortem temporal cortex of depressed suicide victims. *Affect. Disorder* 56:141–151.

Rich, C.L., Dhossche, D.M., Ghani, S., Isacsson, G. 1998. Suicide methods and presence of intoxicating abusable substances: Some clinical and public health implications. *Ann. Clin. Psychiatry* 10:169–175.

Richardson, H.N., Lee, S.Y., O'dDell, L.E., Koob, G.F., Rivier, C.L. 2008. Alcohol self-administration acutely stimulates the hypothalamic–pituitary–adrenal axis, but alcohol dependence leads to a dampened neuroendocrine state. *Eur. J. Neurosci.* 28:1641–1653.

Rodriguez de Fonseca, F., Carrera, M.R., Navarro, M., Koob, G.F., Weiss, F. 1997. Activation of corticotropin-releasing factor in the limbic system during CB withdrawal. *Science* 276:2050–2054.

Rodriguez-Gaztelumendi, A., Rojo, M.L., Pazos, A., Diaz, A. 2009. Altered CB receptor-signaling in prefrontal cortex from an animal model of depression is reversed by chronic fluoxetine. *J. Neurochem.* 108:1423–1433.

Roy, A. 2000. Relation of family history of suicide to suicide attempts in alcoholics. *Am. J. Psychiatry* 157:2050–2051.

Schlaepfer, T.E., Cohen, M.X., Frick, C. et al. 2008. Deep brain stimulation to reward circuitry alleviates anhedonia in refractory major depression. *Neuropsychopharmacology* 33:368–377.

Schmidt, L.G., Samochowiec, J., Finckh, U. et al. 2002. Association of a CB1 cannabinoid receptor gene (*CNR1*) polymorphism with severe alcohol dependence. *Drug Alcohol Depend.* 65:221–224.

Self, D.W., Nestler, E.J. 1998. Relapse to drug-seeking: Neural and molecular mechanisms. *Drug Alcohol Depend.* 51:49–60.

Shearman, L.P., Rosko, K.M., Fleischer, R. et al. 2003. Antidepressant-like and anorectic effects of the cannabinoid CB1 receptor inverse agonist AM251 in mice. *Behav. Pharmacol.* 14:573–582.

Sher, L. 2006. Alcoholism and suicidal behavior: A clinical overview. *Acta Psychiatr. Scand.* 113:13–22.

Sher, L. 2007. The role of the hypothalamic–pituitary–adrenal axis dysfunction in the pathophysiology of alcohol misuse and suicidal behavior in adolescents. *Int. J. Adolesc. Med. Health* 19:3–9.

Sibille, E., Arango, V., Galfalvy, H.C. et al. 2004. Gene expression profiling of depression and suicide in human prefrontal cortex. *Neuropsychopharmacology* 29:351–361.

Simons, J.S., Gaher, R.M., Correia, C.J., Hansen, C.L., Christopher, M.S. 2005. An affective-motivational model of marijuana and alcohol problems among college students. *Psychol. Addict. Behav.* 19:326–334.

Sipe, J.C., Chiang, K., Gerber, A.L., Beutler, E., Cravatt, B.F. 2002. A missense mutation in human fatty acid amide hydrolase associated with problem drug use. *Proc. Natl Acad. Sci. USA* 99:8394–8399.

Solinas, M., Justinova, J., Goldberg, S.R., Tanda, G.J. 2006. Anandamide administration alone and after inhibition of fatty acid amide hydrolase (FAAH) increases dopamine levels in the nucleus accumbens shell in rat. *J. Neurochem.* 98:408–419.

Solinas, M., Tanda, G., Justinova, Z. et al. 2007. The endogenous cannabinoid anandamide produces delta-9-tetrahydrocannabinol-like discriminative and neurochemical effects that are enhanced by inhibition of fatty acid amide hydrolase but not by inhibition of anandamide transport. *J. Pharmacol. Exp. Ther.* 321:370–380.

Soyka, M., Koller, G., Schmidt, P. et al. 2008. Cannabinoid receptor 1 blocker rimonabant (SR 141716) for treatment of alcohol dependence: Results from a placebo-controlled, double-blind trial. *J. Clin. Psychopharmacol.* 28:317–324.

Steiner, M.A., Marsicano, G., Nestler, E.J., Holsboer, F., Lutz, B., Wotjak, C.T. 2008a. Antidepressant-like behavioral effects of impaired cannabinoid receptor type 1 signaling coincide with exaggerated corticosterone secretion in mice. *Psychoneuroendocrinology* 33:54–67.

Steiner, M.A., Wanisch, K., Monory, K. et al. 2008b. Impaired cannabinoid receptor type 1 signaling interferes with stress-coping behavior in mice. *Pharmacogenom. J.* 8:196–208.

Sugiura, T., Kondo, S., Sukagawa, A. et al. 1995. 2-Arachidonoylglycerol: A possible endogenous cannabinoid receptor ligand in brain. *Biochem. Biophys. Res. Commun.* 215:89–97.

Suominen, K., Henriksson, M., Suokas, J., Isometsa, E., Ostamo, A., Lonnqvist, J. 1996. Mental disorders and comorbidity in attempted suicide. *Acta Psychiatr. Scand.* 94:234–240.

Tanda, G., Loddo, P., Di Chiara, G. 1999. Dependence of mesolimbic dopamine transmission on delta9-tetrahydrocannabinol. *Eur. J. Pharmacol.* 376:23–26.

Tremblay, L.K., Naranjo, C.A., Graham, S.J. et al. 2005. Functional neuroanatomical substrates of altered reward processing in major depressive disorder revealed by a dopaminergic probe. *Arch. Gen. Psychiatry* 62:1228–1236.

Tzavara, E.T., Davis, R.J., Perry, K.W. et al. 2003. The CB1 receptor antagonist SR141716A selectively increases monoaminergic neurotransmission in the medial prefrontal cortex: Implications for therapeutic actions. *Br. J. Pharmacol.* 138:544–553.

Uriguen, L., Perez-Rial, S., Ledent, C., Palomo, T., Manzanares, J. 2004. Impaired action of anxiolytic drugs in mice deficient in cannabinoid CB1 receptors. *Neuropharmacology* 46:966–973.

Van Sickle, M.D., Duncan, M., Kingsley, P.J. et al. 2005. Identification and functional characterization of brainstem cannabinoid CB2 receptors. *Science* 310:329–332.

Vinod, K.Y. Hungund, B.L. 2006. The role of endocannabinoid system in depression and suicide. *Trends Pharmacol. Sci.* 27:539–545.

Vinod, K.Y., Arango, V., Xie, S. et al. 2005. Elevated levels of endocannabinoids and CB1 receptor-mediated G-protein signaling in the prefrontal cortex of alcoholic suicide victims. *Biol. Psychiatry* 57:480–486.

Vinod, K.Y., Hungund, B.L. 2005. Endocannabinoid lipids and mediated system: Implications for alcoholism and neuropsychiatric disorders. *Life Sci.* 77:1569–1583.

Vinod, K.Y., Kassir, S.A., Hungund, B.L., Cooper, T.B., Mann, J.J., Arango, V. 2010. Selective alterations of the CB1 receptors and the fatty acid amide hydrolase in the ventral striatum of alcoholics and suicides. *J. Psychiatr. Res.* 44:591–597.

Vinod, K.Y., Maccioni, P., Garcia-Gutierrez, M.S., et al. 2012. Innate difference in the endocannabinoid signaling and its modulation by alcohol consumption in alcohol-preferring sP rats. *Addict. Biol.* 17:62–75.

Vinod, K.Y., Sanguino, E., Yalamanchili, R., Manzanares, J., Hungund, B.L. 2008a. Manipulation of fatty acid amide hydrolase functional activity alters sensitivity and dependence to ethanol. *J. Neurochem.* 104:233–243.

Vinod, K.Y., Yalamanchili, R., Thanos, P.K. et al. 2008b. Genetic and pharmacological manipulations of the CB1 receptor alter ethanol consumption and dependence in ethanol-preferring and ethanol non-preferring mice. *Synapse* 62:574–581.

Wang, L., Liu, J., Harvey-White, J., Zimmer, A., Kunos, G. 2003. Endocannabinoid signaling via cannabinoid receptor 1 is involved in ethanol preference and its age-dependent decline in mice. *Proc. Natl. Acad. Sci. USA* 100:1393–1398.

Wang, W., Sun, D., Pan, B. et al. 2010. Deficiency in endocannabinoid signaling in the nucleus accumbens induced by chronic unpredictable stress. *Neuropsychopharmacology* 35:2249–2261.

Wenger, T., Ledent, C., Tramu, G. 2003. The endogenous cannabinoid, anandamide, activates the hypothalamo–pituitary–adrenal axis in CB1 cannabinoid receptor knockout mice. *Neuroendocrinology* 78:294–300.

Wilson, R.I., Nicoll, R.A. 2002. Endocannabinoid signaling in the brain. *Science* 296:678–682.

Witkin, J.M., Tzavara, E.T., Davis, R.J., Li, X., Nomikos, G.G. 2005. A therapeutic role for cannabinoid CB1 receptor antagonists in major depressive disorders. *Trends Pharmacol. Sci.* 26:609–617.

Zarate, J., Churruca, I., Echevarria, E. et al. 2008. Immunohistochemical localization of CB1 cannabinoid receptors in frontal cortex and related limbic areas in obese Zucker rats: Effects of chronic fluoxetine treatment. *Brain Res.* 1236:57–72.

Zavitsanou, K., Garrick, T., Huang, X.F. 2004. Selective antagonist [^{3}H]SR141716A binding to CB1 receptors is increased in the anterior cingulate cortex in schizophrenia. *Prog. Neuropsychopharmacol. Biol. Psychiatry* 28:355–360.

Zimmer, A., Zimmer, A.M., Hohmann, A.G., Herkenham, M., Bonner, T.I. 1999. Increased mortality, hypoactivity, and hypoalgesia in cannabinoid CB1 receptor knockout mice. *Proc. Natl Acad. Sci. USA* 96:5780–5785.

Zuo, L., Kranzler, H.R., Luo, X., Covault, J., Gelernter, J. 2007. CNR1 variation modulates risk for drug and alcohol dependence. *Biol. Psychiatry* 62:616–626.

6 Stress–Diathesis Model of Suicidal Behavior

Kees van Heeringen

CONTENTS

6.1 INTRODUCTION

Suicide and attempted suicide are complex behaviors, and a large number of proximal and distal risk factors have been identified (Hawton and van Heeringen, 2009). These risk factors can be categorized in explanatory models, which may help to understand suicidal individuals and facilitate the assessment of suicide risk.

Early models have identified key determinants operating during the development of disorders or behavioral problems. For example, psychologists have developed schema models that focused on cognitive characteristics of, for example, depression, anxiety, and personality disorders. This conceptual approach and the empirical research motivated by such models have led to significant insights into these disorders (Ingram and Luxton, 2005). Stress has also been identified as a key determinant of psychopathology, so that a variety of models have featured stress as a primary determinant. Such models suggest that severe enough negative events can precipitate disorders even without reference to individual biological or psychological characteristics.

The stress model of suicidal behavior is an example of such models. It is based on the observation that stressful life events are commonly recognized as triggers of suicidal behavior. A variety of explanatory models, including those applied by lay people, have indeed featured stress as a primary determinant of suicidal behavior.

Such models indicate that negative life events if severe enough can precipitate suicidal behavior even without the existence of individual predisposing psychological or biological characteristics.

Until recently, most studies of suicidal behavior were based on such early models and thus restricted to one domain of possible risk factors, for example, social, psychiatric, or psychological. As pointed out by Mann et al. (1999), such studies are too narrowly focused to estimate the relative importance of different types as risk factors or their interrelationship. A model of suicidal behavior has to take into account proximal and distal risk factors and their potential interaction (Hawton and van Heeringen, 2009). Stress models of suicidal behavior can indeed not explain the observations that even extreme stress does not lead to suicidal behaviors in all exposed individuals. Such observations have led to the recognition that the development of suicidal behavior involves a vulnerability or diathesis as a distal risk factor, which predisposes individuals to such behavior when stress is encountered.

This chapter will review the scientific literature on the stress–diathesis model of suicidal behavior. Preceding this review, general issues regarding the origins, definitions, and components of stress–diathesis models will be addressed. The concluding discussion will point at the advantages of using stress–diathesis models for treating and preventing suicide risk and address issues with regard to future research.

6.2 STRESS–DIATHESIS MODELS: GENERAL ISSUES

Diathetic individuals respond with abnormal or pathological reactions to physiological stimuli or the ordinary conditions of life that are borne by the majority of individuals without injury (Zuckerman, 1999). The concept of diathesis has been intuitively straightforward and discussed intensively in the literature, but few precise definitions are available. (Ingram and Luxton, 2005). A diathesis is commonly conceptualized as a predispositional factor, or a set of factors, that makes possible a disordered state. It reflects a constitutional vulnerability to develop a disorder.

The diathesis concept has a long history in medical terminology. The word diathesis stems from the Greek idea of predisposition, which is related to the humoral theory of temperament and disease (Zuckerman, 1999). The term has been used in a psychiatric context since the 1800s. Theories of schizophrenia brought the stress and diathesis concepts together and the particular terminology of diathesis–stress interaction was developed by Meehl, Bleuler, and Rosenthal in the 1960s (Ingram and Luxton, 2005).

In the modern sense, the biological traits produced by the genetic disposition are the diathesis. The term "diathesis" has, however, been broadened to include cognitive and social predispositions that may make a person vulnerable to a disorder such as depression. In this broader sense, the diathesis is the necessary antecedent condition for the development of a disorder or problem, whether biological or psychological. The "cry of pain" model of suicidal behavior, as described in detail further in this chapter, is a clear example of such a psychological approach to the study of the diathesis to suicidal behavior. In most models, whether biological or psychological, the diathesis alone is not sufficient to produce the disorder but requires other potentiating or releasing factors to become pathogenic. The diathesis, in this case, includes the vulnerability to stress (Zuckerman, 1999).

Most stress–diathesis models presume that all people have some level of diathesis for any given psychiatric disorder (Monroe and Hadjiyannakis, 2002). However, individuals may differ with regard to the point at which they develop a disorder depending on the degree to which predispositional risk factors exist and on the degree of experienced stress. Thus, relatively minor stressors may lead to a disorder in persons who are highly vulnerable. This approach presupposes additivity, that is, the idea that diathesis and stress add together to produce the disorder. Ipsative models more specifically posit an inverse relationship between components such that the greater the presence of one component the less of the other component is needed to bring about the disorder. Thus, for example, minimal stress is needed for depression to occur in individuals with a strongly depressogenic schema (Ingram and Luxton, 2005). Such models assume a dichotomous diathesis, that is, either one has it (a gene, a unique combination of genes, or a particular brain pathology) or one does not have it (Zuckerman, 1999). If the diathesis is absent, there is no effect of stress so that even severe stress will not lead to the development of the disorder. When the diathesis is present, the expression of the disorder will be conditional on the degree of stress: as stress increases so does the risk for the disorder in persons who possess the diathesis (Ingram and Luxton, 2005). However, most disorders in the psychiatric domain probably have a polygenic basis that allows for varying degrees of the diathesis agent, including variations in neurotransmitter activity levels. In this case, the probability of a disorder would increase as a function of both levels of stress and strength of the diathesis.

The conceptualization of a diathesis as dynamic implies that such a diathesis is continuous rather than dichotomous. For example, schema models of depression were commonly regarded as dichotomous models: if an individual possesses a depressogenic schema he or she is at risk of depression when events that activate this schema occur. More recent discussions of the schema model have however pointed at the possibility of a continuous character by describing the depressogenic nature of schemata as ranging from weak or mild to strong.

In line with the possibility of a continuous diathesis it should be noted that the interaction between stress and a diathesis might not be static, and change over time. The diathesis may increase or decrease so that the amount of stress needed for the development of pathology may need to decrease or increase, respectively. The "kindling" phenomenon (Post, 1992) provides an example of the dynamic character of the interaction between stress and vulnerability: repeated occurrences of a disorder may cause neuronal changes that result in more sensitivity to stress. The kindling theory thus proposes that diatheses may change so that more or less stress becomes necessary to activate vulnerability factors (Ingram and Luxton, 2005). It is however not clear whether the diathesis changes under the influence of negative circumstances or whether residua and scarring add to the diathesis and thus increase vulnerability.

Finally, a diathesis may theoretically consist of one single factor or be constituted by multiple components. Polygenic disorders or interpersonal cognitive theories provide examples of diatheses that are composed by multiple factors.

6.3 STRESS–DIATHESIS MODELS OF SUICIDAL BEHAVIOR

Early descriptions of the roles of stress and a diathesis in the development of suicidal behavior were grounded in sociobiology (De Catanzaro, 1980). Further studies focused on cognitive psychological characteristics. For instance, studying a college population Schotte and Clum (1982) described a stress/problem-solving model of suicidal behavior in which poor problem solvers under high life stress are considered to be at risk for depression, hopelessness, and suicidal behavior. Rubinstein (1986) developed a stress–diathesis theory of suicide, in which the effects of specific situational stressors and the categories or predisposing factors of vulnerable individuals in a given culture were integrated in a biocultural model of suicidal behavior. Mann and Arango (1992) then proposed a stress–diathesis model based on the integration of neurobiology and psychopathology, which still forms the basis for much of the current research in suicidology. Particular emphasis was thereby given to changes in the serotonin system and how these may represent a constitutional risk factor as opposed to a state-dependent risk factor for suicidal behavior.

The following sections will focus on the stress component and the diathesis component of stress–diathesis models of suicidal behavior, followed by a description of a number of such models.

6.3.1 Stress Component

Psychosocial crises and psychiatric disorders may constitute the stress component of stress–diathesis models of suicidal behavior (Mann et al., 1999). It is difficult to separate the impact of psychosocial adversity from that of psychiatric illness. Poverty, unemployment, and social isolation have all been implicated in suicide. These factors are clearly not independent from each other or from psychiatric illness. Psychiatric disorders can lead to job loss, to breakup of marriages or relationships, or to the failure to form such relationships. Moreover, psychiatric illness and psychosocial adversity can combine to increase stress on the person (Mann, 2003).

Many studies have focused on state-dependent characteristics of psychiatric disorders, which may be associated particularly with suicide risk. These include the severity of depression, levels of hopelessness and mental pain, and cognitive characteristics. With regard to an effect of severity of depression, study results have not been equivocal as some (e.g., Mann et al., 1999), but not all (e.g., Forman et al., 2004), studies show that the risk of suicide increases with elevated levels of severity of depressive symptoms. There appears to be more agreement about the association between increased levels of hopelessness and an increased risk of suicide in depressed individuals. A substantial number of studies have focused on state-dependent cognitive characteristics of depressive episodes in association with an increased risk of suicide. More particularly, Beck's theory of modes has been shown to offer a framework for conceptualizing suicidal behavior, which is useful for treatment and prevention.

Modes are defined as interconnected networks of cognitive, affective, motivational, physiological, and behavioral schemata that are activated simultaneously by relevant environmental events and result in goal-directed behavior. Thus, suicidal individuals may experience suicide-related cognitions, negative effect, and the motivation to engage in suicidal behavior in the context of a depressive episode and following exposure to triggering life events. Mental pain (or "psychache") thereby appears to be an emotional and motivational characteristic of particular importance (Troister and Holden, 2010).

It appears however that some of these state-dependent characteristics are to be regarded more appropriately as trait dependent and thus as a part of the diathesis. The emergence of cognitive suicidal modes and feelings of hopelessness during suicidal crises may indeed be regarded as activations of trait-dependent vulnerability characteristics.

6.3.2 Diathesis Component

Genetic effects, childhood abuse, and epigenetic mechanisms may be involved in the etiology of the diathesis to suicidal behavior (Mann and Haghgighi, 2010). Clinical studies have indeed shown that reported childhood adversity, such as deprivation and physical or sexual abuse, is a risk factor for psychopathological phenomena in later childhood and adulthood, including depression and suicide. Not all individuals will however develop psychopathology following exposure to childhood adversity, indicating the existence of a diathesis in some but not all individuals. Neuroanatomical, physiological, and genomic alterations may contribute to the long-lasting detrimental effects of exposure to childhood adversity on the risk of psychopathology (Miller et al., 2009). The study by McGowan et al. (2009) provides an intriguing example of how environmental influences may affect the expression of genes. The involvement of serotonin and other neurotransmitters, the (epi)genetics of suicidal behavior, and the role of gene–environment interactions are discussed in Chapters 2, 3, and 10–14. Postmortem and neuroimaging studies have clearly demonstrated structural and functional changes in the brains of individuals with a history of suicidal behavior, which may correlate with components of the diathesis (see Chapter 10; van Heeringen et al., 2011a). Postmortem findings include fewer cortical serotonin neurons in key brain regions such as the dorsal and ventral prefrontal cortex, which also appear to correlate with components of the diathesis (Mann, 2003).

These components may include aggression and/or impulsivity, pessimism and hopelessness, and problem-solving or cognitive rigidity. Some of these characteristics are discussed as intermediate phenotypes of suicidal behavior elsewhere in this book. Recent studies have used neuropsychological approaches to the study of the diathesis, and have focused particularly on decision-making processes (Jollant et al., 2007; Dombrovski et al., 2010; van Heeringen et al., 2011b).

Currently available evidence as reviewed in this chapter suggests that the diathesis to suicidal behavior is continuous. It can be hypothesized that the diathesis becomes more pronounced during the course of the suicidal process that commonly precedes completed suicide (van Heeringen, 2001). Suicide is indeed commonly preceded by nonfatal suicide attempts, which are commonly repeated with an increasing degree of medical severity, suicidal intent, or lethality of the method used. Several studies have provided support for a kindling effect on the occurrence of suicide attempts. Findings from clinical studies point at the possibility that each time such a suicidal mode becomes activated, it becomes increasingly accessible in memory and requires less triggering stimuli to become activated the next time. This phenomenon can be used to explain findings from epidemiological studies in suicide attempters showing that each succeeding suicide attempt is associated with a greater probability of a subsequent suicide attempt (Leon et al., 1990; Oquendo et al., 2004; van Heeringen, 2001).

The concept of a continuous diathesis may explain differences in suicidal behavior between individuals, for example, why individuals differ in their suicidal reaction to similar life events varying from deliberate self-harm with no or minor physical consequences to completed suicide. Repeated exposure to stressors may thus gradually diminish the resilience toward stress, due to which stressors of decreasing severity may lead to suicidal behaviors with increasing suicidal intent. Increasing evidence points at a role of increasing neuropsychological deficits in the medial temporal cortex–hippocampal system, perhaps due to the detrimental effects of stress hormones on serotonergic neurons. As discussed in more detail elsewhere in this book, studies of levels of the serotonin metabolite 5-HIAA in the cerebrospinal fluid of suicide attempters have shown that (1) depressed suicide attempters have lower levels than depressed non-attempters, (2) repeating attempters have lower levels than so-called first-evers, (3) the use of violent methods is associated with lower levels than the use of non-violent methods, and (4) attempted suicide patients with lower levels show a poorer survival in terms of death from suicide (for a review see, van Heeringen, 2001). Such findings point at a possible increase of the vulnerability to suicidal behavior during the suicidal process, which is paralleled by a decrease in serotonergic functioning.

6.4 EXAMPLES OF STRESS–DIATHESIS MODELS OF SUICIDAL BEHAVIOR

6.4.1 Cognitive Stress–Diathesis Model of Suicidal Behavior

Williams and Pollock (2001) have described a diathesis for suicidal behavior in cognitive psychological terms, that is, the "cry of pain" model, which was elaborated in the "differential activation model." According to the "cry of pain" model, suicidal behavior represents the response to a situation that has three components:

1. *Sensitivity to signals of defeat*: Using the "emotional Stroop task," Williams and colleagues clearly demonstrated attentional biases (or so-called perceptual pop-outs) in association with suicidal behavior—an involuntary hypersensitivity to stimuli signaling "loser" status increases the risk that the defeat response will be triggered.
2. *Perceived "no escape"*: Limited problem-solving abilities may indicate to persons that there is no escape from problems or life events. Further study has revealed that such limited abilities correlate with decreases in the specificity of autobiographical memories. To generate potential solutions

to problems, a person apparently needs to have access to the past in some detail. Overgeneral memories prevent the use of strategies, which are sufficiently detailed to solve problems.

3. *Perceived "no rescue"*: The occurrence of suicidal behavior is associated with a limited fluency in coming up with positive events that might happen in the future. This limited fluency is reflected not only by the perception that there is no escape from an aversive situation but also by the judgment that no rescue is possible in the future. It is thereby interesting to note that the fluency of generating positive future events correlates negatively with levels of hopelessness, a core clinical predictor of suicidal behavior. This suggests that hopelessness does not consist of the anticipation of an excess of negative events, but indicates that hopelessness reflects the failure to generate sufficient rescue factors.

The identification of the neuropsychological correlates of the three cognitive components reflects an interesting characteristic of the "cry of pain" model, in addition to its clinical relevance. The authors state that, in this sense, the model fits in life events and biological research. The study of the biological underpinnings of hopelessness and mental pain, as discussed elsewhere in this chapter, indeed suggests that the components of this model can be studied using neurobiological research approaches and thus may contribute to our understanding of the pathophysiology by identifying possible endophenotypes of suicidal behavior.

6.4.2 Clinical Stress–Diathesis Models of Suicidal Behavior

Mann et al. (1999) proposed a stress–diathesis model based on the findings from a clinical study of a large sample of patients admitted to a university psychiatric hospital. When compared to patients without a history of suicide attempts, patients who had attempted suicide show higher scores on subjective depression and suicidal ideation, and reported fewer reasons for living. In addition, suicide attempters show higher rates of lifetime aggression and impulsivity, comorbid borderline personality disorder, substance use disorder or alcoholism, family history of suicidal acts, head injury, smoking, and childhood abuse history. The risk for suicidal acts thus is determined not only by a psychiatric illness (the stressor) but also by a diathesis as reflected by tendencies to experience more suicidal ideation and to be more impulsive and, therefore, more likely to act on suicidal feelings. More in particular, Mann and colleagues describe a predisposition to suicidal acts that appears to be part of a more fundamental predisposition to both externally and self-directed aggression. Aggression, impulsivity, and borderline personality disorder are key characteristics, which may be the result of genetic factors or early life experiences, including a history of physical or sexual abuse. A common underlying genetic or familial factor may therefore explain the association between suicidal behavior with the aggression/impulsivity factor and/or borderline personality disorder, independent of transmission of major depression or psychosis. Suicide risk was also associated with past head injury, and the authors hypothesize that aggressive–impulsive children and adults are more likely to sustain a head injury, which may lead to disinhibition and aggressive behavior. The serotonin neurotransmission system may also play a role. Given the evidence linking low serotonergic activity to suicidal behavior, it is conceivable that such low activity may mediate genetic and developmental effects on suicide, aggression, and alcoholism (Mann et al., 1999).

Based on a review of studies of clinical predictors of suicide, McGirr and Turecki (2007) have provided a second example of a clinical stress–diathesis model. The model is based on the clinical observation that psychopathology, for the most part, appears to be a necessary but not sufficient factor for suicide. Therefore, a promising avenue for improved clinical detection is the elucidation of stable risk factors predating the onset of psychopathology, through which suicidal behavior is salted out. The authors describe personality characteristics as stable risk factors, which can be regarded as reflecting preexisting endophenotypes and which interact with the onset of psychiatric disorders (the stressor) to result in suicide. While the authors acknowledge the potential role of personality characteristics such as neuroticism and introversion in relation to suicide, they focus their review on impulsivity and aggression. Impulsivity is in this context regarded more as a behavioral dimension than as the explosive or instantaneous actions relating to an inability to resist impulses. The behavioral dimension describes behaviors that appear to occur without reflection or consideration of consequences, are often risky or inappropriate to the situation, and are accompanied by undesirable outcomes. They do not necessarily include aggressive behaviors, but high levels of impulsivity correlate with high levels of aggression. A correlation between aggression, impulsivity, and hostility has been confirmed in suicide completers using psychological autopsy approaches. Studies of fatal and nonfatal suicidal behavior have indeed pointed at a role of this behavioral dimension. Impulsivity thus appears to be involved not only in self-harming behaviors without suicidal intent but also in high-lethality and fatal suicidal behaviors.

With respect to aggression, more extensive histories of aggression have been associated with suicide attempts in clinical samples and those meeting criteria for major depression and bipolar disorder. Disruptive aggression appears to distinguish female ideators who attempt suicide from those who do not. More extensive histories of aggression, assault, and irritability have been associated with adolescent suicide completion. In addition, depressed suicides and borderline suicides exhibit higher levels of aggressive behaviors than diseased controls.

Levels of impulsivity thus tend to correlate with those of aggression and hostility. Evidence suggests that these characteristics fall under a superordinate factor relating to the familial transmission of dyscontrol psychopathology.

The involvement of impulsivity and aggression in the diathesis of suicidal behavior has been a matter of debate since many years. The controversy is fueled by, among others, epidemiological observations that many attempted and completed suicides do not appear to be aggressive or impulsive and by theoretical discussions about the multifaceted nature of the aggression and impulsivity concepts.

6.4.3 Neurobiological Model of Suicidal Behavior

Using a functional neuroimaging technique, Jollant et al. (2008) provide an example of a third, that is, neurobiological approach to stress–diathesis models of suicidal behavior. This model of suicidal behavior was investigated by exposing young males with a history of depression to angry, happy, and neutral faces while being euthymic. Findings in young males with a history of attempted suicide were compared to those in young males without such a history. Relative to affective comparison subjects, suicide attempters showed greater activity in the right lateral orbitofrontal cortex (Brodmann area 47) and decreased activity in the right superior frontal gyrus (area 6) in response to prototypical angry versus neutral faces, greater activity in the right anterior cingulate gyrus (area 32 extending to area 10) to mild happy versus neutral faces, and greater activity in the right cerebellum to mild angry versus neutral faces. Thus, suicide attempters were distinguished from non-suicidal patients by responses to angry and happy faces that may suggest increased sensitivity to others' disapproval, higher propensity to act on negative emotions, and reduced attention to mildly positive stimuli. It is concluded that these patterns of neural activity and cognitive processes may represent vulnerability markers of suicidal behavior in men with a history of depression.

6.5 DISCUSSION

Although there are many pathways to suicide, studies in the domains of neuropsychology, cognitive psychology, neurobiology, and clinical psychiatry have provided increasing evidence in support of a stress–diathesis model of suicidal behavior. While depression is the common final pathway to suicidal behavior, the vast majority of depressed individuals neither attempt nor complete suicide. It appears that a diathesis to suicidal behavior differentiates depressed individuals who will kill themselves from other depressed patients. The diathesis may be due to epigenetic effects and childhood adversity and is reflected by a distinct biological, psychological, or clinical profile. This profile may include aggression/impulsivity, pessimism and hopelessness, and deficient problem solving. The involvement of aggression/impulsivity has recently been questioned and it has been suggested that the study of decision making and emotion regulation may help to refine this endophenotype (Brent, 2009).

The application of stress–diathesis models to suicidal behavior has substantial implications for the identification of suicide risk and the prevention of suicidal behavior. The identification of trait-dependent vulnerability factors can be expected to facilitate early recognition of suicide risk. Vulnerability traits are open to modification early in life, and interventions during sensitive periods of development may have durable effects on personality and thereby affect vulnerability to suicide (McGirr and Turecki, 2007). In the context of prediction and prevention, it is important to note that trait-dependent components of the diathesis can be demonstrated and treated beyond depressive episodes. For example, reducing the diathesis for suicidal behavior might be possible as evidenced by the clinical effects of lithium, clozapine, or cognitive behavioral therapy (Mann, 2003). Lithium appears to reduce the rate of suicidal behavior independently of its mood-stabilizing effects in patients with unipolar or bipolar disorder. Clozapine reduces suicidal behavior in schizophrenia independently of its antipsychotic action. The mechanisms that underlie the antisuicidal effects of lithium and clozapine are not known, but both medications affect a component of the diathesis to suicidal behavior, that is, the serotonergic system.

Further research of the applicability of stress–diathesis models to suicidal behavior is however needed. For example, it remains to be demonstrated whether the diathesis to suicidal behavior is continuous or dichotomous, and whether stress–diathesis models of suicidal behavior are additive or interactive. An important issue is the potential interdependence of the stress and diathesis components, as components of the diathesis may increase the probability of exposure to stressors. For example, Jollant et al. (2007) clearly demonstrated that impaired decision making, that is, a potential component of the diathesis to suicidal behavior, increases the risk of problems in affective relationships in suicide attempters. A recent study of a stress–diathesis model of adolescent depression showed that adolescents with a negative cognitive style are more at risk of depression following stressful life events, but also demonstrated that individuals at risk are more likely to report stressors that are at least partly dependent on their behavior. This model suggests a cycle that perpetuates across time, hinting at the mechanisms that may both initiate and maintain or worsen depressive symptoms in adolescence (Kercher and Rapee, 2009). The applicability of a similar stress–diathesis model to suicidal behavior and its implications for our understanding of the dynamic nature of this model remain to be demonstrated. The interdependence of stress and diathesis components would however also mean that interventions targeting the diathesis may also decrease exposure to stressors and suggests that relief of stress effects would enhance the efficacy of therapeutic interventions.

REFERENCES

Brent, D. 2009. In search of endophenotypes for suicidal behavior. *American Journal of Psychiatry* 166:1087–1088.

De Catanzaro, D. 1980. Human suicide: A biological perspective. *Behavioral and Brain Sciences* 3:265–272.

Dombrovski, A.Y., Clark, L., Siegle, G.J., Butters, M.A., Ichikawa, N., Sahakian, B.J., Szanto, K. 2010. Reward/punishment learning in older suicide attempters. *American Journal of Psychiatry* 167:699–707.

Forman, E.M., Berk, M.S., Henriques, G.R., Brown, G.K., Beck, A.T. 2004. History of multiple suicide attempts as a behavioural marker of severe psychopathology. *American Journal of Psychiatry* 161:437–443.

Hawton, K., van Heeringen, K. 2009. Suicide. *The Lancet* 373:1372–1381.

Ingram, R.E., Luxton, D.D. 2005. Vulnerability–stress models. In: *Development of Psychopathology: A Vulnerability–Stress Perspective*. Hankin, B.L., Abela, J.R.Z., Eds. Thousand Oaks, CA: Sage Publications.

Jollant, F., Lawrence, N.S., Giampetro, V., Brammer, M.J., Fullana, M.A., Drapier, D., Courtet, P., Phillips, M.L. 2008. Orbitofrontal cortex response to angry faces in men with histories of suicide attempts. *American Journal of Psychiatry* 165:740–748.

Jollant, F., Guillaume, S., Jaussent, I., Castelanu, D., Malafosse, A., Courtet, P. 2007. Impaired decision making in suicide attempters may increase the risk of problems in affective relationships. *Journal of Affective Disorders* 99:59–62.

Kercher, A., Rapee, R.M. 2009. A test of a cognitive diathesis–stress generation pathway in early adolescent depression. *Journal of Abnormal Child Psychology* 37:845–855.

Leon, A.C., Friedman, R.A., Sweeney, J.A., Brown, R.P., Mann, J.J. 1990. Statistical issues in the identification of risk factors for suicidal behavior: The application of survival analysis. *Psychiatry Research* 31:99–108.

Mann, J.J. 2003. Neurobiology of suicidal behaviour. *Nature Reviews Neuroscience* 4:819–828.

Mann, J.J., Arango, V. 1992. Integration of neurobiology and psychopathology in a unified model of suicidal behavior. *Journal of Clinical Psychopharmacology* 12:S2–S7.

Mann, J.J., Haghgighi, F. 2010. Genes and environment: Multiple pathways to psychopathology. *Biological Psychiatry* 68:403–404.

Mann, J.J., Waternaux, C., Haas, G.L., Malone, K.M. 1999. Toward a clinical model of suicidal behavior in psychiatric patients. *American Journal of Psychiatry* 156:181–189.

McGirr, A., Turecki, G. 2007. The relationship of impulsive aggressiveness to suicidality and other depression-linked behaviors. *Current Psychiatry Reports* 9:460–466.

McGowan, P.O., Sasaki, A., D'Alessio, A.C., Dymov, S., Labonte, B., Szyf, M., Turecki, G., Meaney, M.J. 2009. Epigenetic regulation of the glucocorticoid receptor in human brain associates with childhood abuse. *Nature Neuroscience* 12:342–348.

Miller, J.M., Kinnally, E.L., Ogden, R.T., Oquendo, M.A., Mann, J.J., 2009. Reported childhood abuse is associated with low serotonin binding in vivo in major depressive disorder. *Synapse* 63:565–573.

Monroe, S.M., Hadjiyannakis, H. 2002. The social environment and depression: Focusing on severe life stress. In: *Handbook of Depression*. Gotlib, I.H., Hammen, C.L., Eds. New York: Guilford Press.

Oquendo, M.A., Galfalvy, H., Russo, S., Ellis, S.P., Grunebaum, M.F., Burke, A., Mann, J.J. 2004. Prospective study of clinical predictors of suicidal acts after a major depressive episode in patients with major depressive disorder or bipolar disorder. *American Journal of Psychiatry* 61:1433–1441.

Post, R.M. 1992. Transduction of psychosocial stress into the neurobiology of recurrent affective disorder. *American Journal of Psychiatry* 149:9999–1010.

Rubinstein, D.H. 1986. A stress–diathesis theory of suicide. *Suicide and Life-Threatening Behavior* 16:182–197.

Schotte, D.E., Clum, G.A. 1982. Suicide ideation in a college population. *Journal of Consulting and Clinical Psychology* 50:690–696.

Troister, T., Holden, R.R. 2010. Comparing psychache, depression and hopelessness in their associations with suicidality: A test of Sheidman's theory of suicide. *Personality and Individual Differences* 7:689–693.

van Heeringen, C., Ed. 2001. *Understanding Suicidal Behaviour: The Suicidal Process Approach to Research, Treatment and Prevention*. West Sussex, England: Wiley.

van Heeringen, C., Byttebier, S., Godfrin, K. 2011a. Suicidal brains: A systematic review of structural and functional brain studies in association with suicidal behaviour. *Neuroscience and Biobehavioral Reviews* 35:688–698.

van Heeringen, C., Godfrin, K., Bijttebier, S. 2011b. Understanding the suicidal brain: A review of neuropsychological studies of suicidal ideation and behaviour. In: *The International Handbook of Suicide Prevention: Research, Policy and Practice*. O'Connor, R.C., Platt, S., Gordon, J., Eds. Chichester, U.K.: Wiley.

Williams, J.M.G., Pollock, L. 2001. Psychological aspects of the suicidal process. In: *Understanding Suicidal Behaviour: The Suicidal Process Approach to Research, Treatment and Prevention*. van Heeringen, K., Ed. West Sussex, England: Wiley.

Zuckerman, M. 1999. *Vulnerability to Psychopathology: A Biosocial Model*. Washington, DC: American Psychological Association.

7 Do Serum Cholesterol Values and DST Results Comprise Independent Risk Factors for Suicide?

William Coryell

CONTENTS

7.1 SUICIDE AND HPA AXIS HYPERACTIVITY

7.1.1 General Measures of HPA Axis Hyperactivity

Findings that link hypothalamic–pituitary–adrenal (HPA) axis hyperactivity to risks for later suicide date to 1965 when Bunney and Fawcett[1] described 36 depressed patients who had supplied serial 24 h urine samples for 17-hydroxycorticosteroid (17-OHCS) determinations during extended stays on a research ward. Three committed suicide on pass or shortly after discharge, two made serious attempts, and all five were among the 18 who had had consistently or intermittently high 17-OHCS values (Fisher's exact test, $p = 0.023$). In a subsequent series of 145 patients assessed in similar fashion, the five who committed suicide all had had mean 17-OHCS values that fell in the top 10% of the values for the overall group.[2] At least two more reports also associated high urinary corticosteroid measures with later suicide.[3,4]

Subsequent lines of evidence likewise showed HPA axis hyperactivity to be a risk factor for a completed suicide. In postmortem studies, suicide victims had, in comparison to control subjects, greater adrenal weights,[5–7] greater adrenal cholesterol concentrations,[8] and greater adrenal volumes.[9] Others showed suicide victims to have higher cerebrospinal fluid (CSF) concentrations of corticotropin-releasing hormone (CRH),[10,11] higher amounts of CRH immunoreactivity in specific brain regions,[12,13] and lower numbers of CRF binding sites.[14] Some, though, found no differences in CRF receptor number[15] or in levels of immunoreactivity[16] between individuals dead by suicide and control subjects dead from other causes.

While most postmortem findings have indicated that HPA axis hyperactivity is a risk factor for suicide, a major caveat to these findings is the fact that all comparison groups were comprised of psychiatrically well individuals. The higher levels of HPA axis activity that apparently preceded the suicide deaths may have simply reflected the presence of severe depressive disorder, itself a well-established risk factor for suicide. The use of 24 h urinary free cortisol (UFC) values to, instead, compare depressed patients who later suicide to those who do not therefore speaks more directly to the clinical need to estimate risk for suicide among patients who present with depressive illness and, particularly, among those who also present with suicidal preoccupation or behaviors.

7.1.2 Dexamethasone Suppression Test

Still more relevant, given the impracticalities of 24 h urine collections, is evidence from studies using the more accessible dexamethasone suppression test (DST). In this procedure a low dose of dexamethasone, most commonly 1 mg, is given at 11 p.m. and plasma samples are drawn the following day, most often at 8 a.m. and 4 p.m., but, in some, also at 11 p.m. Recommended cutoffs to indicate an abnormal escape from dexamethasone suppression have varied but 5 µg/dL or more in any post-dexamethasone sample has been the most widely used.

Eleven studies (Table 7.1) have conducted follow-ups of depressed inpatients that underwent 1 or 2 mg DSTs. Not included in this listing is one report that described a diagnostically mixed sample without the detail necessary to isolate those with depressive disorders[17] and another[18] that included patients hospitalized with mania and that overlapped to an unspecified degree those described in a previously published series.[19] These listed studies have varied considerably in sample size, in length of follow-up, and in the proportion that eventually completed suicide, but all found higher rates of nonsuppression among those who later completed suicide. While a number of the group differences did not reach statistical significance, the pooled odds ratio for future suicide in DST nonsuppressors compared to suppressors in these studies is 3.0.

The robustness of the relationship between DST results and subsequent suicide varies considerably across these reports from an odds ratio of 12.6 in one[19] to an odds ratio 1.7 in another.[20] There are no obvious explanations for this variance. It may well be that DST results have more predictive value in some depressive subgroups than in others but few investigators have explored this possibility.

Nearly all of the prospective studies of DST results and later suicide were limited to inpatients or did not separate inpatients from outpatients in their analysis.[21] Coryell et al.[22] included a substantial outpatient cohort, however, and

TABLE 7.1
DST Results and Risk for Suicide among Inpatients with Depressive Disorders

Study	Number Followed	Length of Follow-Up (Years)	Number of Suicides (%)	Number of DST Nonsup-pressors (%)	% Who Suicided among Suppressors	% Who Suicided among Nonsup-pressors
Carroll et al.[23]	~250	?	4 (~1.6)	~125 (~50)	0	3.3
Norman et al.[28]	66	?	13 (19.7)	17 (25.8)	12.2	41.2
Roy et al.[24]	27	1	4 (14.8)	14 (51.8)	7.6	21.4
Yerevanian et al.[21]	101	2	3 (3.0)	34 (33.7)	0	8.8
Coryell and Schlesser[25]	205	2	4 (1.9)	96 (46.8)	0	4.2
Boza et al.[26]	13	4	2 (15.4)	3 (23.1)	0	66.7
Nielsen and Bostwick (personal communication)	114	14	7 (6.1)	58 (50.9)	1.8	10.3
Coryell and Schlesser[19]	78	15	8 (10.3)	32 (41.0)	2.9	26.8
Coryell et al.[103]	54	17	4 (7.4)	26 (48.2)	3.6	11.5
Coryell et al.[22]	184	18	9 (4.8)	112 (60.9)	2.6	8.3
Jokinen et al.[100]	382	18	36 (9.4)	167 (43.7)	7.4	12.0

showed that DST nonsuppression was a risk factor for suicide only among inpatients. Among inpatients, moreover, nonsuppression was predictive only for those patients who were rated, on the basis of suicidal thoughts or behaviors, as having at least a moderate suicide tendency ("often thinks of suicide or has thought of a specific method"). Within this group, 2.6% and 13.2% of suppressors and nonsuppressors, respectively, went on to commit suicide whereas corresponding rates for those without suicide plans or frequent thoughts were 2.6% and 2.9%. In replication, Jokinen et al.[20] found a threefold difference in completed suicide rates between nonsuppressors and suppressors among depressed inpatients who had attempted suicide and no difference among those who had not. These two sets of results thus indicate that HPA axis hyperactivity may be a risk factor for suicide only among depressed patients who have clinically manifested suicidal tendencies. The proportion of such patients is likely to have varied among the studies listed in Table 7.1 and this may account for at least some of the differences in findings between them.

Reported relationships between DST results and risks for suicide attempts have been far less consistent than those concerning DST results and risks for completed suicide.[27] Indeed, in some reports, DST nonsuppression was significantly less likely among suicide attempters than among nonattempters.[18,28] However, the few studies that have grouped suicide attempters by the type of attempt they had made found that

DST nonsuppressors were significantly more likely than normal suppressors to have made a serious attempt,[28–30] or to have achieved high scores on scales designed to quantify the risk of future suicide.[31]

The potential value of the DST as a clinical tool in estimating suicide risks should be viewed in the context of other variables typically considered in this judgment. Sex, age, the availability of support, any history of recent losses, and such clinical dimensions as feelings of hopelessness, anxiety levels, and overall symptom severity are typically taken into account and weighed differently on a case-by-case basis. The variable shown in the largest number of studies to be significantly associated with eventual suicide is a history of suicide plans or attempts[32] and as such, it is probably the most appropriate standard against which to view the performance of DST results. Of the three prospective studies of suicide that have tested both factors as predictors, two found DST suppressor status to be the more strongly associated with later suicide (Table 7.2).

The three reports that have provided survival curves depicting time to suicide by baseline DST suppressor status found the difference in suicide rates between nonsuppressors and suppressors to be larger at 5 years of follow-up then at 1–2 years and still larger at follow-up intervals of 15–25 years.[19,20,22] This suggests that though HPA axis hyperactivity is widely viewed as a state marker, an abnormality that resolves in concert with the underlying depressive episode, an abnormal DST result also appears to identify the type of depressive illness a patient has rather than simply the type of depressive episode. Though DST results have been shown to vary considerably within individuals with depressive illness, both within[33] and across[34] episodes, a positive DST result places an individual at a higher risk for suicide in any subsequent depressive episode. This conclusion is in harmony with earlier findings. Analyses of serial DST results have shown that, in comparison to a single normal result, a single abnormal result is more indicative of specific depressive subtypes[33,34] and of the likelihood of true antidepressant response.[35]

TABLE 7.2
DST Results, Suicide Plans or Attempts, and Eventual Suicide: Comparison of Odds Ratios

	Odds Ratio for Suicide
Coryell and Schlesser[19]	
History of serious attempt vs. no attempt history	3.8
DST nonsuppression vs. suppression	14
Jokinen et al.[20]	
Recent suicide attempt vs. no recent attempt	6.5
DST nonsuppression vs. suppression	1.7
Coryell et al.[103]	
Global rating of suicide threats or behaviors: high vs. low	2.2
DST nonsuppression vs. suppression	3.4

7.2 SUICIDE AND SERUM CHOLESTEROL

7.2.1 Early Findings

Large-scale prevention trials designed to show the effects of lowering cholesterol concentrations on mortality yielded results that first drew attention to low serum cholesterol as a possible risk factor for suicidal behavior. They showed the expected decreases in cardiovascular deaths among groups assigned to the cholesterol-lowering arms but the decreases in overall mortality were absent. A review of earlier trials[36] yielded an odds ratio of 1.76 for death from unnatural causes among subjects assigned to cholesterol-lowering treatment in comparison to control subjects. A later meta-analysis encompassed a larger number of trials and no longer found the risks for non-illness mortality associated with cholesterol-lowering efforts to be significant. Some risk remained, however, with an odds ratio of 1.32 for the 13 cholesterol-lowering trials that used diet and non-statin drugs. The authors speculated that low cholesterol concentrations may increase the risks for suicidal behavior chiefly in the presence of other risk factors and that the exclusion criteria used in later trials may have eliminated a higher proportion of individuals with such factors.

Noninterventional community studies have also addressed the question of whether lower cholesterol concentrations increase risks for unnatural death. Most found that they did[37–44] though some did not.[45] Several found, instead, that such deaths were more common among those with higher rather than lower cholesterol concentrations.[46,47]

Four of the positive studies tested whether the effects of low cholesterol concentrations on suicide rates were time limited. Three[39,41,42] showed that they were not; effects did not diminish over follow-up periods of 7,[42] 12,[41] and 20[39] years. Thus, low cholesterol concentrations appear to comprise a trait marker for elevated suicide risk.

7.2.2 Case–Control Studies

The aforementioned findings have motivated a large number of case–control studies of clinical populations. In most, patients with histories of suicide attempts had significantly lower cholesterol concentrations than did those who had not made suicide attempts.[48–64] Figure 7.1 displays the corresponding group values for mean cholesterol concentrations. Not included are reports that instead described the distribution of suicide attempters across cholesterol concentration groupings. These, however, likewise reported significant associations between low values and histories of suicide attempts.[65,66]

Others have failed to find significant difference between suicide attempt and ill control subjects[67–74] but no retrospective, case-controlled study has reported significantly higher cholesterol values for suicide attempters. Notably, and inexplicably, the only prospective study of baseline cholesterol concentration and subsequent suicide attempts did yield a significantly contrary result.[50] With statistical control for age, depressed patients with cholesterol concentrations above 190 mg/dL were more likely to make a serious suicide attempt during follow-up than were those with lower levels.

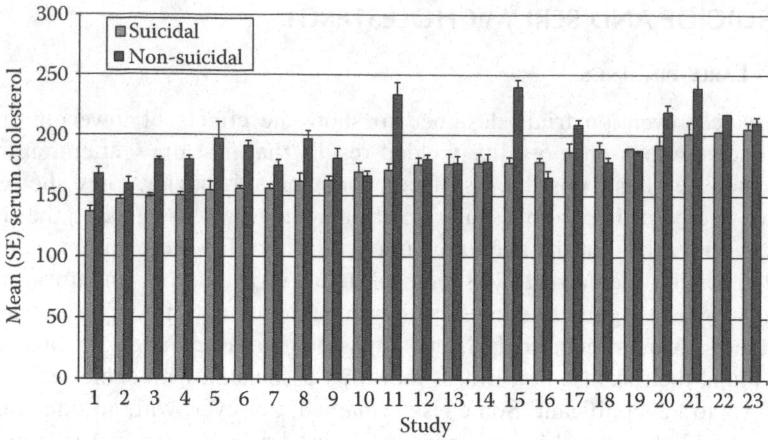

FIGURE 7.1 Suicidal subjects and ill controls.

Comparisons of patients with suicide attempts considered violent or otherwise serious to patients who had made nonserious attempts (Figure 7.2) have been more consistent in finding significantly lower cholesterol levels in the former.[51,53,54,72,75,76] Again, though some differences were not significant,[69,77,78] none described significantly higher cholesterol concentrations in the nonviolent attempters.

In the only postmortem study to compare suicide victims to controls by tissue cholesterol concentrations, Lalovic et al.[79] found no difference but did note that those who had used violent methods had significantly lower concentrations than did those who had used other methods. Further evidence that low cholesterol concentrations are linked to suicide risk through a propensity to aggression or violence derives from a study of criminal offenders in which low cholesterol concentrations were strongly predictive of later suicides among violent offenders but were not at all predictive in nonviolent offenders.[80]

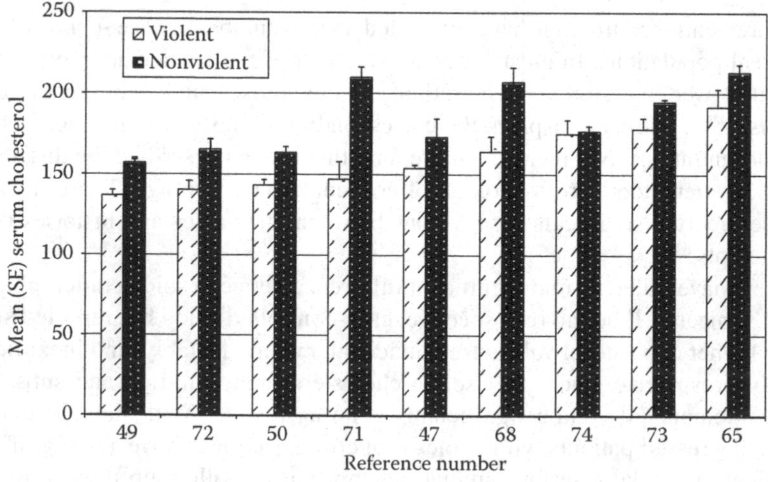

FIGURE 7.2 Violent (serious) attempts and nonviolent attempts.

Other lines of evidence have emerged in support of a low serum cholesterol concentration as a trait-like risk factor for suicidal behavior. In one large series of patients attending a lithium clinic, males in the lowest quartile of cholesterol concentration values were twice as likely to have family histories positive for a completed suicide.[65] Maes et al.,[81] moreover, demonstrated a clear synchrony within the Belgian population between annual fluctuations of serum cholesterol concentrations and suicide. Several other groups have shown that carriers of genetic defects that result in low cholesterol concentrations are at markedly increased risks for attempted and completed suicide.[82,83]

Unfortunately, there have been very few studies in clinical samples of serum cholesterol concentrations and subsequent suicide. One comparison of surviving controls to 41 patients with bipolar disorder who died by suicide after an unspecified interval found no difference in baseline serum cholesterol concentrations. The analysis did not adjust for age, however. In contrast, a follow-up of 78 inpatients with major mood disorders found, after adjusting for age, a significant relationship between cholesterol concentrations and later suicide.[19] The relationship grew stronger as the follow-up interval increased from 5 to 15 years.

7.2.3 Serum Cholesterol, Suicide, and the Serotonin System

The particular association between low serum cholesterol concentrations and suicide attempts or completions by violent means is relevant to the mechanism most frequently proposed as linking these phenomena. Hypofunction of the serotonin system is now a well-established risk factor both for suicidal behavior[27] and for violent behavior generally.[77,84,85] Low serum cholesterol concentrations, in turn, have been linked in a number of ways to serotonin hypofunction. Some reports have described negative correlations between CSF concentrations of 5-HIAA and serum cholesterol in humans[86] and in primates[87] though others have failed to find this.[88–90] Likewise, some have described positive correlations between cholesterol concentrations and platelet 5-HT levels[57,91,92] though, again, some have not.[67,75] Still others have used cortisol and prolactin responses to fenfluramine or MCPP to measure CNS serotonin activity and have reported significant correlations with baseline serum cholesterol concentrations[93–95] or with changes in cholesterol consequent to cholesterol-lowering efforts.[96]

The effects of age and recent alcohol use may confound the apparent correlation of serum cholesterol concentration with suicide risk because these factors can influence both cholesterol concentrations and the propensity to suicidal behavior. A number of the relevant reports have attempted to control for such factors but many did not. The results of a careful study in primates are thus particularly noteworthy because environmental variables could be strictly controlled. Here the investigators assigned cynomolgus monkeys to a "luxury diet" of 0.34 mg cholesterol/calorie or to a "prudent diet" of 0.05 mg cholesterol/calorie.[97] The latter was designed to be equivalent to diets often recommended for humans with significant cardiovascular risk factors. Monkeys assigned the low-cholesterol diet were then observed to exhibit significantly more aggression, to have blunted responses to fenfluramine challenges,[93] and to have lower CSF concentrations of 5-HIAA.[87]

7.3 ARE SERUM CHOLESTEROL AND DST RESULTS ORTHOGONAL PREDICTORS OF SUICIDE?

Fawcett et al.[98] have proposed that both HPA axis hyperactivity and serotonin hypofunction, perhaps as reflected in low serum cholesterol values, increase risks for suicidal behavior but that the two factors take different pathways in doing so. In this model, the HPA axis disturbance drives suicidal behavior through a concomitantly greater severity of anxiety, agitation, and anhedonia, while serotonin hypofunction operates through its correlation with impulsivity and aggression. To the degree that these two factors are biologically independent, the consideration of both together should substantially improve the accuracy of suicide risk estimations. Indeed, the one study to examine this directly yielded results clearly supportive of this model.[99] Fifteen years after index admission, none of the 30 patients who had had both normal DST results and serum cholesterol concentrations above the lower third of values had committed suicide. For the 13 with low cholesterol concentrations but normal DST results, and for the 21 with higher cholesterol concentrations but abnormal DST results, survival analysis indicated cumulative probability of suicide of 10.0% and 20.9%, respectively. For the 10 who had both risk factors, the estimate was 40.9%.

These results are encouraging but await replication. An effort to consider both DST results and CSF concentrations of 5-HIAA had an even longer follow-up period and a similar number of suicides but was not supportive; 9 (28.1%) of the 32 patients who had either low 5-HIAA levels or abnormal DST result, but not both, committed suicide, whereas only 1 (11.1%) of the 9 subjects with both abnormalities did so.[100]

In both of the aforementioned studies, the occurrence of the two risk factors seemed independent of each other. A subsequent analysis of data from the second group, in fact, showed that, among the 25 who had a recent history of suicide attempts, DST nonsuppression was significantly less likely in the group with low 5-HIAA levels (1 of 13 or 7.7%) than in the group with high 5-HIAA levels (7 of 12 or 58.3%) (Fisher's exact test, $p < .01$). Another group, though, described a significantly higher rate of DST nonsuppression in suicide attempters with low 5 HIAA levels in comparison to those with higher 5-HIAA values.[31]

There exists a much larger literature concerning the relationship between HPA axis activity and the serotonin system in depressive disorders but the nature of this relationship remains uncertain. Porter et al.[101] concluded that research so far has not demonstrated a clear link between HPA axis hyperactivity and reduced 5-HT function. Within this review were six studies of post-dexamethasone cortisol concentrations and responses to 5-HTP or fenfluramine challenges. Of these, four showed no significant correlation, one showed a positive correlation, and another showed a negative correlation. On the more direct question of whether serum cholesterol differs by DST result among patients with depressive disorders, the literature is otherwise silent.

7.4 CAVEATS AND CONCLUSIONS

If further studies support serum cholesterol concentrations and DST results as clinically useful tools for the assessment of risks for suicide, important questions would remain regarding thresholds to be used to identify risk status. Jokinen et al.[102]

concluded from an ROC analysis that optimal post-dexamethasone cortisol thresholds for suicide prediction were 3.3 µg/dL for females and 7.3 µg/dL for males. An alternative threshold of 10 µg/dL was tested in two cohorts and appeared to provide a better separation of suicides from survivors in one[103] but not in the other.[22] Jokinen et al.[20] found cortisol values drawn at 4 p.m. to be more predictive than those drawn at 8 a.m. or 11 p.m., but Coryell et al.[22] found that the 11 p.m. results were the most predictive.

An inspection of Figure 7.1 illustrates that the selection of a high-risk threshold for serum cholesterol concentration poses a greater problem. An arrangement of studies in order of mean serum cholesterol concentrations in the suicide attempter groups shows that the mean values for some attempter groups exceed those for nonattempters in other studies. Variance in sample ages and prevailing dietary customs are likely to play a large role in this distribution. One solution would be to use national or regional norms to specify thresholds for low cholesterol levels within age groups. A simpler approach that would sacrifice some sensitivity and specificity would be to simply designate 60 mg/dL as the upper threshold for high-risk status.

This review has summarized the now extensive evidence linking both HPA axis hyperactivity and low serum cholesterol to a heightened risk for suicide. The decision to include either measure in clinical estimations of immediate or long-term risk for suicide must take into account substantial uncertainties as to thresholds for high risk and the subgroups for which these measures are, or are not, meaningful. Answers to these questions will require prospective studies of large samples, and evidence so far indicates that such samples should be drawn from inpatients with major depressive disorder admitted because of suicidal threats or behaviors.

Even if such efforts confirmed the robustness of these risk factors relative to such routinely used considerations as sex, age, symptom severity, and the nature of suicide thoughts and past behaviors, and even if they clarified questions as to risk thresholds, it is likely that widespread skepticism toward their clinical use will exist. Reasons include the lack of precedence for the use of biological measures in psychiatry. While laboratory measures are often used to rule out medical explanations for psychiatric presentations, no biological measures are in wide use for diagnostic or prognostic purposes. In the case of the DST as a clinical tool, reluctance would be compounded by a history in which considerable enthusiasm over its use as a diagnostic test for melancholia was followed by widespread disillusionment.

Also relevant would be the nihilism often expressed over the study of risk factors for suicide. It is often said, "suicide cannot be predicted," but this imbues the concept of prediction with certainty. Nothing can be predicted with certainty but physicians must frequently estimate the likelihood of given outcomes even if such outcomes occur in only small proportions of at-risk individuals. This is no less true of suicide as an outcome since such estimations strongly influence decisions as to hospitalization, discharge, monitoring intensity, and the potential lethality of medications prescribed. To ignore risk factors in making such judgments would be indefensible. The evidence presented here suggests that abnormal DST results and low serum cholesterol may provide additional information that would, in the context of other risk factors, permit a more accurate estimation of true risk status.

REFERENCES

1. Bunney, W. E., Jr. and Fawcett, J. A. 1965. Possibility of a biochemical test for suicidal potential: An analysis of endocrine findings prior to three suicides. *Arch Gen Psychiatry* 13:232–239.
2. Bunney, W. E., Jr., Fawcett, J. A., Davis, J. M., and Gifford, S. 1969. Further evaluation of urinary 17-hydroxycorticosteroids in suicidal patients. *Arch Gen Psychiatry* 21:138–150.
3. Krieger, G. 1970. Biochemical predictors of suicide. *Dis Nerv Syst* 31:478–482.
4. Ostroff, R., Giller, E., Bonese, K., Ebersole, E., Harkness, L., and Mason, J. 1982. Neuroendocrine risk factors of suicidal behavior. *Am J Psychiatry* 139:1323–1325.
5. Dorovini-Zis, K. and Zis, A. P. 1987. Increased adrenal weight in victims of violent suicide. *Am J Psychiatry* 144:1214–1225.
6. Dumser, T., Barocka, A., and Schubert, E. 1998. Weight of adrenal glands may be increased in persons who commit suicide. *Am J Forensic Med Pathol* 19:72–76.
7. Szigethy, E., Conwell, Y., Forbes, N. T., Cox, C., and Caine, E. D. 1994. Adrenal weight and morphology in victims of completed suicide. *Biol Psychiatry* 36:374–380.
8. Hoch-Ligeti, C. 1966. Adrenal cholesterol concentration in cases of suicide. *Br J Exp Pathol* 47:594–598.
9. Willenberg, H. S., Bornstein, S. R., Dumser, T. et al. 1998. Morphological changes in adrenals from victims of suicide in relation to altered apoptosis. *Endocr Res* 24:963–967.
10. Arato, M., Banki, C. M., Bissette, G., and Nemeroff, C. B. 1989. Elevated CSF CRF in suicide victims. *Biol Psychiatry* 25:355–359.
11. Merali, Z., Du, L., Hrdina, P. et al. 2004. Dysregulation in the suicide brain: mRNA expression of corticotropin-releasing hormone receptors and GABA(A) receptor subunits in frontal cortical brain region. *J Neurosci* 24:1478–1485.
12. Austin, M. C., Janosky, J. E., and Murphy, H. A. 2003. Increased corticotropin-releasing hormone immunoreactivity in monoamine-containing pontine nuclei of depressed suicide men. *Mol Psychiatry* 8:324–332.
13. Merali, Z., Kent, P., Du, L. et al. 2006. Corticotropin-releasing hormone, arginine vasopressin, gastrin-releasing peptide, and neuromedin B alterations in stress-relevant brain regions of suicides and control subjects. *Biol Psychiatry* 59:594–602.
14. Nemeroff, C. B., Owens, M. J., Bissette, G., Andorn, A. C., and Stanley, M. 1988. Reduced corticotropin releasing factor binding sites in the frontal cortex of suicide victims. *Arch Gen Psychiatry* 45:577–579.
15. Hucks, D., Lowther, S., Crompton, M. R., Katona, C. L., and Horton, R. W. 1997. Corticotropin-releasing factor binding sites in cortex of depressed suicides. *Psychopharmacology (Berl)* 134:174–178.
16. Charlton, B., Cheetham, S. C., Horton, R. W., Kataona, C. L. E., Pormptom, M. R., and Ferrier, I. N. 1988. Corticotropin-releasing-factor immunoreactivity in post-mortem brain from patients with depressed suicides. *Psychopharmacology (Berl)* 2:13–18.
17. Traskman-Bendz, L., Alling, C., Oreland, L., Regnell, G., Vinge, E., and Ohman, R. 1992. Prediction of suicidal behavior from biologic tests. *J Clin Psychopharmacol* 12:21S–26S.
18. Black, D. W., Monahan, P. O., and Winokur, G. 2002. The relationship between DST results and suicidal behavior. *Ann Clin Psychiatry* 14:83–88.
19. Coryell, W. and Schlesser, M. 2001. The dexamethasone suppression test and suicide prediction. *Am J Psychiatry* 158:748–753.
20. Jokinen, J., Carlborg, A., Martensson, B., Forslund, K., Nordstrom, A. L., and Nordstrom, P. 2007. DST non-suppression predicts suicide after attempted suicide. *Psychiatry Res* 150:297–303.
21. Yerevanian, B. I., Feusner, J. D., Koek, R. J., and Mintz, J. 2004. The dexamethasone suppression test as a predictor of suicidal behavior in unipolar depression. *J Affect Disord* 83:103–108.

22. Coryell, W., Young, E., and Carroll, B. 2006. Hyperactivity of the hypothalamic–pituitary–adrenal axis and mortality in major depressive disorder. *Psychiatry Res* 142:99–104.
23. Carroll, B. J., Greden, J. F., and Feinberg, M. 1980. Suicide, neuroendocrine dysfunction and CSF 5-HIAA concentrations in depression. *Recent advances in Neuropsychopharmacology, Proceedings of 12th CINP Congress*, ed. B. Angrist, Pergamon Press, Oxford, U.K.
24. Roy, A., Agren, H., Picker, D., Linnoila, M., Doran, A. R., Cutlor, N. R., and Paul, S. M. 1986. Reduced concentrations of homovanillic acid and homovanillic acid to 5-hydroxyindoleacetic acid ratios in depressed patients: Relationship to suicidal behavior and dexamethasone nonsuppression. *Am J Psych* 143:1539–1545.
25. Coryell, W. and Schlesser, M. A. 1981. Suicide and the dexamethasone suppression test in unipolar depression. *Am J Psych* 138:1120–1121.
26. Roza, R., Milanes, F., Reisch, J., Slater, V., and Garrigo, L. 1988. The DST and suicide among depressed alcoholic patients. *Am J Psych* 145:266–276.
27. Mann, J. J. and Currier, D. 2007. A review of prospective studies of biologic predictors of suicidal behavior in mood disorders. *Arch Suicide Res* 11:3–16.
28. Norman, W. H., Brown, W. A., Miller, I. W., Keitner, G. I., and Overholser, J. C. 1990. The dexamethasone suppression test and completed suicide. *Acta Psychiatr Scand* 81:120–125.
29. Coryell, W. 1990. DST abnormality as a predictor of course in major depression. *J Affect Disord* 19:163–169.
30. Roy, A. 1992. Hypothalamic–pituitary–adrenal axis function and suicidal behavior in depression. *Biol Psychiatry* 32:812–816.
31. Westrin, A. and Nimeus, A. 2003. The dexamethasone suppression test and CSF-5-HIAA in relation to suicidality and depression in suicide attempters. *Eur Psychiatry* 18:166–171.
32. Coryell, W. and Young, E. A. 2005. Clinical predictors of suicide in primary major depressive disorder. *J Clin Psychiatry* 66:412–417.
33. Coryell, W., Smith, R., Cook, B., Moucharafieh, S., Dunner, F., and House, D. 1983. Serial dexamethasone suppression test results during antidepressant therapy: Relationship to diagnosis and clinical change. *Psychiatry Res* 10:165–174.
34. Coryell, W. and Schlesser, M. A. 1983. Dexamethasone suppression test response in major depression: Stability across hospitalizations. *Psychiatry Res* 8:179–189.
35. Coryell, W., Turner, R. D., and Sherman, A. 1987. Desipramine plasma levels and clinical response: Evidence for a curvilinear relationship. *J Clin Psychopharmacol* 7:138–142.
36. Muldoon, M. F., Manuck, S. B., and Matthews, K. A. 1990. Lowering cholesterol concentrations and mortality: A quantitative review of primary prevention trials. *Br Med J* 301:309–314.
37. Boscarino, J. A., Erlich, P. M., and Hoffman, S. N. 2009. Low serum cholesterol and external-cause mortality: Potential implications for research and surveillance. *J Psychiatr Res* 43:848–854.
38. Chen, Z., Peto, R., Collins, R., Macmahon, S., Lu, J., and Li, W. 1991. Serum cholesterol concentration and coronary heart disease in population with low cholesterol concentrations. *Br Med J* 303:276–282.
39. Ellison, L. F. and Morrison, H. I. 2001. Low serum cholesterol concentration and risk of suicide. *Epidemiology* 12:168–172.
40. Lindberg, G., Rastam, L., Gullberg, B., and Eklund, G. A. 1992. Low serum cholesterol concentration and short term mortality from injuries in men and women. *Br Med J* 305:277–279.
41. Neaton, J. D., Blackburn, H., Jacobs, D. et al. 1992. Serum cholesterol level and mortality findings for men screened in the Multiple Risk Factor Intervention Trial. Multiple Risk Factor Intervention Trial Research Group. *Arch Intern Med* 152:1490–1500.

42. Partonen, T., Haukka, J., Virtamo, J., Taylor, P. R., and Lonnqvist, J. 1999. Association of low serum total cholesterol with major depression and suicide. *Br J Psychiatry* 175:259–262.
43. Tamosiunas, A., Reklaitiene, R., Radisauskas, R., and Jureniene, K. 2005. Prognosis of risk factors and trends in mortality from external causes among middle-aged men in Lithuania. *Scand J Public Health* 33:190–196.
44. Zureik, M., Courbon, D., and Ducimetiere, P. 1996. Serum cholesterol concentration and death from suicide in men: Paris prospective study I. *Br Med J* 313:649–651.
45. Smith, G. D., Shipley, M. J., Marmot, M. G., and Rose, G. 1992. Plasma cholesterol concentration and mortality. The Whitehall Study. *JAMA* 267:70–76.
46. Iribarren, C., Reed, D. M., Wergowske, G., Burchfiel, C. M., and Dwyer, J. H. 1995. Serum cholesterol level and mortality due to suicide and trauma in the Honolulu Heart Program. *Arch Intern Med* 155:695–700.
47. Tanskanen, A., Vartiainen, E., Tuomilehto, J., Viinamaki, H., Lehtonen, J., and Puska, P. 2000. High serum cholesterol and risk of suicide. *Am J Psychiatry* 157:648–650.
48. Atmaca, M., Kuloglu, M., Tezcan, E., and Ustundag, B. 2003. Serum leptin and cholesterol levels in schizophrenic patients with and without suicide attempts. *Acta Psychiatr Scand* 108:208–214.
49. Favaro, A., Caregaro, L., Di Pascoli, L., Brambilla, F., and Santonastaso, P. 2004. Total serum cholesterol and suicidality in anorexia nervosa. *Psychosom Med* 66:548–552.
50. Fiedorowicz, J. G. and Coryell, W. H. 2007. Cholesterol and suicide attempts: A prospective study of depressed inpatients. *Psychiatry Res* 152:11–20.
51. Garland, M., Hickey, D., Corvin, A. et al. 2000. Total serum cholesterol in relation to psychological correlates in parasuicide. *Br J Psychiatry* 177:77–83.
52. Guillem, E., Pelissolo, A., Notides, C., and Lepine, J. P. 2002. Relationship between attempted suicide, serum cholesterol level and novelty seeking in psychiatric in-patients. *Psychiatry Res* 112:83–88.
53. Kim, Y. K., Lee, H. J., Kim, J. Y., Yoon, D. K., Choi, S. H., and Lee, M. S. 2002. Low serum cholesterol is correlated to suicidality in a Korean sample. *Acta Psychiatr Scand* 105:141–148.
54. Kim, Y. K. and Myint, A. M. 2004. Clinical application of low serum cholesterol as an indicator for suicide risk in major depression. *J Affect Disord* 81:161–166.
55. Lee, H. J. and Kim, Y. K. 2003. Serum lipid levels and suicide attempts. *Acta Psychiatr Scand* 108:215–221.
56. Marcinko, D., Marcinko, V., Karlovic, D. et al. 2008. Serum lipid levels and suicidality among male patients with schizoaffective disorder. *Prog Neuropsychopharmacol Biol Psychiatry* 32:193–196.
57. Marcinko, D., Pivac, N., Martinac, M., Jakovljevic, M., Mihaljevic-Peles, A., and Muck-Seler, D. 2007. Platelet serotonin and serum cholesterol concentrations in suicidal and non-suicidal male patients with a first episode of psychosis. *Psychiatry Res* 150:105–108.
58. Marcinko, D., Popovic-Knapic, V., Franic, T. et al. 2008. Association of cholesterol and socio-demographic parameters with suicidality in the male patients with schizophrenia. *Psychiatr Danub* 20:390–395.
59. Modai, I., Valevski, A., Dror, S., and Weizman, A. 1994. Serum cholesterol levels and suicidal tendencies in psychiatric inpatients. *J Clin Psychiatry* 55:252–254.
60. Ozer, O. A., Kutanis, R., Agargun, M. Y. et al. 2004. Serum lipid levels, suicidality, and panic disorder. *Compr Psychiatry* 45:95–98.
61. Plana, T., Gracia, R., Mendez, I., Pintor, L., Lazaro, L., and Castro-Fornieles, J. 2010. Total serum cholesterol levels and suicide attempts in child and adolescent psychiatric inpatients. *Eur Child Adolesc Psychiatry* 19:615–619.
62. Sarchiapone, M., Camardese, G., Roy, A. et al. 2001. Cholesterol and serotonin indices in depressed and suicidal patients. *J Affect Disord* 62:217–219.
63. Sullivan, P. F., Joyce, P. R., Bulik, C. M., Mulder, R. T., and Oakley-Browne, M. 1994. Total cholesterol and suicidality in depression. *Biol Psychiatry* 36:472–477.

64. Vuksan-Cusa, B., Marcinko, D., Nad, S., and Jakovljevic, M. 2009. Differences in cholesterol and metabolic syndrome between bipolar disorder men with and without suicide attempts. *Prog Neuropsychopharmacol Biol Psychiatry* 33:109–112.
65. Bocchetta, A., Chillotti, C., Carboni, G., Oi, A., Ponti, M., and Del Zompo, M. 2001. Association of personal and familial suicide risk with low serum cholesterol concentration in male lithium patients. *Acta Psychiatr Scand* 104:37–41.
66. Golier, J. A., Marzuk, P. M., Leon, A. C., Weiner, C., and Tardiff, K. 1995. Low serum cholesterol level and attempted suicide. *Am J Psychiatry* 152:419–423.
67. Almeida-Montes, L. G., Valles-Sanchez, V., Moreno-Aguilar, J. et al. 2000. Relation of serum cholesterol, lipid, serotonin and tryptophan levels to severity of depression and to suicide attempts. *J Psychiatry Neurosci* 25:371–377.
68. Apter, A., Laufer, N., Bar-Sever, M., Har-Even, D., Ofek, H., and Weizman, A. 1999. Serum cholesterol, suicidal tendencies, impulsivity, aggression, and depression in adolescent psychiatric inpatients. *Biol Psychiatry* 46:532–541.
69. Deisenhammer, E. A., Kramer-Reinstadler, K., Liensberger, D., Kemmler, G., Hinterhuber, H., and Fleischhacker, W. W. 2004. No evidence for an association between serum cholesterol and the course of depression and suicidality. *Psychiatry Res* 121:253–261.
70. Huang, T. and Wu, S. 2000. Serum cholesterol levels in paranoid and non-paranoid schizophrenia associated with physical violence or suicide attempts in Taiwanese. *Psychiatry Res* 96:175–178.
71. Huang, T. L. 2005. Serum lipid profiles in major depression with clinical subtypes, suicide attempts and episodes. *J Affect Disord* 86:75–79.
72. Papassotiropoulos, A., Hawellek, B., Frahnert, C., Rao, G. S., and Rao, M. L. 1999. The risk of acute suicidality in psychiatric inpatients increases with low plasma cholesterol. *Pharmacopsychiatry* 32:1–4.
73. Perez-Rodriguez, M. M., Baca-Garcia, E., Diaz-Sastre, C. et al. 2008. Low serum cholesterol may be associated with suicide attempt history. *J Clin Psychiatry* 69:1920–1927.
74. Roy, A., Gonzalez, B., Marcus, A., and Berman, J. 2001. Serum cholesterol, suicidal behavior and impulsivity in cocaine-dependent patients. *Psychiatry Res* 101:243–247.
75. Alvarez, J. C., Cremniter, D., Gluck, N. et al. 2000. Low serum cholesterol in violent but not in non-violent suicide attempters. *Psychiatry Res* 95:103–108.
76. Atmaca, M., Kuloglu, M., Tezcan, E., and Ustundag, B. 2008. Serum leptin and cholesterol values in violent and non-violent suicide attempters. *Psychiatry Res* 158:87–91.
77. Gallerani, M., Manfredini, R., Caracciolo, S., Scapoli, C., Molinari, S., and Fersini, C. 1995. Serum cholesterol concentrations in parasuicide. *Br Med J* 310:1632–1636.
78. Tripodianakis, J., Markianos, M., Sarantidis, D., and Agouridaki, M. 2002. Biogenic amine turnover and serum cholesterol in suicide attempt. *Eur Arch Psychiatry Clin Neurosci* 252:38–43.
79. Lalovic, A., Levy, E., Luheshi, G. et al. 2007. Cholesterol content in brains of suicide completers. *Int J Neuropsychopharmacol* 10:159–166.
80. Repo-Tiihonen, E., Halonen, P., Tiihonen, J., and Virkkunen, M. 2002. Total serum cholesterol level, violent criminal offences, suicidal behavior, mortality and the appearance of conduct disorder in Finnish male criminal offenders with antisocial personality disorder. *Eur Arch Psychiatry Clin Neurosci* 252:8–11.
81. Maes, M., Scharpe, S., D'hondt, P. et al. 1996. Biochemical, metabolic and immune correlates of seasonal variation in violent suicide: A chronoepidemiologic study. *Eur Psychiatry* 11:21–33.
82. Edgar, P. F., Hooper, A. J., Poa, N. R., and Burnett, J. R. 2007. Violent behavior associated with hypocholesterolemia due to a novel *APOB* gene mutation. *Mol Psychiatry* 12:258–263; 221.
83. Lalovic, A., Merkens, L., Russell, L. et al. 2004. Cholesterol metabolism and suicidality in Smith–Lemli–Opitz syndrome carriers. *Am J Psychiatry* 161:2123–2126.

84. Golomb, B. A. 1998. Cholesterol and violence: Is there a connection? *Ann Intern Med* 128:478–487.
85. Golomb, B. A., Stattin, H., and Mednick, S. 2000. Low cholesterol and violent crime. *J Psychiatr Res* 34:301–309.
86. Asellus, P., Nordstrom, P., and Jokinen, J. 2010. Cholesterol and CSF 5-HIAA in attempted suicide. *J Affect Disord* 125:388–392.
87. Kaplan, J. R., Shively, C. A., Fontenot, M. B. et al. 1994. Demonstration of an association among dietary cholesterol, central serotonergic activity, and social behavior in monkeys. *Psychosom Med* 56:479–484.
88. Engstrom, G., Alsen, M., Regnell, G., and Traskman-Bendz, L. 1995. Serum lipids in suicide attempters. *Suicide Life Threat Behav* 25:393–400.
89. Hibbeln, J. R., Umhau, J. C., George, D. T., Shoaf, S. E., Linnoila, M., and Salem, N., Jr. 2000. Plasma total cholesterol concentrations do not predict cerebrospinal fluid neurotransmitter metabolites: Implications for the biophysical role of highly unsaturated fatty acids. *Am J Clin Nutr* 71:331S–338S.
90. Ringo, D. L., Lindley, S. E., Faull, K. F., and Faustman, W. O. 1994. Cholesterol and serotonin: Seeking a possible link between blood cholesterol and CSF 5-HIAA. *Biol Psychiatry* 35:957–959.
91. Delva, N. J., Matthews, D. R., and Cowen, P. J. 1996. Brain serotonin (5-HT) neuroendocrine function in patients taking cholesterol-lowering drugs. *Biol Psychiatry* 39:100–106.
92. Steegmans, P. H., Fekkes, D., Hoes, A. W., Bak, A. A., Van Der Does, E., and Grobbee, D. E. 1996. Low serum cholesterol concentration and serotonin metabolism in men. *Br Med J* 312:221.
93. Muldoon, M. F., Kaplan, J. R., Manuck, S. B., and Mann, J. J. 1992. Effects of a low-fat diet on brain serotonergic responsivity in cynomolgus monkeys. *Biol Psychiatry* 31:739–742.
94. Papakostas, G. I., Petersen, T., Mischoulon, D. et al. 2003. Serum cholesterol and serotonergic function in major depressive disorder. *Psychiatry Res* 118:137–145.
95. Terao, T., Nakamura, J., Yoshimura, R. et al. 2000. Relationship between serum cholesterol levels and meta-chlorophenylpiperazine-induced cortisol responses in healthy men and women. *Psychiatry Res* 96:167–173.
96. Vevera, J., Fisar, Z., Kvasnicka, T. et al. 2005. Cholesterol-lowering therapy evokes time-limited changes in serotonergic transmission. *Psychiatry Res* 133:197–203.
97. Kaplan, J. R., Manuck, S. B., and Shively, C. 1991. The effects of fat and cholesterol on social behavior in monkeys. *Psychosom Med* 53:634–642.
98. Fawcett, J., Busch, K. A., Jacobs, D., Kravitz, H. M., and Fogg, L. 1997. Suicide: A four-pathway clinical-biochemical model. *Ann N Y Acad Sci* 836:288–301.
99. Coryell, W. and Schlesser, M. 2007. Combined biological tests for suicide prediction. *Psychiatry Res* 150:187–191.
100. Jokinen, J., Nordstrom, A. L., and Nordstrom, P. 2009. CSF 5-HIAA and DST non-suppression–orthogonal biologic risk factors for suicide in male mood disorder inpatients. *Psychiatry Res* 165:96–102.
101. Porter, R. J., Gallagher, P., Watson, S., and Young, A. H. 2004. Corticosteroid-serotonin interactions in depression: A review of the human evidence. *Psychopharmacology (Berl)* 173:1–17.
102. Jokinen, J., Nordstrom, A. L., and Nordstrom, P. 2008. ROC analysis of dexamethasone suppression test threshold in suicide prediction after attempted suicide. *J Affect Disord* 106:145–152.
103. Coryell, W., Fiedorowicz, J., Zimmerman, M., and Young, E. 2008. HPA-axis hyperactivity and mortality in psychotic depressive disorder: Preliminary findings. *Psychoneuroendocrinology* 33:654–658.

8 Brain-Derived Neurotrophic Factor in Suicide Pathophysiology

Yogesh Dwivedi

CONTENTS

8.1 INTRODUCTION

An emerging hypothesis suggests that the pathogenesis of suicidal behavior and depression involves altered neural plasticity (Garcia, 2002), resulting in the inability of the brain to make appropriate adaptive responses to environmental stimuli (Duman et al., 2000; Fossati et al., 2004). This hypothesis is supported by studies showing altered brain structure during stress and in depressed and suicidal patients. These alterations include a reduction in cell number, density, cell body size, and neuronal and glial density in frontal cortical or hippocampal brain areas and a decrease in parahippocampal cortex cortical/laminar thickness (Altshuler et al., 1990; Rajkowska, 1997, 2000, 2002; Ongur et al., 1998; Rosoklija et al., 2000; Cotter et al., 2001, 2002; Miguel-Hidalgo and Rajkowska, 2002). In addition, changes in synaptic circuitry (Aganova and Uranova, 1992), decreased dorsolateral prefrontal cortical (PFC) activity (Dolan et al., 1993; Drevets et al., 1998), impaired synaptic connectivity between the frontal lobe and other brain regions (Andreasen, 1997; Honer, 1999), changes in the number and shape of dendritic spines (Toni et al., 1999; Hajsza et al., 2005), changes in the primary location of synapse formation, altered dendritic morphological characteristics of neurons in the hippocampus, a decrease in length and number of apical dendrites (McEwen, 2000), neuronal atrophy and a decreased volume of the hippocampus (Sheline, 2000; Sala et al., 2004; Frodl et al., 2006), a decreased number of neurons and glia in cortical areas (Rajkowska and Miguel-Hidalgo, 2007), and spatial cognition deficits (Sackeim, 2001) have also been reported during stress and depression. Furthermore, stress, a major factor in suicide, hinders performance on hippocampal-dependent memory tasks and impairs induction of hippocampal long-term potentiation. These studies clearly demonstrate impaired structural and functional plasticity in depression and suicide.

Neurotrophins (NTs) are growth factors that are critical in regulating structural, synaptic, and morphological plasticity and in modulating the strength and number of synaptic connections and neurotransmission (Thoenen, 2000). In addition, the role of NTs in the adult central nervous system is important because they participate in the maintenance of neuronal functions, the structural integrity of neurons, and neurogenesis (Cooper et al., 1996), suggesting their biological role during the entire life span. NTs are homodimeric proteins and are categorized into four different classes: nerve growth factor, brain-derived neurotrophic factor (BDNF), NT-3, and NT-4/5. Most functions of NTs are mediated by the tropomycin receptor kinase (Trk) family of tyrosine kinase receptors. The interaction of NTs with the Trk receptors is specific: nerve growth factor binds with TrkA, BDNF and NT-4 both bind to TrkB, and NT-3 binds to TrkC with the highest affinity but can also bind and mediate its actions via TrkA and TrkB receptors. All NTs can bind to the pan75 NT receptor ($p75^{NTR}$), which plays a role in NT transport, ligand-binding specificity, and Trk functioning (Barbacid, 1994, 1995; Lewin and Barde, 1996; Schweigreiter, 2006). In addition to the full-length TrkB receptor, several noncatalytic truncated TrkB isoforms have also been identified; these isoforms lack the signaling domain, preventing the induction of a signal transduction mechanism. Binding of an NT to the appropriate Trk receptor leads to the dimerization and transphosphorylation of tyrosine residues in the intracellular domain of the Trk receptors and subsequent activation of signaling pathways (Chao et al., 2006; Reichardt, 2006), leading to altered transcription of critical genes.

The most widely distributed member of NT family is BDNF (Huang and Reichardt, 2001). The *BDNF* gene lies on chromosome 11p13 and encodes pro-BDNF, a precursor peptide of mature BDNF (Seidah et al., 1996). BDNF is translated as 30–35 kDa preproproteins consisting of a preprodomain, a prodomain, and a C-terminal mature NT domain. BDNF levels and its intracellular localization in neurons are regulated via several different mechanisms, including BDNF transcripts, messenger RNA (mRNA) and protein transport, and regulated cleavage of pro-BDNF to mature BDNF. The pro-BDNF is produced in the endoplasmic reticulum, which is accumulated in the trans-Golgi network via the Golgi apparatus. Pro-BDNF can be cleaved in the endoplasmic reticulum by furin or in the regulated secretary vesicles by proconvertase enzymes. Pro-BDNF binds to sortilin, an intracellular chaperone that binds to the prodomain of BDNF to traffic it to the regulated secretory pathway, in the Golgi apparatus. This facilitates the correct folding of the mature BDNF domain. The mature domain of BDNF binds to carboxypeptidase E, thereby sorting BDNF to the regulated secretary pathway (Lu et al., 2005). Pro-BDNF can also be processed by serine protease plasmin when pro-BDNF is in the extracellular milieu (Pang et al., 2004). A substitution of valine (Val) to methionine (Met) at codon 66 in the prodomain impairs this sorting of BDNF (Egan et al., 2003). The impact of a

BDNF Val66Met polymorphism has been widely studied in relation to the clinical characteristics of suicidal behavior (Zai et al., 2011).

The expression of the *BDNF* gene is tightly regulated by neuronal activity, through mechanisms dependent on calcium (Mellstrom et al., 2004). In addition to BDNF, the function of a receptor for BDNF (i.e., TrkB) is also regulated in an activity-dependent manner. TrkB is primarily localized in the synaptic sites. Further localization of TrkB occurs at the synaptic sites after neuronal activity (Lu et al., 2005). Neuronal activity, therefore, is critical for synthesis and intracellular targeting of TrkB receptors (Lu et al., 2005). Thus, BDNF release and expression of TrkB receptors in a coordinated fashion are important for optimal synaptic response.

BDNF is directly involved in neurite outgrowth, phenotypic maturation, morphological plasticity, and synthesis of proteins for differentiated functioning of neurons and for synaptic functioning (Huang and Reichardt, 2001). BDNF is also involved in nerve regeneration, structural integrity, and maintenance of neuronal plasticity in adult brain, including regulation of synaptic activity, and in neurotransmitter synthesis (Reichardt, 2006). Thus, a pathological alteration of the BDNF may lead to defects in neural maintenance and regeneration and, therefore, structural abnormalities in the brain and may also reduce neural plasticity and, therefore, impair the individual's ability to adapt to crisis situations. Because of the role played by BDNF in regulating structural, synaptic, and morphological plasticity, there has been great interest in its role in the pathogenic mechanisms of depression (Dwivedi, 2010). The role of BDNF in depression has gained broad attention because many preclinical and clinical studies indicate that depression is associated with decreased expression of BDNF and that antidepressants alleviate depressive behavior by increasing its level (Duman and Monteggia, 2006; Brunoni et al., 2008). More recently, it has been suggested that the relationship of BDNF with depression is not straightforward, although the role of BDNF in the mechanisms of action of antidepressants is consistent (Castrén, 2005).

Although a high rate of depressive disorder is one of the main causes of increased mortality among suicide victims (Marttunen et al., 1991), only a minority of people with such diagnoses commit suicide. This indicates that there is a certain predisposition to suicide that may be independent of the main psychiatric outcome (Mann, 1998, 2002; Blumenthal and Kupfer, 1999; Turecki, 2005). Therefore, in recent years, there has been great interest in examining the role of BDNF in suicide, independent of depression. In this chapter, I discuss the recent findings of a possible role of BDNF in suicide pathogenesis.

8.2 STRESS AND BDNF

Stress plays an important role in suicidal behavior, and several studies demonstrate that stress poses a major risk factor in suicide (reviewed in Mann, 2002). A hyperactive hypothalamic–pituitary–adrenal axis and suicidal behavior are well correlated, such that elevated corticotrophin-releasing hormone levels in the cerebrospinal fluid, reduced corticotrophin-releasing hormone binding sites in the frontal cortex, augmented pro-opiate-melanocortic RNA density in the pituitary gland, large corticotrophic cell size, and alterations in the mineralocorticoid to glucocorticoid receptor mRNA ratio in the hippocampus of subjects who committed suicide

(Nemeroff et al., 1988; Lopez et al., 1992; Szigethy et al., 1994; Dumser et al., 1998) have been demonstrated. Also, a consistent association has been found between subsequently completed suicide and nonsuppression of cortisol using the dexamethasone suppression test (Coryell and Schlesser, 2001; Jokinen et al., 2007, 2009).

The first study showing the role of BDNF in stress was from Smith et al. (1995), who showed that immobilization stress lowers the expression of BDNF in the hippocampus, most notably in the dentate gyrus. Subsequently, this was confirmed by many investigators (Ueyama et al., 1997; Fuchikami et al., 2009). By using a different stress paradigm, Rasmusson et al. (2002) found that exposure to foot shocks decreased BDNF mRNA in rat dentate gyrus. Other stressors, such as social defeat, also decreased BDNF in the hippocampus, but interestingly, this decrease extended to cortical and subcortical regions (Pizarro et al., 2004). Because stress is associated with elevated levels of glucocorticoids, several studies have examined the effect of exogenous glucocorticoids on BDNF expression. For example, corticosterone (CORT) treatment to rats reduces BDNF expression in the hippocampus (Smith et al., 1995; Schaaf et al., 1998). In a recent study, we extensively examined the effects of CORT on BDNF expression in the rat brain and found that implantation of the CORT pellet decreased the expression of BDNF in the hippocampus and in the frontal cortex (Dwivedi et al., 2006). Interestingly, adrenalectomy increased the level of BDNF in the hippocampus (Barbany and Persson, 1992; Chao et al., 1998), whereas supplementation of the synthetic glucocorticoid dexamethasone to adrenalectomized rats restored the level of BDNF to normal (Barbany and Persson, 1992). These studies suggest that CORT plays a critical role in regulating the synthesis of BDNF.

We further examined the molecular basis of decreased BDNF synthesis in response to CORT treatment. The rat *BDNF* gene contains four distinct promoters that are linked to four main transcript forms (Nakayama et al., 1994). Each transcript has four short 5′ noncoding exons (I–IV) containing separate promoters and one shared 3′ exon (exon V) encoding the mature BDNF protein. These transcripts facilitate multilevel regulation of BDNF expression and determine the tissue-specific expression. Although the functions of each BDNF transcript are not clearly known, BDNF transcripts are differentially expressed across brain regions and are differentially regulated (Pattabiraman et al., 2005; Chiaruttini et al., 2008; Wong et al., 2009; Luberg et al., 2010). In general, exons that are closely located in the genome are expressed in a similar manner: exons I, II, and III have brain-enriched expression patterns, and exons IV, V, and VI are widely expressed and are also in nonneural tissues. Exon IV is expressed in the cell body, whereas exon V is expressed in soma and dendrites. Functionally, exon IV is involved in maturation of interneurons through a transsynaptic route, whereas exon V is involved in maturation of excitatory neurons and in dendritic synapse formation.

When we examined whether a decrease in BDNF expression by CORT is associated with alterations in the expression of a specific BDNF transcript, we found that CORT decreased the expression of selective transcripts II and IV, but not transcript I or III, in the rat frontal cortex and hippocampus (Dwivedi et al., 2006). Other studies also suggest that immobilization stress decreased exon IV in the hippocampus (Marmigere et al., 2003) and hypothalamus (Rage et al., 2002), leading to a decrease in total BDNF expression in these brain areas. These studies indicate the possible involvement of CORT in regulating the expression of specific BDNF transcripts and, thus, the expression of BDNF.

Because it has been demonstrated that antidepressants can regulate the levels of glucocorticoids (Carvalho and Pariante, 2008; Fitzsimons et al., 2009; Nikisch, 2009), we examined whether glucocorticoid-mediated down-regulation of BDNF is reversed by antidepressants and, if so, what could be the possible mechanism of BDNF regulation by antidepressants. Initially, we examined how BDNF is regulated in response to antidepressants. We treated rats with different classes of antidepressants (Dwivedi et al., 2006). We observed that desipramine (a norepinephrine blocker) and phenelzine (a monoamine oxidase inhibitor) increased mRNA levels of *BDNF* gene expression in both the frontal cortex and hippocampus, whereas fluoxetine (a serotonin reuptake blocker) increased the mRNA level of BDNF only in the hippocampus. Interestingly, we found that desipramine specifically increased the expression of BDNF transcripts I and III in both the frontal cortex and hippocampus; fluoxetine increased only exon II in the hippocampus; and phenelzine increased exons I and IV in the hippocampus but only exon I in the frontal cortex. We further examined whether antidepressants can reverse the CORT-mediated decrease in BDNF expression and, if so, whether the same BDNF transcripts regulate CORT-mediated down-regulation and antidepressant-mediated up-regulation of the *BDNF* gene (Dwivedi et al., 2006b). It was observed that all the antidepressants normalized the levels of CORT, although the degree of this reversal varied with different antidepressants. When examined, we found that desipramine reversed the CORT-induced decrease in BDNF expression in both the frontal cortex and hippocampus. Fluoxetine only partially reversed such a decrease in the hippocampus, but no effect was found in the frontal cortex. Phenelzine, on the other hand, reversed the CORT-induced decrease in BDNF partially in the frontal cortex and completely in the hippocampus. Interesting results were noted when individual BDNF transcripts were examined after antidepressant treatment to CORT-implanted rats. We found that all the antidepressants increased mRNA levels of those BDNF transcripts that were affected when the respective antidepressant was given to healthy rats without CORT treatment. For example, desipramine increased exons I and III in the frontal cortex and hippocampus, fluoxetine increased exon II in the hippocampus, and phenelzine increased exon I in the frontal cortex and exons I and IV in the hippocampus. Surprisingly, except for an increase in exon II by fluoxetine in the frontal cortex and in exon IV by phenelzine in the hippocampus, the CORT-mediated decrease in exons II and IV persisted even after antidepressant treatment. Interestingly, despite these different effects of CORT and antidepressants on BDNF transcripts, overall, all the antidepressants increased the level of BDNF mRNA in the brain of CORT-treated rats. Although it is difficult to assess the extent of involvement of a particular exon in the regulation of overall BDNF expression, there is complete reversal by desipramine in both the frontal cortex and hippocampus because the increase in exon III was robust in these brain areas. On the other hand, in the hippocampus, fluoxetine was able to reverse the CORT-mediated decrease of only exon II, but not exon IV; therefore, the reversal was partial. However, no effect of fluoxetine on total BDNF expression was observed in the frontal cortex, because fluoxetine was not able to increase either exon II or IV in the frontal cortex. On the other hand, phenelzine was partially effective in the frontal cortex because of its effects on exon II, but complete reversal was noted in the hippocampus because phenelzine increased the levels of both CORT-decreased exons II and IV. Thus, it appears that antidepressants reverse total BDNF expression in CORT-treated rats; however, the mechanisms for the down-regulation of BDNF transcripts by CORT and those that affect their up-regulation by antidepressants are different. Recently, in an attempt to identify potential biomarkers for the onset of antidepressant action in depressive patients, Rojas et al. (2011) examined several molecules, including glucocorticoid receptors and serum BDNF levels during antidepressant treatment. Thirty-four depressed outpatients were treated with venlafaxine, and individuals exhibiting a 50% reduction in their baseline 17-Item Hamilton Depression Rating Scale score by the sixth week of treatment were considered responders. These responders showed an early improvement in parallel with an increase in BDNF levels during the first 2 weeks of treatment. Nonresponders showed increased glucocorticoid receptor levels by the third week and reduced serum BDNF levels by the sixth week of treatment. The authors concluded that levels of BDNF in serum and glucocorticoid receptor levels in lymphocytes may represent biomarkers that could be used to predict responses to venlafaxine treatment.

8.3 BDNF AND SUICIDE

8.3.1 Human Postmortem Brain Studies

Studies in human postmortem tissues provide direct evidence of neurobiological abnormalities in suicide subjects. Our group was the first to examine the role of BDNF in suicide by studying the expression of BDNF in the PFC (Brodmann area 9) and hippocampus of suicide subjects and well-matched nonpsychiatric healthy controls (Dwivedi et al., 2003). We found that the mRNA level of BDNF was significantly lower in both the PFC and the hippocampus of suicide subjects. By using antibody that recognizes mature BDNF, we also found a significantly decreased protein level of BDNF in the PFC and the hippocampus of suicide subjects. The decreased protein level of BDNF was significantly correlated with its mRNA level. This suggests that there is less transcription of BDNF in brains of suicide subjects. Interestingly, when we divided suicide subjects into those who had a history of major depression and those who had other psychiatric disorders, we found no differences in the expression of BDNF in these two groups and the level of BDNF was decreased in all suicide subjects regardless of psychiatric diagnosis. Thus, our findings demonstrate that a reduced level of BDNF is associated with suicidal behavior. More recently, Karege et al. (2005) examined BDNF expression in 30 suicide victims and 24 drug-free nonsuicidal subjects, who were devoid of psychiatric or neurological disease. They found a significant decrease in BDNF level in the hippocampus and PFC, but not in the entorhinal cortex, of suicide victims who were drug free compared with nonsuicidal controls. The decrease was observed in all suicide victims, regardless of diagnosis. This study supports a role of BDNF in the pathophysiological characteristics of suicidal behavior. This study also suggests that a decrease in BDNF may be specific only to brain areas that are related to emotion and cognition. Karege et al. (2005) also found that suicide subjects who were receiving antidepressant treatment did not show any change in the level of BDNF, suggesting that psychotropic drugs normalize the decreased level of BDNF in suicide subjects. The absence of change in BDNF level of drug-treated suicide victims further suggests that BDNF may be a mediator of psychotropic drugs. Interestingly, Kozicz et al. (2008) examined the sex difference in the expression of BDNF in suicide subjects. They found that the BDNF level was much lower in the midbrain of male suicide subjects, whereas female suicide subjects showed an increased level of BDNF in this brain area, suggesting a possible sex effect in the regulation of BDNF expression in suicide subjects. Although the other studies did not find sex-specific changes in BDNF expression in the hippocampus or cortical areas of suicide subjects (Dwivedi et al., 2003; Karege et al., 2005), whether the sex-specific effect in BDNF expression is specific to the midbrain area needs to be further studied.

Because the epiphenomenon of teenage suicide may be different than that of adults, in a recent study, we attempted to delineate the pathogenic mechanisms of adult vs. teenage suicide (Pandey et al., 2008). We found that expression of BDNF mRNA was decreased in the PFC and hippocampus of teenage suicide subjects; however, the protein expression of BDNF was decreased only in the PFC, not in the hippocampus. On the other hand, we previously reported that both mRNA and protein expression are down-regulated in the hippocampus and PFC of adult suicide subjects. Thus, there is a disconnection between mRNA and protein expression of BDNF in the hippocampus of teenage suicide subjects. There is a possibility that differences in expression of BDNF between the PFC and hippocampus of teenage suicide subjects could be associated with defective translation or turnover of BDNF in the hippocampus. Further studies are needed to clarify this issue.

Recently, Keller et al. (2010) tested the hypothesis that alterations of DNA methylation could be involved in the dysregulation of *BDNF* gene expression in the brain of suicide subjects. For this, they examined DNA methylation in 44 suicide completers and 33 nonsuicidal control subjects. They found significant increased DNA methylation at specific CpG sites in BDNF promoter/exon IV. Most of the CpG sites lying in the –300/500 region, on both strands, had low or no methylation, with the exception of a few sites located near the transcriptional start site that had differential methylation, whereas genome-wide methylation levels were comparable among the subjects. The mean methylation degree at the four CpG sites analyzed by pyrosequencing was always <12.9% in the 33 nonsuicidal control subjects, whereas in 13 (30%) of 44 suicide victims, the mean methylation degree ranged between 13.1% and 34.2%. A higher methylation degree corresponded to lower BDNF mRNA levels. This study suggests that epigenetics may play a crucial role in altering the level of BDNF in suicide subjects.

8.3.2 Genetic Linkage Studies

As previously mentioned, the gene encoding human BDNF is localized at chromosome 11p13. In humans, a common single-nucleotide polymorphism at nucleotide 196 within the 5′ pro-BDNF sequence encodes a variant BDNF at codon 66 (Val66Met). This Met66 variant affects activity-dependent BDNF secretion (Egan et al., 2003) and is critical for dendritic trafficking and synaptic localization of BDNF. Interestingly, knockout mice carrying the Val66Met polymorphism show reduced activity-dependent secretion of BDNF (Chen et al., 2004). More interestingly, the BDNF *Met/Met* or *Val/Met* allele is associated with reduced hippocampus volume (Toro et al., 2009). Furthermore, the Val66Met polymorphism in the *BDNF* gene modulates human cortical plasticity and the response to transcranial magnetic stimulation (Cheeran et al., 2008).

Hong et al. (2003) studied the association between the *BDNF* gene Val66Met polymorphism and mood disorders, age of onset, and suicidal behavior in a Chinese population. They found that the genotype and allele frequencies for the *BDNF* gene Val66Met polymorphism were not different between depression groups and control subjects. Also, the *BDNF* gene was not associated with age of onset or suicidal history in patients with a mood disorder. However, recently, Kim et al. (2008) reported that, in a Korean population, although the allelic distributions did not differ between bipolar patients and healthy controls, the rate of suicide attempts among the *Val/Val*, *Val/Met*, and *Met/Met* genotype groups was significantly different. Relative to patients with the *Val/Val* genotype, those with the *Met/Met* genotype had a 4.9-fold higher risk of suicide attempts, suggesting that BDNF *Val/Met* is related to suicidal behavior in bipolar patients. Iga et al. (2007) genotyped the BDNF Val66Met polymorphism in 154 major depressive patients and 154 control subjects. They found that the genotypic distributions and allele frequencies were similar among the patients and control subjects. However, when the relationships of the polymorphism with several clinical variables were examined (i.e., age, sex, age of onset, number of episodes, presence of psychotic features, suicidal behavior, and family history), the dose of *Met* allele had significant effects on psychotic features, suicidal behavior, and family history. More recently, Sarchiapone et al. (2008) studied depressed patients for their history of suicide attempts and BDNF polymorphism. They found that there was a significantly increased risk of suicidal behavior in depressed patients who carried the BDNF Val/Met polymorphism variant (GA + AA). The risk of a suicide attempt was also significantly higher among those reporting higher levels of childhood emotional, physical, and sexual abuse. Secondary analyses suggested that depression severity was a significant risk factor only in the wild-type *BDNF* genotype and that the risk of suicide attempts was more predictable within the wild-type group. In a postmortem brain study of subjects who committed suicide, Zarrilli et al. (2008) found no significant association of the BDNF Val/Met polymorphism and suggested that completed and attempted suicide may have two distinct phenomena and that different molecular genetic components may be involved. Zarrilli et al. (2008) also analyzed two other polymorphisms in the *BDNF* gene, −270C > T and −281C > A, and found their occurrence as <5%. Vincze et al. (2008) genotyped the BDNF polymorphism in bipolar patients and healthy controls. They found *G196* alleles and severity of suicidal behavior in bipolar patients. In another study, Perroud et al. (2008) examined whether a Val/Met BDNF polymorphism could moderate the effect of childhood maltreatment on the onset, number, and violence of suicidal behavior in suicide attempters. They found that childhood sexual abuse was associated with violent suicide attempts in adulthood only among Val/Val individuals, not among Val/Met or Met/Met individuals. The severity of childhood maltreatment was significantly associated with more suicide attempts and with a younger age at onset of suicide attempt. This result suggests that Val/Met modulates the effect of childhood sexual abuse on the violence of suicidal behavior and that BDNF dysfunction may enhance the risk of violent suicidal behavior in adulthood.

Recently, de Luca et al. (2011) performed a family-based association study of the Val66Met polymorphism in nuclear families with at least one subject affected by major psychosis with suicidal behavior and compared allele-specific mRNA levels in postmortem brain samples from suicide and nonsuicidal victims. These investigators reported that allele 3 in the GT repeat polymorphism was transmitted significantly more often to patients who attempted suicide; however, there was no significant difference between maternal and paternal transmission ratios. Also, no significant differences in the ratio of Val/Met-specific mRNA expression between suicide victims and controls were noted. Thus, this study does not support a role for allelic imbalance or parent of origin of BDNF for suicidal behavior in major psychoses. In contrast, Pregelj et al. (2011) evaluated the association between BDNF Val66Met variants and suicide, committed with violent or nonviolent methods, in victims with or without stressful childhood experience and found a role of BDNF in increased vulnerability to suicide. They genotyped the BDNF Val66Met polymorphism on 560 DNA samples from 359 suicide victims and 201 control subjects and subdivided according to sex, method of suicide, and influence of childhood adversity. A similar frequency of BDNF Val66Met variants was found between all included suicide victims and the control groups and also between the male groups. The frequency of the combined *Met/Met* and *Met/Val* genotypes and the homozygous *Val/Val* genotype was significantly different between the female suicide victims and female controls, between the female suicide victims who used violent suicide methods and female controls, and between all included suicide victims with or without stressful life events. The combined *Met/Met* and *Met/Val* genotypes contributed to this significance. This study suggests that combined *Met/Met* and *Met/Val* genotypes of the BDNF Val66Met variant could be the risk factor for violent suicide in female subjects and for suicide in victims exposed to childhood trauma. These results confirm a major role of BDNF in increased vulnerability to suicide.

Schenkel et al. (2010) evaluated the impact of the BDNF Val66Met polymorphism on the clinical characteristics of suicide attempts in a cohort of 120 patients who attempted suicide. They found that sex, *BDNF* genotype, and intent and method of suicide attempt all were risk factors for high lethality in suicide attempts. Furthermore, male sex and the presence of the BDNF *66Met* allele were significantly and independently associated with the high lethality in suicide attempts, suggesting that the BDNF *66Met* allele is an independent predictor of high lethality in suicide attempts of depressed patients and, thus, may allow earlier identification of patients at high risk for suicide.

8.3.3 BDNF Studies in Suicidal Patients and Comparison of Suicidal and Nonsuicidal Depressed Patients

In search for a possible biological marker for suicidal behavior, recently, several studies attempted to examine BDNF levels in the blood cells of suicidal subjects. It has been shown that BDNF may cross the blood–brain barrier and that platelet BDNF postnatally shows changes similar to the brain (Karege et al., 2002), suggesting that there are parallel changes in the blood and brain levels of BDNF. In a group of depressed patients with a recent suicide attempt, nonsuicidal depressed patients, and healthy controls, Kim et al. (2007) measured plasma BDNF levels. They found that the BDNF level was significantly lower in suicidal depressed patients vs. nonsuicidal depressed patients or healthy controls; however, BDNF levels were not different between fatal and nonfatal suicide attempts. Similarly, Lee et al. (2007) found that the plasma BDNF level was significantly decreased in depressed suicidal patients vs. depressed nonsuicidal patients. Interestingly, Dawood et al. (2007) used direct internal jugular vein blood sampling methods to circumvent the issue of whether BDNF is released from sources other than the brain. They examined the relationship between brain BDNF production and suicide risk in patients with depression who were free of medication. They found that the venoarterial BDNF concentration gradient was significantly reduced in patients at medium to high suicide risk and that there was a significant negative correlation between suicide risk and the internal jugular venous venoarterial BDNF concentration gradient. In contrast to these studies, Deveci et al. (2007) reported that the serum BDNF level was lower in both the attempted suicide group and the depressed group vs. the control group. Similarly, the platelet BDNF level was lower in both nonsuicidal and suicidal depressed patients compared with healthy controls (Lee and Kim, 2009). More recently, these investigators (Lee and Kim, 2010) measured BDNF mRNA in peripheral blood mononuclear cells of 30 patients with major depression without recent suicide attempts, 30 patients with major depression with recent suicide attempts, and 30 healthy controls. All depressed patients were either medication naïve or medication free. They found significantly decreased BDNF mRNA expression in peripheral blood mononuclear cells of depressed patients, with or without a history of suicide attempts, when compared with healthy controls. Interestingly, the degree of decrease in BDNF was greater in suicidal depressed patients. This study further suggests an impact of suicidal behavior on the level of BDNF in peripheral cells.

8.4 BDNF RECEPTORS IN SUICIDAL BEHAVIOR

One of the major receptors to which BDNF binds with high affinity and mediates its action is TrkB. Binding of BDNF to the TrkB receptor leads to the dimerization and transphosphorylation of tyrosine residues in the intracellular domain of the TrkB receptors and to subsequent activation of cytoplasmic signaling pathways (Reichardt, 2006). The *TRKB* gene produces two isoforms: the full-length or catalytic form and the truncated or noncatalytic form. The main biological actions of BDNF are mediated via full-length TrkB, which is catalytically active. On the other hand, the truncated TrkB (TrkB.T1) lacks a large part of the intracellular domain and does not display protein–tyrosine kinase activity (Middlemas et al., 1991). Binding with BDNF leads to activation of the full-length TrkB by ligand-induced dimerization and autophosphorylation of tyrosine residues in the intracellular region. The activated TrkB interacts with and phosphorylates several intracellular targets. The truncated TrkB is also a predominant isoform in the adult brain (Armanini et al., 1995) and functions as a cellular adhesion molecule regulating synaptic plasticity and axonal outgrowth, modulating signaling by catalytic TrkB through the formation of heterodimers, and regulating the extracellular availability of its endogenous ligands (Middlemas et al., 1991). The BDNF signaling is impaired as a consequence of the formation of receptor heterodimers (Eide et al., 1996), suggesting that the truncated form of TrkB can also act as a negative modulator of BDNF signaling.

Several preclinical studies have demonstrated that antidepressants up-regulate the expression and/or activation of TrkB (Rantamäki et al., 2007; Kozisek et al., 2008a,b). In fact, ablation of TrkB renders mice behaviorally insensitive to antidepressive treatment in depression- and anxiety-like paradigms (Li et al., 2008). This prompted us to examine whether activation/expression of TrkB is altered in the brain of suicide subjects. We found that the expression of full-length TrkB was significantly lower in the PFC and hippocampus of adult suicide subjects compared with age-matched healthy controls. On the other hand, no significant change was noted in the expression of TrkB.T1. When examined in brains of teenage suicide subjects, we found similar changes in the expression of full-length TrkB (Pandey et al., 2008). Our findings suggest that suicide is associated with a decreased level of BDNF and that functions of BDNF via TrkB are also impaired. In addition, a decrease in full-length TrkB may also affect the supply of BDNF to neurons and, thus, the loss of trophic maintenance of a variety of neuronal types, because the catalytically active full-length TrkB is present predominantly within neuronal axons, cell soma, and dendrites (Fryer et al., 1996). In addition, the presence of truncated TrkB would only exacerbate any effects as a result of the loss of catalytically active full-length TrkB, because truncated TrkB inhibits BDNF-mediated neurite outgrowth via the internalization of BDNF. More recently, we examined the functional status of full-length TrkB in these suicide subjects and observed that tyrosine phosphorylation of TrkB was significantly lower in the brains of suicide subjects (Dwivedi et al., 2009). These studies suggest that both BDNF and TrkB are less expressed and that the functionality of TrkB is impaired in the suicide brain.

Recently, Ernst et al. (2009) studied TrkB.T1 in the postmortem brains of suicide subjects and found that a significant population of suicide completers had a decrease in different probe sets specific to TrkB.T1 in frontal cortical areas. The decrease was specific to the T1 splice variant. There was no effect of genetic variation in a 2500 base pair promoter region or at relevant splice junctions; however, the effect of the methylation state at CpG dinucleotides on TrkB.T1 expression was noted, suggesting that a reduction in TrkB.T1 expression in suicide subjects may be associated with the epigenetic modification of the TrkB.T1. More recently, Keller et al. (2011) investigated whether epigenetic modifications of the *TrkB* gene occur in the Wernicke area of 18 suicide subjects compared with 18 controls. They did not find any correlation between suicidal behavior and TrkB and TrkB.T1 expression and promoter methylation in the Wernicke area. On the other hand, in the same samples, the BDNF promoter IV was significantly hypermethylated in suicide subjects. This study suggests that the expression and methylation state of suicide-related genes may be specific for brain area.

To determine whether the *BDNF* gene or its high-affinity receptor gene *NTRK2* confers risk for suicide attempt and major depression, Kohli et al. (2010) investigated 83 tagging single-nucleotide polymorphisms covering the genetic variability of these loci in European populations in a case–control association design. They studied 394 depressed patients (113 were suicide attempters) and 366 matched healthy control subjects and replicated these studies in 744 German patients with depression and

921 African-American nonpsychiatric clinic patients, of whom 152 and 119 were positive for suicide attempters, respectively. They found that independent single-nucleotide polymorphisms within *NTRK2* were associated with suicide attempters among depressed patients of the discovery sample, which was confirmed in both the German and African-American replication samples. Multi-locus interaction analysis suggested that single-nucleotide polymorphism associations within this locus contributed to the risk of suicide attempts in a multiplicative and interactive fashion. These data clearly demonstrate that a combination of several independent risk alleles within the *NTRK2* locus is associated with suicide attempts in depressed patients and supports a role of NTs in the pathophysiological characteristics of suicide.

Another class of receptor to which BDNF binds is $p75^{NTR}$, which plays an important role in NT transport, ligand-binding specificity, and Trk functioning. The 3.8 kb mRNA for $p75^{NTR}$ encodes a 427 amino acid protein containing a 28 amino acid single peptide, a single transmembrane domain, and a 55 amino acid cytoplasmic domain (Hasegawa et al., 2004). Although $p75^{NTR}$ receptors do not contain a catalytic motif, they interact with several proteins, including Trk receptors, which causes enhancement of ligand specificity and ligand affinities for Trk receptors (Esposito et al., 2001). On the other hand, $p75^{NTR}$ can send negative signals. For example, $p75^{NTR}$ can cause developing hippocampal neuronal death, induced by NTs in the absence of a Trk receptor (Meldolesi et al., 2000). In the adult central nervous system, excitotoxin-induced neuronal apoptosis is accompanied by the induction of $p75^{NTR}$ in the dying neurons (Roux et al., 1999), suggesting that $p75^{NTR}$ may represent a general stress-induced apoptotic mechanism. Interestingly, the apoptotic mechanisms of $p75^{NTR}$ are active only when Trk receptors are less expressed or less active. Moreover, ectopic expression of the Trk receptor can convert a proapoptotic NT to a prosurvival NT. Thus, the ratio of expression levels and/or activation states of Trk receptors and $p75^{NTR}$ is important in NT-mediated functions. Recently, we observed that the expression ratio of $p75^{NTR}$ to Trk receptors is increased in the postmortem brain of suicide subjects. Reduced expression of NTs, together with reduced expression and activation of Trk and concomitant increased expression of $p75^{NTR}$, indicates that the possible consequence is a tipping of the balance away from cell survival, which could be associated with structural abnormalities and reduced neuronal plasticity in the suicide brain. The mechanisms responsible for $p75^{NTR}$-induced apoptotic functions are still not clear. However, several studies demonstrate that $p75^{NTR}$ includes c-Jun kinase signaling, sphingolipid turnover, and association with adaptor proteins, such as the NT receptor–interacting MAGE homolog and $p75^{NTR}$-associated cell death executor, which directly promote cell cycle arrest and apoptosis (Whitfield et al., 2001). In contrast, Trk receptors suppress c-Jun kinase and activation of sphingomyelinase, initiated by $p75^{NTR}$. Sphingomyelinase activation results in the generation of ceramide, which promotes apoptosis by inactivating extracellular signal–regulated kinase and phosphoinositide 3-kinase pathways (Zhou et al., 1998). As discussed later, we reported less-activated extracellular signal–regulated kinase 1/2 (Dwivedi et al., 2001, 2006) and phosphoinositide 3-kinase (Dwivedi et al., 2008) signaling in the postmortem brains of suicide subjects, which could be associated with less activation/expression of Trks.

In addition to Trks, pro-BDNF also plays an important role in p75^{NTR}-mediated apoptosis (Hempstead, 2002). Thus, pro-BDNF and mature BDNF can cause opposite physiological actions through binding to p75^{NTR} and TrkB receptors, respectively (Lu et al., 2005). In a recent study, we observed that the level of pro-BDNF is increased in the postmortem brains of suicide subjects (unpublished data; Dwivedi et al., 2011), whereas a genetic study suggests that the S205L polymorphism, which substitutes a serine with a leucine residue of the *p75^{NTR}* gene, is associated with attempted suicide (McGregor et al., 2007). These studies suggest that mature BDNF, pro-BDNF, and p75^{NTR} may also play an important role in suicide pathogenesis. Further studies are required to determine whether p75^{NTR}-mediated proapoptotic pathways are active in the brain of suicide subjects and how Trk- and p75^{NTR}-mediated signal transduction pathways interplay in the pathophysiological characteristics of suicide.

8.5 CONCLUSION AND FUTURE STUDIES

From the previously described studies, it is clear that BDNF is less expressed in both brain and peripheral tissues of suicide subjects. Interestingly, expression of BDNF is reduced and activation and expression of TrkB, to which BDNF binds and mediates its functions, are lower in the suicide brain. These studies indicate a possible deficit in the functioning of BDNF in suicidal patients. Because depression is the major factor in suicidal behavior, studies have shown that BDNF is down-regulated during depression. Thus, an important point that needs careful consideration is whether the findings of BDNF are linked to depressive symptoms or are specifically associated with suicidal behavior. As previously discussed, several studies have examined BDNF in suicidal and nonsuicidal depressed subjects and have shown that the down-regulation of BDNF is mainly linked to suicide attempts. Genetic studies also support this notion. Nonetheless, more comprehensive studies are required to confirm these findings.

Another important observation is that glucocorticoid affects BDNF expression, which appears to be regulated through modulation of specific BDNF transcripts. These studies suggest that an increase in cortisol/CORT during stress is critical in regulating BDNF expression. These studies also raise the possibility that each transcript may have distinct functions in the brain and may be differentially regulated. Thus, in the future, it will be important to examine the functions of each BDNF transcript and to understand the role of these transcripts in the pathophysiological characteristics of suicide.

Several other questions that need to be examined also arise from these studies. These questions include the following: (1) How does the Val66Met polymorphism induce suicidal behavior? (2) What is the significance of up-regulated pro-BDNF and p75^{NTR} in the brain of suicide subjects? (3) What is the significance of dendritic localization of TrkB and its effects on BDNF signaling in the suicide brain? Interestingly, BDNF induces the expression of Lim kinase 1, a protein kinase whose mRNA translation is inhibited by brain-specific microRNA-134. The microRNA-134 is localized in dendrites, and its overexpression leads to a decrease in spine size through repression of Lim kinase 1 mRNA translation (Schratt et al., 2006). Thus, studying BDNF/TrkB and other interacting proteins in dendrites will further reveal their novel mechanistic roles in the development of suicidal behavior.

ACKNOWLEDGMENTS

The study was supported by grants from the National Institute of Mental Health (R0168777, R21MH081099, and R01MH082802), the National Alliance for Research in Schizophrenia and Depression, and the American Foundation for Suicide Prevention.

REFERENCES

Aganova, E.A., Uranova, N.A. 1992. Morphometric analysis of synaptic contacts in the anterior limbic cortex in the endogenous psychoses. *Neurosci Behav Physiol* 22:59–65.

Altshuler, L.L., Casanova, M.F., Goldberg, T.E., Kleinman, J.E. 1990. The hippocampus and parahippocampus in schizophrenia, suicide, and control brains. *Arch Gen Psychiatry* 47:1029–1034.

Andreasen, N.C. 1997. Linking mind and brain in the study of mental illnesses: A project for a scientific psychopathology. *Science* 275:1586–1593.

Armanini, M.P., McMahon, S.B., Sutherland, J., Shelton, D.L., Philips, H.S. 1995. Truncated and catalytic isoforms of TrkB are co-expressed in neurons of rat and mouse CNS. *Eur J Neurosci* 7:1403–1409.

Barbacid, M. 1994. The Trk family of neurotrophin receptors. *J Neurobiol* 25:1386–1403.

Barbacid, M. 1995. Neurotrophic factors and their receptors. *Curr Opin Cell Biol* 7:148–155.

Barbany, G., Persson, H. 1992. Regulation of neurotrophin mRNA expression in the rat brain by glucocorticoids. *Eur J Neurosci* 4:396–403.

Blumenthal, S.J., Kupfer, D.J. 1999. *Suicide over the Life Cycle: Risk factors, Assessment, and Treatment of Suicidal Patients*. Washington, DC: The American Psychiatric Press.

Brunoni, A.R., Lopes, M., Fregni, F. 2008. A systematic review and meta-analysis of clinical studies on major depression and BDNF levels: Implications for the role of neuroplasticity in depression. *Int J Neuropsychopharmacol* 11:1169–1180.

Carvalho, L.A., Pariante, C.M. 2008. In vitro modulation of the glucocorticoid receptor by antidepressants. *Stress* 11:411–424.

Castrén, E. 2005. Is mood chemistry? *Nat Rev Neurosci* 6:241–246.

Chao, M.V., Rajagopal, R., Lee, F.S. 2006. Neurotrophin signaling in health and disease. *Clin Sci (Lond)* 110:167–173.

Chao, H.M., Sakai, R.R., Ma, L.Y., McEwen, B.S. 1998. Adrenal steroid regulation of neurotrophic factor expression in the rat hippocampus. *Endocrinology* 139:3112–3118.

Cheeran, B., Talelli, P., Mori, F., Koch, G., Suppa, A., Edwards, M., Houlden, H., Bhatia, K., Greenwood, R., Rothwell, J.C. 2008. A common polymorphism in the brain-derived neurotrophic factor gene (*BDNF*) modulates human cortical plasticity and the response to rTMS. *J Physiol* 586:5717–5725.

Chen, Z.Y., Patel, P.D., Sant, G., Meng, C.X., Teng, K.K., Hempstead, B.L., Lee, F.S. 2004. Variant brain-derived neurotrophic factor (BDNF) (Met66) alters the intracellular trafficking and activity-dependent secretion of wild-type BDNF in neurosecretory cells and cortical neurons. *J Neurosci* 24:4401–4411.

Chiaruttini, C., Sonego, M., Baj, G., Simonato, M., Tongiorgi, E. 2008. BDNF mRNA splice variants display activity-dependent targeting to distinct hippocampal laminae. *Mol Cell Neurosci* 37:11–19.

Cooper, J.D., Skepper, J.N., Berzaghi, M.D., Lindholm, D., Sofroniew, M.V. 1996. Delayed death of septal cholinergic neurons after excitotoxic ablation of hippocampal neurons during early postnatal development in the rat. *Exp Neurol* 139:143–155.

Coryell, W., Schlesser, M. 2001. The dexamethasone suppression test and suicide prediction. *Am J Psychiatry* 158:748–753.

Cotter, D., Mackay, D., Chana, G., Beasley, C., Landau, S., Everall, I.P. 2002. Reduced neuronal size and glial cell density in area 9 of the dorsolateral prefrontal cortex in subjects with major depressive disorder. *Cereb Cortex* 12:386–394.

Cotter, D., Mackay, D., Landau, S., Kerwin, R., Everall, I. 2001. Reduced glial cell density and neuronal size in the anterior cingulate cortex in major depressive disorder. *Arch Gen Psychiatry* 58:545–553.

Dawood, T., Anderson, J., Barton, D., Lambert, E., Esler, M., Hotchkin, E., Haikerwal, D., Kaye, D., Lambert, G. 2007. Reduced overflow of BDNF from the brain is linked with suicide risk in depressive illness. *Mol Psychiatry* 12:981–983.

de Luca, V., Souza, R.P., Zai, C.C., Panariello, F., Javaid, N., Strauss, J., Kennedy, J.L., Tallerico, T., Wong, A.H. 2011. Parent of origin effect and differential allelic expression of BDNF Val66Met in suicidal behaviour. *World J Biol Psychiatry* 12:42–47.

Deveci, A., Aydemir, O., Taskin, O., Taneli, F., Esen-Danaci, A. 2007. Serum BDNF levels in suicide attempters related to psychosocial stressors: A comparative study with depression. *Neuropsychobiology* 56:93–97.

Dolan, R.J., Bench, C.J., Liddle, P.F., Friston, K.J., Frith, C.D., Grasby, P.M., Frackowiak, R.S. 1993. Dorsolateral prefrontal cortex dysfunction in the major psychoses; symptom or disease specificity? *J Neurol Neurosurg Psychiatry* 56:1290–1294.

Drevets, W.C., Ongür, D., Price, J.L. 1998. Reduced glucose metabolism in the subgenual prefrontal cortex in unipolar depression. *Mol Psychiatry* 3:190–191.

Duman, R.S., Malberg, J., Nakagawa, S., D'Sa, C. 2000. Neuronal plasticity and survival in mood disorders. *Biol Psychiatry* 48:732–739.

Duman, R.S., Monteggia, L.M. 2006. A neurotrophic model for stress-related mood disorders. *Biol Psychiatry* 59:1116–1127.

Dumser, T., Barocka, A., Schubert, E. 1998. Weight of adrenal glands may be increased in persons who commit suicide. *Am J Forensic Med Pathol* 19:72–76.

Dwivedi, Y. 2010. Brain-derived neurotrophic factor and suicide pathogenesis. *Ann Med* 42:87–96.

Dwivedi, Y., Rizavi, H.S., Conley, R.R., Pandey, G.N. 2006a. ERK MAP kinase signaling in post-mortem brain of suicide subjects: Differential regulation of upstream Raf kinases Raf-1 and B-Raf. *Mol Psychiatry* 11:86–98.

Dwivedi, Y., Rizavi, H.S., Conley, R.R., Roberts, R.C., Tamminga, C.A., Pandey, G.N. 2003. Altered gene expression of brain-derived neurotrophic factor and receptor tyrosine kinase B in postmortem brain of suicide subjects. *Arch Gen Psychiatry* 60:804–815.

Dwivedi, Y., Rizavi, H.S., Pandey, G.N. 2006b. Antidepressants reverse corticosterone-mediated decrease in BDNF expression: Dissociation in regulation of specific exons by antidepressants and corticosterone. *Neuroscience* 139:1017–1029.

Dwivedi, Y., Rizavi, H.S., Roberts, R.C., Conley, R.C., Tamminga, C.A., Pandey, G.N. 2001. Reduced activation and expression of ERK1/2 MAP kinase in the postmortem brain of depressed suicide subjects. *J Neurochem* 77:916–928.

Dwivedi, Y., Rizavi, H.S., Teppen, T., Zhang, H., Mondal, A., Roberts, R.C., Conley, R.R., Pandey, G.N. 2008. Lower phosphoinositide 3-kinase (PI 3-kinase) activity and differential expression levels of selective catalytic and regulatory PI 3-kinase subunit isoforms in prefrontal cortex and hippocampus of suicide subjects. *Neuropsychopharmacology* 33:2324–2340.

Dwivedi, Y., Rizavi, H., Zhang, H., Mondal, A.C., Roberts, R.C., Conley, R.R., Pandey, G.N. 2009. Neurotrophin receptor activation and expression in human postmortem brain: Effect of suicide. *Biol Psychiatry* 65:319–328.

Egan, M.F., Kojima, M., Callicott, J.H., Goldberg, T.E., Kolachana, B.S., Bertolino, A., Zaitsev, E., Gold, B., Goldman, D., Dean, M., Lu, B., Weinberger, D.R. 2003. The BDNF val66met polymorphism affects activity-dependent secretion of BDNF and human memory and hippocampal function. *Cell* 112:257–269.

Eide, E.F., Vining, E.R., Eide, B.L., Zang, K., Wang, X.-Y., Reichardt, L.F. 1996. Naturally occurring truncated TrkB receptors have dominant inhibitory effects on brain-derived neurotrophic factor signaling. *J Neurosci* 16:3123–3129.

Ernst, C., Deleva, V., Deng, X., Sequeira, A., Pomarenski, A., Klempan, T., Ernst, N., Quirion, R., Gratton, A., Szyf, M., Turecki, G. 2009. Alternative splicing, methylation state, and expression profile of tropomyosin-related kinase B in the frontal cortex of suicide completers. *Arch Gen Psychiatry* 66:22–32.

Esposito, D., Patel, P., Stephens, R.M., Perez, P., Chao, M.V., Kaplan, D.R., Hempstead, BL. 2001. The cytoplasmic and transmembrane domains of the p75 and Trk A receptors regulate high affinity binding to nerve growth factor. *J Biol Chem* 276:32687–32695.

Fitzsimons, C.P., van Hooijdonk, L.W., Morrow, J.A., Peeters, B.W., Hamilton, N., Craighead, M., Vreugdenhil, E. 2009. Antiglucocorticoids, neurogenesis and depression. *Mini Rev Med Chem* 9:249–264.

Fossati, P., Radtchenko, A., Boyer, P. 2004. Neuroplasticity: From MRI to depressive symptoms. *Eur Neuropsychopharmacol* 14:S503–S510.

Frodl, T., Schaub, A., Banac, S., Charypar, M., Jäger, M., Kümmler, P., Bottlender, R., Zetzsche, T., Born, C., Leinsinger, G., Reiser, M., Möller, H.J., Meisenzah, E.M. 2006. Reduced hippocampal volume correlates with executive dysfunctioning in depression. *J Psychiatry Neurosci* 31:316–323.

Fryer, R.H., Kaplan, D.R., Feinstein, S.C., Radeke, M.J., Grayson, D.R., Kromer, L.F. 1996. Developmental and mature expression of full-length and truncated TrkB receptors in the rat forebrain. *J Comp Neurol* 374:21–40.

Fuchikami, M., Morinobu, S., Kurata, A., Yamamoto, S., Yamawaki, S. 2009. Single immobilization stress differentially alters the expression profile of transcripts of the brain-derived neurotrophic factor (*BDNF*) gene and histone acetylation at its promoters in the rat hippocampus. *Int J Neuropsychopharmacol* 12:73–82.

Garcia, R. 2002. Stress, synaptic plasticity, and psychopathology. *Rev Neurosci* 13:195–208.

Hajsza, T., MacLusky, N.J., Leranth, C. 2005. Short-term treatment with the antidepressant fluoxetine triggers pyramidal dendritic spine synapse formation in rat hippocampus. *Eur J Neurosci* 21:1299–1303.

Hasegawa, Y., Yamagishi, S., Fujitani M., Yamashita, T. 2004. p75 neurotrophin receptor signaling in the nervous system. *Biotechnol Annu Rev* 10:123–149.

Hempstead, B.L. 2002. The many faces of p75NTR. *Curr Opin Neurobiol* 12:260–267.

Honer, W.G. 1999. Assessing the machinery of mind: Synapses in neuropsychiatric disorders. *J Psychiatry Neurosci* 24:116–121.

Hong, C.J., Huo, S.J., Yen, F.C., Tung, C.L., Pan, G.M., Tsai, S.J. 2003. Association study of a brain-derived neurotrophic-factor genetic polymorphism and mood disorders, age of onset and suicidal behavior. *Neuropsychobiology* 48:186–189.

Huang, E., Reichardt, L.F. 2001. Neurotrophins: Roles in neuronal development and function. *Annu Rev Neurosci* 24:677–736.

Iga, J., Ueno, S., Yamauchi, K., Numata, S., Tayoshi-Shibuya, S., Kinouchi, S., Nakataki, M., Song, H., Hokoishi, K., Tanabe, H., Sano, A., Ohmori, T. 2007. The Val66Met polymorphism of the brain-derived neurotrophic factor gene is associated with psychotic feature and suicidal behavior in Japanese major depressive patients. *Am J Med Genet B Neuropsychiatr Genet* 144B:1003–1006.

Jokinen, J., Carlborg, A., Mårtensson, B., Forslund, K., Nordström, A.L., Nordström, P. 2007. DST non-suppression predicts suicide after attempted suicide. *Psychiatry Res* 150:297–303.

Jokinen, J., Nordström, A.L., Nordström, P. 2009. CSF 5-HIAA and DST non-suppression-orthogonal biologic risk factors for suicide in male mood disorder inpatients. *Psychiatry Res* 165:96–102.

Karege, F., Schwald, M., Cisse, M. 2002. Postnatal developmental profile of brain-derived neurotrophic factor in rat brain and platelets. *Neurosci Lett* 328:261–264.

Karege, F., Vaudan, G., Schwald, M., Perroud, N., La Harpe, R. 2005. Neurotrophin levels in postmortem brains of suicide victims and the effects of antemortem diagnosis and psychotropic drugs. *Brain Res Mol Brain Res* 136:29–37.

Keller, S., Sarchiapone, M., Zarrilli, F., Tomaiuolo, R., Carli, V., Angrisano, T., Videtic, A., Amato, F., Pero, R., di Giannantonio, M., Iosue, M., Lembo, F., Castaldo, G., Chiariotti, L. 2011. TrkB gene expression and DNA methylation state in Wernicke area does not associate with suicidal behavior. *J Affect Disord* 135:400–404.

Keller, S., Sarchiapone, M., Zarrilli, F., Videtic, A., Ferraro, A., Carli, V., Sacchetti, S., Lembo, F., Angiolillo, A., Jovanovic, N., Pisanti, F., Tomaiuolo, R., Monticelli, A., Balazic, J., Roy, A., Marusic, A., Cocozza, S., Fusco, A., Bruni, C.B., Castaldo, G., Chiariotti, L. 2010. Increased BDNF promoter methylation in the Wernicke area of suicide subjects. *Arch Gen Psychiatry* 67:258–267.

Kim, B., Kim, C.Y., Hong, J.P., Kim, S.Y., Lee, C., Joo, Y.H. 2008. Brain-derived neurotrophic factor Val/Met polymorphism and bipolar disorder. Association of the Met allele with suicidal behavior of bipolar patients. *Neuropsychobiology* 582:97–103.

Kim, Y.K., Lee, H.P., Won, S.D., Park, E.Y., Lee, H.Y., Lee, B.H., Lee, S.W., Yoon, D., Han, C., Kim, D.J., Choi, S.H. 2007. Low plasma BDNF is associated with suicidal behavior in depression. *Prog Neuropsychopharmacol Biol Psychiatry* 31:578–579.

Kohli, M.A., Salyakina, D., Pfennig, A., Lucae, S., Horstmann, S., Menke, A., Kloiber, S., Hennings, J., Bradley, B.B., Ressler, K.J., Uhr, M., Müller-Myhsok, B., Holsboer, F., Binder, E.B. 2010. Association of genetic variants in the neurotrophic receptor-encoding gene *NTRK2* and a lifetime history of suicide attempts in depressed patients. *Arch Gen Psychiatry* 67:348–359.

Kozicz, T., Tilburg-Ouwens, D., Faludi, G., Palkovits, M., Roubos, E. 2008. Gender-related urocortin 1 and brain-derived neurotrophic factor expression in the adult human midbrain of suicide victims with depression. *Neuroscience* 152:1015–1023.

Kozisek, M.E., Middlemas, D., Bylund D.B. 2008a. Brain-derived neurotrophic factor and its receptor tropomyosin-related kinase B in the mechanism of action of antidepressant therapies. *Pharmacol Ther* 117:30–51.

Kozisek, M.E., Middlemas, D., Bylund, D.B. 2008b. The differential regulation of BDNF and TrkB levels in juvenile rats after four days of escitalopram and desipramine treatment. *Neuropharmacology* 54:251–257.

Lee, B.H., Kim, Y.K. 2009. Reduced platelet BDNF level in patients with depression. *Prog Neuropsychopharmacol Biol Psychiatry* 33:849–853.

Lee, B.H., Kim, Y.K. 2010. BDNF mRNA expression of peripheral blood mononuclear cells was decreased in depressive patients who had or had not recently attempted suicide. *J Affect Disord* 125:369–373.

Lee, B.H., Kim, H., Park, S.H., Kim, Y.K. 2007. Decreased plasma BDNF level in depressive patients. *J Affect Disord* 101:239–244.

Lewin, G.R., Barde, Y.A. 1996. Physiology of the neurotrophins. *Annu Rev Neurosci* 19:289–317.

Li, Y., Luikart, B.W., Birnbaum, S., Chen, J., Kwon, C.H., Kernie, S.G., Bassel-Duby, R., Parada, L.F. 2008. TrkB regulates hippocampal neurogenesis and governs sensitivity to antidepressive treatment. *Neuron* 59:399–412.

Lopez, J.F., Palkovits, M., Arato, M., Mansour, A., Akil, H., Watson, S.J. 1992. Localization and quantification of proopiomelanocortin mRNA and glucocorticoid receptor mRNA in pituitaries of suicide victims. *Neuroendocrinology* 56:491–501.

Lu, B., Pang, P.T., Woo, N.H. 2005. The yin and yang of neurotrophin action. *Nat Rev Neurosci* 6:603–614.

Luberg, K., Wong, J., Weickert, C.S., Timmusk, T. 2010. Human *TrkB* gene: Novel alternative transcripts, protein isoforms and expression pattern in the prefrontal cerebral cortex during postnatal development. *J Neurochem* 113:952–964.

Mann, J.J. 1998. The neurobiology of suicide. *Nat Med* 4:25–30.

Mann, J.J. 2002. A current perspective of suicide and attempted suicide. *Ann Intern Med* 136:302–311.

Marmigere, F., Givalois, L., Rage, F., Arancibia, S., Tapia-Arancibia, L. 2003. Rapid induction of BDNF expression in the hippocampus during immobilization stress challenge in adult rats. *Hippocampus* 13:646–655.

Marttunen, M.J., Aro, H.M., Henriksson, M.M., Lonnqvist, J.K. 1991. Mental disorders in adolescent suicide. DSM-III-R axes I and II diagnoses in suicides among 13- to 19-year-olds in Finland. *Arch Gen Psychiatry* 48:834–839.

McEwen, B.S. 2000. Effects of adverse experiences for brain structure and function. *Biol Psychiatry* 48:713–714.

McGregor, S., Strauss, J., Bulgin, N., De Luca, V., George, C.J., Kovacs, M., Kennedy, J.L. 2007. *p75(NTR)* gene and suicide attempts in young adults with a history of childhood-onset mood disorder. *Am J Med Genet B Neuropsychiatr Genet* 144B:696–700.

Meldolesi, J., Sciorati, C., Clementi, E. 2000. The p75 receptor: First insights into the transduction mechanisms leading to either cell death or survival. *Trends Pharmacol Sci* 21:242–243.

Mellstrom, B., Torres, B., Link, W.A., Naranjo, J.R. 2004. The *BDNF* gene: Exemplifying complexity in Ca^{2+}-dependent gene expression. *Crit Rev Neurobiol* 16:43–49.

Middlemas, D.S., Lindberg, R.A., Hunter, T. 1991. TrkB, a neural receptor protein-tyrosine kinase: Evidence for a full-length and two truncated receptors. *Mol Cell Biol* 11:143–153.

Miguel-Hidalgo, J., Rajkowska, G. 2002. Morphological brain changes in depression. Can antidepressants reverse them? *CNS Drugs* 16:361–372.

Nakayama, M., Gahara, Y., Kitamura, T., Ohara, O. 1994. Distinctive four promoters collectively direct expression of brain-derived neurotrophic factor gene. *Mol Brain Res* 21:206–218.

Nemeroff, C.B., Owens, M.J., Bissette, G., Andorn, A.C., Stanley, M. 1988. Reduced corticotropin releasing factor binding sites in the frontal cortex of suicide victims. *Arch Gen Psychiatry* 45:577–579.

Nikisch, G. 2009. Involvement and role of antidepressant drugs of the hypothalamic–pituitary–adrenal axis and glucocorticoid receptor function. *Neuro Endocrinol Lett* 30:11–16.

Ongur, D., Drevets, W.C., Price, J.L. 1998. Glial reduction in the subgenual prefrontal cortex in mood disorders. *Proc Natl Acad Sci USA* 95:13290–13295.

Pandey, G.N., Ren, X., Rizavi, H.S., Conley, R.R., Roberts, R.C., Dwivedi, Y. 2008. Brain-derived neurotrophic factor and tyrosine kinase B receptor signalling in post-mortem brain of teenage suicide victims. *Int J Neuropsychopharmacol* 11:1047–1061.

Pang, P.T., Teng, H.K., Zaitsev, E., Woo, N.T., Sakata, K., Zhen, S., Teng, K.K., Yung, W.H., Hempstead, B.L., Lu, B. 2004. Cleavage of proBDNF by tPA/plasmin is essential for long-term hippocampal plasticity. *Science* 306:487–491.

Pattabiraman, P.P., Tropea, D., Chiaruttini, C., Tongiorgi, E., Cattaneo, A., Domenici, L. 2005. Neuronal activity regulates the developmental expression and subcellular localization of cortical BDNF mRNA isoforms in vivo. *Mol Cell Neurosci* 28:556–570,

Perroud, N., Courtet, P., Vincze, I., Jaussent, I., Jollant, F., Bellivier, F., Leboyer, M., Baud, P., Buresi, C., Malafosse, A. 2008. Interaction between BDNF Val66Met and childhood trauma on adult's violent suicide attempt. *Genes Brain Behav* 7:314–322.

Pizarro, J.M., Lumley, L.A., Medina, W., Robison, C.L., Chang, W.E., Alagappan, A., Bah, M.J., Dawood, M.Y., Shah, J.D., Mark, B., Kendall, N., Smith, M.A., Saviolakis, G.A., Meyerhoff, J.L. 2004. Acute social defeat reduces neurotrophin expression in brain cortical and subcortical areas in mice. *Brain Res* 1025:10–20.

Pregelj, P., Nedic, G., Paska, A.V., Zupanc, T., Nikolac, M., Balažic, J., Tomori, M., Komel, R., Seler, D.M., Pivac, N. 2011. The association between brain-derived neurotrophic factor polymorphism (BDNF Val66Met) and suicide. *J Affect Disord* 128:287–290.

Rage, F., Givalois, L., Marmigere, F., Tapia-Arancibia, L., Arancibia, S. 2002. Immobilization stress rapidly modulates BDNF mRNA expression in the hypothalamus of adult male rats. *Neuroscience* 112:309–318.

Rajkowska, G. 1997. Morphometric methods for studying the prefrontal cortex in suicide victims and psychiatric patients. *Ann N Y Acad Sci* 836:253–268.

Rajkowska, G. 2000. Histopathology of the prefrontal cortex in depression: What does it tell us about dysfunctional monoaminergic circuits? *Prog Brain Res* 126:397–412.

Rajkowska, G. 2002. Cell pathology in mood disorders. *Semin Clin Neuropsychiatry* 7:281–292.

Rajkowska, G., Miguel-Hidalgo, J.J. 2007. Gliogenesis and glial pathology in depression. *CNS Neurol Disord Drug Targets* 6:219–233.

Rantamäki, T., Hendolin, P., Kankaanpää, A., Mijatovic, J., Piepponen, P., Domenici, E., Chao, M.V., Männistö, P.T., Castrén, E. 2007. Pharmacologically diverse antidepressants rapidly activate brain-derived neurotrophic factor receptor TrkB and induce phospholipase-C gamma signaling pathways in mouse brain. *Neuropsychopharmacology* 32:2152–2162.

Rasmusson, A.M., Shi, L., Duman, R. 2002. Downregulation of BDNF mRNA in the hippocampal dentate gyrus after re-exposure to cues previously associated with footshock. *Neuropsychopharmacology* 27:133–142.

Reichardt, L.F. 2006. Neurotrophin-regulated signalling pathways. *Philos Trans R Soc Lond B Biol Sci* 361:1545–1564.

Rojas, P.S., Fritsch, R., Rojas, R.A., Jara, P., Fiedler, J.L. 2011. Serum brain-derived neurotrophic factor and glucocorticoid receptor levels in lymphocytes as markers of antidepressant response in major depressive patients: A pilot study. *Psychiatry Res* 189:239–245.

Rosoklija, G., Toomayan, G., Ellis, S.P., Keilp, J., Mann, J.J., Latov, N., Hays, A.P., Dwork, A.J. 2000. Structural abnormalities of subicular dendrites in subjects with schizophrenia and mood disorders: Preliminary findings. *Arch Gen Psychiatry* 57:349–356.

Roux, P.P., Colicos, M.A., Barker, P.A., Kennedy, T.E. 1999. p75 neurotrophin receptor expression is induced in apoptotic neurons after seizure. *J Neurosci* 19:6887–6896.

Sackeim, H.A. 2001. Functional brain circuits in depression and remission. *Arch Gen Psychiatry* 58:649–650.

Sala, M., Perez, J., Soloff, P., Ucelli di Nemi, S., Caverzasi, E., Soares, J.C., Brambilla, P. 2004. Stress and hippocampal abnormalities in psychiatric disorders. *Eur Neuropsychopharmacol* 14:393–405.

Sarchiapone, M., Carli, V., Roy, A., Iacoviello, L., Cuomo, C., Latella, M.C., di Giannantonio, M., Janiri, L., de Gaetano, M., Janal, M.N. 2008. Association of polymorphism (Val66Met) of brain-derived neurotrophic factor with suicide attempts in depressed patients. *Neuropsychobiology* 57:139–145.

Schaaf, M.J.M., de Jong, J., de Kloet, E.R., Vreugdenhil, E. 1998. Downregulation of BDNF mRNA and protein in the rat hippocampus by corticosterone. *Brain Res* 813:112–120.

Schenkel, L.C., Segal, J., Becker, J.A., Manfro, G.G., Bianchin, M.M., Leistner-Segal, S. 2010. The BDNF Val66Met polymorphism is an independent risk factor for high lethality in suicide attempts of depressed patients. *Prog Neuropsychopharmacol Biol Psychiatry* 34:940–944.

Schratt, G.M., Tuebing, F., Nigh, E.A., Kane, C.G., Sabatini, M.E., Kiebler, M., Greenberg, M.E. 2006. A brain-specific microRNA regulates dendritic spine development. *Nature* 439:283–289.

Schweigreiter, R. 2006. The dual nature of neurotrophins. *Bioessays* 28:583–594.

Seidah, N.G., Benjannet, S., Pareek, S., Chrétien, M., Murphy, R.A. 1996. Cellular processing of the neurotrophin precursors of NT3 and BDNF by the mammalian proprotein convertases. *FEBS Lett* 379:247–250.

Sheline, Y.I. 2000. 3D MRI studies of neuroanatomic changes in unipolar depression: The role of stress and medical comorbidity. *Biol Psychiatry* 48:791–800.

Smith, M.A., Makino, S., Kvetnansky, R., Post, R.M. 1995. Stress and glucocorticoids affect the expression of brain-derived neurotrophic factor and neurotrophin-3 mRNAs in the hippocampus. *J Neurosci* 15:1768–1777.

Szigethy, E., Conwell, Y., Forbes, N.T., Cox, C., Caine, E.D. 1994. Adrenal weight and morphology in victims of completed suicide. *Biol Psychiatry* 36:374–380.

Thoenen, H. 2000. Neurotrophins and activity-dependent plasticity. *Prog Brain Res* 128:183–191.

Toni, N., Buchs, P.A., Nikonenko, I., Bron, C.R., Muller, D. 1999. LTP promotes formation of multiple spine synapses between a single axon terminal and a dendrite. *Nature* 402:421–425.

Toro, R., Chupin, M., Garnero, L., Leonard, G., Perron, M., Pike, B., Pitiot, A., Richer, L., Veillette, S., Pausova, Z., Paus, T. 2009. Brain volumes and Val66Met polymorphism of the *BDNF* gene: Local or global effects? *Brain Struct Funct* 213:501–509.

Turecki, G. 2005. Dissecting the suicide phenotype: The role of impulsive–aggressive behaviors. *J Psychiatry Neurosci* 30:398–408.

Ueyama, T., Kawai, Y., Nemoto, K., Sekimoto, M., Toné, S., Senba, E. 1997. Immobilization stress reduced the expression of neurotrophins and their receptors in the rat brain. *Neurosci Res* 28:103–110.

Vincze, I., Perroud, N., Buresi, C., Baud, P., Bellivier, F., Etain, B., Fournier, C., Karege, F., Matthey, M.L., Preisig, M., Leboyer, M., Malafosse, A. 2008. Association between brain-derived neurotrophic factor gene and a severe form of bipolar disorder, but no interaction with the serotonin transporter gene. *Bipolar Disorder* 10:580–587.

Whitfield, J., Neame, S.J., Paquet, L., Bernard, O., Ham, J. 2001. Dominant-negative c-Jun promotes neuronal survival by reducing BIM expression and inhibiting mitochondrial cytochrome c release. *Neuron* 29:629–643.

Wong, J., Webster, M.J., Cassano, H., Weickert, C.S. 2009. Changes in alternative brain-derived neurotrophic factor transcript expression in the developing human prefrontal cortex. *Eur J Neurosci* 29:1311–1322.

Zai, C.C., Manchia, M., De Luca, V., Tiwari, A.K., Chowdhury, N.I., Zai, G.C., Tong, R.P., Yilmaz, Z., Shaikh, S.A., Strauss, J., Kennedy, J.L. 2011. The brain-derived neurotrophic factor gene in suicidal behaviour: A meta-analysis. *Int J Neuropsychopharmacol* 30:1–6.

Zarrilli, F., Angiolillo, A., Castaldo, G., Chiariotti, L., Keller, S., Sacchetti, S., Marusic, A., Zagar, T., Carli, V., Roy, A., Sarchiapone, M. 2008. Brain derived neurotrophic factor (BDNF) genetic polymorphism (Val66Met) in suicide: A study of 512 cases. *Am J Med Genet B Neuropsychiatr Genet* 150:599–600.

Zhou, H., Summers, S.A., Birnbaum, M.J., Pittman, R.N. 1998. Inhibition of Akt kinase by cell-permeable ceramide and its implications for ceramide-induced apoptosis. *J Biol Chem* 273:16568–16575.

9 Neuroimaging High Risk States for Suicide

Jeffrey H. Meyer

CONTENTS

9.1 OVERVIEW OF NEUROIMAGING STRATEGIES

Neuroimaging methods provide a great opportunity to improve our understanding of the pathway between adverse environmental conditions and suicide (see Figure 9.1). Since neuroimaging is completed *in vivo*, it is tailored to investigate the intermediary links between neuropathology and the symptoms or traits associated with suicide since such symptoms and traits may be measured at the time of brain scanning.

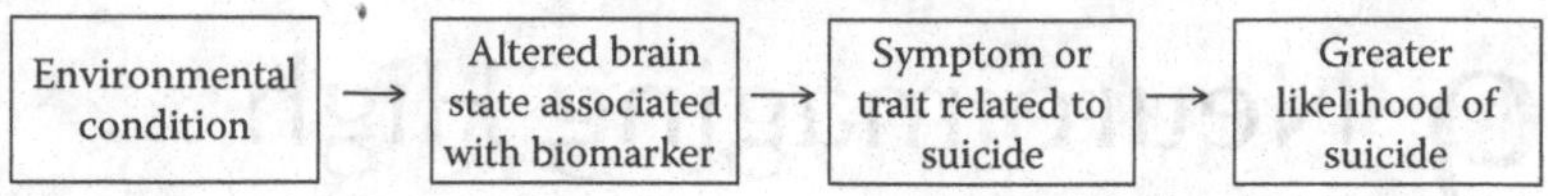

FIGURE 9.1 The strength of neuroimaging methods is to elucidate steps as outlined in the above diagram. *In vivo* imaging offers the opportunity to link environmental conditions with biomarker changes and in turn link these biomarkers with symptoms or traits implicated in risk for suicide.

As well, links between environmental conditions and neuropathology associated with suicide may be investigated since neuroimaging may be conducted during different environmental conditions.

In this chapter, the neurobiological targets related to suicide presented include serotonin 5-HT_{2A} receptors, serotonin transporters (5-HTT), monoamine oxidase A, D_2 receptor binding, and μ-opioid receptors. Recent discoveries in the area of 5-HT_{2A} receptor, serotonin transporter, and MAO-A imaging have led to a pathophysiological understanding for the mechanism of pessimism in major depressive disorder (MDD). Neuroimaging of MAO-A, D_2 and μ-opioid imaging are leading toward better understanding of mechanisms of impulsivity. An important ongoing direction is determining how environmental conditions lead to alterations in brain neurochemistry that can increase the risk for suicide and this will be illustrated in an example of neuroimaging MAO-A in early postpartum.

9.2 5-HT_{2A} RECEPTORS AND SUICIDE

9.2.1 Greater Prefrontal Cortex 5-HT_{2A} Binding: Most Frequent Abnormality in Suicide

In the year 2000, it could be argued that the most consistent postmortem biological abnormality of suicide was increased $serotonin_{2A}$ receptor binding in the prefrontal cortex. Beginning in 1984, a number of investigations examined $serotonin_{2A}$ receptor binding in suicide victims and it was often reported that this binding was increased in the prefrontal cortex, most commonly in Brodmann area 9.[1–11] At the time of study, these were reported as alterations in $serotonin_2$ receptor binding but it is generally believed that these studies represented investigations of $serotonin_{2A}$ receptors given that ligand binding to 5-HT_{2C} receptors in cortex is extremely low[12,13] and mRNA of 5-HT_{2B} receptors is extremely low in cortex.[14] These findings were reported in studies in which diagnosis of the suicide victim was unrestricted. However, these findings were more consistent in the subsample of studies of depressed suicide victims and medication-free suicide victims.[4,7]

It was unclear whether this finding was associated with a specific diagnosis, a symptom cluster associated with suicide, or represented a direct marker of suicidal ideation. It had been proposed that this finding was somewhat more common in studies of suicide victims who died of violent means,[1,3,5,6,15,16] but it was unclear how this explanation linked this abnormality to symptoms, life events, or personality traits. For example, it could be argued that selection of violent means of suicide represented greater planning of suicide or a predisposition toward violent behavior.

However, one psychological autopsy investigation suggested that the latter association was possible.[17] Since it is much easier to recruit subjects with specific diagnoses *in vivo*, most neuroimaging studies of 5-HT_{2A} receptors focused upon three diagnostic areas in relation to 5-HT_{2A} receptors: MDD, borderline personality disorder, and aggressive behavior.

9.2.2 Greater Prefrontal Cortex 5-HT_{2A} Binding: A Diagnostic and Symptom-Specific Finding

If one examines the list of neuroimaging studies of 5-HT_{2A} receptors that purely compare depressed and healthy samples as found in Table 9.1, they appear to contradict postmortem findings as many of these between-group comparisons reported a regional decrease in 5-HT_{2A} binding. The discrepancy can be resolved substantially through the observation that studies in which subjects recently received selective serotonin reuptake inhibitor (SSRI) treatment usually report decreased regional 5-HT_{2A} receptor binding.[18–20,23,26,29,30] Since the initial studies in the field sampled people with recent antidepressant treatment, the initial impression derived from this work was that regional 5-HT_{2A} binding tended to be reduced in MDD.

However, the first two studies not sampling subjects who recently had antidepressant treatment found no difference between depressed and healthy subjects.[21,22] The first study of medication-free subjects by Meyer et al. applied [^{18}F]setoperone. [^{18}F]Setoperone is a very good radioligand for imaging 5-HT_{2A} receptors owing to its specific binding in cortex, reversibility, and favorable ratio of specific binding to free and nonspecific binding[31–35] (see Meyer for a review of neuroimaging radioligands that bind to 5-HT_{2A} receptor[28]). The study by Meyer et al. sampled medication-free (>6 months) subjects in the midst of a major depressive episode (MDE) from early onset MDD. Subjects also had no comorbid psychiatric illnesses. No difference in prefrontal cortex 5-HT_{2A} binding was found as compared to healthy controls.[21] The second study of medication-free subjects applied [^{18}F]altanserin positron emission tomography (PET) in older depressed subjects and found no difference in 5-HT_{2A} binding between patients and healthy controls.[22] [^{18}F]Altanserin is a reasonable technique as it has high specific binding ratio but there are radioactive metabolites that cross the blood–brain barrier, so groups that use this technique currently apply it with a bolus plus infusion approach.[36] After considering medication-free status, there still was a discrepancy, but a lesser one, between studies of suicide victims and neuroimaging studies of MDEs: 5-HT_{2A} density was often elevated in prefrontal cortex in postmortem studies of suicide victims yet 5-HT_{2A} binding in prefrontal cortex was unchanged in medication-free depressed subjects.

One interpretation of the lack of difference in prefrontal cortex 5-HT_{2A} binding between medication-free depressed and control subjects is that there is no relationship of this analysis to the investigations of suicide victims. This seems unlikely given that half of suicide victims have a diagnosis of MDD[37,38] and that some of the postmortem studies that reported elevated 5-HT_{2A} density in prefrontal cortex sampled subjects with MDD.[37,38]

An alternative perspective is that a subgroup of MDE subjects have the biological abnormality reported in suicide victims. To investigate this question, one could take

TABLE 9.1
Imaging Studies of 5-HT$_{2A}$ Receptors in Major Depressive Disorder

Study	Method	Number of Subjects	Medication-Free Status	Result
D'Haenen et al.[18]	[^{123}I]Ketanserin SPECT	19 Depressed 10 Healthy	7 days	Greater in parietal cortex
Biver et al.[19]	[^{18}F]Altanserin PET	8 Depressed 22 Healthy	10 days	Lower in orbitofrontal cortex
Attar Levy et al.[20]	[^{18}F]Setoperone PET	7 Depressed 7 Healthy	Taking benzodiazepines	Lower in prefrontal cortex
Meyer et al.[21]	[^{18}F]Setoperone PET	14 Depressed 14 Healthy	3 months plus 5 half-lives	No difference
Meltzer et al.[22]	[^{18}F]Altanserin PET	11 Depressed 11 Healthy	"Untreated"	No difference
Yatham et al.[23]	[^{18}F]Setoperone PET	20 Depressed 20 Healthy	2 weeks	Decrease in all cortex
Messa et al.[24]	[^{18}F]Setoperone PET	19 Depressed 19 Healthy	Taking benzodiazepines	Decrease in all cortex
Meyer et al.[25,a]	[^{18}F]Setoperone PET	22 Depressed 22 Healthy	6 months	Positive association with dysfunctional attitude severity in cortex
Mintun et al.[26,b]	[^{18}F]Altanserin PET	46 Depressed 29 Healthy	4 weeks	Decrease in hippocampus
Bhagwagar et al.[27]	[^{11}C]MDL100907	20 Recovered depressed 20 Healthy	6 months	Positive association with dysfunctional attitude severity in prefrontal cortex Elevation in most cortex regions

Source: Updated from Meyer, J.H., *Semin. Nucl. Med.*, 38, 287, 2008.

[a] Subjects enrolled in the study by Meyer et al.[21] were also included in the expanded study by Meyer et al.[25] of 5-HT$_{2A}$ receptors and dysfunctional attitudes in subjects with depression as well as subjects with borderline personality disorder.

[b] Findings appear largely driven by a single healthy subject with very high 5-HT$_{2A}$ BP_{ND}.

advantage of the issue that 5-HT$_{2A}$ receptor density has an inverse relationship with extracellular serotonin levels such that the density of 5-HT$_{2A}$ receptors in cortex increases after chronic serotonin depletion and decreases after chronically raising extracellular serotonin.[39–42] Based upon this relationship, the candidate subgroup are depressed subjects suspected of having low extracellular serotonin in prefrontal cortex.

The symptom used to identify this subgroup was the elevated pessimism (dysfunctional attitudes) observed during MDEs. There is a modest level of dysfunctional attitudes in health, which increase to a variable extent during depressive episodes.[43–45] Greater pessimism during MDEs is an important symptom that contributes to the generation of sad mood and is targeted by cognitive therapy.[43–45] This symptom also abates when SSRI treatment is successful.[44,45] The rationale for choosing this symptom of elevated dysfunctional attitudes is that raising extracellular serotonin after administration of intravenous D-fenfluramine is associated with a strong shift in dysfunctional attitudes toward optimism 1 h later in healthy individuals.[25] This argues that among the many roles of serotonin, one of them is to modulate dysfunctional attitudes in humans. Both the anterior cingulate cortex and subregions of prefrontal cortex (dorsolateral and medial prefrontal cortex) participate in functions related to optimism and pessimism.[46–49] It is convenient that dysfunctional attitudes can be measured with the dysfunctional attitudes scale (DAS), a measure sensitive for detecting negativistic thinking in the midst of depressive episodes,[50,51] with very good internal consistency (Cronbach's $\alpha = 0.85–0.87$)[52,53] and high test–retest reliability.[43,53]

A strong correlation was observed between severity of dysfunctional attitudes (pessimism) and the elevation in cortex 5-HT_{2A} binding potential reflecting specific binding relative to free and nonspecific binding (BP_{ND}) was discovered. Moreover, cortex 5-HT_{2A} BP_{ND} was significantly elevated in subjects with MDE and severe pessimism.[25] For example, in the prefrontal cortex region centered on Brodmann area 9, 5-HT_{2A} BP_{ND} was elevated 29% in depression subjects with dysfunctional attitude scores higher (more pessimistic) than the median for the group. There was a strong, significant correlation between severity of pessimism and prefrontal cortex 5-HT_{2A} BP_{ND} (see Figure 9.2). A recent study by Bhagwagar et al. replicated this relationship between dysfunctional attitudes severity and prefrontal cortex 5-HT_{2A} BP_{ND} in recovered depressed subjects.[27] In a separate study of a large sample of healthy subjects, two personality facets related to pessimism, vulnerability, and anxiety were also positively correlated with prefrontal cortex, temporal cortex, and left insula 5-HT_{2A} BP_{ND}.[54]

The investigations correlating severity of dysfunctional attitudes with greater 5-HT_{2A} BP_{ND}[25,27] are highly consistent with postmortem investigations reporting greater 5-HT_{2A} receptor density in the prefrontal cortex of suicide victims. Fifty percent of suicide victims have MDD.[37,38] The DAS is highly correlated with hopelessness measured with the Beck Hopelessness Scale.[55–58] Given that hopelessness is a risk factor for suicide,[59,60] it is plausible that investigations of suicide victims reporting increased 5-HT_{2A} BP_{ND} sampled depressed subjects with greater severity of pessimism. See Figure 9.3 that represents the relationship of sampling studies for MDD, dysfunctional attitudes, and suicide.

Two other conditions that may be sampled in the studies of suicide victims are borderline personality and antisocial personality disorder.[37,38,61] In two investigations of 5-HT_{2A} binding in medication-free borderline personality disorder, no difference was found in the prefrontal cortex, although the second study reported an increase in 5-HT_{2A} binding in hippocampus in a *post hoc* analysis.[25,62] Meyer et al. investigated that 5-HT_{2A} binding in aggressive individuals found no group difference in any region, but did report an age interaction such that 5-HT_{2A} binding tended to increase considerably more with age in individuals with aggression.[63] Some theories of aggression

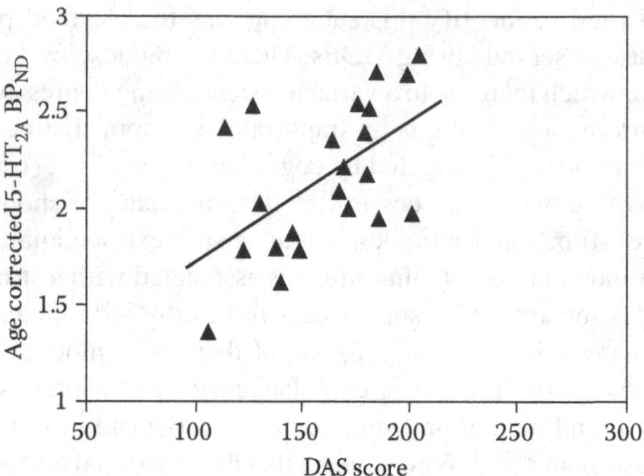

FIGURE 9.2 5-HT_{2A} Binding potential in prefrontal cortex is associated with dysfunctional attitudes in depressed subjects. Age corrected 5-HT_{2A} receptor binding potential (5-HT_{2A} BP_{ND}) within bilateral prefrontal cortex (Brodmann area 9) in depressed subjects is plotted against dysfunctional attitudes scale (DAS) score. When controlling for age, the correlation coefficient between 5-HT_{2A} BP and DAS is 0.56, $p = 0.009$. The age corrected 5-HT_{2A} BP_{ND} was calculated by applying a linear regression with predictor variables age and DAS to the 5-HT_{2A} BP_{ND}. The slope of the line for the age predictor was used to normalize each subject's 5-HT_{2A} BP_{ND} to that expected for a 30 year old subject. (Updated from Meyer JH et al., *Am J Psychiatry*, 160(1), 90, 2003.)

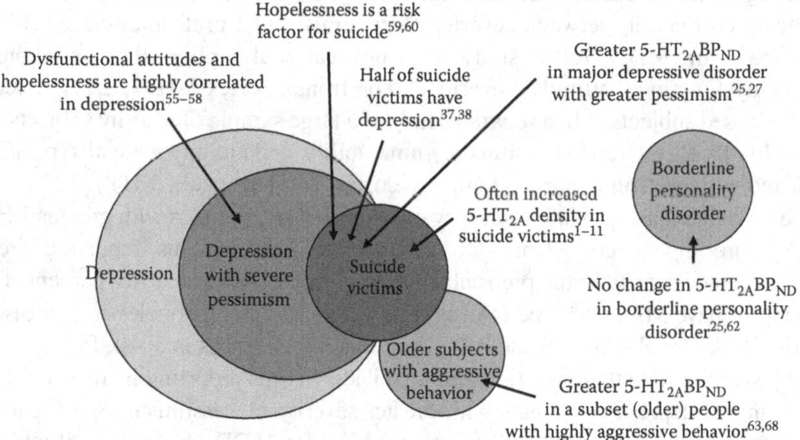

FIGURE 9.3 Elevated prefrontal cortex 5-HT_{2A} density in suicide reflects sampling from major depressive disorder with pessimism and from aggressive behavior but not from borderline personality disorder.

and impulsivity consider the mechanisms of inadequate pruning and it is possible that this age-related effect might represent an index of differential change in pyramidal cell loss, since most 5-HT_{2A} receptors are found in apical dendrites of pyramidal cell neurons and the pattern of loss of pyramidal cells matches the loss in 5-HT_{2A} receptors with age.[64–67] A recent study of 5-HT_{2A} receptors by Rosell et al. reported greater binding in orbitofrontal cortex in a group of aggressive individuals who met current criteria for intermittent explosive personality disorder, some of whom had personality disorders.[68] A limitation of the latter study is that study subjects did not receive urine screening for substance abuse, a condition commonly comorbid with aggression, and such substances may bias 5-HT_{2A} receptor binding.[69,70] Hence, among the investigations of 5-HT_{2A} binding there is support for increased binding in older individuals with aggression throughout cortex and in orbitofrontal cortex in people with aggression.

The findings of greater 5-HT_{2A} density in the prefrontal cortex of suicide victims in postmortem study can be reinterpreted in light of diagnostic and symptom-specific findings from neuroimaging studies of 5-HT_{2A} binding. MDD with high levels of pessimism is associated with greater prefrontal cortex 5-HT_{2A} binding, is sampled heavily in studies of suicide victims because half of the suicide victims have MDD and hopelessness (a strong correlate of pessimism), which is associated with the risk of suicide.[37,38,55–60] The diagnosis of borderline personality disorder does not contribute to the finding of greater prefrontal cortex 5-HT_{2A} density in suicide victims since no change in prefrontal 5-HT_{2A} BP_{ND} was found in this group.[25,62] In those with aggressive behavior, at least a subgroup of subjects older than age 34 show a similar neurobiological finding and would be expected to also contribute to the original findings of greater prefrontal 5-HT_{2A} density in suicide victims.[63,68] See Figure 9.3 for a Venn diagram that represents the sampling contributions of the various diagnoses and symptoms to the finding of greater 5-HT_{2A} density in suicide victims.

9.3 SEROTONIN TRANSPORTER BINDING AND SUICIDE

9.3.1 Discovery of Quantitative Method to Measure Serotonin Transporter Binding

In 2000, after 20 years of attempts, a major discovery in the field of serotonin transporter neuroimaging occurred with the development of [^{11}C]-3-amino-4-(2-dimethylaminomethyl-phenylsulfanyl)-benzonitrile ([^{11}C]DASB). The two previous radiotracers applied for *in vivo* imaging had limitations: 2-β-carbomethoxy-3-β-(4-iodophenyl)-tropane (β-CIT) single photon emission tomography (SPECT) was the first technique developed and the specific binding signal could be differentiated in the midbrain,[71–73] but it has almost equal affinity for the dopamine transporter as compared to the serotonin transporter.[74,75] As dopamine transporter density is high in the *substantia nigra*,[76] one cannot determine the relative contributions of dopamine and serotonin transporter binding in midbrain (the location where this radiotracer technique is applied for an index of serotonin transporter binding). The other early radiotracer, [^{11}C](+)McN5652 has a low ratio of specific binding relative to free and nonspecific binding, which in combination with modest reversibility makes valid and reliable quantitation difficult in regions other than the thalamus, and impossible in human cortex.[77–80] The radiotracer [^{11}C] (DASB, 3-amino-4-(2-dimethylaminomethylphenylsulfanyl)-benzonitrile) was a major advance owing to its selectivity, reversibility, greater specific binding relative to free and nonspecific binding, and reliability.[81–89] This radiotracer was three orders of magnitude more selective for the 5-HTT over other monoamine transporters and highly selective for the 5-HTT in comparison to a number of other targets screened.[88,89] Selectivity was further confirmed as 92%–95% of the specific binding to 5-HTT was displaceable by 5-HTT binding medications in animal models.[88,89] [^{11}C]DASB has very good brain uptake in humans.[81,82] In humans, its ratio of specific binding relative to free and nonspecific is good and its free and nonspecific binding has low between-subject variability.[81,83] Multiple brain regions may be assessed with noninvasive methods[81–89] and reliability of regional 5-HTT BP_{ND} measures is very good.[85–87,90] The 5-HTT BP_{ND} measures are low in cortex, but with standardized region of interest methods, very good reliability of 5-HTT BP_{ND} in human cortex was achieved.[85–87,90] In summary, [^{11}C]DASB PET imaging was a discovery that created a new opportunity for quantifying 5-HTT binding in humans.

Given that depleting serotonin in humans can lead to sad mood, as evidenced in the tryptophan depletion paradigm,[91–95] that 5-HT_{2A} receptor binding is often elevated in samples of suicide victims[1–11] and during pessimism with MDD,[25,27] and that 5-HT_{2A} receptor binding shows an inverse relationship to extracellular serotonin,[39–42] understanding whether extracellular serotonin levels may be lower in suicide is an important question. Direct evidence that serotonin is low in suicide is difficult to obtain: brain serotonin cannot be directly measured *in vivo* and it is likely, based upon animal simulations of postmortem delay, that serotonin levels are very unstable, even within 24h of death.[96] Moreover, postmortem investigations (previously listed by Mann et al.[97] and Stockmeier[11]) have difficultly sampling medication-free subjects.

9.3.2 Serotonin Transporter Binding and Disease Models

Measurement of serotonin transporter binding has considerable value in several different models of disease that may affect extracellular serotonin levels. There are four plausible models to consider in regard to how serotonin transporter binding could be altered in a disease that lowers extracellular brain serotonin.[70] These are referred to here as models one through four. Model one is a lesion model that reduces monoamine-releasing neurons. In a lesion model, reductions in binding occur. Model two is a model of secondary change in serotonin transporter binding consequent to serotonin lowering via a different process. Model three is increased clearance of extracellular monoamine via greater monoamine transporter density. In model three, greater available serotonin transporter binding leads to greater clearance of monoamines from extracellular locations. Model four is endogenous displacement and is dependent upon the properties of the radioligand. Endogenous displacement is the property of a few radioligands to express different binding after short-term manipulations of their endogenous neurotransmitter. Abnormalities in serotonin transporter binding during MDEs may be discussed in the context of these models.

Model two is unlikely to be relevant for serotonin transporter binding. Acute reductions in serotonin have repeatedly shown reductions in 5-HTT mRNA.[98–100] However, longer-term reductions or elevations in serotonin typically show no effect upon regional 5-HTT density.[101–103] However, it is worth noting that available evidence suggests that the different monoamine transporters do not regulate in the same fashion after chronic depletion of their endogenous monoamine. In contrast, for dopamine transporters in striatum, the evidence to support a relationship between long-term reductions in extracellular dopamine and a lowering of striatal dopamine transporter density is fairly strong.[104–107] It is reported that norepinephrine density decreases in the locus coeruleus after chronic norepinephrine depletion.[108]

Model four is unlikely for [^{11}C]DASB, but it is unclear whether this model would complicate interpretation of other serotonin transporter binding radiotracers because it has not been investigated for other serotonin transporter binding radiotracers. Endogenous displacement refers to the property, found in a minority of PET radiotracers under physiological conditions, to have increased binding potential measures after an acute reduction in endogenous neurotransmitter.[109] The phenomenon described is that the neurotransmitter itself prevents access of the radiotracer to receptors through competition. For [^{11}C]DASB, endogenous displacement may occur with exceptionally large magnitude changes in extracellular 5-HT but this would not be expected to occur with extracellular 5-HT changes that are physiologically tolerable for humans. In an animal study with [^{11}C]DASB, after an intraperitoneal injection of 10 mg/kg of the MAO-A/B inhibitor tranylcypromine, which raises extracellular serotonin several hundred to thousand percent,[110–112] a reduction in 5-HTT BP_{ND} was observed.[113] This has been replicated in animals with similar doses of tranylcypromine.[114] However, humans cannot tolerate one-tenth of this dose of tranylcypromine even with lengthy titrations and oral administration. Thus, this magnitude of serotonin change likely exceeds what is physiologically tolerable in humans. In accord with this perspective, we found no effect of tryptophan depletion upon 5-HTT BP_{ND} in 14 humans, demonstrating that endogenous serotonin occupancy is unlikely to appreciably influence [^{11}C]DASB binding under physiologically tolerable conditions.[87] Talbot et al. reported similar results in eight humans.[115] For other PET radiotracers such as [^{11}C]-N,N-dimethyl-2-(2′-amino-4′-hydroxymethylphenylthio)benzylamine ([^{11}C]HOMADAM), [^{11}C] MADAM, trans-1,2,3,5,6,10-β-Hexahydro-6-[4-(methylthio) phenyl]-pyrrolo-[2,1-a]-isoquinoline ([^{11}C](+)McN5652), or SPECT radiotracers, single photon emission tomography 2-beta-carbomethoxy-3-beta-(4-iodophenyl)-tropane (β CIT SPECT), or (2-([2-([dimethylamino]methyl)phenyl]thio)-5-^{123}I-iodophenylamine) ([^{123}I]ADAM), it is it is unknown whether endogenous levels of serotonin influence binding levels. This fourth model is unlikely to apply to PET imaging studies with [^{11}C]DASB in humans but it is unclear as to whether this model applies to other serotonin transporter radiotracers since the question has not been tested.

9.3.3 Relationship of Serotonin Transporter Imaging to Dysfunctional Attitudes and Major Depressive Disorder

Dysfunctional attitude is an important symptom of MDD, particularly because it is strongly related to suicide. Hopelessness[59,60] (and difficulty seeing positive reasons for living[116]) is an important risk factor for suicide and it has been clearly established in four separate samples of depressed subjects that greater hopelessness is associated with greater severity of dysfunctional attitudes as measured with the DAS.[55–58] The DAS[43] is a sensitive measure for detecting pessimistic thinking in the midst of MDE[50,51] that has very good internal consistency (Cronbach's $\alpha = 0.85–0.87$)[52,53] and high test–retest reliability.[43,53]

Low extracellular serotonin in prefrontal and anterior cingulate cortex during MDD is a highly plausible explanation for the strong correlation between prefrontal and anterior cingulate cortex $5\text{-}HT_{2A}$ receptor binding and dysfunctional attitudes.

Key subregions of prefrontal cortex (particularly the medial prefrontal cortex and dorsolateral prefrontal cortex) as well as the anterior cingulate cortex participate in cognitive functions related to optimism/pessimism.[46–49]

Modulation of extracellular serotonin can shift dysfunctional attitudes: Elevating extracellular serotonin abruptly in healthy humans via intravenous D-fenfluramine administration (in contrast to control condition) shifted perspective toward optimism as measured by the DAS.[25] The interpretation of this shift toward optimism after D-fenfluramine was that one of the functions of extracellular serotonin in humans is to reduce pessimism.[25] A property of 5-HT_{2A} receptors is that 5-HT_{2A} receptor density has an inverse relationship with extracellular serotonin levels such that the density of 5-HT_{2A} receptors in cortex increases after chronic serotonin depletion and decreases after chronically raising extracellular serotonin.[39–42] Therefore, an interpretation of the finding that prefrontal and anterior cingulate cortex 5-HT_{2A} BP_{ND} was significantly elevated in subjects with MDE and severe pessimism,[25] and that 5-HT_{2A} BP_{ND} in these regions is positively associated with severity of dysfunctional attitudes,[25,27] is that extracellular serotonin is low in MDD with severe pessimism.

There are only two postmortem investigations of 5-HTT density in subjects with recent symptoms of depressive episodes.[117,118] In these investigations, one of which concurrently sampled suicide victims reported no changes in 5-HTT density in the dorsal raphe or the locus coeruleus. Other postmortem investigations of 5-HTT density sampled subjects with a history of a depressive episode and these investigations usually studied the prefrontal cortex and/or dorsal raphe nucleus. Findings ranged from decreased 5-HTT density[119–123] to no difference in 5-HTT density.[124–128] In several of these studies, subjects were medication free,[121,124,125] and for many of these investigations, average postmortem delay was less than a day.[117–119,121,123,127] For greater detail, the reader is referred to the review of Stockmeier.[11] Other sampling issues that may influence postmortem investigations are effects of additionally sampling patients with bipolar disorder and possible differences between early versus late onset MDD. None of the postmortem studies investigated the relationship between 5-HTT binding to indices of hopelessness and pessimism.

The first application of [^{11}C]DASB PET imaging to investigate MDD investigated the relationship of 5-HTT BP_{ND} to severity of dysfunctional attitudes and presence of a MDE. In this study, Meyer et al. sampled 20 subjects with MDEs (from early onset MDD) and 20 healthy controls.[84] Subjects were medication free for at least 3 months, had no other comorbid axis I illnesses, were nonsmoking, and had early onset depression. There was no evidence for a difference in regional 5-HTT BP_{ND} in the group with MDEs and those in the midst of health. However, subjects with MDEs and severely pessimistic dysfunctional attitudes had significantly higher 5-HTT BP_{ND}, compared to healthy in brain regions sampling serotonin nerve terminals (dorsolateral and medial prefrontal cortex, anterior cingulated cortex, thalamus, bilateral caudate, and bilateral putamen). On average, 5-HTT BP_{ND} was 21% greater in these regions in depressed subjects with severely pessimistic dysfunctional attitudes. Moreover, within the MDE group, greater 5-HTT BP_{ND} was strongly associated with more negativistic dysfunctional attitudes in the same brain regions (see Figure 9.4). The interpretation was that serotonin transporters have an important role in influencing extracellular serotonin during MDEs: Greater regional 5-HTT levels can provide greater vulnerability to low

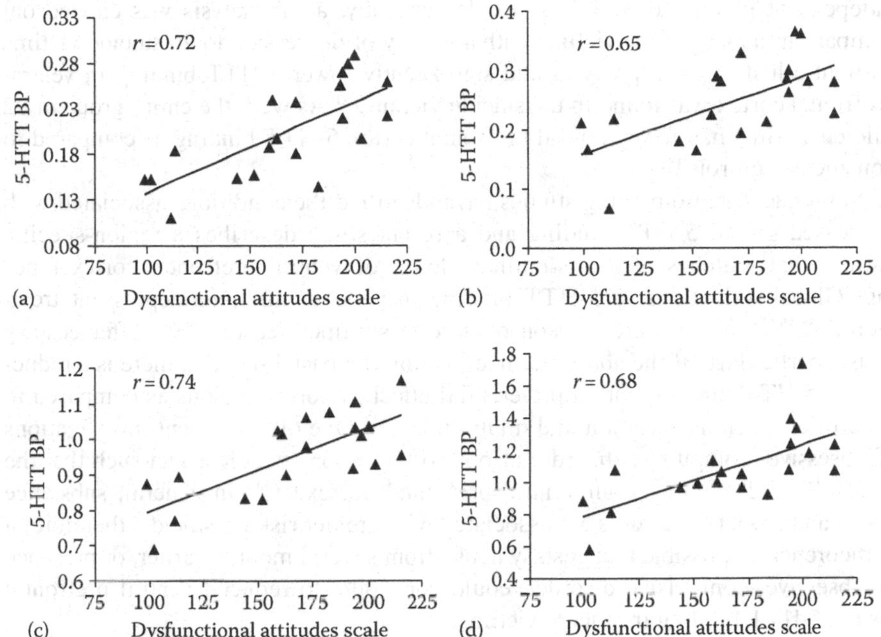

FIGURE 9.4 Correlations between dysfunctional attitudes scale (DAS) and serotonin transporter binding potential (5-HTT BP) in some of the larger regions in depressed subjects. Highly significant correlations were found: (a) dorsolateral prefrontal cortex ($p = 0.0004$), (b) anterior cingulated cortex ($p = 0.002$), (c) bilateral putamen ($p = 0.0002$), and (d) bilateral thalamus ($p = 0.001$). (Reprinted from Meyer, J.H. et al., *Arch. Gen. Psychiatry*, 61(12), 1271, 2004.)

extracellular 5-HT and symptoms of extremely negativistic dysfunctional attitudes. This interpretation, in subjects with high levels of pessimism during MDE, corresponds to model number three (see earlier under Section 9.3.2).

In general, neuroimaging studies that sample subjects who have early onset MDD, are medication free for greater than 2 months, are nonsmoking, and do not have comorbid axis I disorders and apply better quality radiotracer technology tend to find either no change in regional 5-HTT binding or an increase in regional 5-HTT binding.[84,129–131] Investigations that include sampling of subjects with late onset MDD, comorbid axis I psychiatric disorders, recent antidepressant use, and current cigarette smoking and do not apply a selective radiotracer are more likely to report a reduction in regional 5-HTT binding.[132–136] Only the first study of [^{11}C]DASB PET concurrently investigated a measure of pessimism or hopelessness.

9.3.4 Relationship of Serotonin Transporter Imaging to Postmortem Investigations of Suicide Victims

In addition to the series of postmortem studies of 5-HTT binding in depression (with or without concurrent suicide), a pair of studies, both from the same group, reported decreased 5-HTT binding in the ventral prefrontal cortex associated with suicide, independent of diagnosis.[121,137] In the larger study, a subanalysis was carried out comparing the 43 suicide victims with a history of depression to 20 suicide victims without a history of depression and significantly lower 5-HTT binding in ventral prefrontal cortex was found in the suicide victims.[121] As well, the entire group of 82 suicide victims had lower ventral prefrontal cortex 5-HTT binding as compared to nonsuicide controls.[121]

Subsequent neuroimaging studies have identified the conditions associated with decreased global 5-HTT binding and a recent study describes a region-specific pattern that matches of a predominant lower of ventral prefrontal cortex binding. Global reductions in 5-HTT binding may occur after antidepressant treatment,[78,85,86,138] or in winter season relative to summer season.[139,140] After ecstasy abuse, particularly if the abuse occurred within the past 4 months, there is a reduction in 5-HTT binding with a preferential effect in cortex regions as compared to subcortical regions (striatum and midbrain).[141–143] One of two recent investigations of obsessive–compulsive disorder also report a region-specific effect such that the lowest 5-HTT binding is in ventral prefrontal cortex.[144,145] In general, substance abuse and anxiety disorders are associated with greater risk for suicide; therefore, it is theoretically possible that ecstasy abuse from several months earlier, or presence of obsessive–compulsive disorder, could contribute to reduced ventral prefrontal cortex 5-HTT binding in suicide victims.[121]

Another possibility that requires further investigation is whether 5-HTT binding in impulsivity disorder or other aspects related to risk for suicide is related to the reduction in 5-HTT binding. Interestingly, a loss of 5-HTT binding occurred in peer-reared rhesus monkeys with early maternal separation (who later develop aggressiveness and impulsivity), but the pattern of 5-HTT binding loss, which included midbrain, thalamus, caudate, putamen, and anterior cingulate cortex, did not extend into the prefrontal cortex.[146] Takano et al. reported that neuroticism, which can include impulsivity, was positively associated with 5-HTT binding in the thalamus and Kalbitzer et al. reported a negative correlation between neuroticism and openness to values, the latter which then correlated negatively with 5-HTT binding in most subcortical (but not cortical) regions.[147,148] There is a study of 5-HTT binding in aggression applying [^{11}C]McN5652 PET reporting decreased 5-HTT binding in the anterior cingulate cortex[149] and a study of [^{123}I]ADAM reporting greater 5-HTT binding in brain stem of people with borderline personality disorder,[150] but, unfortunately, neither of these two radiotracers has sufficient specific binding signal to assess whether such binding changes occur in prefrontal cortex. A recent study of 5-HTT binding in alcohol abuse was negative.[151] To date, neuroimaging studies of 5-HTT binding in impulsivity (or conditions related to impulsivity) have not matched the ventral prefrontal cortex pattern of 5-HTT binding loss described in the two postmortem studies that specifically focused upon this region.[121,137]

9.4 MONOAMINE OXIDASE A AND SUICIDALITY

Monoamine oxidase A (MAO-A) is a target of interest in research of suicidality given its role in metabolizing monoamines and its widespread distribution throughout the brain. In the brain, the predominant location for this enzyme is on the outer mitochondria membranes in neurons,[152] and monoamine oxidase A density is highest in brain stem (locus coeruleus), lower in the hippocampus, cortex, striatum, and minimal in white matter tissue.[152,153] Monoamine oxidase A metabolizes serotonin given that serotonin is a high affinity substrate for MAO-A,[154–157] MAO-A is detectable in serotonin-releasing neurons,[158,159] and that MAO-A clearly influences extracellular serotonin as administration of MAO-A inhibitors increase extracellular serotonin from 20% to 200%, depending upon drug, dose, and region.[110,160–165] Norepinephrine[157,166] and dopamine[154–156] are also high affinity substrates for MAO-A,[157,166] and there is evidence that under MAO-A inhibition, extracellular concentrations of these monoamines[160,167–171] are raised. MAO-A is easily detectable in cells that synthesize norepinephrine[152,158,159,172] but more difficult to detect in dopamine-synthesizing neurons.[159,173] In knockout models of MAO-A, extracellular serotonin, norepinephrine, and dopamine are also raised substantively (100%–200%) in prefrontal cortex, hippocampus, and superior raphe nuclei.[174]

As of 2005, based upon postmortem study, it was unknown as to whether brain MAO-A binding or activity is abnormal during at risk states for suicide such as MDEs because each previous investigation of brain MAO-A had at least two important limitations that created significant heterogeneity in the samples[175–180] such as nonspecificity of technique for MAO-A versus monoamine oxidase B, enrollment of subjects who recently took medication, unclear diagnosis of suicide victims, small sample size, and lack of categorizing between early onset depression and late onset depression. In contrast to the common, early onset depression prior to age 40, late onset depression likely has a different pathophysiology attributable to lesions and/or degenerative disease.[181]

Neuroimaging radioligands to measure an index of MAO-A levels in humans include [^{11}C]clorgyline, deuterium-labeled [^{11}C]clorgyline, [^{11}C]harmine, and [^{11}C] befloxatone.[182–187] The latter two have an useful advantage in terms of having more rapid kinetics and greater reversibility. [^{11}C]Harmine was also modeled in humans and it possesses high affinity for the MAO-A site, as well as high selectivity, excellent specific binding relative to free and nonspecific binding ratios, and high brain uptake.[184,186,188–190] Given these properties and validation, it is the lead radiotracer for quantitating brain MAO-A binding in humans at this time.

9.4.1 Monoamine Oxidase A and Major Depressive Disorder

Prior to 2006, there were no studies of MAO-A density, activity, or mRNA in early onset MDD. In 2006, MAO-A V_S, an index of MAO-A density, was measured in 17 MDE and 17 healthy subjects with [^{11}C]harmine PET.[190] All subjects were otherwise healthy. Depressed subjects had early onset depression (before age 40) and were drug free for at least 5 months although most were antidepressant naïve. Depressed subjects were aged 18–50, met DSM-IV diagnosis of current MDE and MDD verified by SCID for DSM IV,[191] were nonsmoking and had a reasonable severity of depression with a score greater than 17 on the 17-item HDRS. The MAO-A V_S was highly significantly elevated ($p < 0.001$ each region, average magnitude 34% [or two standard deviations]) in the depressed subjects (see Figure 9.5). The study was

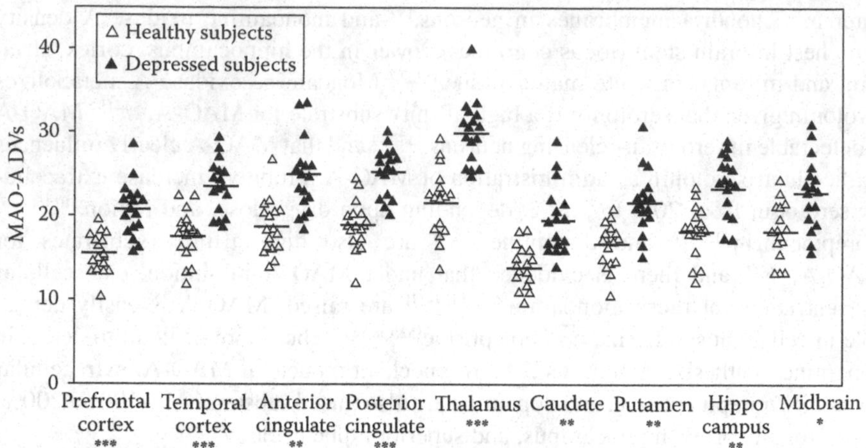

FIGURE 9.5 Comparison of MAO-A DVs between depressed and healthy subjects. On average MAO-A DVs was elevated by 34%, or two standard deviations, in depressed individuals. Differences between groups were highly significant in each region: *$p = 0.001$, **$p < 0.0001$, ***$p < 0.00001$. (Reprinted from Meyer J.H. et al., *Arch Gen Psychiatry*, 61(12), 1271, 2004.)

considered definitive for showing that MAO-A binding is elevated in early onset depression because the magnitude was large, the sample was carefully defined, and the method is selective for MAO-A.

This finding was subsequently replicated by the same group and the relationship between MAO-A binding and state of MDD was also investigated.[192] During remission from MDD, MAO-A binding was elevated in most brain regions such as prefrontal cortex, anterior cingulate cortex, striatum, hippocampus, thalamus, and midbrain. Elevated MAO-A binding may be viewed as an index of a monoamine-lowering process and a historical set of observations were known that had linked chronic monoamine-lowering processes with subsequent MDEs: During the development of reserpine-based antihypertensives in the 1950s, subsequent onset of MDE occurred typically 2 weeks to 4 months later.[193] Consistent with this line of observation, those recovered MDD subjects who had recurrence of their MDEs in the subsequent 6 months had the highest levels MAO-A binding in the prefrontal cortex and anterior cingulate cortex at the time of scanning.[192]

The raised MAO-A binding during MDEs adds to a pathophysiological model of extracellular serotonin loss contributing to pessimism during MDEs. Greater MAO-A binding can be viewed as an index of greater MAO-A levels. When MAO-A levels are greater in cell lines or in brain homogenates, MAO-A activity shows a positively correlated parallel, and typically linear relationship.[153,194–196] Hence, the index of greater MAO-A binding may be viewed as a mechanism of excessive serotonin removal. The inverse relationship between extracellular serotonin levels and dysfunctional attitudes as demonstrated by the shift in dysfunctional attitudes after intravenous D-fenfluramine and the reduction in dysfunctional attitudes after SSRI treatment suggests a role for extracellular serotonin in modulating optimism/pessimism.[25] Brain regions most strongly associated for a role in optimism/pessimism include the anterior cingulate cortex, and the medial and dorsolateral subdivisions

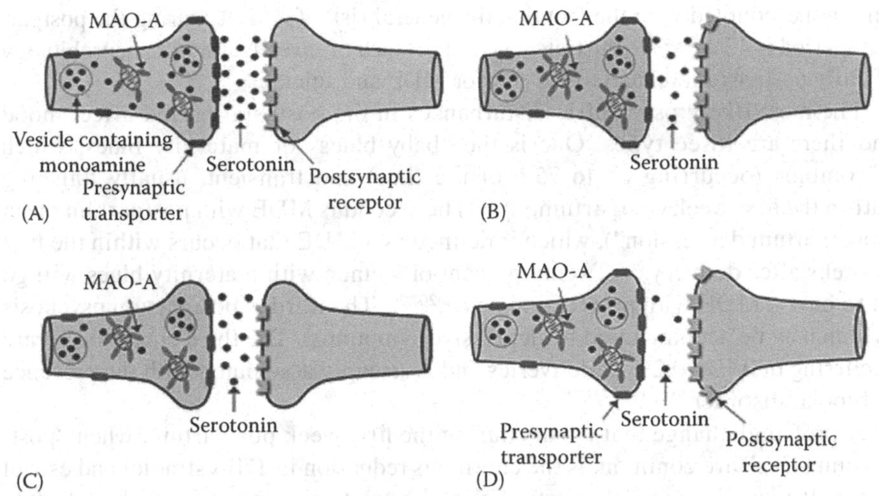

FIGURE 9.6 **(See color insert.)** Modern model of excessive extracellular serotonin loss during major depressive disorder (A) serotonin release in a synapse in health. (B) During a major depressive episode, monoamine oxidase A (MAO-A) density is elevated resulting in greater metabolism of serotonin. Outcomes range from (C) to (D). (C) If the serotonin transporter density is low during a major depressive episode, the effect of elevated MAO-A upon reducing extracellular serotonin is attenuated resulting in a moderate loss of serotonin. This eventually results in a moderate rise in dysfunctional attitudes. (D) If the serotonin transporter density is not low during a major depressive episode, then there is no protection against the effect of elevated MAO-A. The extracellular concentration of serotonin is severely reduced and rise in dysfunctional attitudes is severe. This model is most applicable in regions implicated in the generation of optimism and pessimism such as the anterior cingulate cortex and medial/dorsolateral subregions of the prefrontal cortex. MAO-A in cells that do not release serotonin (such as norepinephrine-releasing neurons, glia, and astrocytes) may also play a role in the metabolism of serotonin. (Adapted from Meyer, J.H. et al., *Arch Gen Psychiatry*, 63(11), 1209, 2006.)

of the prefrontal cortex.[46,197] 5-HT$_{2A}$ receptor density has an inverse relationship to extracellular serotonin levels,[39–42] and greater 5-HT$_{2A}$ receptor binding was observed in these regions during MDD with higher levels of pessimism.[25,27] 5-HTT has a role in clearing extracellular serotonin, and a positive correlation between total 5-HTT binding and severity of dysfunctional attitudes was found in the medial prefrontal cortex, the dorsolateral prefrontal cortex, and the anterior cingulate cortex.[84] Collectively, the findings suggest that pessimism in the anterior cingulate cortex and prefrontal cortex is associated with greater 5-HT$_{2A}$ binding, which may reflect ongoing loss of extracellular serotonin in these regions, and potential sources of loss are excessive 5-HTT binding, and elevated MAO-A levels (see Figure 9.6).

9.4.2 Postpartum, Suicide, and Monoamine Oxidase A

On average, during pregnancy, the risk of suicide is lower and it elevates during postpartum, yet is still less than other times in life.[198] However, when psychiatric illness such as postpartum depression (PPD) occurs, the risk of suicide elevates >10-fold.[199,200]

This issue, coupled with the fact that the general risk of a MDE during the postpartum period is 13%,[201–205] illustrates the importance of investigating the neurobiology of early postpartum related to the risk for MDE and suicide.

The most likely psychiatric disturbances in the postnatal period affect mood and there are three types. One is the "baby blues" or maternity blues, which is common (occurring up to 75% of the time) and transient, usually finishing within the first week postpartum.[206,207] The second is MDE with postpartum onset ("postpartum depression"), which is defined as a MDE that occurs within the first 4 weeks after delivery.[208] Twenty percent of women with maternity blues will go on to have a MDE with postpartum onset.[207,209] The third is postpartum psychosis (which may be accompanied by depressive symptoms). The third condition is rare occurring in 0.1%–0.2% of deliveries and is strongly associated with the presence of bipolar disorder.[210,211]

A profound change that occurs during the first week postpartum, when "postpartum blues" are common, is the enormous reduction in 17β-estradiol and estriol levels. It is well known that estradiol and estriol are produced by the placenta during pregnancy and that plasma levels raise 100-fold and 1000-fold, respectively.[212] It is also well known that estradiol and estriol levels decrease abruptly during postpartum with loss of the placenta. Most of the decline is in the first 4 days with a modest decline thereafter.[207,213,214] It is also a time of reduction in progesterone levels.

Declines in estrogen and progesterone have been implicated in generating sad mood. In an experimental paradigm involving a reduction in estrogen and progesterone, Bloch et al.[215] found that women with a history of PPD had lowering of mood. In this paradigm, eight women with a history of PPD and eight women without a history of PPD were given a gonadotrophin-releasing agonist for 8 weeks and then the agonist was withdrawn (to simulate delivery and loss of placenta). Five of the eight women who had a history of PPD had sustained lowered mood, whereas none of the women who had no history of PPD had sustained lowered mood. Additional support for an inverse relationship between estrogen and mood was supported by the fact that estrogen (17β-estradiol) administration has reduced mood symptoms in some preliminary clinical trials as a treatment for MDE with postpartum onset and MDE in postmenopausal women.[216–223] These findings implicate an inverse relationship between estrogen change and mood; however, it was unclear how this might link to the pathophysiology of MDD.

A candidate link between estrogen loss and the pathophysiology of MDD was the relationship between changes in estrogen levels and changes in MAO-A density. Within regions of high MAO-A density, repeated (increased) estrogen administration is associated with reductions in MAO-A density, mRNA, and/or activity in contrast to estrogen depletion, which is associated with increases in MAO-A density, mRNA, and/or activity: Estradiol administration for 1–3 weeks consistently lowers MAO-A activity in high density areas (amygdala and hypothalamus) in ovariectomized rats.[224–226] In ovariectomized macaque monkeys, 1 month of estradiol administration reduced MAO-A mRNA in the dorsal raphe nucleus.[227] In a separate study of ovariectomized macaque monkeys, 1 month of estradiol administration reduced MAO-A protein in the dorsal raphe region by ~50%.[228] Repeated estradiol administration is associated with a reduction in MAO-A activity in neuroblastoma cell lines. In a neuroblastoma cell line that expresses the human estrogen receptor, exposure to 17β-estradiol for 12 days lowers MAO-A activity.[229] This finding was replicated with exposure to physiological levels of 17β-estradiol for 10 days.[230] Although changes in several indices of MAO-A levels had been reported in relation to estrogen loss, MAO-A binding, mRNA, or activity had never been studied in any species in early postpartum.

Given the 100- to 1000-fold decline in estrogen over the first 4 days postpartum,[201,213,214] the relationship between estrogen decline and elevation in MAO-A synthesis, and the link between greater MAO-A levels and lower mood, there was a strong rationale to neuroimage MAO-A binding in brain regions involved in affect regulation during early postpartum (during days 4–6). Applying [^{11}C]harmine PET neuroimaging, we completed the first investigation of MAO-A binding in any species during the immediate postpartum period. The main finding is a significant elevation of MAO-A binding postpartum (magnitude of 43%) throughout all brain regions assayed (prefrontal cortex, anterior cingulate cortex, hippocampus, striatum, and thalamus) in the immediate postpartum period in healthy women (see Figure 9.7). A voxel-based analysis confirmed that the elevation in MAO-A binding was present throughout the gray matter of the brain (MAO-A density is minimal in white matter[152,153]).

This discovery provides a neurobiological model of postpartum blues in humans, involving a rapid decline in estrogen, followed by a rapid rise in MAO-A levels in affect-modulating structures in the brain, with subsequent sad mood and symptoms

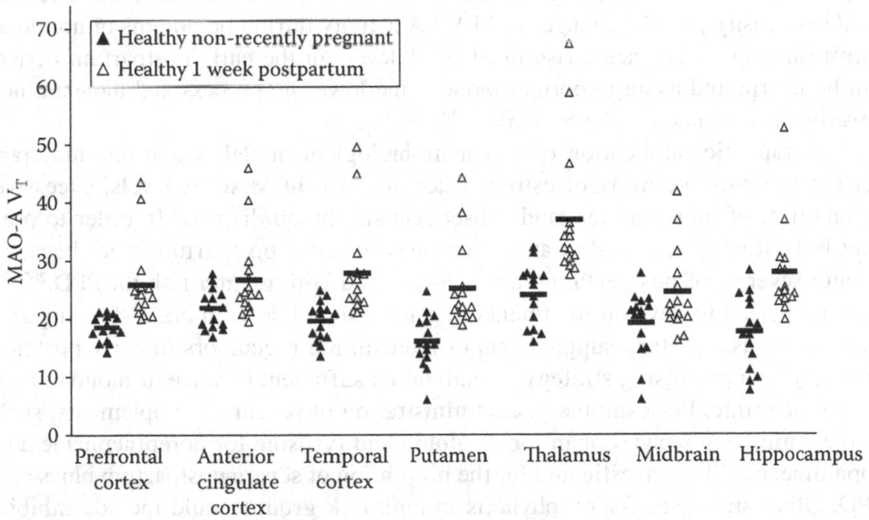

FIGURE 9.7 Monoamine oxidase A binding in the immediate postpartum period. MAO-A V_T is an index of harmine binding to tissue at equilibrium and is an index of MAO-A level. On average regional MAO-A, V_T was elevated by 43% during the early postpartum period. Differences were statistically significant in each brain region ($p < 0.001$ to $p = 0.04$). (From Sacher, J. et al., *Arch. Gen. Psychiatry*, 67(5), 468, 2010.)

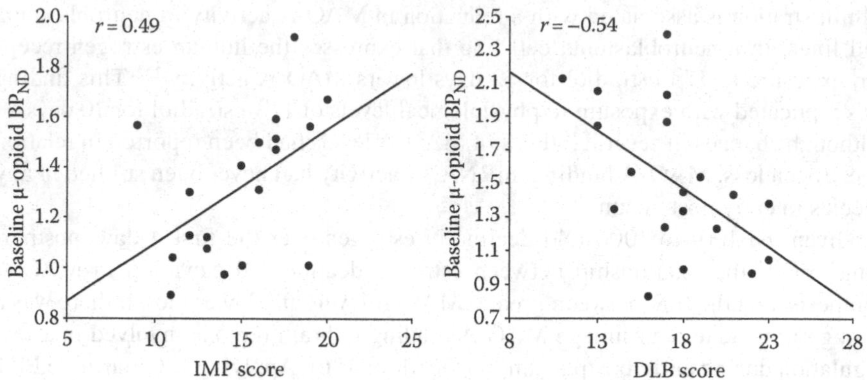

FIGURE 9.8 Association between impulsiveness (IMP) and deliberation (DLB) scores and baseline μ-opioid receptor availability. Areas in which significant differences in μ-receptor availability *in vivo* were observed between individuals with high and low IMP scores (left) and DLB scores (right). (From Love, T.M. et al., *Arch. Gen. Psychiatry*, 66(10), 1124, 2009.)

of postpartum blues. As mentioned earlier, estrogen removal has been associated with a rise in indexes of MAO-A levels,[224,227,228,230,231] and the findings shown in Figure 9.8 argue that this relationship applies strongly during the immediate postpartum period in humans. Consistent with this neurobiological model, MAO-A binding was highest on day 5 postpartum, the day most strongly associated with the lowest mood in most investigations of postpartum blues.[233] MAO-A V_T measured in early postpartum can be viewed as an index for MAO-A levels. Changes in levels of MAO-A density parallel changes in MAO-A activity during paradigms of hormone administration.[196] The acute rise in MAO-A levels in the early postpartum period can be interpreted as an important monoamine-lowering process and monoamine-lowering is associated with sad mood.[91,92,95,233–237]

A therapeutic implication of this neurobiological model is that one may target the proposed pathway of estrogen decline, rise in MAO-A levels, excessive metabolism of monoamines, and subsequent sad mood/distress. In order to prevent PPD, it may be useful to attenuate the severity of postpartum blues because greater severity of postpartum blues is associated with greater risk for PPD.[207,238] Given the need to develop treatments that are compatible with breast-feeding and widespread use, dietary supplements of monoamine precursors in early postpartum may be a promising strategy to maintain a sufficient balance of monoamines during this time. For example, the administration of precursor supplements, such as the amino acids tryptophan for serotonin and tyrosine for norepinephrine and dopamine, could be investigated for the prevention of severe postpartum blues and PPD. Other strategies for prophylaxis in high risk groups could include inhibiting MAO-A, or raising multiple monoamines (that are metabolized by MAO-A with antidepressants). Less invasive strategies such as administration of amino acid precursors have potential for women at low risk of developing PPD, whereas interventions such as MAO-A inhibitors could be considered for those at high risk for PPD.

9.5 NEUROCHEMICAL ABNORMALITIES ASSOCIATED WITH PERSONALITY TRAITS OF IMPULSIVITY

Since the etiology of suicidal behavioral is multifactorial, one useful contribution from neuroimaging is the ability to quantitate the neurochemistry associated with specific risk factors for suicide. Earlier in this chapter, the neurochemistry of dysfunctional attitudes, a construct related closely to hopelessness, was presented. Similarly, greater impulsivity represents another factor associated with greater risk for death by suicide. Personality assessment instruments have been applied in combination with brain imaging and to date have identified strong relationships between impulsivity and MAO-A binding (in all brain regions assessed), striatal dopamine release, and ventral striatal μ-opioid binding.

9.5.1 Monoamine Oxidase A, Impulsivity, and Deliberation

There is a significant body of work to argue that reductions in MAO-A function during development lead to impulsivity in rodents. In rodents, both MAO-A KO and MAO-A inhibitor administration during development are associated with greater aggression, decreased latency to attack in the resident–intruder paradigm, and enhanced fear conditioning.[239–243] MAO-A-deficient animals also show less positive social exploration and fewer investigative behaviors such as time spent exploring novelty, nose-poking, and social investigations of cage intruders.[240,242,243]

It has been argued that excess serotonin during neurodevelopment, which may occur during reduced MAO-A function,[110,154–157,160–165] is important in developing impulsive behavior. Excess brain serotonin (5-HT) that occurs during development disrupts the organization of somatosensory thalamocortical afferents and retinal afferents in the dorsal lateral geniculate nucleus and superior colliculus.[244,245] The abnormal development of thalamocortical afferents is also associated with abnormalities in the somatosensory cortex barrel fields.[245] It has been proposed that impulsive aggression observed in MAO-A KO mice is consequent to impairment in organizing sensory input secondary to the loss of barrel field structure in the somatosensory cortex.[246,247]

Although experimentally induced reductions in MAO-A function during development lead to impulsivity in animals, it need not necessarily be an influential mechanism in humans. However, there was reason to suspect that these animal models have relevance for some subpopulations of people. Men who are deficient in MAO-A, consequent to an X-chromosome defect at the vicinity of the *MAO-A* gene, often show aggressive behavior.[248] Also, a genotype associated with low levels of MAO-A from *in vitro* assay was associated with greater risk of developing antisocial behavior.[249,250] Low 5-hydroxyindoleacetic acid (5-HIAA) levels in cerebrospinal fluid (CSF) are associated with greater likelihood of developing antisocial personality disorder and aggressive criminal behavior.[251–253] While this dataset has been traditionally interpreted as a crude surrogate of extracellular serotonin levels, it is possible that it represents a reduction in MAO-A levels since lower 5-HIAA levels can occur when MAO-A is pharmacologically inhibited.[254,255]

With respect to phenotype, it was shown that reductions in MAO-A function during development are associated with greater impulsivity in animals, and it was shown that in rare cases in humans when there is a severe deficiency in MAO-A there is aggressive behavior.[248] However, this does not necessarily mean that the phenotypic relationship applies frequently, that is, it does not necessarily hold that MAO-A levels have a relationship with impulsivity levels in the general population. Two neuroimaging studies have investigated this question and demonstrated that there is a strong relationship between reductions in MAO-A binding and levels of aggression and impulsivity. In the first study, applying [^{11}C]clorgyline PET, Alia-Klein et al. demonstrated that reduced prefrontal cortex MAO-A binding is associated with self-reported aggression in males.[256] The second study investigated both genders, and in addition to replicating the relationship between prefrontal cortex MAO-A binding and the impulsive/aggressive facets of personality, it also identified a second covariation between MAO-A and deliberation, such that lower levels of deliberation were associated with lower MAO-A binding. In the latter study, 38% of the variance in prefrontal cortex MAO-A binding was accounted for by the personality facets of angry hostility and deliberation, clearly implicating a strong relationship between these two personality facets and MAO-A binding.[257] In both studies, the correlations between the personality traits and MAO-A binding were present throughout the brain regions assessed; however, this is probably due to the issue that MAO-A binding in different brain regions covaries and it is suspected that MAO-A binding in particular brain regions that participate in impulsivity and planning, such as the orbitofrontal cortex and dorsolateral prefrontal cortex,[258] is where the relationship between these two measures is most important.

9.5.2 Dopamine and Impulsivity

Dopamine release in the nucleus accumbens is associated with exposures to rewarding stimuli[259–261] and it was postulated that a greater biological tendency toward the reward from salient stimuli contributes toward impulsivity.[262] In a recent study applying the D_2/D_3 radiotracer [^{18}F]fallypride with PET, Buckholtz et al. discovered that trait impulsivity inversely associated with D_2/D_3 binding in the *substantia nigra*/ventral tegmental area.[263] They also found that trait impulsivity positively associated with striatal responsivity to D-amphetamine. In an earlier study, Dalley et al. found that reduced ventral striatal D_2/D_3 binding with [^{18}F]fallypride PET was also associated with greater levels of impulsivity.[262] Since reduced D_2/D_3 binding may occur when extracellular dopamine is increased, one interpretation of these findings is that in humans, lesser availability of autoreceptors in the *substantia nigra*/ventral tegmental area leads to lesser inhibitory function predisposing toward dopamine release, which, in the context of salient stimuli, leads to impulsiveness.

9.5.3 μ-Opioid Receptors and Impulsivity

Some behavioral studies in humans and rodents have supported a link between pharmacological challenges upon the μ-opioid receptor and impulsive behavior. Some studies report greater impulsive behavior in the context of rewarding circumstances when agonists are present and reduced impulsive behavior in similar context with some antagonists.[264–266] Furthermore, motivation and positive behavioral responses to rewards are enhanced by administration of μ-opioid agonists in the ventral striatum/nucleus accumbens.[268] Love et al. recently reported that high impulsiveness and low deliberation, two orthogonal facets of the NEO Personality Inventory Revised,[268] were associated with higher μ-opioid receptor binding in the ventral striatum, as well as in the anterior cingulate cortex and medial prefrontal cortex.[269] In addition, in this study, low deliberation was associated with greater right prefrontal cortex and thalamus binding. There is one study of μ-opioid binding, applying [^{3}H]DAGO autoradiography in suicide victims reporting increased binding in the inferior prefrontal cortex, cingulate gyrus, postcentral gyrus, and medial temporal gyrus, although binding was increased in every region assessed, and at least by 20%.[270] Given these findings, it would be interesting to assess in future neuroimaging work whether μ-opioid receptor binding and impulsivity are related in samples with psychiatric illnesses that have both impulsivity and greater risk for suicide.

9.6 CONCLUSIONS

While neuroimaging is limited by the range in biomarkers available, its ability for *in vivo* measurement enables discoveries bridging neurochemistry to specific clinic states. For example, the finding of greater 5-HT_{2A} binding in the dorsolateral prefrontal cortex of suicide victims can be translated into diagnostic and symptom specificity: 5-HT_{2A} binding is greater throughout the prefrontal cortex in MDD with high levels of pessimism (a symptom that creates risk for suicide).[25,27] The mechanism for this finding is best explained by excessive serotonin loss consequent to greater levels of functional serotonin transporter and high levels of monoamine oxidase A (leading to greater metabolism).[190,84] Greater monoamine oxidase A binding occurs throughout the brain in early postpartum and may explain the vulnerability to psychiatric illness, in particular MDD, which occurs in postpartum.[271] Since MDD during early postpartum is associated with more than a 10-fold greater risk of suicide, preventing PPD is important.[199,200] Potential biological strategies for prevention of PPD may involve suppressing the pathway between estrogen decline and MAO-A rise or dietary compensations for high MAO-A.

The findings of reduced ventral prefrontal 5-HTT binding in suicide victims are also present with similar regional specificity in a neuroimaging study of obsessive–compulsive disorder,[144,145] but reduced 5-HTT binding may also be present in this region in ecstasy abuse within the previous 4 months, and in the winter relative to summer.[139–143] Neuroimaging of impulsive macaque monkeys shows a pattern of brain 5-HTT loss that spares the ventral prefrontal region.[146] Further neuroimaging studies are needed to understand the diagnostic and symptom specificity of 5-HTT change related to suicide.

Another advantage of *in vivo* imaging measurement is that it can be related to clinical measures associated with suicide, that are also sampled *in vivo*, proximal to the time of brain scanning. For example, impulsivity is associated with particular neurochemical findings: The association of greater impulsivity with reduced regional brain MAO-A binding in humans[256,257] is implicated in altered neurodevelopment leading to altered neural architecture and organization of sensory input. Impulsivity in humans is also associated with neurochemical markers implicated in greater responsivity to reward such as enhanced ventral striatal dopamine release, reduced D_2/D_3 autoreceptor binding,[263] and greater ventral striatal μ-opioid receptor binding.[269] In the future, it is possible that these findings will represent new therapeutic targets for impulsivity-reducing treatments so as to reduce the risk of suicide.

REFERENCES

1. Arango V, Ernsberger P, Marzuk PM et al. Autoradiographic demonstration of increased serotonin 5-HT_2 and beta-adrenergic receptor binding sites in the brain of suicide victims. *Arch Gen Psychiatry*. 1990;47(11):1038–1047.
2. Arango V, Underwood M, Mann J. Alterations in monoamine receptors in the brain of suicide victims. *J Clin Psychopharm*. 1992;12(2):8S–12S.
3. Arora RC, Meltzer HY. Serotonergic measures in the brains of suicide victims: 5-HT_2 binding sites in the frontal cortex of suicide victims and control subjects. *Am J Psychiatry*. 1989;146(6):730–736.
4. Hrdina P, Vu T. Chronic fluoxetine treatment upregulates 5-HT uptake sites and 5-HT_2 receptors in rat brain: An autoradiographic study. *Synapse*. 1993;14:324–331.
5. Mann JJ, Stanley M, McBride PA, McEwen BS. Increased serotonin2 and beta-adrenergic receptor binding in the frontal cortices of suicide victims. *Arch Gen Psychiatry*. 1986;43(10):954–959.
6. Stanley M, Mann JJ. Increased serotonin-2 binding sites in frontal cortex of suicide victims. *Lancet*. 1983;1(8318):214–216.
7. Yates M, Leake A, Candy JM, Fairbairn AF, McKeith IG, Ferrier IN. 5HT_2 receptor changes in major depression. *Biol Psychiatry*. 1990;27(5):489–496.
8. Stockmeier CA, Dilley GE, Shapiro LA, Overholser JC, Thompson PA, Meltzer HY. Serotonin receptors in suicide victims with major depression. *Neuropsychopharmacology*. 1997;16(2):162–173.
9. Turecki G, Briere R, Dewar K et al. Prediction of level of serotonin 2A receptor binding by serotonin receptor 2A genetic variation in postmortem brain samples from subjects who did or did not commit suicide. *Am J Psychiatry*. 1999;156(9):1456–1458.
10. Pandey GN, Dwivedi Y, Rizavi HS et al. Higher expression of serotonin 5-HT(2A) receptors in the postmortem brains of teenage suicide victims. *Am J Psychiatry*. 2002;159(3):419–429.
11. Stockmeier CA. Involvement of serotonin in depression: Evidence from postmortem and imaging studies of serotonin receptors and the serotonin transporter. *J Psychiatr Res*. 2003;37(5):357–373.
12. Hoyer D, Pazos A, Probst A, Palacios JM. Serotonin receptors in the human brain. II. Characterization and autoradiographic localization of 5-HT_{1C} and 5-HT_2 recognition sites. *Brain Res*. 1986;376(1):97–107.
13. Marazziti D, Rossi A, Giannaccini G et al. Distribution and characterization of [^{3}H] mesulergine binding in human brain postmortem. *Eur Neuropsychopharmacol*. 1999;10(1):21–26.
14. Schmuck K, Ullmer C, Engels P, Lubbert H. Cloning and functional characterization of the human 5-HT_{2B} serotonin receptor. *FEBS Lett*. 1994;342(1):85–90.
15. Mann J. Postmortem studies of suicide victims. In: Watson S, ed. *Biology of Schizophrenia and Affective Disease*. London, U.K.: American Psychiatric Press Inc.; 1996, pp. 197–221.

16. Hrdina PD, Demeter E, Vu TB, Sotonyi P, Palkovits M. 5-HT uptake sites and 5-HT_2 receptors in brain of antidepressant-free suicide victims/depressives: Increase in 5-HT_2 sites in cortex and amygdala. *Brain Res.* 1993;614(1–2):37–44.
17. Dumais A, Lesage AD, Lalovic A et al. Is violent method of suicide a behavioral marker of lifetime aggression? *Am J Psychiatry.* 2005;162(7):1375–1378.
18. Dhaenen H, Bossuyt A, Mertens J, Bossuyt-Piron C, Gijesmans M, Kaufman L. SPECT imaging of serotonin2 receptors in depression. *Psychiatry Res Neuroimaging.* 1992;45:227–237.
19. Biver F, Wikler D, Lotstra F, Damhaut P, Goldman S, Mendlewicz J. Serotonin 5-HT_2 receptor imaging in major depression: Focal changes in orbito-insular cortex. *Br J Psychiatry.* 1997;171:444–448.
20. Attar-Levy D, Martinot J-L, Blin J et al. The cortical serotonin2 receptors studied with positron emission tomography and [^{18}F]-setoperone during depressive illness and antidepressant treatment with clomipramine. *Biol Psychiatry.* 1999;45:180–186.
21. Meyer J, Kapur S, Houle S et al. Prefrontal cortex 5-HT_2 receptors in depression: A [^{18}F] setoperone PET imaging study. *Am J Psychiatry.* 1999;156:1029–1034.
22. Meltzer C, Price J, Mathis C et al. PET imaging of serotonin type 2A receptors in late-life neuropsychiatric disorders. *Am J Psychiatry.* 1999;156(12):1871–1878.
23. Yatham LN, Liddle PF, Shiah IS et al. Brain serotonin2 receptors in major depression: A positron emission tomography study. *Arch Gen Psychiatry.* 2000;57(9):850–858.
24. Messa C, Colombo C, Moresco RM et al. 5-HT(2A) receptor binding is reduced in drug-naive and unchanged in SSRI-responder depressed patients compared to healthy controls: A PET study. *Psychopharmacology (Berl).* 2003;167(1):72–78.
25. Meyer JH, McMain S, Kennedy SH et al. Dysfunctional attitudes and 5-HT(2) receptors during depression and self-harm. *Am J Psychiatry.* 2003;160(1):90–99.
26. Mintun MA, Sheline YI, Moerlein SM, Vlassenko AG, Huang Y, Snyder AZ. Decreased hippocampal 5-HT_{2A} receptor binding in major depressive disorder: In vivo measurement with [^{18}F]altanserin positron emission tomography. *Biol Psychiatry.* 2004;55(3):217–224.
27. Bhagwagar Z, Hinz R, Taylor M, Fancy S, Cowen P, Grasby P. Increased 5-HT(2A) receptor binding in euthymic, medication-free patients recovered from depression: A positron emission study with [(11)C]MDL 100,907. *Am J Psychiatry.* 2006;163(9):1580–1587.
28. Meyer JH. Applying neuroimaging ligands to study major depressive disorder. *Semin Nucl Med.* 2008;38(4):287–304.
29. van Heeringen C, Audenaert K, Van Laere K et al. Prefrontal 5-HT_{2a} receptor binding index, hopelessness and personality characteristics in attempted suicide. *J Affect Disord.* 2003;74(2):149–158.
30. Audenaert K, Van Laere K, Dumont F et al. Decreased frontal serotonin 5-HT 2a receptor binding index in deliberate self-harm patients. *Eur J Nucl Med.* 2001;28(2):175–182.
31. Blin J, Pappata S, Kiyosawa M, Crouzel C, Baron J. [^{18}F]Setoperone: A new high-affinity ligand for positron emission tomography study of the serotonin-2 receptors in baboon brain in vivo. *Eur J Pharmacol.* 1988;147:73–82.
32. Blin J, Sette G, Fiorelli M et al. A method for the in vivo investigation of the serotonergic 5-HT_2 receptors in the human cerebral cortex using positron emission tomography and 18F-labeled setoperone. *J Neurochem.* 1990;54(5):1744–1754.
33. Petit-Taboue MC, Landeau B, Osmont A, Tillet I, Barre L, Baron JC. Estimation of neocortical serotonin-2 receptor binding potential by single-dose fluorine-18-setoperone kinetic PET data analysis. *J Nucl Med.* 1996;37(1):95–104.
34. Maziere B, Crouzel C, Venet M et al. Synthesis, affinity and specificity of 18F-setoperone, a potential ligand for in-vivo imaging of cortical serotonin receptors. *Nucl Med Biol.* 1988;15(4):463–468.

35. Fischman A, Bonab A, Babich J et al. Positron emission tomographic analysis of central 5-hydroxytryptamine 2 receptor occupancy in healthy volunteers treated with the novel antipsychotic agent ziprasidone. *J Pharmacol Exp Ther.* 1996;279(2):939–947.
36. Pinborg LH, Adams KH, Svarer C et al. Quantification of 5-HT_{2A} receptors in the human brain using [^{18}F]altanserin-PET and the bolus/infusion approach. *J Cereb Blood Flow Metab.* 2003;23(8):985–996.
37. Barraclough B, Bunch J, Nelson B, Sainsbury P. A hundred cases of suicide: Clinical aspects. *Br J Psychiatry.* 1974;125:355–373.
38. Robins E, Murphy G, Wilkinson R, Gassner S, Kayes J. Some clinical considerations in the prevention of suicide based on a study of 134 successful suicides. *Am J Public Health.* 1959;49(7):888–899.
39. O'Regan D, Kwok RP, Yu PH, Bailey BA, Greenshaw AJ, Boulton AA. A behavioural and neurochemical analysis of chronic and selective monoamine oxidase inhibition. *Psychopharmacology.* 1987;92(1):42–47.
40. Roth B, McLean S, Zhu X, Chuang D. Characterization of two [^{3}H]ketanserin recognition sites in rat striatum. *J Neurochem.* 1987;49(6):1833–1838.
41. Stockmeier CA, Kellar KJ. In vivo regulation of the serotonin-2 receptor in rat brain. *Life Sci.* 1986;38(2):117–127.
42. Todd KG, McManus DJ, Baker GB. Chronic administration of the antidepressants phenelzine, desipramine, clomipramine, or maprotiline decreases binding to 5-hydroxytryptamine2A receptors without affecting benzodiazepine binding sites in rat brain. *Cell Mol Neurobiol.* 1995;15(3):361–370.
43. Weissman A. The dysfunctional attitude scale: A validation study. *Dissert Abstr Int.* 1979;40:1389B–1390B.
44. Simons AD, Murphy GE, Levine JL, Wetzel RD. Cognitive therapy and pharmacotherapy for depression. Sustained improvement over one year. *Arch Gen Psychiatry.* 1986;43(1):43–48.
45. Fava M, Bless E, Otto M, Pava J, Rosenbaum J. Dysfunctional attitudes in major depression changes with pharmacotherapy. *J Nerv Ment Dis.* 1994;182(1):45–49.
46. Sharot T, Riccardi AM, Raio CM, Phelps EA. Neural mechanisms mediating optimism bias. *Nature.* 2007;450(7166):102–105.
47. Mitterschiffthaler MT, Williams SC, Walsh ND et al. Neural basis of the emotional Stroop interference effect in major depression. *Psychol Med.* 2008;38(2):247–256.
48. Elliott R, Rubinsztein JS, Sahakian BJ, Dolan RJ. The neural basis of mood-congruent processing biases in depression. *Arch Gen Psychiatry.* 2002;59(7):597–604.
49. Taylor Tavares JV, Clark L, Furey ML, Williams GB, Sahakian BJ, Drevets WC. Neural basis of abnormal response to negative feedback in unmedicated mood disorders. *Neuroimage.* 2008;42(3):1118–1126.
50. Dohr K, Rush A, Bernstein I. Cognitive biases and depression. *J Abnorm Psychol.* 1989;98(3):263–267.
51. Simons AD, Garfield SL, Murphy GE. The process of change in cognitive therapy and pharmacotherapy for depression. Changes in mood and cognition. *Arch Gen Psychiatry.* 1984;41(1):45–51.
52. Cane D, Olinger L, Gotlib I, Kuiper N. Factor structure of the dysfunctional attitude scale in a student population. *J Clin Psychol.* 1986;42:307–309.
53. Oliver J, Baumgart E. The dysfunctional attitude scale: Psychometric properties and relation to depression in an unselected adult population. *Cog Ther Res.* 1985;9:161–167.
54. Frokjaer VG, Mortensen EL, Nielsen FA et al. Frontolimbic serotonin 2A receptor binding in healthy subjects is associated with personality risk factors for affective disorder. *Biol Psychiatry.* 2008;63(6):569–576.

55. Bouvard M, Charles S, Guerin J, Aimard G, Cottraux J. [Study of Beck's hopelessness scale. Validation and factor analysis]. *Encephale*. 1992;18(3):237–240.
56. Cannon B, Mulroy R, Otto MW, Rosenbaum JF, Fava M, Nierenberg AA. Dysfunctional attitudes and poor problem solving skills predict hopelessness in major depression. *J Affect Disord*. 1999;55(1):45–49.
57. DeRubeis RJ, Evans MD, Hollon SD, Garvey MJ, Grove WM, Tuason VB. How does cognitive therapy work? Cognitive change and symptom change in cognitive therapy and pharmacotherapy for depression. *J Consult Clin Psychol*. 1990;58(6):862–869.
58. Norman W, Miller I, Dow M. Characteristics of depressed patients with elevated levels of dysfunctional cognitions. *Cog Ther Res*. 1988;12:39–51.
59. Beck A, Steer R, Kovacs M, Garrison B. Hopelessness and eventual suicide: A 10-year prospective study of patients hospitalized with suicidal ideation. *Am J Psychiatry*. 1985;142(5):559–563.
60. Beck AT, Brown G, Steer RA. Prediction of eventual suicide in psychiatric inpatients by clinical ratings of hopelessness. *J Consult Clin Psychol*. 1989;57(2):309–310.
61. Brodsky BS, Malone KM, Ellis SP, Dulit RA, Mann JJ. Characteristics of borderline personality disorder associated with suicidal behavior. *Am J Psychiatry*. 1997;154(12):1715–1719.
62. Soloff PH, Price JC, Meltzer CC, Fabio A, Frank GK, Kaye WH. $5HT_{2A}$ receptor binding is increased in borderline personality disorder. *Biol Psychiatry*. 2007;62(6):580–587.
63. Meyer JH, Wilson AA, Rusjan P et al. Serotonin2A receptor binding potential in people with aggressive and violent behaviour. *J Psychiatry Neurosci*. 2008;33(6):499–508.
64. Jakab R, Goldman-Rakic P. 5-Hydroxytryptamine2A serotonin receptors in the primate cerebral cortex: Possible site of action of hallucinogenic and antipsychotic drugs in pyramidal cell apical dendrites. *Proc Natl Acad Sci U. S. A*. 1998;95:735–740.
65. Santana N, Bortolozzi A, Serrats J, Mengod G, Artigas F. Expression of serotonin1A and serotonin2A receptors in pyramidal and GABAergic neurons of the rat prefrontal cortex. *Cereb Cortex*. 2004;14(10):1100–1109.
66. Wu C, Singh SK, Dias P, Kumar S, Mann DM. Activated astrocytes display increased $5\text{-}HT_{2a}$ receptor expression in pathological states. *Exp Neurol*. 1999;158(2):529–533.
67. Jacobs B, Driscoll L, Schall M. Life-span dendritic and spine changes in areas 10 and 18 of human cortex: A quantitative Golgi study. *J Comp Neurol*. 1997;386(4):661–680.
68. Rosell DR, Thompson JL, Slifstein M et al. Increased serotonin 2A receptor availability in the orbitofrontal cortex of physically aggressive personality disordered patients. *Biol Psychiatry*. 2010;67(12):1154–1162.
69. Bubar MJ, Cunningham KA. Prospects for serotonin $5\text{-}HT_{2R}$ pharmacotherapy in psychostimulant abuse. *Prog Brain Res*. 2008;172:319–346.
70. Meyer JH. Imaging the serotonin transporter during major depressive disorder and antidepressant treatment. *J Psychiatry Neurosci*. 2007;32(2):86–102.
71. Brucke T, Kornhuber J, Angelberger P, Asenbaum S, Frassine H, Podreka I. SPECT imaging of dopamine and serotonin transporters with [^{123}I]beta-CIT. Binding kinetics in the human brain. *J Neural Transm Gen Sect*. 1993;94(2):137–146.
72. Innis RB, Seibyl JP, Scanley BE et al. Single photon emission computed tomographic imaging demonstrates loss of striatal dopamine transporters in Parkinson disease. *Proc Natl Acad Sci U. S. A*. 1993;90(24):11965–11969.
73. Kuikka JT, Bergstrom KA, Vanninen E, Laulumaa V, Hartikainen P, Lansimies E. Initial experience with single-photon emission tomography using iodine-123-labelled 2 beta-carbomethoxy-3 beta-(4-iodophenyl) tropane in human brain. *Eur J Nucl Med*. 1993;20(9):783–786.

74. Carroll FI, Kotian P, Dehghani A et al. Cocaine and 3 beta-(4′-substituted phenyl)tropane-2 beta-carboxylic acid ester and amide analogues. New high-affinity and selective compounds for the dopamine transporter. *J Med Chem*. 1995;38(2):379–388.
75. Laruelle M, Giddings SS, Zea-Ponce Y et al. Methyl 3 beta-(4-[^{125}I]iodophenyl)tropane-2 beta-carboxylate in vitro binding to dopamine and serotonin transporters under "physiological" conditions. *J Neurochem*. 1994;62(3):978–986.
76. Ciliax BJ, Drash GW, Staley JK et al. Immunocytochemical localization of the dopamine transporter in human brain. *J Comp Neurol*. 1999;409(1):38–56.
77. Parsey RV, Kegeles LS, Hwang DR et al. In vivo quantification of brain serotonin transporters in humans using [^{11}C]McN 5652. *J Nucl Med*. 2000;41(9):1465–1477.
78. Kent JM, Coplan JD, Lombardo I et al. Occupancy of brain serotonin transporters during treatment with paroxetine in patients with social phobia: A positron emission tomography study with ^{11}C McN 5652. *Psychopharmacology (Berl)*. 2002;164(4):341–348.
79. Ikoma Y, Suhara T, Toyama H et al. Quantitative analysis for estimating binding potential of the brain serotonin transporter with [^{11}C]McN5652. *J Cereb Blood Flow Metab*. 2002;22(4):490–501.
80. Buck A, Gucker PM, Schonbachler RD et al. Evaluation of serotonergic transporters using PET and [^{11}C](+)McN-5652: Assessment of methods. *J Cereb Blood Flow Metab*. 2000;20(2):253–262.
81. Ginovart N, Wilson AA, Meyer JH, Hussey D, Houle S. Positron emission tomography quantification of [(11)C]-DASB binding to the human serotonin transporter: Modeling strategies. *J Cereb Blood Flow Metab*. 2001;21(11):1342–1353.
82. Houle S, Ginovart N, Hussey D, Meyer J, Wilson A. Imaging the serotonin transporter with positron emission tomography: Initial human studies with [^{11}C]DAPP and [^{11}C] DASB. *Eur J Nucl Med*. 2000;27(11):1719–1722.
83. Ichise M, Liow JS, Lu JQ et al. Linearized reference tissue parametric imaging methods: Application to [^{11}C]DASB positron emission tomography studies of the serotonin transporter in human brain. *J Cereb Blood Flow Metab*. 2003;23(9):1096–1112.
84. Meyer JH, Houle S, Sagrati S et al. Brain serotonin transporter binding potential measured with carbon 11-labeled DASB positron emission tomography: Effects of major depressive episodes and severity of dysfunctional attitudes. *Arch Gen Psychiatry*. 2004;61(12):1271–1279.
85. Meyer JH, Wilson AA, Ginovart N et al. Occupancy of serotonin transporters by paroxetine and citalopram during treatment of depression: A [(11)C]DASB PET imaging study. *Am J Psychiatry*. 2001;158(11):1843–1849.
86. Meyer JH, Wilson AA, Sagrati S et al. Serotonin transporter occupancy of five selective serotonin reuptake inhibitors at different doses: An [^{11}C]DASB positron emission tomography study. *Am J Psychiatry*. 2004;161(5):826–835.
87. Praschak-Rieder N, Wilson AA, Hussey D et al. Effects of tryptophan depletion on the serotonin transporter in healthy humans. *Biol Psychiatry*. 2005;58(10):825–830.
88. Wilson A, Schmidt M, Ginovart N, Meyer J, Houle S. Novel radiotracers for imaging the serotonin transporter by positron emission tomography: Synthesis, radiosynthesis, in vitro and ex vivo evaluation of [^{11}C]-labelled 2-(phenylthio) araalkylamines. *J Med Chem*. 2000;43(16):3103–3110.
89. Wilson AA, Ginovart N, Hussey D, Meyer J, Houle S. In vitro and in vivo characterisation of [^{11}C]-DASB: A probe for in vivo measurements of the serotonin transporter by positron emission tomography. *Nucl Med Biol*. 2002;29(5):509–515.
90. Takano A, Suhara T, Ichimiya T, Yasuno F, Suzuki K. Time course of in vivo 5-HTT transporter occupancy by fluvoxamine. *J Clin Psychopharmacol*. 2006;26(2):188–191.
91. Leyton M, Young SN, Blier P et al. The effect of tryptophan depletion on mood in medication-free, former patients with major affective disorder. *Neuropsychopharmacology*. 1997;16(4):294–297.

92. Neumeister A, Nugent AC, Waldeck T et al. Neural and behavioral responses to tryptophan depletion in unmedicated patients with remitted major depressive disorder and controls. *Arch Gen Psychiatry*. 2004;61(8):765–773.
93. Neumeister A, Praschak-Rieder N, Hesselmann B et al. Effects of tryptophan depletion in drug-free depressed patients who responded to total sleep deprivation. *Arch Gen Psychiatry*. 1998;55(2):167–172.
94. Smith K, Clifford E, Hockney R, Clark M, Cowen P. Effect of tryptophan depletion on mood in male and female volunteers: A pilot study. *Hum Psychopharm*. 1997;12:111–117.
95. Young SN, Smith SE, Pihl RO, Ervin FR. Tryptophan depletion causes a rapid lowering of mood in normal males. *Psychopharmacology*. 1985;87(2):173–177.
96. Kontur PJ, al-Tikriti M, Innis RB, Roth RH. Postmortem stability of monoamines, their metabolites, and receptor binding in rat brain regions. *J Neurochem*. 1994;62(1):282–290.
97. Mann J, Underwood M, Arango V. Postmortem studies of suicide victims. *Biology of Schizophrenia and Affective Disorders*. Washington, DC: American Psychiatric Press; 1996, pp. 197–221.
98. Linnet K, Koed K, Wiborg O, Gregersen N. Serotonin depletion decreases serotonin transporter mRNA levels in rat brain. *Brain Res*. 1995;697(1–2):251–253.
99. Xiao Q, Pawlyk A, Tejani-Butt SM. Reserpine modulates serotonin transporter mRNA levels in the rat brain. *Life Sci*. 1999;64(1):63–68.
100. Yu A, Yang J, Pawlyk AC, Tejani-Butt SM. Acute depletion of serotonin down-regulates serotonin transporter mRNA in raphe neurons. *Brain Res*. 1995;688(1–2):209–212.
101. Benmansour S, Cecchi M, Morilak D et al. Effects of chronic antidepressant treatments on serotonin transporter function, density and mRNA level. *J Neurosci*. 1999;19(23):10494–10501.
102. Dewar KM, Grondin L, Carli M, Lima L, Reader TA. [^{3}H]Paroxetine binding and serotonin content of rat cortical areas, hippocampus, neostriatum, ventral mesencephalic tegmentum, and midbrain raphe nuclei region following *p*-chlorophenylalanine and *p*-chloroamphetamine treatment. *J Neurochem*. 1992;58(1):250–257.
103. Graham D, Tahraoui L, Langer SZ. Effect of chronic treatment with selective monoamine oxidase inhibitors and specific 5-hydroxytryptamine uptake inhibitors on [^{3}H] paroxetine binding to cerebral cortical membranes of the rat. *Neuropharmacology*. 1987;26(8):1087–1092.
104. Gordon I, Weizman R, Rehavi M. Modulatory effect of agents active in the presynaptic dopaminergic system on the striatal dopamine transporter. *Eur J Pharmacol*. 1996;298(1):27–30.
105. Han S, Rowell PP, Carr LA. D2 Autoreceptors are not involved in the down-regulation of the striatal dopamine transporter caused by alpha-methyl-*p*-tyrosine. *Res Commun Mol Pathol Pharmacol*. 1999;104(3):331–338.
106. Ikawa K, Watanabe A, Kaneno S, Toru M. Modulation of [^{3}H]mazindol binding sites in rat striatum by dopaminergic agents. *Eur J Pharmacol*. 1993;250(2):261–266.
107. Kilbourn MR, Sherman PS, Pisani T. Repeated reserpine administration reduces in vivo [^{18}F]GBR 13119 binding to the dopamine uptake site. *Eur J Pharmacol*. 1992;216(1):109–112.
108. Lee CM, Javitch JA, Snyder SH. Recognition sites for norepinephrine uptake: Regulation by neurotransmitter. *Science*. 1983;220(4597):626–629.
109. Laruelle M. Imaging synaptic neurotransmission with in vivo binding competition techniques: A critical review. *J Cereb Blood Flow Metab*. 2000;20(3):423–451.
110. Celada P, Artigas F. Monoamine oxidase inhibitors increase preferentially extracellular 5-hydroxytryptamine in the midbrain raphe nuclei. A brain microdialysis study in the awake rat. *Naunyn Schmiedebergs Arch Pharmacol*. 1993;347(6):583–590.

111. Malyszko J, Urano T, Serizawa K et al. Serotonergic measures in blood and brain and their correlations in rats treated with tranylcypromine, a monoamine oxidase inhibitor. *Jpn J Physiol.* 1993;43(5):613–626.
112. Ferrer A, Artigas F. Effects of single and chronic treatment with tranylcypromine on extracellular serotonin in rat brain. *Eur J Pharmacol.* 1994;263(3):227–234.
113. Ginovart N, Wilson AA, Meyer JH, Hussey D, Houle S. [^{11}C]-DASB, a tool for in vivo measurement of SSRI-induced occupancy of the serotonin transporter: PET characterization and evaluation in cats. *Synapse.* 2003;47(2):123–133.
114. Lundquist P, Wilking H, Hoglund AU et al. Potential of [^{11}C]DASB for measuring endogenous serotonin with PET: Binding studies. *Nucl Med Biol.* 2005;32(2):129–136.
115. Talbot PS, Frankle WG, Hwang DR et al. Effects of reduced endogenous 5-HT on the in vivo binding of the serotonin transporter radioligand 11C-DASB in healthy humans. *Synapse.* 2005;55(3):164–175.
116. Malone KM, Oquendo MA, Haas GL, Ellis SP, Li S, Mann JJ. Protective factors against suicidal acts in major depression: Reasons for living. *Am J Psychiatry.* 2000;157(7):1084–1088.
117. Bligh-Glover W, Kolli TN, Shapiro-Kulnane L et al. The serotonin transporter in the midbrain of suicide victims with major depression. *Biol Psychiatry.* 2000;47(12):1015–1024.
118. Klimek V, Roberson G, Stockmeier CA, Ordway GA. Serotonin transporter and MAO-B levels in monoamine nuclei of the human brainstem are normal in major depression. *J Psychiatr Res.* 2003;37(5):387–397.
119. Perry EK, Marshall EF, Blessed G, Tomlinson BE, Perry RH. Decreased imipramine binding in the brains of patients with depressive illness. *Br J Psychiatry.* 1983;142:188–192.
120. Crow TJ, Cross AJ, Cooper SJ et al. Neurotransmitter receptors and monoamine metabolites in the brains of patients with Alzheimer-type dementia and depression, and suicides. *Neuropharmacology.* 1984;23(12B):1561–1569.
121. Mann JJ, Huang YY, Underwood MD et al. A serotonin transporter gene promoter polymorphism (5-HTTLPR) and prefrontal cortical binding in major depression and suicide [see comments]. *Arch Gen Psychiatry.* 2000;57(8):729–738.
122. Arango V, Underwood MD, Boldrini M et al. Serotonin 1A receptors, serotonin transporter binding and serotonin transporter mRNA expression in the brainstem of depressed suicide victims. *Neuropsychopharmacology.* 2001;25(6):892–903.
123. Austin MC, Whitehead RE, Edgar CL, Janosky JE, Lewis DA. Localized decrease in serotonin transporter-immunoreactive axons in the prefrontal cortex of depressed subjects committing suicide. *Neuroscience.* 2002;114(3):807–815.
124. Hrdina P, Foy B, Hepner A, Summers R. Antidepressant binding sites in brain: Autoradiographic comparison of [^{3}H]paroxetine and [^{3}H]imipramine localization and relationship to serotonin transporter. *J Pharm Exp Ther.* 1990;252(1):410–418.
125. Lawrence KM, De Paermentier F, Cheetham SC, Crompton MR, Katona CL, Horton RW. Brain 5-HT uptake sites, labelled with [^{3}H]paroxetine, in antidepressant-free depressed suicides. *Brain Res.* 1990;526(1):17–22.
126. Leake A, Fairbairn AF, McKeith IG, Ferrier IN. Studies on the serotonin uptake binding site in major depressive disorder and control post-mortem brain: Neurochemical and clinical correlates. *Psychiatry Res.* 1991;39(2):155–165.
127. Little KY, McLauglin DP, Ranc J et al. Serotonin transporter binding sites and mRNA levels in depressed persons committing suicide. *Biol Psychiatry.* 1997;41(12):1156–1164.
128. Hendricksen M, Thomas AJ, Ferrier IN, Ince P, O'Brien JT. Neuropathological study of the dorsal raphe nuclei in late-life depression and Alzheimer's disease with and without depression. *Am J Psychiatry.* 2004;161(6):1096–1102.

129. Herold N, Uebelhack K, Franke L et al. Imaging of serotonin transporters and its blockade by citalopram in patients with major depression using a novel SPECT ligand [(123) I]-ADAM. *J Neural Transm*. 2006;113(5):659–670.
130. Ichimiya T, Suhara T, Sudo Y et al. Serotonin transporter binding in patients with mood disorders: A PET study with [^{11}C](+)McN5652. *Biol Psychiatry*. 2002;51(9):715–722.
131. Cannon DM, Ichise M, Rollis D et al. Elevated serotonin transporter binding in major depressive disorder assessed using positron emission tomography and [(11)C]DASB; Comparison with bipolar disorder. *Biol Psychiatry*. 2007;62(8):870–877.
132. Malison RT, Price LH, Berman R et al. Reduced brain serotonin transporter availability in major depression as measured by [^{123}I]-2 beta-carbomethoxy-3 beta-(4-iodophenyl)tropane and single photon emission computed tomography [see comments]. *Biol Psychiatry*. 1998;44(11):1090–1098.
133. Newberg AB, Amsterdam JD, Wintering N et al. ^{123}I-ADAM binding to serotonin transporters in patients with major depression and healthy controls: A preliminary study. *J Nucl Med*. 2005;46(6):973–977.
134. Parsey RV, Hastings RS, Oquendo MA et al. Lower serotonin transporter binding potential in the human brain during major depressive episodes. *Am J Psychiatry*. 2006;163(1):52–58.
135. Selvaraj S, Venkatesha Murthy N, Bhagwagar Z et al. Diminished brain 5-HT transporter binding in major depression: A positron emission tomography study with [(11)C] DASB. *Psychopharmacology (Berl)*. 2011;213(2–3):555–562.
136. Joensuu M, Tolmunen T, Saarinen PI et al. Reduced midbrain serotonin transporter availability in drug-naive patients with depression measured by SERT-specific [(123)I] nor-beta-CIT SPECT imaging. *Psychiatry Res*. 2007;154(2):125–131.
137. Arango V, Underwood MD, Gubbi AV, Mann JJ. Localized alterations in pre- and postsynaptic serotonin binding sites in the ventrolateral prefrontal cortex of suicide victims. *Brain Res*. 1995;688(1–2):121–133.
138. Suhara T, Takano A, Sudo Y et al. High levels of serotonin transporter occupancy with low-dose clomipramine in comparative occupancy study with fluvoxamine using positron emission tomography. *Arch Gen Psychiatry*. 2003;60(4):386–391.
139. Kalbitzer J, Erritzoe D, Holst KK et al. Seasonal changes in brain serotonin transporter binding in short serotonin transporter linked polymorphic region-allele carriers but not in long-allele homozygotes. *Biol Psychiatry*. 2010;67(11):1033–1039.
140. Praschak-Rieder N, Willeit M, Wilson AA, Houle S, Meyer JH. Seasonal variation in human brain serotonin transporter binding. *Arch Gen Psychiatry*. 2008;65(9):1072–1078.
141. McCann UD, Szabo Z, Seckin E et al. Quantitative PET studies of the serotonin transporter in MDMA users and controls using [^{11}C]McN5652 and [^{11}C]DASB. *Neuropsychopharmacology*. 2005;30(9):1741–1750.
142. Selvaraj S, Hoshi R, Bhagwagar Z et al. Brain serotonin transporter binding in former users of MDMA ('ecstasy'). *Br J Psychiatry*. 2009;194(4):355–359.
143. Kish SJ, Lerch J, Furukawa Y et al. Decreased cerebral cortical serotonin transporter binding in ecstasy users: A positron emission tomography/[(11)C]DASB and structural brain imaging study. *Brain*. 2010;133(Pt 6):1779–1797.
144. Matsumoto R, Ichise M, Ito H et al. Reduced serotonin transporter binding in the insular cortex in patients with obsessive–compulsive disorder: A [^{11}C]DASB PET study. *Neuroimage*. 2010;49(1):121–126.
145. Reimold M, Smolka MN, Zimmer A et al. Reduced availability of serotonin transporters in obsessive–compulsive disorder correlates with symptom severity—A [^{11}C]DASB PET study. *J Neural Transm*. 2007;114(12):1603–1609.

146. Ichise M, Vines DC, Gura T et al. Effects of early life stress on [^{11}C]DASB positron emission tomography imaging of serotonin transporters in adolescent peer- and mother-reared rhesus monkeys. *J Neurosci*. 2006;26(17):4638–4643.
147. Takano A, Arakawa R, Hayashi M, Takahashi H, Ito H, Suhara T. Relationship between neuroticism personality trait and serotonin transporter binding. *Biol Psychiatry*. 2007;62(6):588–592.
148. Kalbitzer J, Frokjaer VG, Erritzoe D et al. The personality trait openness is related to cerebral 5-HTT levels. *Neuroimage*. 2009;45(2):280–285.
149. Frankle WG, Lombardo I, New AS et al. Brain serotonin transporter distribution in subjects with impulsive aggressivity: A positron emission study with [^{11}C]McN 5652. *Am J Psychiatry*. 2005;162(5):915–923.
150. Koch W, Schaaff N, Popperl G et al. [I-123]ADAM and SPECT in patients with borderline personality disorder and healthy control subjects. *J Psychiatry Neurosci*. 2007;32(4):234–240.
151. Brown AK, George DT, Fujita M et al. PET [^{11}C]DASB imaging of serotonin transporters in patients with alcoholism. *Alcohol Clin Exp Res*. 2007;31(1):28–32.
152. Saura J, Bleuel Z, Ulrich J et al. Molecular neuroanatomy of human monoamine oxidases A and B revealed by quantitative enzyme radioautography and in situ hybridization histochemistry. *Neuroscience*. 1996;70(3):755–774.
153. Saura J, Kettler R, Da Prada M, Richards JG. Quantitative enzyme radioautography with 3H-Ro 41-1049 and 3H-Ro 19-6327 in vitro: Localization and abundance of MAO-A and MAO-B in rat CNS, peripheral organs, and human brain. *J Neurosci*. 1992;12(5):1977–1999.
154. Fowler C, Oreland L. Substrate-selective interaction between monoamine oxidase and oxygen. In: Singer T, Von Korff R, Murphy D, eds. *Monoamine Oxidase: Structure, Function, and Altered Functions*. New York: Academic Press, Inc.; 1979, pp. 145–151.
155. Kinemuchi H, Fowler C, Tipton K. Substrate specificities of the two forms of monoamine oxidase. In: Tipton K, Dostert P, Strolin-Benedetti M, eds. *Monoamine Oxidase and Disease: Prospects for Therapy with Reversible Inhibitors*. New York: Academic Press, Inc.; 1984, pp. 53–62.
156. Schoepp DD, Azzaro AJ. Specificity of endogenous substrates for types A and B monoamine oxidase in rat striatum. *J Neurochem*. 1981;36(6):2025–2031.
157. White H, Tansik R. Characterization of multiple substrate binding sites of MAO. In: Singer T, Von Korff R, Murphy D, eds. *Monoamine Oxidase: Structure, Function and Altered Functions*. New York: Academic Press, Inc.; 1979, pp. 129–144.
158. Konradi C, Svoma E, Jellinger K, Riederer P, Denney R, Thibault J. Topographic immunocytochemical mapping of monoamine oxidase-A, monoamine oxidase-B and tyrosine hydroxylase in human post mortem brain stem. *Neuroscience*. 1988;26(3):791–802.
159. Luque JM, Kwan SW, Abell CW, Da Prada M, Richards JG. Cellular expression of mRNAs encoding monoamine oxidases A and B in the rat central nervous system. *J Comp Neurol*. 1995;363(4):665–680.
160. Fagervall I, Ross SB. A and B forms of monoamine oxidase within the monoaminergic neurons of the rat brain. *J Neurochem*. 1986;47(2):569–576.
161. Adell A, Biggs TA, Myers RD. Action of harman (1-methyl-beta-carboline) on the brain: Body temperature and in vivo efflux of 5-HT from hippocampus of the rat. *Neuropharmacology*. 1996;35(8):1101–1107.
162. Bel N, Artigas F. In vivo evidence for the reversible action of the monoamine oxidase inhibitor brofaromine on 5-hydroxytryptamine release in rat brain. *Naunyn Schmiedebergs Arch Pharmacol*. 1995;351(5):475–482.
163. Celada P, Bel N, Artigas F. The effects of brofaromine, a reversible MAO-A inhibitor, on extracellular serotonin in the raphe nuclei and frontal cortex of freely moving rats. *J Neural Transm Suppl*. 1994;41:357–363.

164. Curet O, Damoiseau-Ovens G, Sauvage C et al. Preclinical profile of befloxatone, a new reversible MAO-A inhibitor. *J Affect Disord*. 1998;51(3):287–303.
165. Haefely W, Burkard WP, Cesura AM et al. Biochemistry and pharmacology of moclobemide, a prototype RIMA. *Psychopharmacology (Berl)*. 1992;106(Suppl):S6–S14.
166. Houslay MD, Tipton KF. A kinetic evaluation of monoamine oxidase activity in rat liver mitochondrial outer membranes. *Biochem J*. 1974;139(3):645–652.
167. Finberg JP, Pacak K, Goldstein DS, Kopin IJ. Modification of cerebral cortical noradrenaline release by chronic inhibition of MAO-A. *J Neural Transm Suppl*. 1994;41:123–125.
168. Finberg JP, Pacak K, Kopin IJ, Goldstein DS. Chronic inhibition of monoamine oxidase type A increases noradrenaline release in rat frontal cortex. *Naunyn Schmiedebergs Arch Pharmacol*. 1993;347(5):500–505.
169. Colzi A, d'Agostini F, Kettler R, Borroni E, Da Prada M. Effect of selective and reversible MAO inhibitors on dopamine outflow in rat striatum: A microdialysis study. *J Neural Transm Suppl*. 1990;32:79–84.
170. Arbuthnott GW, Fairbrother IS, Butcher SP. Dopamine release and metabolism in the rat striatum: An analysis by 'in vivo' brain microdialysis. *Pharmacol Ther*. 1990;48(3):281–293.
171. Butcher SP, Fairbrother IS, Kelly JS, Arbuthnott GW. Effects of selective monoamine oxidase inhibitors on the in vivo release and metabolism of dopamine in the rat striatum. *J Neurochem*. 1990;55(3):981–988.
172. Konradi C, Kornhuber J, Froelich L et al. Demonstration of monoamine oxidase-A and -B in the human brainstem by a histochemical technique. *Neuroscience*. 1989;33(2):383–400.
173. Moll G, Moll R, Riederer P, Gsell W, Heinsen H, Denney RM. Immunofluorescence cytochemistry on thin frozen sections of human substantia nigra for staining of monoamine oxidase A and monoamine oxidase B: A pilot study. *J Neural Transm Suppl*. 1990;32:67–77.
174. Evrard A, Malagie I, Laporte AM et al. Altered regulation of the 5-HT system in the brain of MAO-A knock-out mice. *Eur J Neurosci*. 2002;15(5):841–851.
175. Grote SS, Moses SG, Robins E, Hudgens RW, Croninger AB. A study of selected catecholamine metabolizing enzymes: A comparison of depressive suicides and alcoholic suicides with controls. *J Neurochem*. 1974;23(4):791–802.
176. Gottfries CG, Oreland L, Wiberg A, Winblad B. Lowered monoamine oxidase activity in brains from alcoholic suicides. *J Neurochem*. 1975;25(5):667–673.
177. Mann JJ, Stanley M. Postmortem monoamine oxidase enzyme kinetics in the frontal cortex of suicide victims and controls. *Acta Psychiatr Scand*. 1984;69(2):135–139.
178. Sherif F, Marcusson J, Oreland L. Brain gamma-aminobutyrate transaminase and monoamine oxidase activities in suicide victims. *Eur Arch Psychiatry Clin Neurosci*. 1991;241(3):139–144.
179. Ordway GA, Farley JT, Dilley GE et al. Quantitative distribution of monoamine oxidase A in brainstem monoamine nuclei is normal in major depression. *Brain Res*. 1999;847(1):71–79.
180. Galva MD, Bondiolotti GP, Olasmaa M, Picotti GB. Effect of aging on lazabemide binding, monoamine oxidase activity and monoamine metabolites in human frontal cortex. *J Neural Transm Gen Sect*. 1995;101(1–3):83–94.
181. Krishnan KR. Biological risk factors in late life depression. *Biol Psychiatry*. 2002;52(3):185–192.
182. Fowler JS, Volkow ND, Wang GJ et al. Brain monoamine oxidase A inhibition in cigarette smokers. *Proc Natl Acad Sci U. S. A*. 1996;93(24):14065–14069.
183. Fowler JS, Logan J, Ding YS et al. Non-MAO A binding of clorgyline in white matter in human brain. *J Neurochem*. 2001;79(5):1039–1046.

184. Bergstrom M, Westerberg G, Nemeth G et al. MAO-A inhibition in brain after dosing with esuprone, moclobemide and placebo in healthy volunteers: In vivo studies with positron emission tomography. *Eur J Clin Pharmacol*. 1997;52(2):121–128.
185. Bottlaender M, Dolle F, Guenther I et al. Mapping the cerebral monoamine oxidase type A: Positron emission tomography characterization of the reversible selective inhibitor [^{11}C]befloxatone. *J Pharmacol Exp Ther*. 2003;305(2):467–473.
186. Ginovart N, Meyer JH, Boovariwala A et al. Positron emission tomography quantification of [^{11}C]-harmine binding to monoamine oxidase-A in the human brain. *J Cereb Blood Flow Metab*. 2006;26(3):330–344.
187. Dolle F, Valette H, Bramoulle Y et al. Synthesis and in vivo imaging properties of [^{11}C] befloxatone: A novel highly potent positron emission tomography ligand for monoamine oxidase-A. *Bioorg Med Chem Lett*. 2003;13(10):1771–1775.
188. Bergstrom M, Westerberg G, Langstrom B. ^{11}C-Harmine as a tracer for monoamine oxidase A (MAO-A): In vitro and in vivo studies. *Nucl Med Biol*. 1997;24(4):287–293.
189. Tweedie DJ, Burke MD. Metabolism of the beta-carbolines, harmine and harmol, by liver microsomes from phenobarbitone- or 3-methylcholanthrene-treated mice. Identification and quantitation of two novel harmine metabolites. *Drug Metab Dispos*. 1987;15(1):74–81.
190. Meyer JH, Ginovart N, Boovariwala A et al. Elevated monoamine oxidase a levels in the brain: An explanation for the monoamine imbalance of major depression. *Arch Gen Psychiatry*. 2006;63(11):1209–1216.
191. First M, Spitzer R, Williams J, Gibbon M. *Structured Clinical Interview for DSM-IV-Non-Patient Edition (SCID-NP, Version 1.0)*. Washington, DC: American Psychiatric Press; 1995.
192. Meyer JH, Wilson AA, Sagrati S et al. Brain monoamine oxidase A binding in major depressive disorder: Relationship to selective serotonin reuptake inhibitor treatment, recovery, and recurrence. *Arch Gen Psychiatry*. 2009;66(12):1304–1312.
193. Freis ED. Mental depression in hypertensive patients treated for long periods with large doses of reserpine. *N Engl J Med*. 1954;251(25):1006–1008.
194. Nelson DL, Herbet A, Glowinski J, Hamon M. [^{3}H]Harmaline as a specific ligand of MAO A—II. Measurement of the turnover rates of MAO A during ontogenesis in the rat brain. *J Neurochem*. 1979;32(6):1829–1836.
195. Nelson DL, Herbet A, Petillot Y, Pichat L, Glowinski J, Hamon M. [^{3}H]Harmaline as a specific ligand of MAO A—I. Properties of the active site of MAO A from rat and bovine brains. *J Neurochem*. 1979;32(6):1817–1827.
196. Edelstein SB, Breakefield XO. Monoamine oxidases A and B are differentially regulated by glucocorticoids and "aging" in human skin fibroblasts. *Cell Mol Neurobiol*. 1986;6(2):121–150.
197. Knutson B, Taylor J, Kaufman M, Peterson R, Glover G. Distributed neural representation of expected value. *J Neurosci*. 2005;25(19):4806–4812.
198. Samandari G, Martin SL, Kupper LL, Schiro S, Norwood T, Avery M. Are pregnant and postpartum women: At increased risk for violent death? Suicide and homicide findings from North Carolina. *Matern Child Health J*. 2011;15(5):660–669.
199. Comtois KA, Schiff MA, Grossman DC. Psychiatric risk factors associated with postpartum suicide attempt in Washington State, 1992–2001. *Am J Obstet Gynecol*. 2008;199(2):120.e1–120.e5.
200. Lewis G. *Why Mothers Die 1997–1999, Confidential Enquiry in Maternal Deaths*. London, U.K.: Royal College of Obstetricians and Gynaecologists; 2001.
201. O'Hara MW, Swain A. Rates and risk of postpartum depression—A meta analysis. *Int Rev Psychiatry*. 1996;8:37–54.
202. Steiner M. Perinatal mood disorders: Position paper. *Psychopharmacol Bull*. 1998;34(3):301–306.

203. Pop VJ, Essed GG, de Geus CA, van Son MM, Komproe IH. Prevalence of post partum depression—Or is it post-puerperium depression? *Acta Obstet Gynecol Scand.* 1993;72(5):354–358.
204. Carothers AD, Murray L. Estimating psychiatric morbidity by logistic regression: Application to post-natal depression in a community sample. *Psychol Med.* 1990;20(3):695–702.
205. Kornstein SG. The evaluation and management of depression in women across the life span. *J Clin Psychiatry.* 2001;62(Suppl 24):11–17.
206. O'Hara M. *Postpartum Depression: Causes and Consequences.* New York: Springer-Verlag; 1994.
207. O'Hara MW, Schlechte JA, Lewis DA, Varner MW. Controlled prospective study of postpartum mood disorders: Psychological, environmental, and hormonal variables. *J Abnorm Psychol.* 1991;100(1):63–73.
208. Association AP. *Diagnostic and Statistical Manual of Mental Disorders*, 4th edn. Washington, DC: American Psychiatric Association; 1994.
209. Campbell S, Cohn J, Flanagan C, Popper S, Meyers T. Course and correlates of postpartum depression during the transition to parenthood. *Dev Psychopathol.* 1992;4:29–47.
210. Brockington IF, Cernik KF, Schofield EM, Downing AR, Francis AF, Keelan C. Puerperal psychosis. Phenomena and diagnosis. *Arch Gen Psychiatry.* 1981;38(7):829–833.
211. Kendell RE, Chalmers JC, Platz C. Epidemiology of puerperal psychoses. *Br J Psychiatry.* 1987;150:662–673.
212. Hendrick V, Altshuler LL, Suri R. Hormonal changes in the postpartum and implications for postpartum depression. *Psychosomatics.* 1998;39(2):93–101.
213. Nott PN, Franklin M, Armitage C, Gelder MG. Hormonal changes and mood in the puerperium. *Br J Psychiatry.* 1976;128:379–383.
214. O'Hara MW, Schlechte JA, Lewis DA, Wright EJ. Prospective study of postpartum blues. Biologic and psychosocial factors. *Arch Gen Psychiatry.* 1991;48(9):801–806.
215. Bloch M, Schmidt PJ, Danaceau M, Murphy J, Nieman L, Rubinow DR. Effects of gonadal steroids in women with a history of postpartum depression. *Am J Psychiatry.* 2000;157(6):924–930.
216. Sichel DA, Cohen LS, Robertson LM, Ruttenberg A, Rosenbaum JF. Prophylactic estrogen in recurrent postpartum affective disorder. *Biol Psychiatry.* 1995;38(12):814–818.
217. Gregoire AJ, Kumar R, Everitt B, Henderson AF, Studd JW. Transdermal oestrogen for treatment of severe postnatal depression. *Lancet.* 1996;347(9006):930–933.
218. Ahokas A, Kaukoranta J, Wahlbeck K, Aito M. Estrogen deficiency in severe postpartum depression: Successful treatment with sublingual physiologic 17beta-estradiol: A preliminary study. *J Clin Psychiatry.* 2001;62(5):332–336.
219. Schneider LS, Small GW, Hamilton SH, Bystritsky A, Nemeroff CB, Meyers BS. Estrogen replacement and response to fluoxetine in a multicenter geriatric depression trial. Fluoxetine Collaborative Study Group. *Am J Geriatr Psychiatry.* 1997;5(2):97–106.
220. Rasgon NL, Altshuler LL, Fairbanks LA et al. Estrogen replacement therapy in the treatment of major depressive disorder in perimenopausal women. *J Clin Psychiatry.* 2002;63(Suppl 7):45–48.
221. Schneider LS, Small GW, Clary CM. Estrogen replacement therapy and antidepressant response to sertraline in older depressed women. *Am J Geriatr Psychiatry.* 2001;9(4):393–399.
222. Amsterdam J, Garcia-Espana F, Fawcett J et al. Fluoxetine efficacy in menopausal women with and without estrogen replacement. *J Affect Disord.* 1999;55(1):11–17.
223. Grigoriadis S, Kennedy SH, Srinivisan J, McIntyre RS, Fulton K. Antidepressant augmentation with raloxifene. *J Clin Psychopharmacol.* 2005;25(1):96–98.
224. Luine VN, McEwen BS. Effect of oestradiol on turnover of type A monoamine oxidase in brain. *J Neurochem.* 1977;28(6):1221–1227.

225. Leung TK, Lai JC, Marr W, Lim L. The activities of the A and B forms of monoamine oxidase in liver, hypothalamus and cerebral cortex of the female rat: Effects of administration of ethinyloestradiol and the progestogens norethisterone acetate and D-norgestrel. *Biochem Soc Trans*. 1980;8(5):607–608.
226. Holschneider DP, Kumazawa T, Chen K, Shih JC. Tissue-specific effects of estrogen on monoamine oxidase A and B in the rat. *Life Sci*. 1998;63(3):155–160.
227. Gundlah C, Lu NZ, Bethea CL. Ovarian steroid regulation of monoamine oxidase-A and -B mRNAs in the macaque dorsal raphe and hypothalamic nuclei. *Psychopharmacology (Berl)*. 2002;160(3):271–282.
228. Smith LJ, Henderson JA, Abell CW, Bethea CL. Effects of ovarian steroids and raloxifene on proteins that synthesize, transport, and degrade serotonin in the raphe region of macaques. *Neuropsychopharmacology*. 2004;29(11):2035–2045.
229. Ma ZQ, Bondiolotti GP, Olasmaa M et al. Estrogen modulation of catecholamine synthesis and monoamine oxidase A activity in the human neuroblastoma cell line SK-ER3. *J Steroid Biochem Mol Biol*. 1993;47(1–6):207–211.
230. Ma ZQ, Violani E, Villa F, Picotti GB, Maggi A. Estrogenic control of monoamine oxidase A activity in human neuroblastoma cells expressing physiological concentrations of estrogen receptor. *Eur J Pharmacol*. 1995;284(1–2):171–176.
231. Chevillard C, Barden N, Saavedra JM. Estradiol treatment decreases type A and increases type B monoamine oxidase in specific brain stem areas and cerebellum of ovariectomized rats. *Brain Res*. 1981;222(1):177–181.
232. Harris B, Lovett L, Newcombe RG, Read GF, Walker R, Riad-Fahmy D. Maternity blues and major endocrine changes: Cardiff puerperal mood and hormone study II. *Br Med J*. 1994;308(6934):949–953.
233. Neumeister A. Tryptophan depletion, serotonin, and depression: Where do we stand? *Psychopharmacol Bull*. 2003;37(4):99–115.
234. Leyton M, Young SN, Pihl RO et al. Effects on mood of acute phenylalanine/tyrosine depletion in healthy women. *Neuropsychopharmacology*. 2000;22(1):52–63.
235. Verhoeff NP, Kapur S, Hussey D et al. A simple method to measure baseline occupancy of neostriatal dopamine d(2) receptors by dopamine in vivo in healthy subjects. *Neuropsychopharmacology*. 2001;25(2):213–223.
236. Laruelle M, D'Souza CD, Baldwin RM et al. Imaging D2 receptor occupancy by endogenous dopamine in humans. *Neuropsychopharmacology*. 1997;17(3):162–174.
237. Hasler G, Fromm S, Carlson PJ et al. Neural response to catecholamine depletion in unmedicated subjects with major depressive disorder in remission and healthy subjects. *Arch Gen Psychiatry*. 2008;65(5):521–531.
238. Adewuya AO. Early postpartum mood as a risk factor for postnatal depression in Nigerian women. *Am J Psychiatry*. 2006;163(8):1435–1437.
239. Kim JJ, Shih JC, Chen K et al. Selective enhancement of emotional, but not motor, learning in monoamine oxidase A-deficient mice. *Proc Natl Acad Sci U. S. A*. 1997;94(11):5929–5933.
240. Cases O, Seif I, Grimsby J et al. Aggressive behavior and altered amounts of brain serotonin and norepinephrine in mice lacking MAOA. *Science*. 1995;268(5218):1763–1766.
241. Shih JC, Ridd MJ, Chen K et al. Ketanserin and tetrabenazine abolish aggression in mice lacking monoamine oxidase A. *Brain Res*. 1999;835(2):104–112.
242. Popova NK, Skrinskaya YA, Amstislavskaya TG, Vishnivetskaya GB, Seif I, de Meier E. Behavioral characteristics of mice with genetic knockout of monoamine oxidase type A. *Neurosci Behav Physiol*. 2001;31(6):597–602.
243. Popova NK, Vishnivetskaya GB, Ivanova EA, Skrinskaya JA, Seif I. Altered behavior and alcohol tolerance in transgenic mice lacking MAO A: A comparison with effects of MAO A inhibitor clorgyline. *Pharmacol Biochem Behav*. 2000;67(4):719–727.

244. Upton AL, Salichon N, Lebrand C et al. Excess of serotonin (5-HT) alters the segregation of ipsilateral and contralateral retinal projections in monoamine oxidase A knockout mice: Possible role of 5-HT uptake in retinal ganglion cells during development. *J Neurosci*. 1999;19(16):7007–7024.
245. Cases O, Vitalis T, Seif I, De Maeyer E, Sotelo C, Gaspar P. Lack of barrels in the somatosensory cortex of monoamine oxidase A-deficient mice: Role of a serotonin excess during the critical period. *Neuron*. 1996;16(2):297–307.
246. Bortolato M, Chen K, Shih JC, Bortolato M, Chen K, Shih JC. Monoamine oxidase inactivation: From pathophysiology to therapeutics. *Adv Drug Deliv Rev*. 2008;60(13–14):1527–1533.
247. Chen K, Cases O, Rebrin I et al. Forebrain-specific expression of monoamine oxidase A reduces neurotransmitter levels, restores the brain structure, and rescues aggressive behavior in monoamine oxidase A-deficient mice. *J Biol Chem*. 2007;282(1):115–123.
248. Brunner HG, Nelen M, Breakefield XO, Ropers HH, van Oost BA. Abnormal behavior associated with a point mutation in the structural gene for monoamine oxidase A. *Science*. 1993;262(5133):578–580.
249. Caspi A, McClay J, Moffitt TE et al. Role of genotype in the cycle of violence in maltreated children [see comment]. *Science*. 2002;297(5582):851–854.
250. Kim-Cohen J, Caspi A, Taylor A et al. MAOA, maltreatment, and gene–environment interaction predicting children's mental health: New evidence and a meta-analysis. *Mol Psychiatry*. 2006;11(10):903–913.
251. Brown GL, Goodwin FK, Ballenger JC, Goyer PF, Major LF. Aggression in humans correlates with cerebrospinal fluid amine metabolites. *Psychiatry Res*. 1979;1(2):131–139.
252. Linnoila M, Virkkunen M, Scheinin M, Nuutila A, Rimon R, Goodwin FK. Low cerebrospinal fluid 5-hydroxyindoleacetic acid concentration differentiates impulsive from nonimpulsive violent behavior. *Life Sci*. 1983;33(26):2609–2614.
253. Linnoila VM, Virkkunen M. Aggression, suicidality, and serotonin. *J Clin Psychiatry*. 1992;53(Suppl):46–51.
254. Waldmeier PC, Baumann PA. Effects of CGP 11305 A, a new reversible and selective inhibitor of MAO A, on biogenic amine levels and metabolism in the rat brain. *Naunyn Schmiedebergs Arch Pharmacol*. 1983;324(1):20–26.
255. Garrick NA, Seppala T, Linnoila M, Murphy DL. Rhesus monkey cerebrospinal fluid amine metabolite changes following treatment with the reversible monoamine oxidase type-A inhibitor cimoxatone. *Psychopharmacology (Berl)*. 1985;86(3):265–269.
256. Alia-Klein N, Goldstein RZ, Kriplani A et al. Brain monoamine oxidase A activity predicts trait aggression. *J Neurosci*. 2008;28(19):5099–5104.
257. Soliman A, Bagby RM, Wilson AA et al. Relationship of monoamine oxidase A binding to adaptive and maladaptive personality traits. *Psychol Med*. 2011;41(5):1051–1060.
258. Bechara A, Damasio H, Damasio AR. Emotion, decision making and the orbitofrontal cortex. *Cereb Cortex*. 2000;10(3):295–307.
259. Di Chiara G, Imperato A. Drugs abused by humans preferentially increase synaptic dopamine concentrations in the mesolimbic system of freely moving rats. *Proc Natl Acad Sci U. S. A*. 1988;85(14):5274–5278.
260. Vaccarino FJ. Nucleus accumbens dopamine–CCK interactions in psychostimulant reward and related behaviors. *Neurosci Biobehav Rev*. 1994;18(2):207–214.
261. Schultz W, Dayan P, Montague PR. A neural substrate of prediction and reward. *Science*. 1997;275(5306):1593–1599.
262. Dalley JW, Fryer TD, Brichard L et al. Nucleus accumbens D2/3 receptors predict trait impulsivity and cocaine reinforcement. *Science*. 2007;315(5816):1267–1270.
263. Buckholtz JW, Treadway MT, Cowan RL et al. Dopaminergic network differences in human impulsivity. *Science*. 30;329(5991):532.

264. Mitchell JM, Tavares VC, Fields HL, D'Esposito M, Boettiger CA. Endogenous opioid blockade and impulsive responding in alcoholics and healthy controls. *Neuropsychopharmacology*. 2007;32(2):439–449.
265. Kieres AK, Hausknecht KA, Farrar AM, Acheson A, de Wit H, Richards JB. Effects of morphine and naltrexone on impulsive decision making in rats. *Psychopharmacology (Berl)*. 2004;173(1–2):167–174.
266. Pattij T, Schetters D, Janssen MC, Wiskerke J, Schoffelmeer AN. Acute effects of morphine on distinct forms of impulsive behavior in rats. *Psychopharmacology (Berl)*. 2009;205(3):489–502.
267. Pecina S, Berridge KC. Hedonic hot spot in nucleus accumbens shell: Where do mu-opioids cause increased hedonic impact of sweetness? *J Neurosci*. 2005;25(50):11777–11786.
268. Young MS, Schinka JA. Research Validity Scales for the NEO-PI-R: Additional evidence for reliability and validity. *J Pers Assess*. 2001;76(3):412–420.
269. Love TM, Stohler CS, Zubieta JK. Positron emission tomography measures of endogenous opioid neurotransmission and impulsiveness traits in humans. *Arch Gen Psychiatry*. 2009;66(10):1124–1134.
270. Gabilondo AM, Meana JJ, Garcia-Sevilla JA. Increased density of mu-opioid receptors in the postmortem brain of suicide victims. *Brain Res*. 1995;682(1–2):245–250.
271. Sacher J, Wilson A, Houle S et al. Elevated brain monoamine oxidase A binding in early postpartum. *Arch Gen Psychiatry*. 2010;67(5):468–474.

10 Gene–Environment Interaction and Suicidal Behavior

Alec Roy

CONTENTS

This chapter reviews the possibility that gene–environment (GXE) interaction may play a role in suicidal behavior. Moffit et al. (2005) stated that "a GXE occurs when the effect of exposure to an environmental pathogen on health is conditional on a person's genotype." Thus, some introductory remarks about the environment and genes, as they relate to suicide, will provide a background for the GXE studies to be discussed.

10.1 CURRENT ENVIRONMENTAL STRESS AND SUICIDAL BEHAVIOR

Stress in relation to suicidal behavior has largely been considered as the recent stress in the weeks and months before the attempt or completed suicide. Thus, initial research efforts were directed toward examining adverse life events proximal to the suicidal behavior. For example, Paykel et al. (1975) first demonstrated that suicide attempters reported four times as many life events in 6 months before the attempt than controls. Their conclusion that there is strong and immediate relationship between suicide attempts and recent life stress has now been replicated in different countries and age groups as well as in completed suicides (Beatrais et al., 1997; Crane et al., 2007; Heikkinen et al., 2007; Osvath et al., 2004). However, it is noteworthy that the great majority of individuals who experience current stress or adverse life events do not attempt or commit suicide.

10.2 EARLY ENVIRONMENTAL STRESS AND SUICIDAL BEHAVIOR

Early life events have also been of interest in relation to environmental stress and suicidal behavior. In particular, the early environmental stress of experiencing childhood trauma has been much investigated in early life event. In this regard, it is noteworthy that childhood trauma is common: a general population survey revealed that 31% of males and 21% of females reported a history of childhood physical abuse while childhood sexual abuse was reported by 13% of females and 4% of males (Holmes and Slap, 1998; MacMillan et al., 1997).

The results of both general population and clinical studies have led to the recognition that exposure to childhood trauma is an independent risk factor for suicidal behavior. For example, results from the Epidemiologic Catchment Area (ECA) Study and the U.S. National Comorbidity Survey demonstrated a strong association between childhood sexual abuse and suicidal behavior (Davidson et al., 1996; Molnar et al., 2001). Similarly, an Australian community study of twins found that a history of childhood trauma significantly increased the risk of a suicide attempt (Nelson et al., 2002). Childhood trauma has been reported to raise the risk of attempting suicide among 13,494 adults attending a Health Maintenance Organization (HMO). Other clinical studies have similarly reported that childhood trauma is associated with suicide attempts in patients with various psychiatric disorders (Roy, 2001a,b, 2005a,b; Sarchiapone et al., 2007). Recent data from the U.S. National Comorbidity Survey Replication indicate that exposure to childhood physical abuse or sexual abuse or witnessing domestic violence accounts for 16% and 50% of suicidal ideation and attempts, respectively, among women and 21% and 33% of ideation and attempts among men (Afifi et al., 2008).

10.3 INDIVIDUAL PREDISPOSITION TO ENVIRONMENTAL STRESS AND SUICIDAL BEHAVIOR

As the great majority of individuals who experience either current or early environmental stress do not exhibit suicidal behavior, those who do so are presumed to have predisposing suicide risk factors. In the stress–diathesis model, suicide risk factors may be either distal or proximal. Distal risk factors create a predisposing diathesis and determine an individual's response to a stressor. They include personality, biological, and genetic variables. They affect the threshold for suicide and increase an individual's risk when he or she experiences a proximal risk factor. Proximal, or trigger, factors are more closely related to the suicidal behavior and act as precipitants. They include life events, stress, acute episodes of mental illness, and acute alcohol or substance abuse. Suicidal individuals differ from nonsuicidal individuals in distal risk factors, for example, genetic factors, and may be moved toward suicidal behaviors by proximal risk factors like recent stress (Mann et al., 1999).

10.4 GENETIC FACTORS AND SUICIDAL BEHAVIOR

Genetic factors are now recognized as one of the main predisposing factors for suicidal behavior. Suicide has always been considered to be a multidetermined act with social, psychiatric, psychodynamic, biological, and personality determinants. However, over the past two decades information from family, twin, and adoption studies has strongly suggested that genetic factors also play a role in suicidal behavior (reviewed in Roy et al., 2000). More recent genetic–epidemiology studies have estimated that at least one-third of the variance for suicidal behavior is genetic (Fu et al., 2002; Glowinski et al., 2001; Statham et al., 1998). Molecular genetic studies have reported that polymorphisms in serotonergic and other genes have been associated with both attempts at suicide and completed suicide (reviewed in chapters 11, 12, and 14). Thus, genetic factors are an important predisposing factor for suicidal behavior for an individual who experiences environmental stress—either current or early adverse life events. This chapter discusses the three genes for which GXE interactions have been reported in relation to attempting suicide.

10.5 SEROTONIN TRANSPORTER PROMOTER (5-HTTLPR)

The serotonin transporter (5-HTT) is located on presynaptic serotonin neurons and its protein, which is responsible for the reuptake of serotonin, is encoded by single gene on chromosome 17. A functional polymorphism in the 5′ regulatory promoter region of the *5-HTT* gene (5-HTTLPR) involves two common alleles (0.4–0.6) that regulate transcription of the gene. The allele containing 14 copies of an imperfect 22 bp repeat (s allele) reduces transcription efficiency of the *5-HTT* gene, while the 16-repeat allele (l allele) does not (Lesch et al., 1996).

10.6 FIRST GXE STUDY OF SUICIDALITY

A GXE interaction between life events and 5-HTTLPR polymorphism and suicidal behavior was first reported by Caspi et al. (2003). This Dunedin longitudinal study is best known for the finding of a significant interaction between the s allele and life events occurring after 21 years of age in producing depressive symptoms from the age of 21–26 years. However, less well known is that Caspi et al. also found that a significant GXE interaction between stressful life events and 5-HTTLPR genotype also predicted suicidal ideation or suicide attempts, but again only among subjects with an s allele. Caspi et al. went on to hypothesize that if 5-HTTLPR genotype moderates the influence of current stressful life events it should also moderate the effect of earlier life events. When they examined this, they found a further GXE interaction as childhood maltreatment also predicted adult depression in subjects with the s allele but not among l/l homozygotes.

These positive GXE results are particularly relevant given the negative results of a review of the 18 controlled studies of 5-HTTLPR genotype in relation to suicidal behavior. No significant difference for 5-HTTLPR distribution between cases and controls was found (Lin and Tsai, 2004). Significant heterogeneity among the odds ratios of the 18 studies was observed and the authors suggested the presence of some moderating variables that might account for this heterogeneity. Such moderating variables may include the environmental ones of childhood trauma and recent life events—as found by Caspi et al. (2003).

10.7 SUBSEQUENT 5-HTTLPR GXE STUDY OF SUICIDE ATTEMPTS IN SUBSTANCE ABUSERS

Patients with substance dependence are a group at increased risk of suicidal behavior. Therefore, we genotype a large group of abstinent substance-dependent patients and controls for the triallelic 5-HTTLPR genotype (Roy et al., 2007). Having experienced early life events, childhood trauma was assessed as subjects completed the 34-item Childhood Trauma Questionnaire (CTQ) of Bernstein and Fink (1998).

Attempters were similar to nonattempters in distribution of 5-HTTLPR genotypes. The sample was divided using the median split for CTQ scores and using the three functional categories of triallelic HTTLPR genotypes: high, intermediate, and low expressing. For CTQ subscales of emotional abuse (χ^2 (1) = 4.46, p = .035), physical abuse (χ^2 (1) = 8.58, p = .003), emotional neglect (χ^2 (1) = 8.58, p = .003), physical neglect (χ^2 (1) = 5.84, p = .016), and sexual abuse (χ^2 (1) = 2.57, p = .11), higher levels of childhood trauma were associated with increased rates of suicide attempts in the low-expressing genotype group. Rates of suicide attempts in the intermediate and high-expression groups were in general unaffected by increasing levels of childhood trauma (Figure 10.1).

Logistic regression showed a GXE interaction for both childhood physical abuse and emotional neglect. For both of these environmental factors, there was a significant increment in suicide attempt risk in the context of the low-expressing 5-HTTLPR genotype. Thus, childhood trauma interacted with low-expressing 5-HTTLPR genotype

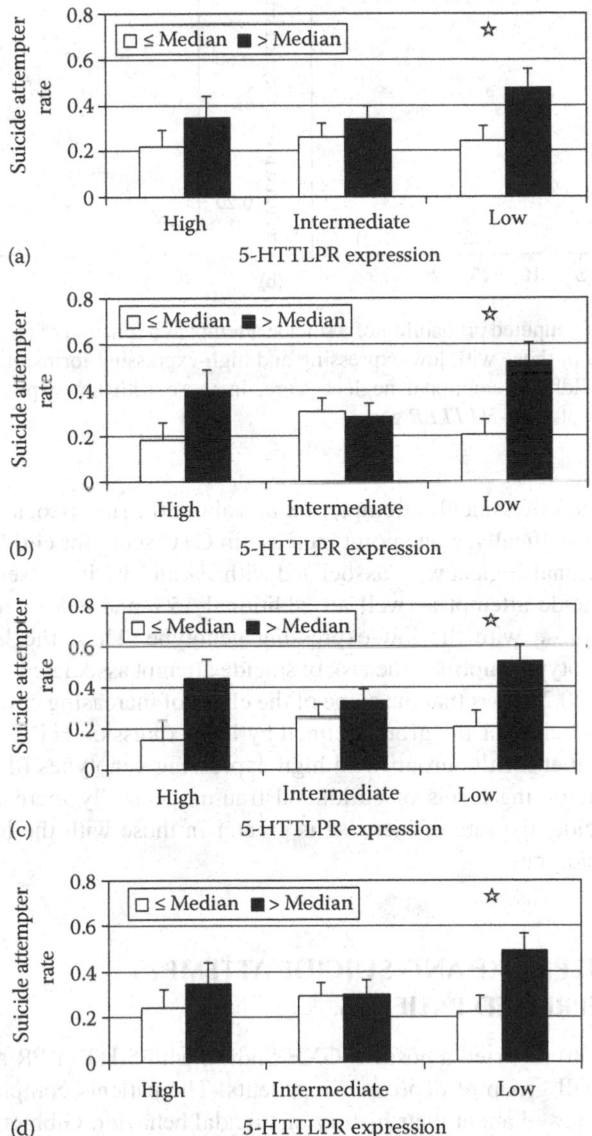

FIGURE 10.1 Suicide attempter rates (±SD) as a function of median CTQ scores: (a) emotional abuse; (b) physical abuse; (c) emotional neglect; and (d) physical neglect, and high-, intermediate-, and low-expression groupings of the 5-HTTLPR genotype. ☆ indicates a statistically significant difference in suicide attempt rate.

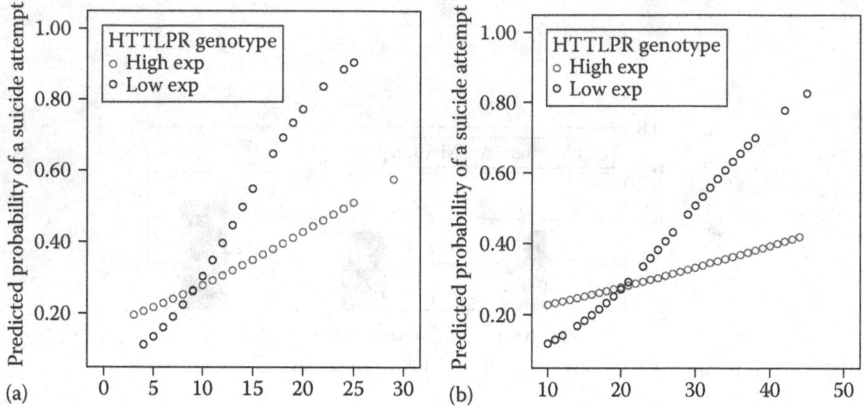

FIGURE 10.2 Computed probability of a suicide attempt as a function of (a) childhood physical abuse score, in those with low-expressing and high-expressing forms of the *5-HTTLPR* gene and (b) childhood emotional neglect score, in those with low-expressing and high-expressing forms of the *5-HTTLPR* gene.

to increase the risk of a suicide attempt, over and above the risk associated with childhood trauma. Specifically, a one point increase in CTQ score for childhood physical abuse and emotional neglect were associated with 7% and 3% increases, respectively, in risk of a suicide attempt as well an additional 15% and 8% increases, respectively, among those with the low-expressing genotype. Thus, the low-expression 5-HTTLPR genotype amplified the risk of suicide attempt associated with childhood trauma. Figure 10.2 shows that the slope of the effect of increasing childhood trauma on attempting suicide for the group defined by low-expression HTTLPR genotypes is steeper than that for the group with high-expressing genotypes (the interaction). Thus, while increasing levels of childhood trauma generally increase the risk of attempting suicide, the rate of increase is greater in those with the low-expressing 5-HTTLPR genotype.

10.8 5-HTTLPR GXE AND SUICIDE ATTEMPTS IN DEPRESSED PATIENTS

Gibb et al. (2006) reported a positive GXE finding with 5-HTTLPR and childhood trauma in a small group of depressed inpatients. The patients completed the CTQ and were interviewed about their history of suicidal behavior. Gibb et al. found that both childhood physical abuse and sexual abuse were significantly associated with suicide attempts among patients with one or two copies of the s allele but not among those who were homozygous for the long allele. They concluded, similar to the aforementioned study in substance abuse patients, that the 5-HTTLPR genotype moderated the link between childhood physical and sexual abuse and suicide attempts in depressed patients.

10.9 PRIMATE GXE HTTLPR STUDIES RELEVANT TO SUICIDAL BEHAVIOR

Low central serotonin function is implicated in suicidal behavior (Asberg, 1997). Thus, GXE studies in primates showing that the interaction of 5-HTTLPR genotype with early experience affects central serotonin are of interest (Bennett et al., 2002). Bennett et al. compared monkeys reared with their mothers with monkeys separated from their mothers at birth and peer reared. The measure of central serotonin function used was cisternal cerebrospinal fluid 5-hydroxy indoleacetic acid (CSF 5-HIAA) concentrations. They found that CSF 5-HIAA was significantly influenced by genotype but only in the peer-reared monkeys. Peer-reared 5-HTTLPR s/l heterozygous monkeys had significantly lower CSF 5-HIAA than peer-raised homozygous monkeys with two l alleles. However, there was no difference in CSF 5-HIAA between the s/l and l/l genotypes in mother-reared monkeys. Only monkeys with early deleterious rearing experiences showed the genotype-predicted difference in CSF 5-HIAA.

Also relevant are earlier findings that monkeys with low CSF 5-HIAA were significantly more impulsive and aggressive than monkeys with CSF 5-HIAA within the normal range (Higley et al., 1996). The trait of impulsive aggression is associated with low central serotonergic activity and is thought to be an intermediate phenotype for suicidal behavior in humans (Higley et al., 1996; Roy and Linnoila, 1985).

10.10 GXE WITH *BDNF* GENE INCREASES RISK OF ATTEMPTING SUICIDE

Perroud et al. (2008) noted that brain-derived neurotrophic factor (BDNF) plays an important role in the growth of serotonergic neurons during development. They studied 813 suicide attempters who completed the CTQ and were genotyped for BDNF. The suicide attempts were classified as either violent or nonviolent. Perroud et al. found that childhood sexual abuse was significantly associated with violent suicide attempts among Val/Val patients but not among Met/Met patients. This led Perroud et al. to conclude that Val66Met modulates the effect of childhood abuse on the violence of a suicide attempt. They suggested the possibility that childhood sexual abuse may lead to brain structural modifications through BDNF dysfunction, which may predispose to the risk of violent suicidal behavior as an adult.

10.11 GXE WITH HYPOTHALAMIC–PITUITARY–ADRENAL AXIS GENES AND ATTEMPTING SUICIDE

10.11.1 Hypothalamic–Pituitary–Adrenal Axis Dysregulation Occurs in Suicidal Individuals

Changes in the dexamethasone suppression test (DST), 24 h urinary free cortisol outputs, salivary cortisol, CSF levels of corticotrophin-releasing hormone (CRH), and CRH-binding sites in frontal cortex have been reported in individuals exhibiting suicidal behavior (Arato et al., 1989; Bunney and Fawcett, 1965; Bunney et al., 1969; Coryell and

Schlesser, 2001; Pfennig et al., 2005; Jokinen and Nordstrom, 2009; Lindqvist et al., 2008a,b; Nemeroff et al., 1988; Roy, 1992; Roy et al., 1986). Dexamethasone nonsuppression on the DST has been found to be significantly associated with predicting suicide in depressed patients, particularly among those who have previously attempted suicide (Coryell and Schlesser, 2001, Coryell et al., 2006; Jokinen et al., 2007). A meta-analysis estimated that the odds ratio for suicide was 4.5 times greater for DST nonsuppressors compared to suppressors (Mann et al., 2005).

10.11.2 *CRH* Gene and Environmental Stress Studies

There have been two studies that have reported relationship between environmental stress and hypothalamic–pituitary–adrenal (HPA) axis genes in relation to suicidal behavior. The mechanism of HPA axis dysregulation includes dysregulation of the hypothalamic peptides CRH and arginine vasopressin (AVP) leading to increased release of plasma ACTH and cortisol. Wasserman et al. (2008a,b), in a large series of family trios with suicide attempt offspring, found that a SNP of a CRH receptor 1 gene—CRHRI SNP rs 4792887—was significantly associated with suicide attempters exposed to low levels of stress, most of whom were depressed males. Wasserman et al. (2008a,b) went on to further report that two other CRHR1 SNPs—rs 110402 and rs 12936511—were also associated with depression among suicidal males.

10.11.3 FKBP5: A Stress-Related Gene and HPA Axis

Childhood trauma has been shown to impact stress reactivity in adulthood by altering HPA axis function (De Bellis et al., 1994; Heim et al., 2004; Nemeroff, 2004; Roy, 2002). Similarly, preclinical studies in both primates and rodents demonstrate that early life trauma impacts HPA axis function in the offspring (Meaney, 2001; Sánchez et al., 2005; Weaver et al., 2005). HPA axis dysregulation may be coupled with glucocorticoid receptor insensitivity that results in impairment of the negative feedback loop. Glucocorticoid receptor activation and ligand binding is moderated by a large molecular complex that includes FKBP5, a co-chaperone of hsp-90. Upon ligand binding, the glucocorticoid receptor dissociates from the chaperone complex and migrates to the nucleus where it initiates transcription in the target genes, *CRH* and *AVP* (Binder, 2009).

10.11.4 GXE with *FKBP5* Gene and Suicide Attempts

As FKBP5 may moderate the sensitization by childhood trauma of the stress-responsive HPA pathway, we hypothesized that GXE interaction between FKBP5 and childhood trauma would be associated with a raised risk of attempting suicide. We genotyped for FKBP5 a sample of 830 substance abusers and controls who had completed the CTQ and been interviewed about suicide attempts (Roy et al., 2010).

10.11.5 GXE Interaction with Four SNP Haplotype Analyses

The primary analyses were conducted with the four SNPs found by Binder et al. (2008) to interact with childhood trauma to predict PTSD symptoms. There were two major yin yang haplotypes, 2122 and 1211, that account for 90% of the haplotype

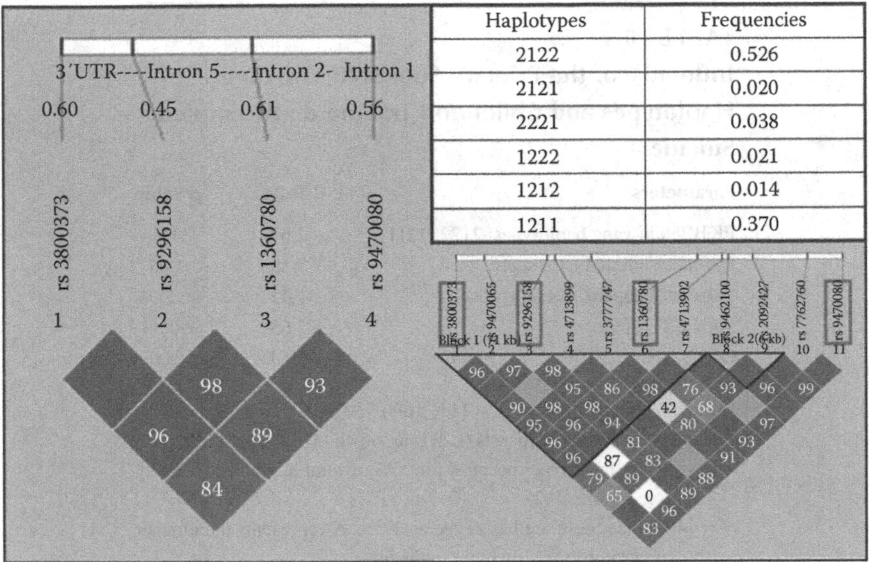

Haplotypes	Frequencies
2122	0.526
2121	0.020
2221	0.038
1222	0.021
1212	0.014
1211	0.370

FIGURE 10.3 FKBP5 11 SNP haplotype block structure and 4 SNP haplotype block and haplotypes. The numbers in the squares refer to pairwise linkage disequilibrium (LD) measured as D′. Haplotype blocks were defined using a setting of average pairwise D′ within block of ≥0.80. The left panel shows the haplotype block structure for the four SNPs highlighted in the right panel; the four SNP haplotypes are listed and the yin yang haplotypes are indicated in bold.

diversity (Figure 10.3). Logistic regression showed that the continuous total CTQ score had a significant effect on suicide attempt but there was no main effect of the yin yang haplotypes. However, the interaction between the yin yang haplotypes and CTQ score significantly influenced the risk of attempted suicide. The total model shown in Table 10.1 was significant ($c^2 = 111$, 5 df, $p \leq .0001$) and contributed 13% of the variance in attempted suicide.

Dichotomous CTQ high/low variables were also examined. Of the individuals who had experienced a high level of childhood trauma, 51% of those with two copies of the 2122 haplotype had attempted suicide compared with 36% of those with one copy of 2122 and 20% of those with no copies of the risk haplotype ($c^2 = 8.5$, 2 df, $p = .014$) (Figure 10.4).

10.11.6 GXE Interaction between Childhood Trauma and 11 SNP Haplotypes

Using a denser SNP coverage, an 11 SNP haplotype analyses showed two relevant haplotypes. The logistic regression analysis revealed no main effect of H2 or H5 haplotypes on suicide attempt. Instead, a GXE interaction was apparent. The total model was significant ($c^2 = 43$, 5 df, $p < .0001$) and contributed 19% of the variance in attempted suicide (Table 10.2).

TABLE 10.1
Influence of the Primary Four SNP Yin Yang Haplotypes and Childhood Trauma on Attempted Suicide

Parameters	L-R Tests c^2	*p* Value
FKBP5 yin yang haplotypes: 2122, 1211	2.63	.11
European ethnic factor score	7.59	.006
Substance dependence diagnosis	31.42	<.0001
Total CTQ score	27.63	<.0001
GXE	4.53	.006

Notes: Whole model test: $c^2 = 111$; 5 df; $p < .0001$, $r^2 = 0.13$.
The effect likelihood ratio tests (L-R) are shown (1 df for each test).
The yin yang haplotypes account for 90% of total haplotype diversity (see Figure 10.2).
$N = 749$. Since these are haplotype analyses, N represents the number of chromosomes (two per individual).
The total childhood trauma questionnaire (CTQ) score is a continuous measure.
GXE indicates the interaction between yin yang haplotypes and total CTQ score.

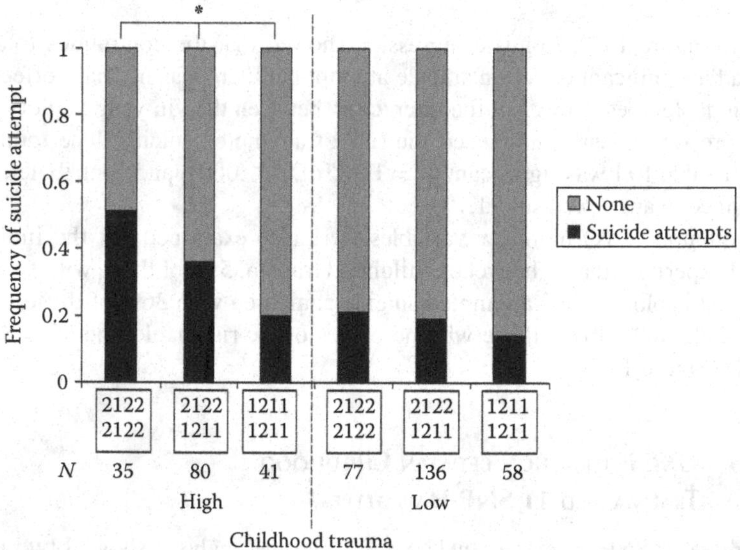

FIGURE 10.4 Interaction of *FKBP5* yin yang diplotypes with high levels of childhood trauma predicts suicide attempts. High level of childhood trauma = total CTQ score ≥1 SD (≥48) above mean of non-substance-dependent/no suicide attempt participants. Low level of childhood trauma = total CTQ score <48. Total $N = 427$. *$c^2 = 8.5$, 2 df, $p < .014$.

TABLE 10.2
Influence of Two 11 SNP Haplotypes, H2 and H5, and Childhood Trauma on Attempted Suicide

Parameters	L-R Tests c^2	p Value
11 SNP haplotypes: H2 and H5	1.38	.239
European ethnic factor score	5.20	.023
Substance dependence diagnosis	17.24	<.0001
Total CTQ score	6.83	.009
GXE	6.89	.009

Notes: Whole model test: $c^2 = 43$; 5 df; $p < .0001$, $r^2 = 0.19$.
The effect likelihood ratio tests (L-R) are shown (1 df for each test).
$N = 230$. Since these are haplotype analyses, N represents the number of chromosomes (two per individual). Only carriers of haplotypes H2 (frequency = 17%) and H5 (frequency = 9%) were included in these analyses. Haplotype H2: 11212111211; haplotype H5: 21111221112.
The total childhood trauma questionnaire (CTQ) score is a continuous measure.
GXE indicates the interaction between haplotypes (H2 and H5) and total CTQ score.

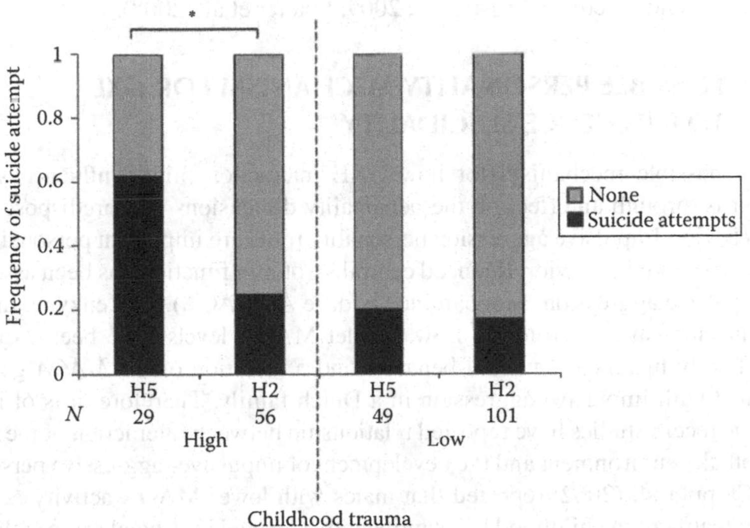

FIGURE 10.5 Interaction between *FKBP5* haplotypes and childhood trauma; influence on suicide attempt. Haplotype H2 (11212111211), frequency = 17%; haplotype H5 (21111221112), frequency = 9%. High level of childhood trauma = total CTQ score ≥1 SD (≥48) above mean of non-substance-dependent/no suicide attempt participants. Low level of childhood trauma = total CTQ score < 48. $N = 235$. *$c^2 = 11.2$, 1 df, $p = .0008$.

Of the individuals in the total group who had experienced a high level of childhood trauma, 62% of those with the H5 haplotype had attempted suicide compared with 25% of those with the H2 haplotype ($c^2 = 11.2$, 1 df, $p = .0008$) (Figure 10.5). In contrast, haplotype H5 conferred no risk to the individuals who had been exposed to low levels of childhood trauma.

10.12 POSSIBLE EPIGENETIC MECHANISMS OF GXE AND SUICIDAL BEHAVIOR

Are there possible explanations in terms of specific etiologic pathways by which the interaction of *genes* and childhood trauma might increase the risk of an individual attempting suicide? Epigenetics refers to processes that regulate the expression of genes in certain cells and/or at specific developmental points (Autry and Monteggia, 2009). In the case of *FKBP5*, a gene known to impact glucocorticoid sensitivity, one might speculate that one mechanism underlying its interaction might be the epigenetic effects of childhood trauma. This is because a recent postmortem study has reported a significant effect of childhood abuse on the genetic regulation of glucocorticoid receptors in suicide victims (McGowan et al., 2009).

In relation to epigenetics, it is also noteworthy that early maltreatment in infant rats leads to persisting changes in methylation of BDNF DNA and altered *BDNF* gene expression in the adult rat's prefrontal cortex (Roth et al., 2009). A postmortem study of suicide victims found significantly increased DNA methylation in the *BDNF* promoter gene in Wernicke's area (Keller et al., 2010). Other studies have also reported hypermethylation in both the GABAA and Trk.T1 receptor promoter regions of suicide victims (Ernst et al., 2009; Poulter et al., 2008).

10.13 POSSIBLE PERSONALITY MECHANISM FOR GXE TO INFLUENCE SUICIDALITY

Another possible mechanism for how GXE interaction might influence suicidal behavior is through an effect on the personality dimensions that predispose to suicidal behavior. Impulsive aggression personality traits are important personality risk factors for suicidal behavior. Reduced central serotonin function has been associated with impulsive aggression. Monoamine Oxidase A (MAOA) is an enzyme involved in the metabolism of serotonin. Low platelet MAOA levels have been associated with a family history of suicidal behavior and a mutation of the *MAOA* gene was associated with impulsive aggression in a Dutch family. Therefore, it is of interest that some recent studies have reported relationship between interaction of the *MAOA* gene with the environment and the development of impulsive–aggressive personality traits. Caspi et al. (2002) reported that males with lower MAOA activity genotype and maltreatment in childhood had significantly elevated antisocial scores relative to their lower MAOA counterparts who were not maltreated in childhood. This finding has been replicated (Foley et al., 2004). Also, females with an early abuse history have been reported to have an association of the low-expression MAOA genotype with suicide attempts as adults (Huang et al., 2004).

Most recently, Perroud et al. (2010) tested the interaction of nine candidate genes and scores on the CTQ in modulating anger-related responses on the State-trait Anger Expression Inventory in 875 suicide attempters. They found that the functional polymorphism of the catecholamine-*O*-methyl transferase (*COMT*) gene Val158Met significantly modulated the association between childhood sexual abuse and anger trait level. Specifically, attempters with the Val high-activity allele showed greater disposition toward anger than attempters homozygous for the Met allele. Interestingly, none of the serotonin-related genes significantly modulated the interaction of childhood abuse on anger traits. They concluded that anger trait level is impacted by the interaction of childhood abuse and functional polymorphism in the *COMT* gene.

10.14 CONCLUSIONS

This chapter has noted the importance of both the environment and genes in suicidal behavior and has reviewed the few published clinical studies of GXE interaction in relation to suicide attempts. Three main points seem noteworthy. The first point is that these few studies all showed no direct effect of genotype on suicidality but did show positive GXE results. These positive GXE results in relationship to suicidal behavior support the conclusion of Moffit et al. that "in psychiatric genetics, ignoring nurture handicaps the field's capacity to make new discoveries about nature," that is, suicidal behavior. Thus, it would seem imperative for there to be more GXE studies of suicidal behavior.

The second point that is noteworthy is that most of these few GXE studies of suicidal behavior have studied the interaction of genes with the early life event of childhood trauma. This is noteworthy because of Rutter (2010) conclusion that if "GXE may bring about biological changes through epigenetic mechanisms operating mostly in early life … it would seem to favor maltreatment rather than current life events as the key environmental hazard." Thus, it would seem that further GXE studies of suicidal behavior might consider including a measure of childhood trauma.

The third point about the GXE studies reviewed in this chapter is that they all had a specific hypothesis that they tested. This is relevant as it echoes the advice of Owen et al. (2010) that GXE studies "to constrain the opportunity for false positives through multiple testing, should be based on strong prior hypotheses concerning at least 1 of the variables, genetic or environmental." Thus, it would seem that further GXE studies of suicidal behavior might set out to test a defined hypothesis.

REFERENCES

Afifi, T., Enns, M., Cox, B., Asmundson, G., Stein, M., Sareen, J. 2008. Population attributable fractions of psychiatric disorders and suicide ideation and attempts associated with adverse childhood experiences. *Am J Pub Health* 98:946–952.

Arato, M., Banki, C., Bissette, G., Nemeroff, C. 1989. Elevated CSF CRH in suicide victims. *Biol Psychiatry* 25:255–259.

Asberg, M. 1997. Neurotransmitters and suicidal behavior: The evidence from cerebrospinal fluid studies. *Ann N Y Acad Sci* 836:158–181.

Autry, A., Monteggia, L. 2009. Epigenetics in suicide and depression. *Biol Psychiatry* 66:812–813.

Beatrais, A., Joyce, P., Mulder, R. 1997. Precipitating factors and life events in serious suicide attempters among youth. *J Am Acad Child Adol Psychiatry* 36:1543–1551.

Bennett, A., Lesch, P., Heils, A., Long, J., Shoaf, S., Champoux, M., Soumi, S., Linnoila, M., Highley, D. 2002. Early experience and serotonin gene variation interact to influence primate CNS function. *Mol Psychiatry* 7:118–122.

Bernstein, D., Fink, L. 1998. *Childhood Trauma Questionnaire: A Retrospective Self-Report Manual.* The Psychological Corporation, San Antonio, TX.

Binder, E.B. 2009. The role of FKBP5, a co-chaperone of the glucocorticoid receptor in the pathogenesis and therapy of affective and anxiety disorders. *Psychoneuroendocrinology* 34:S186–S195.

Binder, E.B., Bradley, R.G., Liu, W., Epstein, M.P., Deveau, T.C., Mercer, K.B., Tang, Y., Gillespie, C.F., Heim, C.M., Nemeroff, C.B., Schwartz, A.C., Cubells, J.F., Ressler, K.J. 2008. Association of FKBP5 polymorphisms and childhood abuse with risk of posttraumatic stress disorder symptoms in adults. *JAMA* 299:1291–1305.

Bunney, W., Fawcett, J. 1965. Possibility of a biochemical test for suicidal potential. *Arch Gen Psychiatry* 17:317–332.

Bunney, W., Fawcett, J., Davis, J., Gifford, S. 1969. Further evaluation of urinary 17-hydroxycorticosteroids in suicidal patients. *Arch Gen Psychiatry* 21:138–150.

Caspi, A., Mcclay, J., Moffitt, T., Mill, J., Martin, J., Craig, I.W., Taylor, A., Poulton, R. 2002. Role of genotype in the maltreated children. *Science* 297:851–854.

Caspi, A., Sugden, K., Moffit T., Taylor, A., Craig, I.W., Harrington, H.L., McClay, J., Mill, J., Martin, J., Braithwaite, A., Poulton, R. 2003. Influence of life stress on depression: Moderation by a polymorphism in the *5-HTT* gene. *Science* 301(5631):386–389.

Coryell, W., Schlesser, M. 2001. The dexamethasone suppression test and suicide prediction. *Am J Psychiatry* 158:748–753.

Coryell, W., Young, E., Carroll, B. 2006. Hyperactivity of the hypothalamic–pituitary–adrenal axis and mortality in major depressive disorder. *Psych Res* 142:99–104.

Crane, C., Williams, J., Hawton, K., Arensman, E., Hjelmeland, H., Bille-Brahe U, Corcoran, P., De Leo, D., Fekete, S., Grad, O., Haring, C., Kerkhof, A.J., Lonngvist, A.J., Michel, K., Salander-Renberg, E., Schmidtke, A., van Heeringen., C., Wasserman, D. 2007. The association between life events and suicide intent in self-poisoners with and without a history of self-harm: A preliminary study. *Suicide Life Threat Behav* 37:367–378.

Davidson, J., Hughes, D., George, L., Blazer, D. 1996. The association of sexual assault and attempted suicide in the community. *Arch Gen Psychiatry* 53:550–555.

De Bellis, M.D., Chrousos, G.P., Dorn, L.D., Burke, L., Helmers, K., Kling, M.A., Trickett, P.K., Putnam, F.W. 1994. Hypothalamic–pituitary–adrenal axis dysregulation in sexually abused girls. *J Clin Endocrinol Metab* 78:249–255.

Ernst, C., Delva, V., Deng, X., Sequeira, A., Pomarenski, A., Klempen, T., Ernst, N., Quirion, R., Gratton, A., Szyf, M., Turecki, G. 2009. Alternative splicing, methylation state, and expression profile of tropomysin-related kinase B in the frontal cortex of suicide completers. *Arch Gen Psychiatry* 66:22–32.

Foley, D., Eaves, L., Wormley, B., Silberg, J., Maes, H., Kuhn, J., Riley, B. 2004. Childhood adversity, monoamine oxidase A genotype, and risk for conduct disorder. *Arch Gen Psychiatry* 61:738–744.

Fu, Q., Heath, Bucholz, K., Nelson, E., Glowinski, A., Golberg, J., Lyons, M., Tsuang, M., Jacob, T., True, M., Eisen, S. 2002. A twin study of genetic and environmental influences on suicidality in men. *Psych Med* 32:11–24.

Gibb, B.E., McGeary, J.E., Beevers, C.G., Miller, I.W. 2006. Serotonin transporter (5-HTTLPR) genotype, childhood abuse, and suicide attempts in adult psychiatric inpatients. *Suicide Life Threat Behavior* 36:687–693.

Glowinski, A., Bucholz, K., Nelson, E., Fu, Q., Madden, P.A., Reich, W, Heath, A.C. 2001. Suicide attempts in an adolescent female twin sample. *J Am Acad Child Adol Psychiatry* 40:1300–1307.

Heikkinen, M., Aro, H., Lonnqvist, J. 2007. Recent life events, social support and suicide. *Acta Psychiatr Scand* 89:65–67.

Heim, C., Plotsky, P., Nemeroff, C. 2004. Importance of studying the contribution of early adverse experiences to neurobiological finding in depression. *Neuropsychopharmacology* 29:641–648.

Higley, D., King, S., Hasert, M., Champoux, M., Suomi, S., Linnoila, M. 1996. Stability of interindividual differences in serotonin function and its relationship to severe aggression and competent social behavior in rhesus macaque females. *Neuropsychopharmacology* 14:67–76.

Holmes, W., Slap, G. 1998. Sexual abuse of boys: Definition, prevalence, correlates, sequelae, and management. *JAMA* 280:1855–1862.

Huang, Y., Cate, S., Battistuzzi, C., Oquendo, M., Brent, D., Mann, J. 2004. An association between a functional polymorphism in the monoamine oxidase A gene promoter, impulsive traits and early abuse experience. *Neuropsychopharmacology* 29:1498–1505.

Jokinen, J., Carlborg, A., Martensson, B., Forsland, K., Nordstrom, A.-L., Nordstrom, P. 2007. DST non-suppression predicts suicide after attempted suicide. *Psych Res* 150:297–303.

Jokinen, J., Nordstrom, P. 2009. HPA axis hyperactivity and attempted suicide in young adult mood disorder inpatients. *J Affect Disord* 116:117–120.

Keller, S., Sarchiapone, M., Zarrilli, F., Videtic, A., Ferrero, A., Carli, V., Sacchetti, S., Lembo, F., Angiolillo, A., Jovanovic, N., Pisanti, F., Tomaiuolo, R., Monticelli, A., Roy, A., Marusic, A., Cocozza, S., Fusco, A., Bruni, C., Casaldo, G., Chiariotti, L. 2010. Increased BDNF promoter methylation in Wernicke's area of suicide subjects. *Arch Gen Psychiatry* 67:258–267.

Lesch, P., Benjel, D., Heils, A., Sabol, S., Greenberg, D., Petri, S., Benjamin, J., Müller, C.R., Hamer, D.H., Murphy, D.L. 1996. Association of anxiety-related trails with a polymorphism in the serotonin transporter gene regulatory region. *Science* 274:1527–1531.

Lin, P.-Y., Tsai, G. 2004. Association between serotonin transporter gene promoter polymorphism and suicide: Results of a meta-analysis. *Biol Psychiatry* 55:1023–1030.

Lindqvist, D., Isaksson, A., Träskman-Bendz, L., Brundin, L. 2008a. Salivary cortisol and suicidal behavior—A follow up study. *Psychoneuroendocrinology* 33:1061–1068.

Lindqvist, D., Traskman-Bendz, L., Vang, F. 2008b. Suicidal intent and the HPA-axis characteristics of suicide attempters with major depressive disorder and adjustment disorders. *Arch Suicide Res* 12:197–207.

MacMillan, H., Fleming, J., Trocme, N., Boyle, M., Wong, M., Racine, Y.A., Beardslee, W.R., Offord, D.R. 1997. Prevalence of child physical and sexual abuse in the community. Results from the Ontario Health Supplement. *JAMA* 278:131–135.

Mann, J.J., Currier, D., Stanley, B., Oquendo, M., Amsel, L., Ellis, S. 2005. Can biological tests assist prediction of suicide in mood disorders? *International Journal of Neuropsychopharmacology* 21:1–10.

Mann, J.J., Waternaux, C., Haas, G., Malone, K. 1999. Towards a clinical model of suicidal behavior in psychiatric patients. *Am J Psychiatry* 156:181–189.

McGowan, P., Sasaki, A., D'Alessio, C., Dymov, S., Labonte, B., Szyf, M., Turecki, G., Meaney, M.J. 2009. Epigenetic regulation of the glucocorticoid receptor in human brain associates with childhood abuse. *Nat Neurosci* 12:342–348.

Meaney, M. 2001. Maternal care, gene expression, and the transmission of individual differences in stress reactivity across generations. *Annu Rev Neurosci* 24:1161–1192.

Moffit, T., Caspi, A., Rutter, M. 2005. Strategies for investigating interactions between measured genes and measured environments. *Arch Gen Psychiatry* 62:473–481.

Molnar, B., Berkman, L., Buka, S. 2001. Psychopathology, childhood sexual abuse, and other childhood adversities: Relative links to subsequent suicidal behavior in the US. *Psychol Med* 31:963–977.

Nelson, E., Health, A., Madden, P., Cooper, L., Dinwiddle, S., Bucholz, K.K., Glowinski, A., McLaughlin, T., Dunne, M.P., Statham, D.J., Martin, N.G. 2002. Association between self-reported childhood sexual abuse and adverse psychosocial outcomes. Results from a twin study. *Arch Gen Psychiatry* 59:139–145.

Nemeroff, C. 2004. Neurobiological consequences of childhood trauma. *J Clin Psychiatry* 65:18–28.

Nemeroff, C.B., Owens, M.J., Bissette, G., Andorn, A.C., Stanley, M. 1988. Reduced corticotrophin releasing factor binding sites in the frontal cortex of suicide victims. *Arch Gen Psychiatry* 45:577–579.

Osvath, P., Voros, V., Fekete, S. 2004. Life events and psychopathology in a group of suicide attempters. *Psychopathology* 37:36–40.

Owen, M., Craddock, N., O'Donovan, M. 2010. Suggestion of roles for both common and rare risk variants in genome-wide studies of schizophrenia. *Arch Gen Psychiatry* 67:667–673.

Paykel, G., Prusoff, B., Myers, J. 1975. Suicide attempts and recent life events: A controlled comparison. *Arch Gen Psychiatry* 32:327–333.

Perroud, N., Courtet, P., Vincze, I., Jaussent, I., Jollant, F., Bellivier, F., Leboyer, M., Baud, P., Buresi, C., Malafosse, A. 2008. Interaction between BDNF Val66Met and childhood trauma on adult's violent suicide attempt. *Genes Brain Behav* 7:314–322.

Perroud, N., Jaussent, I., Guillaume, S., Bellivier, F., Baud, P., Jollant, F., Bellivier, F., Leboyer, M., Baud, P., Malafosse, A. 2010. COMT but not serotonin-related genes modulates the influence of childhood abuse on anger traits. *Genes Brain Behav* 9:193–202.

Pfennig, A., Kunzel, H., Kern, N., Ising, M., Majer, M., Fuchs, B., Ernst, G., Holsboer, F., Binder, E.B. 2005. Hypothalamus–pituitary–adrenal system regulation and suicidal behavior in depression. *Biol Psychiatry* 57:336–342.

Poulter, M., Du, L., Weaver, I., Palkovits, M., Faludi, G., Merali, Z., Szyf, M., Anisman, H. 2008. GABAA receptor promoter hypermethylation in suicide brain: Implications for the involvement of epigenetic processes. *Biol Psychiatry* 64:645–652.

Roth, T., Lubin, F., Funk, A., Sweatt, D. 2009. Lasting epigenetic influence of early-life adversity on the *BDNF* gene. *Biol Psychiatry* 65:760–769.

Roy, A. 1992. Hypothalamic–pituitary–adrenal axis function and suicidal behavior in depression. *Biol Psychiatry* 32:812–816.

Roy, A. 2001a. Childhood trauma and attempted suicide in alcoholics. *J Nerv Mental Dis* 189:120–121.

Roy, A. 2001b. Characteristics of cocaine dependent patients who attempt suicide. *Am J Psychiatry* 158:1214–1219.

Roy, A. 2002. Urinary-free cortisol and childhood trauma in cocaine dependent adults. *J Psych Res* 36:173–177.

Roy, A. 2005a. Reported childhood trauma and suicide attempts in schizophrenic patients. *Suicide Life Threat Behav* 35:690–693.

Roy, A. 2005b. Childhood trauma and suicidal behavior in male cocaine dependent patients. *Suicide Life Threat Behav* 31:194–196.

Roy, A., Agren, H., Pickar, D., Linnoila, M., Doran, A., Cutler, N., Paul, S.M. 1986. Reduced CSF concentrations of homovanillic acid and homovanillic acid to 5-hydroyindoleacetic acid ratios in depressed patients: Relationship to suicidal behavior and dexamethasone nonsuppression. *Am J Psychiatry* 143:1539–1545.

Roy, A., Gorodetsky, E., Yuan, Q., Goldman, D., Enoch, M.A. 2010. The interaction of FKBP5, a stress related gene, with childhood trauma increases the risk for attempting suicide. *Neuropsychopharmacology* 35:1674–1683.

Roy, A., Hu, X.-Z., Janal, M., Goldman, D. 2007. Interaction between childhood trauma and serotonin gene variation in attempting suicide. *Neuropsychopharmacology* 32:2046–2052.

Roy, A., Linnoila, M. 1985. Suicidal behavior, impulsiveness and serotonin. *Acta Psych Scand* 78:529–535.

Roy, A., Nielsen, D., Rylander, G., Sarchiapone, M. 2000. The genetics of suicidal behavior, *The International Handbook of Suicide*. Hawton, K., Van Heeringen, K., Wiley, J., Eds. Wiley, England, U.K., pp. 209–222.

Rutter, M. 2010. Gene–environment interplay. *Depress Anxiety* 27:1–4.

Sánchez, M., Noble, P., Lyon, C., Plotsky, P., Davis, M., Nemeroff, C., Winslow, J. 2005. Alterations in diurnal cortisol rhythm and acoustic startle response in nonhuman primates with adverse rearing. *Biol Psychiatry* 57:373–381.

Sarchiapone, M., Carli, V., Cuomo, C., Roy, A. 2007. Childhood trauma and suicide attempts in unipolar depressed patients. *Depress Anxiety* 24:268–272.

Statham, D.J., Heath, A.C., Madden, P.A., Bucholz, K.K., Bierut, L., Dinwiddie, S.H., Slutske, W.S., Dunne, M.P., Martin, N.G. 1998. Suicidal behavior: An epidemiological and genetic study. *Psychol Med* 28(4):839–855.

Wasserman, D., Sokolowski, M., Rozanov, V., Wasserman, J. 2008a. The *CRH1* gene: A marker for suicidality in depressed males exposed to low stress. *Genes Brain Behav* 7:14–19.

Wasserman, D., Waserman, J., Rozanov, V., Sokolwski, M. 2008b. Depression in suicidal males: Genetic risk variants in the *CRHR1* gene. *Genes Brain Behav* 8(1):72–79.

Weaver, I.C., Champagne, F.A., Brown, S.E., Dymov, S., Sharma, S., Meaney, M.J., Szyf, M. 2005. Reversal of maternal programming of stress responses in adult offspring through methyl supplementation: Altering epigenetic marking later in life. *J Neurosci* 25:11045–11054.

11 Genetic Factors and Suicidal Behavior

Clement C. Zai, Vincenzo de Luca, John Strauss, Ryan P. Tong, Isaac Sakinofsky, and James L. Kennedy

CONTENTS

11.1 INTRODUCTION

Approximately 1 million people worldwide die by suicide each year. With a prevalence rate of 0.0145% and suicide accounting for 1.5% of death by all causes, it is the 10th leading cause of mortality worldwide (Hawton and van Heeringen 2009). Suicide is complex, multifactorial behavioral phenotype. Suicide is also familial: a family history of suicide increases risk of suicide attempts and completed suicide. In this chapter, we will examine the family, twin, and adoption studies that establish the existence of both genetic and environmental bases of suicidal behavior. We will then review the major candidate gene findings. Lastly, we will discuss recent developments in genetic studies of suicide and comment on future experiments that may help resolve the challenges that are hindering genetic research into the pathophysiology of suicide.

11.2 FAMILY, TWIN, AND ADOPTION STUDIES

11.2.1 Family Studies

Family studies of suicide are designed to explore the extent to which suicidal behavior permeates families. A number of family studies have indicated familial aggregation of suicidal behavior. Most studies have shown a higher rate of suicidal behavior in relatives of suicide victims or attempters compared to relatives of non-suicidal controls (Tsuang 1983; Pfeffer et al. 1994; Malone et al. 1995; Brent et al. 1996; Johnson et al. 1998). Population registries in Denmark and Sweden have provided large datasets to explore the familiality of suicide. Four such studies have been carried out, all reporting increased rates of suicide in offspring of suicidal parents compared to offspring of non-suicidal parents (Agerbo et al. 2002; Qin et al. 2002, 2003; Runeson and Asberg 2003). One of the investigations provides evidence that this observed increase may not be due to grief, since suicide rates of individuals who lost their parent(s) to suicide were higher compared to individuals who lost their parent(s) to homicides or accidents (Runeson and Asberg 2003).

While population registry studies have the advantage of high statistical power, they may lack details in psychiatric diagnoses and suicidal behavior. In a study of suicide records dating from 1880 to 1980 in an Amish community, Egeland and Sussex (1985) found 26 reported suicides that aggregated within four families who also had a high incidence of mood disorders. The authors also found other families that were affected by multiple mood disorders but had no history of suicidal behavior. This suggests that the presence of mood disorders may be one risk factor for suicide, but additional factors likely play a role. Similarly, Tsai et al. (2002) showed that there was an increased risk for suicide in relatives of bipolar suicide victims compared to relatives of bipolar disorder patients who were not suicidal. Tsuang et al. (1985) and Powell et al. (2000) both found higher rates of suicides within families of psychiatric inpatients who completed suicide than within families of psychiatric inpatients who were not suicidal, regardless of the inpatient's psychiatric diagnosis. These results are in agreement with the finding that the suicide attempt rate in families of suicide attempters is higher compared to families of nonattempters, despite both attempters and nonattempters having similar mood disorder profiles (Brent et al. 2002, 2003). These studies also suggest that the trend for familiality of suicide completion may be at least partially independent of the familiality of psychiatric diagnoses.

Brent et al. (1996) examined adolescent suicide victims and matched non-suicidal controls from chart reviews. They found an increased rate of suicide attempts in first-degree relatives of suicide victims compared to those of non-suicidal controls. Fourteen additional studies also reported increased rate of suicide attempts or completion in families of suicide completers (Gould et al. 1996; Cheng et al. 2000; Kim et al. 2005) and probands with a history of suicide attempt (Garfinkel et al. 1982; Roy 1983, 2001, 2002, 2003; Linkowski et al. 1985; Mitterauer 1990; Pfeffer et al. 1994; Bridge et al. 1997; Johnson et al. 1998; Goodwin et al. 2004; Mann et al. 2005). These studies suggested the existence of a common suicide phenotype that includes both attempt and completion. Suicidal ideation, in contrast, may present as a separate suicide phenotype. The rate of suicidal ideation in relatives of suicidal probands was not increased after controlling for the presence of psychiatric disorders in relatives (Brent et al. 1996; Kim et al. 2005). The rate of familial suicide attempts was higher in suicide attempters compared to nonattempters, but the rate of suicide attempts was not higher in families of suicidal ideators compared to nonideators (Pfeffer et al. 1994). These observations suggest that suicidal ideation may segregate more with psychiatric diagnoses than suicide attempt/completion. Suicidal ideation and suicide attempt/completion may be partly independent phenotypes and do not fit strictly on a severity scale of suicidal behaviors.

The family studies we have reviewed are retrospective studies. Lieb et al. (2005) carried out a prospective study to examine the risk of suicidal ideation or attempts in the offspring of depressed mothers who had attempted or contemplated suicide. They found a greater than 50% increase in the risk for suicidal ideation or attempt relative to offspring whose mothers had never attempted suicide. From the aforementioned discussion, family studies support a genetic component of suicidal behavior. They provide evidence that the inheritance of suicidal behavior include both suicide attempt and suicide completion. Also, they show that the familial transmission may be independent of that of psychiatric disorders despite the fact that most suicide attempters/completers have underlying neuropsychiatric diagnoses.

11.2.2 Twin Studies

Twin studies are designed to evaluate the magnitude by which genetic and environmental factors influence a phenotype in a population. For the purpose of this review, twin studies investigate the risk of a twin exhibiting suicidal behavior given that the co-twin completed suicide. They also compare the suicide risk between monozygotic twin (MZ) pairs to dizygotic twin (DZ) pairs while assuming that environment factors are similar between MZ and DZ. The earliest studies were case reports from the 1800s, including concordant twins described by Dr. Benjamin Rush in 1812. Nonetheless, these findings cannot be confirmed because the state of monozygosity and (same sex) dizygosity was not ascertained until the 1930s. Additional case reports were published regarding suicide concordance rates in MZ and DZ twin pairs. Kallman (1953) reported a MZ concordance rate of 5.6% (1/18) versus a DZ concordance rate of 0% (0/21). Harvald and Hauge (1965) examined the Danish national twin registry and found 4 concordant MZ pairs out of 21 compared to no concordance within 75 DZ pairs. Kaprio et al. (1995), in comparison, reported no concordance in suicide within 34 MZ twin pairs and 2 concordant DZ pairs out of 119 in the Finnish national twin registry. In the United States, Roy et al. (1991) reported 7 concordant MZ pairs for suicide completion out of 62 compared to 2 concordant DZ twin pairs out of 114 collected from a national twin registry, the Minnesota Twin Loss Study, and other sources.

Roy et al. (1995) furthered the findings to include suicide attempts in the analysis. They found 10 of 26 surviving MZ co-twins had attempted suicide, while none of the nine surviving DZ co-twins had a history of suicide attempts. This suggests that suicide attempts and completions may share a common genetic component. Roy and Segal (2001) continued on to collect a sample from the United States and Canada consisting of 13 MZ and 15 DZ pairs. The difference between the MZ (4/13) and the DZ (0/15) concordance rates for suicide completion or attempt was significant. Unfortunately, these studies did not include psychiatric conditions and many cases of suicide attempts and suicidal ideation may have been missed. Nevertheless, the variability in reported heritability may likely have resulted from the small size of individual samples, which would have limited the extent of these analyses. The latest estimate for the heritability of suicide attempt/completion is 43% (95% confidence interval [CI]: 27%–60%; McGuffin et al. 2010).

Larger population-based twin studies have been conducted. A telephone interview study (Statham et al. 1998) where the authors included 5995 respondents of European ancestry showed increased concordance within MZ pairs compared to DZ pairs across suicide phenotypes that include the presence of any suicidal thoughts, persistent thoughts/plan/minor attempt, and serious suicide attempt. They estimated the heritability of suicidal ideation to be 43%, suicide plan/attempt 44%, and that of serious suicide attempt to be 55%. Fu et al. (2002) took into consideration psychopathology and reported the heritability of suicidal ideation to be 36% and that of suicide attempt to be 17% from 3372 male twin pairs belonging to the Vietnam Era Twin Registry. Glowinski et al. (2001) carried out a similar study on suicides attempted by 3416 female adolescent twins from a U.S. twin registry and found lifetime suicide attempt concordance rates to be 25% and 13% for MZ and DZ pairs, respectively. After taking into account shared and non-shared environmental factors, genetic factors were found to explain 48% of the variance in suicide attempt. More recently, Cho et al. (2006) investigated the rates of self-reported suicidal ideation or attempt on U.S. adolescent twins within 12 months prior to the interviews. As with previous reports, they found higher concordance rates for either suicidal ideation or suicide attempt in MZ pairs (23% and 38%, respectively) than in DZ pairs (17% and

17%, respectively). The investigators also found that aggression and depression may explain more of the variance in suicidality in females compared to males, while alcohol use and binge drinking seem to explain a larger portion of the variance in suicidal behavior in males than in females (Cho et al. 2006).

In all of the aforementioned four large-scale studies, the authors consistently found that psychiatric diagnoses (including major depressive disorder, conduct disorder, anxiety disorder, and alcohol dependence) and environmental factors (including childhood sexual and physical abuse) were associated with higher risk of attempting suicide. As with family studies, twin studies support a genetic basis of suicidal behavior and suggest that the genetic component of suicidal behavior may be independent of the heritability of psychiatric disorders. Furthermore, at least part of the heritability of suicidal behavior may overlap with suicidal ideation, suicide attempt, and suicide completion.

11.2.3 Adoption Studies

There are relatively few adoption studies that investigate the effect of environment on suicidal behavior. Using the Danish Adoption Registry, Schulsinger et al. (1979) matched 57 adoptees who committed suicide with 57 adoptees without history of suicidal behavior. Of the 269 biological relatives of suicidal adoptees, 12 committed suicide, while only 2 of the 269 biological relatives of non-suicidal adoptees committed suicide. None of the relatives of the adopted families in the study were affected by suicide. These observations support a genetic component of suicide. From the same registry, Wender et al. (1986) compared 71 adoptees with affective disorder to 71 adoptees who did not report significant affective disorder symptoms. They found higher frequencies of suicide in those related to affective disorder adoptees, especially those of the "affect reaction" subtype, compared to the relatives of unaffected adoptees. This suggests that the mechanism of suicide may involve the regulation of impulsive behavior.

11.3 REVIEW OF CANDIDATE GENES

Candidate gene studies have been the predominant methodology used in published genetic studies of suicidal behavior. Most of the published studies are recorded in Suicidal Behaviours: Genetic Association Studies Database (SBGAS) maintained by the McGill Group for Suicide Studies (URL: http://gmes.mcgill.ca/).

11.3.1 Serotonergic System

The serotonin neurotransmission system has received the most attention in candidate gene studies of suicidal behavior due largely to the role of serotonin in mood regulation and studies showing altered serotonin function in suicide victims (Virkkunen et al. 1989; Koller et al. 2003). Specifically, low cerebrospinal fluid concentration of the serotonin metabolite 5-hydroxyindoleacetic acid (5HIAA) has been found in numerous studies of suicidal patients (Asberg 1997), especially in high-lethality attempters (Placidi et al. 2001).

11.3.1.1 Tryptophan Hydroxylases *TPH1* and *TPH2*

The synthesis of serotonin (5-hydroxytryptamine [5-HT]) is catalyzed by the rate-limiting enzymes, the more broadly expressed TPH1 and the neuron-specific TPH2. The 29 kb human *TPH1* gene is localized on chromosome 11p15.3-p14 (Boularand et al. 1995). Several common single-nucleotide polymorphisms (SNPs) have been described in the upstream regulatory region and in intron 7 (Rotondo et al. 1999). Although functional variants have not been reported in the coding region of the *TPH1* gene (Han et al. 1999), the A (U) allele of the intron-7 A779C (rs1799913) marker has been associated with higher cerebrospinal fluid levels of the serotonin metabolite 5HIAA (Nielsen et al. 1994), and the CC genotype of the intron-7 A218C (rs1800532) marker has been associated with lower TPH1 protein levels in the prefrontal cortex (Ono et al. 2002). TPH1 expression in the anterior pituitary among other brain areas suggests a role in the hypothalamic–pituitary–adrenal (HPA) axis (Zill et al. 2009).

Since the first published finding by Nielsen et al. (1994) that the C-allele of the A779C polymorphism was associated with suicide attempts in alcoholic violent offenders/arsonists, over 40 papers have investigated this gene for association with suicide attempt, suicide completion, and suicidal ideation. Rujescu et al. (2003) conducted the first meta-analysis of A218C genotype data from seven studies on Caucasians and found that the A-allele (odds ratio [OR] = 1.33; 95% CI: 1.17–1.51) and both the AA and AC genotypes (OR = 1.48; 95% CI: 1.22–1.79) conferred risk for suicide attempt/completion. However, the suicide attempters have a mixture of different psychiatric diagnoses. Thus, the influence of other psychiatric diagnoses on these findings remained uncertain. More recently, Saetre et al. (2010) carried out a meta-analysis of studies in which the suicide cases were matched with controls for psychiatric diagnoses. They did not find A218C/A779C to be significantly associated with suicide (OR = 0.96; 95% CI: 0.80–1.16). The reason behind the discrepant results between these two meta-analyses could be that the first meta-analysis utilized healthy controls, while the latest meta-analysis employed controls matched with suicide cases for psychiatric disorders (e.g., schizophrenia suicide attempters versus schizophrenia nonattempters). Saetre et al. (2010) instead found a significant association between A218C with schizophrenia, suggesting that the association detected in the previous meta-analysis may have been confounded by the suicide cases' comorbid psychiatric conditions. It may also explain the lack of significant difference in TPH1 expression between suicide and non-suicidal groups (Zill et al. 2009; Perroud et al. 2010a). Despite numerous studies of *TPH1* polymorphisms in suicidal behavior, taken together, the results have not been meaningful. Nonetheless, the role of *TPH1* gene in suicide cannot be dismissed without examining additional polymorphisms, including those in its upstream regulatory region (Sun et al. 2005).

Unlike TPH1 being expressed in both the central nervous system and the periphery, TPH2 is the predominant form of tryptophan hydroxylase expressed in the brain (Walther et al. 2003). There have been conflicting results in the literature over the difference in expression of the more recently discovered, neuron-specific TPH2 between non-suicidal controls and suicide completers. Some papers have reported an increased expression of the *TPH2* gene in suicide victims (Bach-Mizrachi et al. 2008;

Perroud et al. 2010a), while others did not observe this difference in TPH2 expression between suicide victims and non-suicidal controls (De Luca et al. 2006a). The different brain regions examined could have contributed to the mixed gene expression findings. The 93.6 kb gene, located at 12q21.1, has been examined frequently since the first report of the G-allele of the rs1386494 marker being nominally associated with suicide completion in a German sample (Zill et al. 2004). Many subsequent studies examining multiple polymorphisms yielded mostly negative findings (De Luca et al. 2004a, 2005b; Zill et al. 2007; Mann et al. 2008; Mouri et al. 2009; Must et al. 2009), except for studies in depressed patients (Ke et al. 2006; Lopez de Lara et al. 2007; Yoon and Kim 2009). Unfortunately, even though multiple *TPH2* markers have been investigated, the analyzed polymorphisms often do not overlap among the studies, making interpretation of the findings difficult. It might be possible that *TPH2* gene expression changes with depression status, but not with suicide behavior *per se*. It might also be possible that the brain region–specific expression changes observed in suicide victims may be influenced by epigenetic changes in the *TPH2* gene promoter region.

11.3.1.2 Serotonin Transporter 5HTT (*SLC6A4* Gene)

The gene coding for 5HTT (*SLC6A4*, 37.8 kb at 17q11.1-q12) is another frequently examined candidate for studying the genetics of suicide. It is the primary target of many commonly prescribed antidepressant medications. 5HTT expression has been shown to be decreased in prefrontal cortical regions of suicide completers (Arango et al. 1995; Mann et al. 1996; Austin et al. 2002), but a number of study findings disputed these earlier reports (Mann et al. 1996; Little et al. 1997; Perroud et al. 2010a). Underlying psychiatric conditions and medications could have contributed to the mixed findings.

Two polymorphisms, in particular, a 44 nucleotide insertion/deletion polymorphism (serotonin transporter linked polymorphic region, HTTLPR) in the promoter region ~1 kb from the transcription start site and a 17 base pair (bp) variable number tandem repeat (VNTR) polymorphism in intron 2 (STin2), affect the function of 5HTT. More specifically, the HTTLPR genotypes carrying the short (S) allele (SS or LS) are associated with decreased transcriptional activity of the promoter (Collier et al. 1996) and reduced reuptake of serotonin compared to the long–long (LL) genotype (Heils et al. 1996). Also, the STin2 region has been shown to be an enhancer element in transgenic reporter mouse embryos, where the 12-repeat allele of STin2 was associated with increased reporter gene expression (MacKenzie and Quinn 1999). Since the first published report of *5HTT* gene and suicide attempt by Bellivier et al. (1997), over 40 papers have been published on this topic. Included is a recent comprehensive meta-analysis of studies up to January 2006 showing the long allele to be associated with decreased risk for suicide in the entire sample (OR = 0.88; 95% CI: 0.80–0.97) (Li and He 2007). The authors did not find STin2 to be associated with suicidal behavior from pooling five study samples and stratifying the meta-analysis by sex, ethnicity, diagnostic groups, and method of case–control pairing (i.e., whether suicide cases and controls were matched for psychiatric diagnosis) (Li and He 2007). In nine studies where the suicide attempters were paired with suicide nonattempters with the same psychiatric conditions, carriers of the long allele were associated with decreased risk for suicide (OR = 0.83; 95% CI: 0.73–0.95). The results from the meta-analysis of four studies found that the HTTLPR short allele carrying genotypes were not significantly associated with suicide completion (OR = 1.07; 95% CI: 0.48–2.37). This could possibly be due to reduced sample sizes and the fact that the control groups consisted of nonpsychiatric subjects. More recent studies of 5HTT in suicide found the short allele to be associated with violent suicide attempts (Wasserman et al. 2007; Neves et al. 2008, 2010).

It is important to note that the HTTLPR polymorphism has been associated with numerous neuropsychiatric disorders including depression (Kiyohara and Yoshimasu 2010), bipolar disorder (Cho et al. 2005), child aggression (Beitchman et al. 2006), and alcohol dependence (Feinn et al. 2005) as well as neuroticism (Sen et al. 2004). All of these disorders have in turn been associated with increased suicidality (Persson et al. 1999b). Similarly, STin2 has been associated with schizophrenia (Fan and Sklar 2005). In addition, the *5HTT* gene has been associated with response to antidepressants (Smits et al. 2004; Porcelli et al. 2011). Thus, it would be crucial to include the administration of serotonin reuptake inhibitors as a covariate when analyzing this gene in suicide. Also, the SNP rs25531 has been linked with the long allele of HTTLPR. It involves an A to G substitution that has been associated with lower 5HTT expression (Nakamura et al. 2000; Hu et al. 2006). This additional level of complexity at this locus may have contributed to some of the mixed findings in earlier genetic studies involving this gene. Studies including the rs25531 polymorphism, however, did not find significant association between the HTTLPR and suicidal behavior (De Luca et al. 2006c, 2008a; Chen et al. 2007; Segal et al. 2009). In the presence of childhood trauma, the short allele has been associated with suicide (Gibb et al. 2006; Roy et al. 2007), suggesting that gene–environment interaction could play an important role in suicidal behavior. Functional effects of the 10-repeat, 12-repeat, and other alleles of STin2 as well as its interaction with HTTLPR need to be verified in humans (Ali et al. 2010). The interaction between STin2 and childhood trauma should be explored. Additional polymorphisms across the *5HTT* gene, especially those in the HTTLPR region (Nakamura et al. 2000; Sakai et al. 2002), have been largely unexplored in most published studies (De Luca et al. 2006c; Zhang et al. 2008).

11.3.1.3 Serotonin Receptors (*HTR1A, HTR1B, HTR2A, HTR2C*, etc.)

Over 10 studies have reported negative findings with respect to *HTR1A* in suicidal behavior, beginning with Nishiguchi et al. (2002) who reported nonsignificant results with two infrequent missense variants in Japanese suicide completers versus healthy controls. Most studies since then have concentrated on the promoter C-1019G (rs6295) polymorphism, which is located within the consensus sequence for a transcriptional repressor called nuclear DEAF-1 related (Lemonde et al. 2003), with the C-allele having higher binding affinity than the G-allele. Thus, the G-allele represents the un-repressed or high-expression variant. Lemonde et al. (2003) also found the high-expression G-allele to be overrepresented in suicide completers of French Canadian ancestry compared to healthy controls. This may explain the low serotonin levels observed in suicide completers. These positive findings were not replicated however in other samples of suicide completers (Huang et al. 2004;

Ohtani et al. 2004; Serretti et al. 2007b; Videtic et al. 2009b) or suicide attempters (Huang et al. 2004; Serretti et al. 2007b; Wang et al. 2009).

The 1.17 kb intronless *HTR1B* gene located at 6q13 is also another frequently studied gene in suicide research due to the aggressive and impulsive behavioral phenotypes observed in knockout mice lacking the expression of the homologous *Htr1b* gene (Saudou et al. 1994; Brunner and Hen 1997; Zhuang et al. 1999; Bouwknecht et al. 2001). It has also been associated with antisocial behavior in alcohol-dependent individuals (Soyka et al. 2004). Since high level of impulsive aggression has been reported in psychological autopsy of suicide victims compared to psychiatric controls (Dumais et al. 2005; McGirr et al. 2008), *HTR1B* was considered a candidate gene for suicidal behavior. A meta-analysis of six studies with various study designs did not yield a significant finding of suicidal behavior with the G861C polymorphism (rs6296) (Kia-Keating et al. 2007).

Later publications did not find G861C to be associated with suicidal behaviors. The T-261G (rs11568817) and A-161T (rs130058) polymorphisms, which have been shown to alter 5HT1B expression in cell lines (Sun et al. 2002; Duan et al. 2003a), were found to be associated with suicidal ideation in depressed patients (Wang et al. 2009). However, other investigations into these and other polymorphisms, namely, A1180G, C129T (Huang et al. 1999; Zouk et al. 2007), T-261G (Zouk et al. 2007), A-161T (Hong et al. 2004; Tsai et al. 2004; Videtic et al. 2006; Zouk et al. 2007), and G371T (De Luca et al. 2008a), did not yield significant results. An exception was a marginal overrepresentation of the A-161T T-allele in suicide victims compared to healthy controls (Zouk et al. 2007). The different study designs and different markers investigated in each study made comparison among these studies challenging. For instance, Zouk et al. (2007) examined suicide victims versus healthy controls while Wang et al. (2009) examined the severity of suicidal behavior within a group of patients. The suicide cases in Zouk et al. (2007) were Caucasians of various psychiatric diagnoses, while the sample recruited by Wang et al. (2009) consisted of a group of Han Chinese major depressive patients without history of alcoholism, drug abuse, drug-induced bipolar disorder, or depressive disorders. It remains to be explored whether the presence of certain psychiatric conditions affects the association of *HTR1B* with suicidal behavior. Of interest, obsessive–compulsive disorder and bipolar disorder have been associated with *HTR1B* in several studies (Mundo et al. 2000, 2001, 2002). Nonetheless, the results from the literature thus far suggest that *HTR1B* may not play a major role in suicidal behavior.

The serotonin 2A receptor (*HTR2A*) gene is localized to 13q14-q21 and is 62.66 kb in length. The majority of studies to date reported 5HT2A expression or binding to be increased in suicide victims (Cheetham et al. 1988; Lowther et al. 1994; Turecki et al. 1999; Arango et al. 2003). Increased 5HT2A levels have also been reported in platelets of suicidal patients (Pandey 1997). A SNP in the promoter region of *HTR2A*, A-1438G (rs6311), as well as an exon 1 SNP, T102C (Ser34Ser, rs6313), have been extensively studied in numerous psychiatric disorders. The C102 allele has been associated with a 20% decrease in 5HT2A receptor levels in the temporal cortex (Polesskaya and Sokolov 2002). Li et al. (2006) carried out a meta-analysis of T102C of 25 studies published up to July 2005 on suicidal behavior, and they did not find significant association with suicide attempt versus non-suicidal patients (seven studies; OR = 0.98; 95% CI: 0.83–1.16). Other matching strategies yielded positive findings where the T102 was found to be a protective allele. These methods included comparing individuals with suicidal ideation to a combined group of patients lacking ideation and healthy controls (five studies; OR = 0.77; 95% CI: 0.62–0.95), or comparing individuals with suicidal ideation and attempts to psychiatric controls without attempt or ideation and healthy controls (20 studies; OR = 0.88; 95% CI: 0.77–1.00). In the same paper, the meta-analysis of A-1438G of seven studies in suicidal behavior showed that A-allele carrying genotypes were protective against suicide (OR = 0.67; 95% CI: 0.50–0.89) (Li et al. 2006). More recent studies with T102C and A-1438G yielded mostly negative findings. Only one study thus far has investigated the tag SNPs, SNPs selected to minimize overlapping genetic information, spanning *HTR2A* for association with suicidal ideation. This study used a sample of 270 families affected by schizophrenia or schizoaffective disorder (Fanous et al. 2009). They did not observe significant association between any of the tag SNPs or their haplotypes and suicidal ideation. Phenotypic heterogeneity from analyzing these tag SNPs in suicide attempters and completers may have contributed to the mixed results for this gene.

Suicidal behavior has been hypothesized to be partly X-linked. This is because of the seemingly different rates of suicidal ideation (2 females:1 male), suicide attempts (4 females:1 male), and suicide completion (1 female:3 males) between the two sexes (Stefulj et al. 2004a,b; Bondy et al. 2006). Turecki et al. (2003) first investigated the X-linked *HTR2C* gene where the authors did not find the G-995A (rs3813928) polymorphism to be significant in suicide completion. Subsequent studies focused on a missense polymorphism Ser23Cys (rs6318, G68C) in the coding region of the gene and most did not find this polymorphism significant. However, Videtic et al. (2009a) found the G-allele to be associated with risk for suicide in their Slovenian sample. The *HTR2C* gene is large (326.1 kb), and thus additional polymorphisms need to be investigated. Also, changes in mRNA editing have been reported in suicide victims (Gurevich et al. 2002). In addition to the *HTR2C* gene, Turecki et al. (2003) also explored for possible association between suicide completion and the *HTR1D*, *HTR1E*, *HTR1F*, *HTR5A*, and *HTR6* genes. They did not find any of the tested markers in these genes to be associated with suicide risk.

11.3.1.4 Monoamine Oxidases–*MAOA* and *MAOB*

MAOA is a mitochondrial enzyme that degrades monoamines including dopamine, norepinephrine, and serotonin. It is the major target of the monoamine oxidase inhibitor (MAOI) class of antidepressants. Elevated activity of the MAOA enzyme in the hypothalamic region of suicide victims has been reported (Sherif et al. 1991). The *MAOA* gene is localized to Xp11.23-p11.4. A 30 bp repeat in the promoter region has been associated with levels of expression (Sabol et al. 1998) where alleles 2 (with 3.5 repeats) and 3 (with four repeats) were associated with higher *in vitro* transcriptional activity. These alleles have also been associated with higher 5HIAA levels (Jonsson et al. 2000; Williams et al. 2003) and lower response to serotonin (as measured by increase in circulating prolactin levels upon fenfluramine administration) (Manuck et al. 2000). Two additional polymorphisms, EcoRV (rs1137070) and Fnu4HI (rs6323), have been associated with changes in enzyme activity (Hotamisligil and Breakefield 1991). The promoter VNTR has been examined in more than eight studies, with Ho et al. (2000) finding the 132 bp allele 3 to be associated with risk for suicide attempts in bipolar disorder patients. The authors also analyzed and found the Fnu4HI (rs6323) T-allele to be associated with suicide risk in the same sample (Ho et al. 2000). Courtet et al. (2005) found in a sample of hospitalized suicide attempters with various diagnoses that alleles 2 and 3 were overrepresented in violent attempters. Other polymorphisms in this 90.66 kb gene should be interrogated for possible association with suicidal behavior. Only one study (Brezo et al. 2010) explored the 115.8 kb *MAOB* gene, which is adjacent to *MAOA*, in association with suicide attempts. The authors did not report any significant findings with two examined *MAOB* polymorphisms.

11.3.2 Dopaminergic System

The dopamine system has not been a major target for suicide genetic research. There have been few studies looking into altered dopamine (and norepinephrine) levels in brain tissues of suicide victims (Arango et al. 1993) and cerebrospinal fluid (CSF) of suicide attempters (Roy et al. 1986; Jones et al. 1990; Lester 1995). Dopamine has also been associated with impulsivity, a personality trait that is implicated in suicide (van Gaalen et al. 2006; Oswald et al. 2007).

11.3.2.1 Tyrosine Hydroxylase

Tyrosine hydroxylase (*TH*) is a small 7.9 kb gene located on chromosome 11p15.5. A nominally significant finding was reported between the putatively functional 252 bp eight-repeat allele of the tetranucleotide repeat (Albanese et al. 2001) in the promoter region of *TH* and suicide attempters compared to healthy control subjects (Persson et al. 1997). Another research group did not replicate these positive findings with suicide attempt in bipolar disorder patients (Ho et al. 2000). Giegling et al. (2008) investigated three SNPs in the DOPA decarboxylase (*DDC*) gene in addition to the Val81Met (rs6356) and a 3′ polymorphism (rs3842727) in the *TH* gene. They did not find positive results with either gene. Gerra et al. (2005) explored the relationship between the gene coding for the dopamine transporter *SLC6A3* and did not find the VNTR in the 3′ untranslated region of the gene to be associated with male suicide-attempting heroin addicts compared to controls. The VNTR in exon 3 of the dopamine *DRD4* gene has also been investigated in two studies (Persson et al. 1999a; Zalsman et al. 2004) with negative findings.

11.3.2.2 Catechol-*O*-Methyltransferase

Catechol-*O*-methyltransferase (*COMT*) deactivates dopamine and norepinephrine by the addition of a methyl group from *S*-adenosylmethionine. The 27.2 kb gene is mapped to 22q11.1-q11.2. The Val158Met (rs4680) polymorphism has been investigated in many psychiatric disorders. The Met allele has been associated with lower thermostability and resultant lower COMT enzymatic activity (Weinshilboum and Raymond 1977; Lachman et al. 1998). Two meta-analyses have been conducted. Kia-Keating et al. (2007), using data from six previous studies (Ohara et al. 1998a; Nolan et al. 2000; Russ et al. 2000; Liou et al. 2001; Rujescu et al. 2003;

Ono et al. 2004), found the low-functioning Met allele to be associated with suicide risk. However, the results could be influenced by the different sex ratios among the studies. It is important to note that several groups have reported the sex-specific association of suicidal behavior with Val158Met (Nolan et al. 2000; Ono et al. 2004). Calati et al. (2011) presented an updated meta-analysis with four additional samples (De Luca et al. 2005a, 2006b; Baud et al. 2007; Zalsman et al. 2008), and they did not report Val158Met to be significant. Similar to the Kia-Keating et al.'s (2007) paper, this meta-analysis could be influenced by different case–control matching strategies employed by each included study or insufficient number of studies. In addition, both meta-analyses reported a significant effect of sex ratios (Kia-Keating et al. 2007; Calati et al. 2011). Sex-specific analysis of this and additional *COMT* polymorphisms (e.g., Nackley et al. 2006) in suicide are warranted.

11.3.2.3 Dopamine Receptor *DRD2*

Four suicide phenotype studies of the dopamine receptor *DRD2* gene (11q22-q23) have been published. Finckh et al. (1997) reported an exon 8 polymorphism in the 3′ untranslated region where the AA genotype appeared to be overrepresented in the alcoholic patients with history of suicide attempts. Subsequent studies investigated a putatively functional promoter insertion/deletion polymorphism (−141 Ins/Del, rs1799732; Arinami et al. 1997), with Johann et al. (2005) reporting the Del allele conferring risk for suicide attempt or ideation in alcoholics. Suda et al. (2009) recently reported the Ins allele as well as the A2 allele of the TaqIA polymorphism to be overrepresented in the suicide attempters compared to the healthy controls. A third paper by Ho et al. (2000) did not find −141C Ins/Del to be associated with suicide attempt in their sample of bipolar and unipolar disorder patients. Additional *DRD2* polymorphisms, particularly the putatively functional C957T (rs6277, Duan et al. 2003b), rs12364238, and rs1076560 (Zhang et al. 2007) markers, should be explored in suicidal behavior. Caution must be given when analyzing *DRD2* markers as, similar to *5HTT*, many of these *DRD2* markers have been implicated in a number of psychiatric disorders (Noble 2003).

11.3.3 Hypothalamic–Pituitary–Adrenal Axis

The Hypothalamic Pituitary Adrenal (HPA) axis is a very important component of the stress response system. HPA axis abnormalities, as indicated by nonsuppression in the dexamethasone suppression test, have been implicated in suicidal behavior (McGirr et al. 2010; Pompili et al. 2010). One study has explored two polymorphisms (rs1870393 and rs3176921) in the corticotrophin-release hormone (*CRH*) gene in suicide attempt using a family-based study design (Wasserman et al. 2008). The authors reported negative findings. Wasserman et al. (2008) also investigated two polymorphisms in the gene coding for the CRH receptor (*CRHR1*) (rs4792887 and rs1396862) and found the T-allele of rs4782887 to confer risk for suicide attempts, particularly in males who had experienced low-level stressful life events. In the same year, Papiol et al. (2007) reported a similar study genotyping two different *CRHR1* SNPs (rs110402 and rs242937) in their sample of suicide attempters and healthy controls. They did not find these SNPs to be associated with suicide attempt. Papiol et al. (2007) also reported negative findings with two SNPs (rs2270007 and rs2240403) in the CRH receptor 2 gene (*CRHR2*), in contrast to a nominal association of a GT-repeat polymorphism in *CRHR2* with severity of suicidal behavior measured in bipolar disorder subjects in a nuclear family sample (De Luca et al. 2007). Papiol et al. (2007) also did not find one putative functional SNP (rs1360780; Binder et al. 2004) in FK506-binding protein (*FKBP5*) gene to be associated with suicide attempt, but Willour et al. (2009) reported significant association of four markers in *FKBP5* with bipolar disorder depending on suicide attempt status. Papiol et al. (2007) did not find two SNPs (rs7728378 and rs1875999) in the gene encoding CRH-binding protein (*CRHBP*) to be significant, the results that are not consistent with a more recent report of an association between *CRHBP* and suicide attempt history in schizophrenia patients (De Luca et al. 2010). Although our group (De Luca et al. 2010) did not find the glucocorticoid receptor gene (*NR3C1*) to be associated with suicide attempt history in schizophrenia patients, increased methylation of specific *NR3C1* promoter CpG sites have been reported in the hippocampi of suicide victims with history of childhood trauma (McGowan et al. 2009), the results that may account for the observed decreased expression of glucocorticoid receptors in these brain regions (Labonte and Turecki 2010).

11.3.4 Brain-Derived Neurotrophic Factor and Tropomyosin-Related Kinase B

Decreased brain-derived neurotrophic factor (BDNF) levels have been reported in suicide victims (Dwivedi et al. 2003; Karege et al. 2005; Kim et al. 2007; Dwivedi 2010). Of the regional SNPs, the Val66Met (rs6265) polymorphism has received more attention in genetic studies of suicide. Initial studies of this functional polymorphism did not find it to be associated with suicide attempts (Hong et al. 2003; Huang and Lee 2007). However, Vincze et al. (2008) found an overrepresentation of the Val allele in violent bipolar suicide attempters versus healthy controls. The majority of recent studies reported the Met allele to be associated with suicide attempt in the context of various psychiatric diagnoses, including schizophrenia (Huang and Lee 2007), bipolar disorder (Kim et al. 2008), and depression (Iga et al. 2007; Sarchiapone et al. 2008). Additional studies did not report the Val66Met to be associated with suicide (Zarrilli et al. 2009; Kohli et al. 2010; Spalletta et al. 2010) or worsening of suicidal behavior during a drug trial in treatment-resistant depressed adolescents (Brent et al. 2010). Overall, a recent meta-analysis by our group of 11 published studies found the Met allele to be associated with risk for suicide ($p = 0.032$; $OR_{Met} = 1.16$; 95% CI: 1.01–1.32; Zai et al. 2011).

The association of *BDNF* may be influenced by its downstream receptor tropomyosin-related kinase B (*NTRK2*; Kunugi et al. 2004). Kunugi et al. (2004) was first to report an association between Ser205Leu of *NTRK2* and depressed suicide. McGregor et al. (2007) did not replicate this finding in young adults who had childhood depression and had been prospectively followed for over 20 years. More recently, multiple polymorphisms in the *NTRK2* gene have been consistently associated with suicidal behavior. Perroud et al. (2009) found *NTRK2* to be associated with suicidal ideation in the Genome-based Therapeutic Drugs for Depression (GENDEP) study and Kohli et al. (2010) recently supported these findings by showing that multiple polymorphisms in *NTRK2* were associated with suicide attempt history in two ethnically distinct depression samples. Gene–gene interaction may be able to refine the association findings in suicide and identify pathways involved in this devastating outcome. In addition, the observed changes in BDNF expression levels in suicide could be epigenetically determined. Keller et al. (2010) reported hypermethylation at specific CpG sites in the *BDNF* promoter region that correlated with decreased BDNF expression. Similarly, hypermethylation at specific CpG dinucleotides in the *NTRK2* gene promoter have been linked to reduced frontal cortical expression of a TrkB protein isoform in postmortem brain studies in suicide victims (Ernst et al. 2009). Further discussion of methylation and other epigenetic mechanisms is provided later.

11.3.5 Adrenergic Receptor A2A

The levels of adrenergic α-2A receptor have been found increased in the postmortem brain tissues of depressed suicide victims (De Paermentier et al. 1997; Garcia-Sevilla et al. 1999). However, these findings need further replication (Gross-Isseroff et al. 2000). Its gene, adrenergic receptor A2A (*ADRA2A*), is 3.65 kb in length and mapped to chromosomal region 10q24-q26. The Asn251Lys (rs1800035) polymorphism has been reported as functional because the Lys allele increases the ligand-mediated G-protein coupling and enhances downstream signaling in the inhibition of adenylate cyclase and activation of MAPK (Small et al. 2000). Neither Ohara et al. (1998b) nor Sequeira et al. (2004) found significant results with a promoter polymorphism (C-1291G, rs1800544) in Japanese suicide attempters with mood disorders and French Canadian suicide victims, respectively. A more recent study by Fukutake et al. (2008), however, reported the C-1291 allele to be associated with risk for female suicide completion in their Japanese sample.

11.3.6 Other Genes

Poulter et al. (2008) found increased anterior prefrontal *GABRA1* gene promoter methylation that was consistent with their previous finding of decreased anterior prefrontal GABRA1 levels in major depressive suicide victims (Merali et al. 2004). There is increasing evidence of altered expression of GABA (γ-aminobutyric acid) system components in suicide, yet genetic studies of GABA system genes in suicide are still lacking. One study investigating polymorphisms in the $GABA_A$ receptor α3 subunit gene *GABRA3* (Baca-Garcia et al. 2004), and another study on the glutamate decarboxylase genes *GAD1* and *GAD2*, did not find a significant association with suicidal behavior (De Luca et al. 2004b). Nonetheless, recent reports of altered GABA receptor subunit mRNA expression in depressed suicide victims (Sequeira et al. 2009; Poulter et al. 2010) encourage exploration of additional GABA system genes in suicide (Lee et al. 2009). In particular, the altered expression levels point to examination of polymorphisms in regulatory regions.

Signaling molecules have been investigated in suicidal behavior. The rs1130214 and rs2494746 markers in the gene coding for the intracellular signaling molecule protein kinase B (AKT1) were associated with suicide attempt and violent attempt, respectively (Magno et al. 2010). Previously, angiotensin-receptor blockers were identified as a possible risk factor for suicide (OR = 3.52) (Callreus et al. 2007). Four studies have explored the role of the angiotensin-converting enzyme (*ACE*) gene in suicidal behavior. Two studies found the low-functioning insertion allele (Rigat et al. 1990) to be associated with suicide completion (Hishimoto et al. 2006; Fudalej et al. 2009). One study found the deletion allele to be overrepresented in suicide attempters and completers compared to controls (Sparks et al. 2009), while another did not find this marker associated with suicide attempt history in depressed patients (Hong et al. 2002). Other genes that have been implicated in suicidal behavior and are awaiting replication include apolipoprotein E (*APOE*) (Hwang et al. 2006), cholecystokinin (*CCK*; Shindo and Yoshioka 2005), 14-3-3 epsilon (Yanagi et al. 2005), spermine/spermidine *N*-acetyltransferase (*SAT1*; Sequeira et al. 2006), regulator of G-protein signaling (*RGS2*; Cui et al. 2008), and neuronal nitric oxide synthase (*NOS1*; Rujescu et al. 2008; Cui et al. 2010).

11.3.7 Genome-Wide Studies of Suicide

Zubenko and coworkers reported a genome-wide linkage study of suicide attempts in 81 families where the probands were affected by recurrent early-onset major depression (Zubenko et al. 2004). Using 389 microsatellite markers, they reported significant linkage in chromosomal regions 2p, 5q, 6q, 8p, 11q, and Xq. Hesselbrock et al. (2004) reported another genome scan of suicide using 336 microsatellite markers across the genomes of multiplex families affected by alcoholism from the Collaborative Study on the Genetics of Alcoholism (COGA). When they conducted a nonparametric analysis on 59 pairs of siblings who had both attempted suicide(s), they found the marker D2S1790 near chromosomal region 2p11 to be linked (Hesselbrock et al. 2004). The findings on 2p12 were later replicated in a sample of 162 bipolar disorder pedigrees (Willour et al. 2007). This study on bipolar disorder patients also found linkage with markers D6S1035, D6S1277, and D6S1027 on chromosome 6q25-q26 (Willour et al. 2007). The marker D6S1035 was found to be linked to bipolar disorder/recurrent major depressive disorder diagnosis in a large sample of 154 families from the National Institute of Mental Health genetics initiative (Cheng et al. 2006); however, two other markers at 6q24-q25, D6S1848 and D6S2436, were found to be linked to suicide completion in this bipolar disorder sample. Recent advances have allowed for large-scale genotyping of SNPs across the genome in large samples to carry out genome-wide hypothesis-generating studies. Unfortunately, using genome-wide significance threshold of 5×10^{-8} did not yield significant findings from any of the published studies thus far, indicating that the genetic architecture underlying suicidal behavior is complex with multiple genes of moderate effect. Perroud et al. (2012) published a genome-wide association study of treatment-associated suicidality on the GENDEP sample of 706 major depression patients treated with escitalopram or nortriptyline. These patients were assessed for increased suicidality during 12 weeks of treatment using a composite score calculated from the 3rd item of the HDRS-17, the 10th item of the Montgomery-Asberg Depression Rating Scale, and the 9th item of the Beck Depression Inventory. The authors found rs11143230 downstream of the guanine deaminase (cypin) (*GDA*) gene to be most significantly associated with increased suicidality during antidepressant treatment. They also found rs358592 in the voltage-gated potassium channel (*KCNIP4*) gene and rs4732812 upstream of the elongation protein 3 homolog (*ELP3*) gene to be most significantly associated with increased suicidality during escitalopram treatment. The rs6812841 marker was most significantly associated with increased suicidality during nortriptyline treatment. Since cypin interacts with the postsynaptic density protein-95 (PSD-95), and PSD-95 is involved in glutamatergic neurotransmission, glutamate signaling may be involved in suicidality. The authors also extracted genotypes for SNPs in 33 candidate genes for suicide, and they found the *NTRK2* gene to be the most significant in the escitalopram-treated group and the *CRHR2* gene to be the most significant in the nortriptyline-treated group (Perroud et al. 2010b). A genome-wide association study was conducted using two bipolar disorder and two major depressive disorder patient conglomerate samples totaling 8737 patients of which 2805 had a lifetime history of at least one suicide attempt (Perroud et al. 2010b). The strongest association signal came from the intergenic marker rs1466846 for suicide attempt in the bipolar disorder discovery sample, and rs2576377 in the Abl-interactor family member 3 binding protein (*ABI3BP/TARSH*) gene for suicide attempt in major depression discovery sample. However, these findings were not replicated in the bipolar disorder or major depression replication samples. The authors selected 19 candidate genes for suicide and found only nominal results with *FKBP5* and *NGFR* genes (Perlis et al. 2010). The authors further conducted a random effects meta-analysis of markers with $p < 1 \times 10^{-3}$ across all four mood disorder samples, half of the 10 most significant markers reside in the gene coding for sorbin and SH3-domain containing-1 (SORBS1), which has been implicated in insulin signaling. Willour et al. (2011) reported a genome-wide association study using the same samples of bipolar disorder patients. After meta-analysis of markers with $p < 1 \times 10^{-3}$ from the first sample in both bipolar disorder samples, they found the most significant signal with rs300774 in an intergenic region at 2p25 in linkage disequilibrium with the *SH3YL1*, *ACP1*, and *FAM150B* genes. Further investigations in postmortem prefrontal cortical brain samples from suicide completers revealed significantly higher ACP1 expression in suicide victims compared to non-suicidal victims (Willour et al. 2011). More recently, another genome-wide association study of suicidality was reported on major depression sample from the RADIANT study (Schosser et al. 2011). In addition to analyzing the discrete variable of serious suicide attempt, genotypes were also analyzed using the SCAN interview, which captures suicide severity from suicide ideation to attempt. None of the top findings from the RADIANT sample were replicated in the German replication sample. However, meta-analysis of top findings from the RADIANT and Sequenced Treatment Alternatives to Relieve Depression (STAR*D) samples revealed an association with rs1377287 in the solute carrier family 4 member 4 (*SLC4A4*) gene. Recent advances in genotyping technologies have enabled us to collect large amounts of genetic information. However, more work needs to be done in terms of better characterization of suicidal behavior and clearer understanding of how gene markers work in pathways to clarify these preliminary whole-genome findings.

11.4 METHODOLOGICAL AND PHENOTYPIC CONSIDERATIONS

With most of the genetic findings of suicidal behavior being nominal or mixed, it has become apparent that there are numerous issues that need to be resolved as we move toward more comprehensive and large-scale genome-wide studies. These issues of suicide genetic studies, as well as more recent research strategies, are discussed later.

11.4.1 Gene–Gene Interactions

Genes rarely work alone. Relatively low ORs derived from meta-analyses of various candidate genes discussed earlier suggest that multiple genes are involved in the complex phenotype of suicide. Only a handful of studies looked at the combined effect of two or more genes in suicidal behavior. Our group (De Luca et al. 2010) reported a nominally significant interaction between *CRHR1* rs16940665 and *CRHBP* rs1875999 in the severity of suicidal behavior in a sample of 231 schizophrenia patients. Investigation into the interaction between *MAOA* and *COMT* polymorphisms in suicide attempt did not yield significant findings in schizophrenia or bipolar disorder patients (De Luca et al. 2005a, 2006b). The interaction between the X-linked genes *MAOA* and *HTR2C* also was not significant (De Luca et al. 2008b). Vincze et al. (2008) found *BDNF* and *5HTT* polymorphisms to be independently associated with violent suicide in their bipolar disorder sample, but they did not find a significant interaction between these two genes in suicidal behavior. Perroud et al. (2009) recently found a significant interaction between *BDNF* and *NTRK2* polymorphisms in suicidal ideation in their GENDEP clinical trial sample. Additional interactions between genes should be explored, using software including the generalized multifactor dimensionality reduction (GMDR; Lou et al. 2007) and HELIXTREE (Golden Helix, Inc.; e.g., Zai et al. 2009). With large datasets from genome-wide association and gene expression microarray studies, pathway analysis using ingenuity pathway analysis (IPA; e.g., Inada et al. 2008; Charlesworth et al. 2010) could be a viable option for more comprehensive analysis of gene pathways in suicidal behavior.

11.4.2 Gene–Environment Interactions

Suicidal behavior is considered a complex phenotype that involves both genetic predispositions and distal/proximal environmental factors (Roy et al. 2009). Personal history of childhood abuse has been repeatedly implicated as a risk factor for suicidal behavior (Brodsky et al. 1997; Melhem et al. 2007; Brezo et al. 2008b; Carballo et al. 2008). Some epidemiological studies have estimated that sexual abuse may explain 20% of the risk variance in suicide (Brent and Melhem 2008). Using data collected for the Dunedin Multidisciplinary Health and Development study, Caspi et al. (2003) found that individuals carrying the HTTLPR short (S) allele and who experienced childhood (during 3–11 years of age) maltreatment or increasing number of stressful life events (at 21–26 years of age), assessed using the life-history calendar, were more prone to suicidal ideation or attempts. Gibb et al. (2006) reported similar findings of increased risk of suicide attempts in 30 psychiatric inpatients who were HTTLPR S-allele carriers and who had experienced physical or sexual abuse but not emotional abuse. Roy et al. (2007) also reported in a sample of African-American substance-dependent patients that those carrying the low-expressing *5HTT* genotypes (SS, SLg, and LgLg) and having experienced higher levels of childhood trauma (as assessed by the Childhood Trauma Questionnaire) were at increased risk for suicide attempts. The interaction between S-allele carrying genotypes and maltreatment was also observed in suicidal ideation in low-income children (Cicchetti et al. 2010). A recent study did not replicate the findings on suicidal ideation in 3243 subjects from 2230 families with HTTLPR (Coventry et al. 2010). Possible explanation for the discrepancy between this study and the others could be that stressful life events assessed for this study were restricted to 12 months leading up to the study rather than history of childhood trauma as assessed in most of the other studies. Nonetheless, further investigations using various study designs are required to resolve these mixed findings. Also, measurement of life events needs to be standardized and likely more information will be obtained if the life events are quantified (Risch et al. 2009; Uher et al. 2011; Mandelli et al. unpublished).

While individuals carrying the *MAOA* VNTR low-functioning allele were more prone to antisocial behavior and conduct disorder (Caspi et al. 2002), they were not prone to suicidal behaviors (Caspi et al. 2003). Brezo et al. (2010) recently reported on a large longitudinal study wherein a cohort of 1255 youths was followed for 22 years and assessed for the development of mood disorders and suicidal behaviors. The authors then examined variants within 11 serotonin system genes for association with these psychiatric phenotypic outcomes. They also explored the possible interaction between these gene variants and childhood physical/sexual abuse on these outcomes. They reported that the *TPH1* rs10488683 marker was associated directly with suicide attempt and three markers in *HTR2A* were associated with suicide attempt with history of abuse (Brezo et al. 2010). Thus, given these initial reports, it appears that incorporation of stressful life events and/or maltreatment is a useful strategy for enhancing genetic studies of suicide.

11.4.2.1 Effects of Medication or Substance Use

Bipolar subjects with comorbid substance use disorders (SUD) had a 39.5% rate of lifetime attempted suicide, compared to those without SUD at a rate of 23.8% (Dalton et al. 2003). A forensic study on suicide victims with respect to alcohol intake prior to suicide showed that while HTTLPR, *TPH1* A218C, and *ACE* insertion/deletion variants were not associated with the presence of alcohol in the blood of suicide victims at time of death, *TPH2* rs1386483 TT genotype appeared underrepresented in the group of suicide victims where alcohol was detected. This suggests that the *TPH2* marker may protect against suicide related to alcohol consumption (Fudalej et al. 2009).

The majority of suicidal behavior occurs in depressed patients, but the role of antidepressants in suicide remains controversial. Some reports suggest that certain antidepressants increase the risk of suicide (Teicher et al. 1990; Khan et al. 2003), but others argue against it (Gibbons et al. 2007). Perroud et al. (2009) published the first paper examining the effects of candidate gene variants on the emergence or worsening of suicidal behavior, as measured by the Beck Depression Inventory (BDI) and Montgomery-Åsberg Depression Rating Scale (MADRS), during 12 weeks of antidepressant treatment in the GENDEP clinical trial on 796 major depressive disorder patients. The authors analyzed genes implicated in the mechanism of the antidepressants escitalopram and nortriptyline, namely, *TPH1*, *TPH2*, *HTR1A*, *HTR2A*, *5HTT*, *ADRA2A*, *SLC6A2*, *BDNF*, and *NTRK2*. The *BDNF* markers rs962369 and rs11030102 were significantly associated with suicide ideation, as was the rs1439050 marker in the BDNF receptor gene *NTRK2*. They also found haplotypes containing Val66 to be associated with an increase in suicidal ideation. In male patients taking nortriptyline, the authors found *ADRA2A* rs11195419 A-allele carriers to have increased suicidal ideation at a 12 week follow-up compared to GG genotype carriers. This finding highlights the importance of genetic findings in the context of medication treatment. Brent et al. (2010) also looked at the changes in suicidality in treatment-resistant depressed adolescents after 12 weeks of being on another antidepressant medication alone or in combination with cognitive behavioral therapy. They found *FKBP5* to be associated with suicide events (worsening or emergence of suicidal behavior during the study treatment period). Laje et al. (2007) reported findings from screening of SNPs within 68 candidate genes; they found markers in the glutamate receptor genes glutamate receptor, ionotropic, kainite 2 (*GRIK2*) and glutamate receptor, ionotropic, alpha-amino-3-hydroxy-5-methyl-4-isoxazole propionate 3 (*GRIA3*) to be associated with suicide ideation emerging during selective serotonin reuptake inhibitor (SSRI) treatment in patients from the STAR*D clinical trial, while the genes *BDNF* or *NTRK2* were not associated. In the same sample, Perlis et al. (2007) found the cAMP response element–binding protein (*CREB1*), which acts upstream of *BDNF*, to be associated in males only. Laje et al. (2009) followed up with the first genome-wide association study (GWAS) in the same sample using the Illumina Human-1 BeadChip that samples 109,365 SNPs. They reported the proteoglycan-like sulfated glycoprotein papilin gene *PAPLN* and the interleukin receptor gene *IL28RA* to be genome-wide significant (Laje et al. 2009). An important limitation of the GWAS approach is the extensive correction required for massive multiple testing. In the statistical effort to reduce false positive findings, important true positives may be missed.

11.4.2.2 Epigenetic Mechanism

We have noted earlier in this chapter that there is a strong genetic component in suicidal behavior. It is possible that part of this strong genetic component is determined by DNA modification, that is, epigenetics (Petronis 2010). The epigenetic factors may include modifications of histone proteins and cytosine residues at CpG dinucleotides within promoter regions of genes. While epigenetic signatures are heritable, they can be modified by the environment, as demonstrated in rodent models of maternal nursing behavior (Weaver et al. 2004) and physical exercise (Collins et al. 2009). In the maternal behavior model, maternal licking and grooming of the pups was shown to alter the epigenetic profile at the glucocorticoid receptor (*NR3C1*) gene promoter, as well as change the binding of a specific transcription factor (EGR1) to the *NR3C1* promoter, alter hippocampal NR3C1 expression, and alter HPA stress response in the pup. Physical exercise has also been shown to alter histone modification patterns in the dentate gyrus (Collins et al. 2009). These epigenetic modifications in turn change promoter activity via differential binding to transcription factors and silencers. In a human postmortem study, McGowan et al. (2009) extended these findings to humans, wherein suicide victims who had experienced child abuse had increased cytosine methylation at the binding site for EGR1 compared to suicide victims without history of abuse. The authors went on to demonstrate in cell culture experiments that the observed specific methylation interfered with EGR1 binding, leading to decreased *NR3C1* promoter activity (McGowan et al. 2009). In view of these mechanisms, the gene–environment interactions noted earlier could be partly mediated by an epigenetic mechanism (Tsankova et al. 2007). Other postmortem brain studies have shown hypermethylation in CpG islands of promoters of other genes, including *rRNA* (McGowan et al. 2008), *GABRA1* (Poulter et al. 2008), *NTRK2* (Ernst et al. 2009), spermine oxidase (*SMOX*; Fiori and Turecki 2010), and *BDNF* (Keller et al. 2010). Given that hypermethylation leads to decreased promoter activity, the observations could explain the decreased expression of these gene products reported in suicide victims. Thus, epigenetic investigation of other suicide candidate genes may clarify some of the inconsistencies in genetic findings of suicidal behavior. Further investigations including differential allele expression should also be carried out (De Luca et al. 2011).

11.4.3 Better Characterization of Suicidal Behavior

As previously mentioned, suicidal behaviors range from suicidal ideation through attempts of varying degrees of intentionality and lethality to suicide completion. Even though family studies have pointed to overlapping genetic basis among these three behaviors, their differences may explain in part the inconsistencies in the results reported in genetic studies.

11.4.3.1 Consideration of the Psychiatric Disorder Context

Another possible explanation for the mixed candidate gene findings could be that different research groups used different case–control matching strategies. Many studies matched suicide victims with healthy control subjects. These studies would not be able to distinguish possible associations between the genetic variants and psychiatric disorders from associations between these same genetic variants and suicide. Preexisting psychiatric disorders likely play a significant role (Arsenault-Lapierre et al. 2004). For instance, over 90% of a random sample of suicide victims have at least one Axis I psychiatric diagnosis upon psychological autopsy (Henriksson et al. 1993; Cavanagh et al. 2003). Over half of the suicide victims suffered from a depressive disorder though the prevalence is higher in females compared to males. Alcohol dependence, on the other hand, was observed more often in male victims (39%) than in females (18%). Eighty-eight percent of these victims had two or more diagnoses. Ernst et al. (2004) studied the 10% of suicide completers with no Axis I diagnoses (n = 16) detected in their psychological autopsy sample and compared them with living controls. They concluded that most of the individuals who committed suicide and appeared psychiatrically normal after a psychological autopsy may have had an underlying psychiatric process that the psychological autopsy method, as commonly carried out, failed to detect.

A possible way to overcome this methodological problem would be to match suicide attempters with suicide nonattempters who have the same underlying psychiatric disorders, but the comparability of these groups is based on the assumption that the genetic mechanism of suicide attempts is the same for all psychiatric disorders.

Both borderline personality disorder and major depressive disorder have high rates of suicide, but suicide attempters with borderline personality disorder display significantly more impulsive aggression than attempters with major depression alone. The former group also made their first suicide attempts at significantly younger ages than the latter group of patients (Soloff et al. 2000). Although not without its challenges, patients with suicide attempts and borderline personality disorder can be matched with borderline personality disordered patients without suicide attempts. One challenge in this regard is the fact that recurrent suicidal behavior is regarded as a common diagnostic feature of borderline personality disorder; thus, patients without suicide attempts are less common and it is therefore more difficult to obtain a large sample size.

11.4.3.2 Intermediate Phenotypes

Gene expression profiling will be increasingly used to look for new candidate genes for suicide. Klempan et al. (2009) recently published replicated findings of reduced *SAT1* gene expression across 12 postmortem cortical regions of depressed suicide victims compared to nonpsychiatric controls from the Quebec Suicide Brain Bank. Sequeira et al. (2009) reported findings from the first gene expression microarray analysis on samples from the same brain bank. Down-regulation of the metabotropic glutamate receptor (GRM3) was found in the prefrontal and parietal cortices of suicide victims with or without depression compared to controls. They also found altered expression of multiple subunits of the $GABA_A$ receptor in suicide victims with depression compared to non-suicidal controls. $GABA_A$ receptor expression was also altered in depressed suicide victims compared to suicide victims without history of depression. These findings suggest that the altered expression of $GABA_A$ subunit genes may be specific for major depression and not directly related to suicide (Sequeira et al. 2009). If these results are replicated in independent samples, expression changes may serve as a candidate intermediate phenotype that may disentangle the confounds between suicide and depression.

van Heeringen et al. (2003) found reduced 5HT2A receptor binding in the frontal cortical region of a sample of suicide attempters compared to normal controls. Also, impulsivity was correlated to 5HTT binding in suicide attempters, but not in controls (Lindstrom et al. 2004). A single-photon emission computed tomography (SPECT) split-dose activation study undertaken after a verbal fluency test found decreased prefrontal cortical activation in recent depressed suicide attempters compared to healthy controls (Audenaert et al. 2002). *BDNF* Val66Met was recently associated with 5HTT binding in multiple brain regions, with the ValVal genotype associated with higher 5HTT availability than the Met-containing genotypes in both male suicide attempters and healthy controls (Henningsson et al. 2009). It appears that genetic studies of brain imaging data, as well as postmortem studies (e.g., Hercher et al. 2009, 2010) with respect to suicidal behavior, should take into account potential confounders, including underlying psychiatric diagnoses and medication.

The mechanism of suicidal attempt and completion may be mediated by personality traits (e.g., Brezo et al. 2006, 2008a). In particular, family studies have pointed to impulsive aggression or Cluster B personality disorder to be an intermediate phenotype of suicide (Brent et al. 1996; Johnson et al. 1998; Kim et al. 2005; Zouk et al. 2006; Diaconu and Turecki 2009; McGirr et al. 2009). The more suicidal behaviors aggregate within families, the higher the level of aggressive behavior observed in both the probands and their offspring (Brent et al. 2003; Melhem et al. 2007). Twin studies have demonstrated the genetic heritability of aggressive trait to be ~45% (Rushton et al. 1986; Coccaro et al. 1997). Impulsivity has been associated with a promoter polymorphism in *HTR1A* (Benko et al. 2010) and *HTR2A* (Nomura et al. 2006). *HTR2A* (Giegling et al. 2006; Serretti et al. 2007a) and *MAOA* (Manuck et al. 2000) have also been associated with anger and aggression. Whether or not previous associations reported between suicide and *HTR1B* (Zouk et al. 2007), *HTR2A* (Preuss et al. 2001; Giegling et al. 2006), and *MAOA* (Manuck et al. 2000; Courtet et al. 2005) were due to an underlying association between these genes and aggressive behavior remains to be further scrutinized.

Several personality traits, namely, neuroticism, hopelessness, and extroversion as measured using standardized instruments including the NEO-PI-R, have been found to be associated with suicide ideation, attempt, and completion (Brezo et al. 2006; Heisel et al. 2006; Stankovic et al. 2006). Within the domain of neuroticism, the depression facet was associated while self-consciousness was inversely related to suicidal ideation (Chioqueta and Stiles 2005). Using available follow-up data for up to 7 years from the Collaborative Longitudinal Personality Disorders study, Yen et al. (2009) found negative affect to be a significant predictor of suicide attempt, even after accounting for gender, childhood sex abuse, major depressive disorder, and SUD. Genetic studies of these suicide-related personality traits may clarify some of the mixed findings in suicide.

Suicidal behavior has also been suggested to be partly associated with impaired problem solving (Sakinofsky et al. 1990) and learning and decision making. As such, Jollant et al. (2005) compared performance on the Iowa Gambling Task (Bechara et al. 1999) between suicide attempters and nonattempters, and found attempters to perform significantly worse than nonattempters, irrespective of psychiatric disorder or violence of the attempts. Subsequently, Jollant et al. (2007) investigated polymorphisms in the serotonergic system (*TPH2*, *5HTT*, *TPH1*, and *MAOA*) in the Iowa Gambling Taskin suicide attempters. While overall performance on the task was not significantly different among the genotype groups of each polymorphism, the suicide risk genotype carriers improved significantly less than the other genotype carriers over the course of the task (Jollant et al. 2007). The results suggest that these genetic risk variants influence the rigidity in the decision-making process and the inability to learn from negative outcomes from earlier decisions in suicide attempters. It is important to note that performance on the Iowa Gambling Task was correlated to emotional lability and anger expression but not impulsivity (Jollant et al. 2005). Interestingly, the AA genotype of the A218C marker in *TPH1* has been associated with reduced anger control in suicide attempters (Baud et al. 2009). The recently published modified Stroop test may offer more accurate predictive information for suicidal behavior (Malloy-Diniz et al. 2009; Cha et al. 2010); however, the long-term predictive value of this test remains to be evaluated.

11.5 CONCLUDING REMARKS

Family, twin, and adoption studies have established a genetic basis of suicidal behavior. However, suicide candidate gene studies have been plagued by inconsistent findings. To move forward, a consensus needs to be reached for the definition of different types of suicidal behaviors. Researchers are increasingly using the Columbia Classification Algorithm of Suicide Assessment (C-CASA) to better categorize suicidal behaviors. This instrument requires the suicidal act to be self-directed with intent to die (Posner et al. 2007). Intermediate phenotypes, including impulsive aggression, major depression, borderline personality disorder, cognitive inflexibility, and stress sensitivity, should be used to clarify the various aspects of suicidal behavior (Mann et al. 2009). Study design should control for sex (e.g., in analysis of *COMT*) and underlying psychiatric disorders. Genes should be studied more comprehensively by analyzing coding and promoter polymorphisms as well as tag markers across each gene to assess their role in suicidal behavior and related phenotypes. The traditional candidate gene approach remains very important since understanding the biology starts with one gene at a time (Neale and Sham 2004).

With the rapid development of genome-wide technology, GWAS are turning out some novel genes as in the study by Laje et al. (2009). Analysis of genetic findings should include genes that are functionally related, using software such as FORGE (Pedroso et al. in press), GMDR, and IPA. Key environmental contributors need to be considered as part of the experimental design. These would include distal childhood trauma (sexual abuse, adverse life events), more proximal acute stress, as well as support systems. The effect of environmental stressors may be mediated through epigenetic regulation of gene expression (with microarrays analyzed using weighted correlation network analysis (WGCNA); Zhang and Horvath 2005) including methylation and hydroxymethylation of CpG sites in gene promoter regions (Irizarry et al. 2009). Thus, epigenetic examination of risk gene promoters will provide complementary data that may account for part of the variance in gene expression levels and suicide susceptibility that cannot be explained by polymorphisms alone.

Overall, suicide is a complex phenotype with multiple contributing genetic and environmental factors. Multidisciplinary examination into suicidal behavior will enable us to elucidate the pathophysiological mechanism underlying suicide. Understanding this mechanism may lead to better treatments and prevention in those at risk.

REFERENCES

Agerbo, E., Nordentoft, M., and Mortensen, P. B. 2002. Familial, psychiatric, and socioeconomic risk factors for suicide in young people: Nested case–control study. *Br Med J* 325(7355): 74.

Albanese, V., Biguet, N. F., Kiefer, H., Bayard, E., Mallet, J., and Meloni, R. 2001. Quantitative effects on gene silencing by allelic variation at a tetranucleotide microsatellite. *Hum Mol Genet* 10(17): 1785–1792.

Ali, F. R., Vasiliou, S. A., Haddley, K., Paredes, U. M., Roberts, J. C., Miyajima, F. et al. 2010. Combinatorial interaction between two human serotonin transporter gene variable number tandem repeats and their regulation by CTCF. *J Neurochem* 112(1): 296–306.

Arango, V., Ernsberger, P., Sved, A. F., and Mann, J. J. 1993. Quantitative autoradiography of alpha 1- and alpha 2-adrenergic receptors in the cerebral cortex of controls and suicide victims. *Brain Res* 630(1–2): 271–282.

Arango, V., Huang, Y. Y., Underwood, M. D., and Mann, J. J. 2003. Genetics of the serotonergic system in suicidal behavior. *J Psychiatr Res* 37(5): 375–386.

Arango, V., Underwood, M. D., Gubbi, A. V., and Mann, J. J. 1995. Localized alterations in pre- and postsynaptic serotonin binding sites in the ventrolateral prefrontal cortex of suicide victims. *Brain Res* 688(1–2): 121–133.

Arinami, T., Gao, M., Hamaguchi, H., and Toru, M. 1997. A functional polymorphism in the promoter region of the dopamine D2 receptor gene is associated with schizophrenia. *Hum Mol Genet* 6(4): 577–582.

Arsenault-Lapierre, G., Kim, C., and Turecki, G. 2004. Psychiatric diagnoses in 3275 suicides: A meta-analysis. *BMC Psychiatry* 4: 37.

Asberg, M. 1997. Neurotransmitters and suicidal behavior. The evidence from cerebrospinal fluid studies. *Ann N Y Acad Sci* 836: 158–181.

Audenaert, K., Goethals, I., Van Laere, K., Lahorte, P., Brans, B., Versijpt, J. et al. 2002. SPECT neuropsychological activation procedure with the Verbal Fluency Test in attempted suicide patients. *Nucl Med Commun* 23(9): 907–916.

Austin, M. C., Whitehead, R. E., Edgar, C. L., Janosky, J. E., and Lewis, D. A. 2002. Localized decrease in serotonin transporter-immunoreactive axons in the prefrontal cortex of depressed subjects committing suicide. *Neuroscience* 114(3): 807–815.

Baca-Garcia, E., Vaquero, C., Diaz-Sastre, C., Jimenez-Trevino, L., Saiz-Ruiz, J., Fernandez-Piqueras, J. et al. 2004. Lack of association between polymorphic variations in the alpha 3 subunit GABA receptor gene (*GABRA3*) and suicide attempts. *Prog Neuropsychopharmacol Biol Psychiatry* 28(2): 409–412.

Bach-Mizrachi, H., Underwood, M. D., Tin, A., Ellis, S. P., Mann, J. J., and Arango, V. 2008. Elevated expression of tryptophan hydroxylase-2 mRNA at the neuronal level in the dorsal and median raphe nuclei of depressed suicides. *Mol Psychiatry* 13(5): 507–513, 465.

Baud, P., Courtet, P., Perroud, N., Jollant, F., Buresi, C., and Malafosse, A. 2007. Catechol-*O*-methyltransferase polymorphism (COMT) in suicide attempters: A possible gender effect on anger traits. *Am J Med Genet B Neuropsychiatr Genet* 144B(8): 1042–1047.

Baud, P., Perroud, N., Courtet, P., Jaussent, I., Relecom, C., Jollant, F. et al. 2009. Modulation of anger control in suicide attempters by TPH-1. *Genes Brain Behav* 8(1): 97–100.

Bechara, A., Damasio, H., Damasio, A. R., and Lee, G. P. 1999. Different contributions of the human amygdala and ventromedial prefrontal cortex to decision-making. *J Neurosci* 19(13): 5473–5481.

Beitchman, J. H., Baldassarra, L., Mik, H., De Luca, V., King, N., Bender, D. et al. 2006. Serotonin transporter polymorphisms and persistent, pervasive childhood aggression. *Am J Psychiatry* 163(6): 1103–1105.

Bellivier, F., Laplanche, J. L., Leboyer, M., Feingold, J., Bottos, C., Allilaire, J. F. et al. 1997. Serotonin transporter gene and manic depressive illness: An association study. *Biol Psychiatry* 41(6): 750–752.

Benko, A., Lazary, J., Molnar, E., Gonda, X., Tothfalusi, L., Pap, D. et al. 2010. Significant association between the C(−1019)G functional polymorphism of the *HTR1A* gene and impulsivity. *Am J Med Genet B Neuropsychiatr Genet* 153B(2): 592–599.

Binder, E. B., Salyakina, D., Lichtner, P., Wochnik, G. M., Ising, M., Putz, B. et al. 2004. Polymorphisms in FKBP5 are associated with increased recurrence of depressive episodes and rapid response to antidepressant treatment. *Nat Genet* 36(12): 1319–1325.

Bondy, B., Buettner, A., and Zill, P. 2006. Genetics of suicide. *Mol Psychiatry* 11(4): 336–351.

Boularand, S., Darmon, M. C., Ravassard, P., and Mallet, J. 1995. Characterization of the human tryptophan hydroxylase gene promoter. Transcriptional regulation by cAMP requires a new motif distinct from the cAMP-responsive element. *J Biol Chem* 270(8): 3757–3764.

Bouwknecht, J. A., Hijzen, T. H., van der Gugten, J., Maes, R. A., Hen, R., and Olivier, B. 2001. Absence of 5-HT(1B) receptors is associated with impaired impulse control in male 5-HT(1B) knockout mice. *Biol Psychiatry* 49(7): 557–568.

Brent, D. A., Bridge, J., Johnson, B. A., and Connolly, J. 1996. Suicidal behavior runs in families. A controlled family study of adolescent suicide victims. *Arch Gen Psychiatry* 53(12): 1145–1152.

Brent, D. A. and Melhem, N. 2008. Familial transmission of suicidal behavior. *Psychiatr Clin North Am* 31(2): 157–177.

Brent, D., Melhem, N., and Turecki, G. 2010. Pharmacogenomics of suicidal events. *Pharmacogenomics* 11(6): 793–807.

Brent, D. A., Oquendo, M., Birmaher, B., Greenhill, L., Kolko, D., Stanley, B. et al. 2002. Familial pathways to early-onset suicide attempt: Risk for suicidal behavior in offspring of mood-disordered suicide attempters. *Arch Gen Psychiatry* 59(9): 801–807.

Brent, D. A., Oquendo, M., Birmaher, B., Greenhill, L., Kolko, D., Stanley, B. et al. 2003. Peripubertal suicide attempts in offspring of suicide attempters with siblings concordant for suicidal behavior. *Am J Psychiatry* 160(8): 1486–1493.

Brezo, J., Bureau, A., Merette, C., Jomphe, V., Barker, E. D., Vitaro, F. et al. 2010. Differences and similarities in the serotonergic diathesis for suicide attempts and mood disorders: A 22-year longitudinal gene-environment study. *Mol Psychiatry* 15(8): 831–843.

Brezo, J., Paris, J., Hebert, M., Vitaro, F., Tremblay, R., and Turecki, G. 2008a. Broad and narrow personality traits as markers of one-time and repeated suicide attempts: A population-based study. *BMC Psychiatry* 8: 15.

Brezo, J., Paris, J., and Turecki, G. 2006. Personality traits as correlates of suicidal ideation, suicide attempts, and suicide completions: A systematic review. *Acta Psychiatr Scand* 113(3): 180–206.

Brezo, J., Paris, J., Vitaro, F., Hebert, M., Tremblay, R. E., and Turecki, G. 2008b. Predicting suicide attempts in young adults with histories of childhood abuse. *Br J Psychiatry* 193(2): 134–139.

Bridge, J. A., Brent, D., Johnson, B. A., and Connolly, J. 1997. Familial aggregation of psychiatric disorders in a community sample of adolescents. *J Am Acad Child Adolesc Psychiatry* 36(5): 628–636.

Brodsky, B. S., Malone, K. M., Ellis, S. P., Dulit, R. A., and Mann, J. J. 1997. Characteristics of borderline personality disorder associated with suicidal behavior. *Am J Psychiatry* 154(12): 1715–1719.

Brunner, D. and Hen, R. 1997. Insights into the neurobiology of impulsive behavior from serotonin receptor knockout mice. *Ann N Y Acad Sci* 836: 81–105.

Calati, R., Porcelli, S., Giegling, I., Hartmann, A. M., Moller, H. J., De Ronchi, D. et al. 2011. Catechol-*o*-methyltransferase gene modulation on suicidal behavior and personality traits: Review, meta-analysis and association study. *J Psychiatr Res* 45(3): 309–321.

Callreus, T., Agerskov Andersen, U., Hallas, J., and Andersen, M. 2007. Cardiovascular drugs and the risk of suicide: A nested case-control study. *Eur J Clin Pharmacol* 63(6): 591–596.

Carballo, J. J., Harkavy-Friedman, J., Burke, A. K., Sher, L., Baca-Garcia, E., Sullivan, G. M. et al. 2008. Family history of suicidal behavior and early traumatic experiences: Additive effect on suicidality and course of bipolar illness? *J Affect Disord* 109(1–2): 57–63.

Caspi, A., McClay, J., Moffitt, T. E., Mill, J., Martin, J., Craig, I. W. et al. 2002. Role of genotype in the cycle of violence in maltreated children. *Science* 297(5582): 851–854.

Caspi, A., Sugden, K., Moffitt, T. E., Taylor, A., Craig, I. W., Harrington, H. et al. 2003. Influence of life stress on depression: Moderation by a polymorphism in the *5-HTT* gene. *Science* 301(5631): 386–389.

Cavanagh, J. T., Carson, A. J., Sharpe, M., and Lawrie, S. M. 2003. Psychological autopsy studies of suicide: A systematic review. *Psychol Med* 33(3): 395–405.

Cha, C. B., Najmi, S., Park, J. M., Finn, C. T., and Nock, M. K. 2010. Attentional bias toward suicide-related stimuli predicts suicidal behavior. *J Abnorm Psychol* 119(3): 616–622.

Charlesworth, J. C., Curran, J. E., Johnson, M. P., Goring, H. H., Dyer, T. D., Diego, V. P. et al. 2010. Transcriptomic epidemiology of smoking: The effect of smoking on gene expression in lymphocytes. *BMC Med Genomics* 3: 29.

Cheetham, S. C., Crompton, M. R., Katona, C. L., and Horton, R. W. 1988. Brain 5-HT2 receptor binding sites in depressed suicide victims. *Brain Res* 443(1–2): 272–280.

Chen, C. K., Lin, S. K., Huang, M. C., Su, L. W., Hsiao, C. C., Chiang, Y. L. et al. 2007. Analysis of association of clinical correlates and 5-HTTLPR polymorphism with suicidal behavior among Chinese methamphetamine abusers. *Psychiatry Clin Neurosci* 61(5): 479–486.

Cheng, A. T., Chen, T. H., Chen, C. C., and Jenkins, R. 2000. Psychosocial and psychiatric risk factors for suicide. Case–control psychological autopsy study. *Br J Psychiatry* 177: 360–365.

Cheng, R., Juo, S. H., Loth, J. E., Nee, J., Iossifov, I., Blumenthal, R. et al. 2006. Genome-wide linkage scan in a large bipolar disorder sample from the National Institute of Mental Health genetics initiative suggests putative loci for bipolar disorder, psychosis, suicide, and panic disorder. *Mol Psychiatry* 11(3): 252–260.

Chioqueta, A. P. and Stiles, T. C. 2005. Personality traits and the development of depression, hopelessness, and suicide ideation. *Pers Individual Diff* 38(6): 1283–1291.

Cho, H., Guo, G., Iritani, B. J., and Hallfors, D. D. 2006. Genetic contribution to suicidal behaviors and associated risk factors among adolescents in the U.S. *Prev Sci* 7(3): 303–311.

Cho, H. J., Meira-Lima, I., Cordeiro, Q., Michelon, L., Sham, P., Vallada, H. et al. 2005. Population-based and family-based studies on the serotonin transporter gene polymorphisms and bipolar disorder: A systematic review and meta-analysis. *Mol Psychiatry* 10(8): 771–781.

Cicchetti, D., Rogosch, F. A., Sturge-Apple, M., and Toth, S. L. 2010. Interaction of child maltreatment and 5-HTT polymorphisms: Suicidal ideation among children from low-SES backgrounds. *J Pediatr Psychol* 35(5): 536–546.

Coccaro, E. F., Bergeman, C. S., Kavoussi, R. J., and Seroczynski, A. D. 1997. Heritability of aggression and irritability: A twin study of the Buss–Durkee aggression scales in adult male subjects. *Biol Psychiatry* 41(3): 273–284.

Collier, D. A., Stober, G., Li, T., Heils, A., Catalano, M., Di Bella, D. et al. 1996. A novel functional polymorphism within the promoter of the serotonin transporter gene: Possible role in susceptibility to affective disorders. *Mol Psychiatry* 1(6): 453–460.

Collins, A., Hill, L. E., Chandramohan, Y., Whitcomb, D., Droste, S. K., and Reul, J. M. 2009. Exercise improves cognitive responses to psychological stress through enhancement of epigenetic mechanisms and gene expression in the dentate gyrus. *PLoS One* 4(1): e4330.

Courtet, P., Jollant, F., Buresi, C., Castelnau, D., Mouthon, D., and Malafosse, A. 2005. The monoamine oxidase A gene may influence the means used in suicide attempts. *Psychiatr Genet* 15(3): 189–193.

Coventry, W. L., James, M. R., Eaves, L. J., Gordon, S. D., Gillespie, N. A., Ryan, L. et al. 2010. Do 5HTTLPR and stress interact in risk for depression and suicidality? Item response analyses of a large sample. *Am J Med Genet B Neuropsychiatr Genet* 153B(3): 757–765.

Cui, H., Nishiguchi, N., Ivleva, E., Yanagi, M., Fukutake, M., Nushida, H. et al. 2008. Association of *RGS2* gene polymorphisms with suicide and increased RGS2 immunoreactivity in the postmortem brain of suicide victims. *Neuropsychopharmacology* 33(7): 1537–1544.

Cui, H., Supriyanto, I., Asano, M., Ueno, Y., Nagasaki, Y., Nishiguchi, N. et al. 2010. A common polymorphism in the 3′-UTR of the *NOS1* gene was associated with completed suicides in Japanese male population. *Prog Neuropsychopharmacol Biol Psychiatry* 34(6): 992–996.

Dalton, E. J., Cate-Carter, T. D., Mundo, E., Parikh, S. V., and Kennedy, J. L. 2003. Suicide risk in bipolar patients: The role of co-morbid substance use disorders. *Bipolar Disord* 5(1): 58–61.

De Luca, V., Likhodi, O., Van Tol, H. H., Kennedy, J. L., and Wong, A. H. 2006a. Gene expression of tryptophan hydroxylase 2 in post-mortem brain of suicide subjects. *Int J Neuropsychopharmacol* 9(1): 21–25.

De Luca, V., Mueller, D. J., Tharmalingam, S., King, N., and Kennedy, J. L. 2004a. Analysis of the novel *TPH2* gene in bipolar disorder and suicidality. *Mol Psychiatry* 9(10): 896–897.

De Luca, V., Muglia, P., Masellis, M., Jane Dalton, E., Wong, G. W., and Kennedy, J. L. 2004b. Polymorphisms in glutamate decarboxylase genes: Analysis in schizophrenia. *Psychiatr Genet* 14(1): 39–42.

De Luca, V., Souza, R. P., Zai, C. C., Panariello, F., Javaid, N., Strauss, J. et al. 2011. Parent of origin effect and differential allelic expression of BDNF Val66Met in suicidal behaviour. *World J Biol Psychiatry* 12(1): 42–47.

De Luca, V., Strauss, J., and Kennedy, J. L. 2008a. Power based association analysis (PBAT) of serotonergic and noradrenergic polymorphisms in bipolar patients with suicidal behaviour. *Prog Neuropsychopharmacol Biol Psychiatry* 32(1): 197–203.

De Luca, V., Tharmalingam, S., and Kennedy, J. L. 2007. Association study between the corticotropin-releasing hormone receptor 2 gene and suicidality in bipolar disorder. *Eur Psychiatry* 22(5): 282–287.

De Luca, V., Tharmalingam, S., Muller, D. J., Wong, G., de Bartolomeis, A., and Kennedy, J. L. 2006b. Gene–gene interaction between *MAOA* and *COMT* in suicidal behavior: Analysis in schizophrenia. *Brain Res* 1097(1): 26–30.

De Luca, V., Tharmalingam, S., Sicard, T., and Kennedy, J. L. 2005a. Gene–gene interaction between *MAOA* and *COMT* in suicidal behavior. *Neurosci Lett* 383(1–2): 151–154.

De Luca, V., Tharmaligam, S., Strauss, J., and Kennedy, J. L. 2008b. 5-HT2C receptor and MAO-A interaction analysis: No association with suicidal behaviour in bipolar patients. *Eur Arch Psychiatry Clin Neurosci* 258(7): 428–433.

De Luca, V., Tharmalingam, S., Zai, C., Potapova, N., Strauss, J., Vincent, J. et al. 2010. Association of HPA axis genes with suicidal behaviour in schizophrenia. *J Psychopharmacol* 24(5): 677–682.

De Luca, V., Voineskos, D., Wong, G. W., Shinkai, T., Rothe, C., Strauss, J. et al. 2005b. Promoter polymorphism of second tryptophan hydroxylase isoform (TPH2) in schizophrenia and suicidality. *Psychiatry Res* 134(2): 195–198.

De Luca, V., Zai, G., Tharmalingam, S., de Bartolomeis, A., Wong, G., and Kennedy, J. L. 2006c. Association study between the novel functional polymorphism of the serotonin transporter gene and suicidal behaviour in schizophrenia. *Eur Neuropsychopharmacol* 16(4): 268–271.

De Paermentier, F., Mauger, J. M., Lowther, S., Crompton, M. R., Katona, C. L., and Horton, R. W. 1997. Brain alpha-adrenoceptors in depressed suicides. *Brain Res* 757(1): 60–68.

Diaconu, G. and Turecki, G. 2009. Family history of suicidal behavior predicts impulsive–aggressive behavior levels in psychiatric outpatients. *J Affect Disord* 113(1–2): 172–178.

Duan, J., Sanders, A. R., Molen, J. E., Martinolich, L., Mowry, B. J., Levinson, D. F. et al. 2003a. Polymorphisms in the 5′-untranslated region of the human serotonin receptor 1B (*HTR1B*) gene affect gene expression. *Mol Psychiatry* 8(11): 901–910.

Duan, J., Wainwright, M. S., Comeron, J. M., Saitou, N., Sanders, A. R., Gelernter, J. et al. 2003b. Synonymous mutations in the human dopamine receptor D2 (DRD2) affect mRNA stability and synthesis of the receptor. *Hum Mol Genet* 12(3): 205–216.

Dumais, A., Lesage, A. D., Alda, M., Rouleau, G., Dumont, M., Chawky, N. et al. 2005. Risk factors for suicide completion in major depression: A case–control study of impulsive and aggressive behaviors in men. *Am J Psychiatry* 162(11): 2116–2124.

Dwivedi, Y. 2010. Brain-derived neurotrophic factor and suicide pathogenesis. *Ann Med* 42(2): 87–96.

Dwivedi, Y., Rizavi, H. S., Conley, R. R., Roberts, R. C., Tamminga, C. A., and Pandey, G. N. 2003. Altered gene expression of brain-derived neurotrophic factor and receptor tyrosine kinase B in postmortem brain of suicide subjects. *Arch Gen Psychiatry* 60(8): 804–815.

Egeland, J. A. and Sussex, J. N. 1985. Suicide and family loading for affective disorders. *JAMA* 254(7): 915–918.

Ernst, C., Deleva, V., Deng, X., Sequeira, A., Pomarenski, A., Klempan, T. et al. 2009. Alternative splicing, methylation state, and expression profile of tropomyosin-related kinase B in the frontal cortex of suicide completers. *Arch Gen Psychiatry* 66(1): 22–32.

Ernst, C., Lalovic, A., Lesage, A., Seguin, M., Tousignant, M., and Turecki, G. 2004. Suicide and no axis I psychopathology. *BMC Psychiatry* 4: 7.

Fan, J. B. and Sklar, P. 2005. Meta-analysis reveals association between serotonin transporter gene STin2 VNTR polymorphism and schizophrenia. *Mol Psychiatry* 10(10): 928–938, 891.

Fanous, A. H., Chen, X., Wang, X., Amdur, R., O'Neill, F. A., Walsh, D. et al. 2009. Genetic variation in the serotonin 2A receptor and suicidal ideation in a sample of 270 Irish high-density schizophrenia families. *Am J Med Genet B Neuropsychiatr Genet* 150B(3): 411–417.

Feinn, R., Nellissery, M., and Kranzler, H. R. 2005. Meta-analysis of the association of a functional serotonin transporter promoter polymorphism with alcohol dependence. *Am J Med Genet B Neuropsychiatr Genet* 133B(1): 79–84.

Finckh, U., Rommelspacher, H., Kuhn, S., Dufeu, P., Otto, G., Heinz, A. et al. 1997. Influence of the dopamine D2 receptor (DRD2) genotype on neuroadaptive effects of alcohol and the clinical outcome of alcoholism. *Pharmacogenetics* 7(4): 271–281.

Fiori, L. M. and Turecki, G. 2010. Genetic and epigenetic influences on expression of spermine synthase and spermine oxidase in suicide completers. *Int J Neuropsychopharmacol* 13(6): 725–736.

Fu, Q., Heath, A. C., Bucholz, K. K., Nelson, E. C., Glowinski, A. L., Goldberg, J. et al. 2002. A twin study of genetic and environmental influences on suicidality in men. *Psychol Med* 32(1): 11–24.

Fudalej, S., Fudalej, M., Kostrzewa, G., Kuzniar, P., Franaszczyk, M., Wojnar, M. et al. 2009. Angiotensin-converting enzyme polymorphism and completed suicide: An association in Caucasians and evidence for a link with a method of self-injury. *Neuropsychobiology* 59(3): 151–158.

Fukutake, M., Hishimoto, A., Nishiguchi, N., Nushida, H., Ueno, Y., Shirakawa, O. et al. 2008. Association of alpha2A-adrenergic receptor gene polymorphism with susceptibility to suicide in Japanese females. *Prog Neuropsychopharmacol Biol Psychiatry* 32(6): 1428–1433.

Garcia-Sevilla, J. A., Escriba, P. V., Ozaita, A., La Harpe, R., Walzer, C., Eytan, A. et al. 1999. Up-regulation of immunolabeled alpha2A-adrenoceptors, Gi coupling proteins, and regulatory receptor kinases in the prefrontal cortex of depressed suicides. *J Neurochem* 72(1): 282–291.

Garfinkel, B. D., Froese, A., and Hood, J. 1982. Suicide attempts in children and adolescents. *Am J Psychiatry* 139(10): 1257–1261.

Gerra, G., Garofano, L., Pellegrini, C., Bosari, S., Zaimovic, A., Moi, G. et al. 2005. Allelic association of a dopamine transporter gene polymorphism with antisocial behaviour in heroin-dependent patients. *Addict Biol* 10(3): 275–281.

Gibb, B. E., McGeary, J. E., Beevers, C. G., and Miller, I. W. 2006. Serotonin transporter (5-HTTLPR) genotype, childhood abuse, and suicide attempts in adult psychiatric inpatients. *Suicide Life Threat Behav* 36(6): 687–693.

Gibbons, R. D., Brown, C. H., Hur, K., Marcus, S. M., Bhaumik, D. K., and Mann, J. J. 2007. Relationship between antidepressants and suicide attempts: An analysis of the Veterans Health Administration data sets. *Am J Psychiatry* 164(7): 1044–1049.

Giegling, I., Hartmann, A. M., Moller, H. J., and Rujescu, D. 2006. Anger- and aggression-related traits are associated with polymorphisms in the *5-HT-2A* gene. *J Affect Disord* 96(1–2): 75–81.

Giegling, I., Moreno-De-Luca, D., Rujescu, D., Schneider, B., Hartmann, A. M., Schnabel, A. et al. 2008. Dopa decarboxylase and tyrosine hydroxylase gene variants in suicidal behavior. *Am J Med Genet B Neuropsychiatr Genet* 147(3): 308–315.

Glowinski, A. L., Bucholz, K. K., Nelson, E. C., Fu, Q., Madden, P. A., Reich, W. et al. 2001. Suicide attempts in an adolescent female twin sample. *J Am Acad Child Adolesc Psychiatry* 40(11): 1300–1307.

Goodwin, R. D., Beautrais, A. L., and Fergusson, D. M. 2004. Familial transmission of suicidal ideation and suicide attempts: Evidence from a general population sample. *Psychiatry Res* 126(2): 159–165.

Gould, M. S., Fisher, P., Parides, M., Flory, M., and Shaffer, D. 1996. Psychosocial risk factors of child and adolescent completed suicide. *Arch Gen Psychiatry* 53(12): 1155–1162.

Gross-Isseroff, R., Weizman, A., Fieldust, S. J., Israeli, M., and Biegon, A. 2000. Unaltered alpha(2)-noradrenergic/imidazoline receptors in suicide victims: A postmortem brain autoradiographic analysis. *Eur Neuropsychopharmacol* 10(4): 265–271.

Gurevich, I., Tamir, H., Arango, V., Dwork, A. J., Mann, J. J., and Schmauss, C. 2002. Altered editing of serotonin 2C receptor pre-mRNA in the prefrontal cortex of depressed suicide victims. *Neuron* 34(3): 349–356.

Han, L., Nielsen, D. A., Rosenthal, N. E., Jefferson, K., Kaye, W., Murphy, D. et al. 1999. No coding variant of the tryptophan hydroxylase gene detected in seasonal affective disorder, obsessive–compulsive disorder, anorexia nervosa, and alcoholism. *Biol Psychiatry* 45(5): 615–619.

Harvald, B. and Hauge, M. (1965). Hereditary factors elucidated by twin studies. *Genetics and the Epidemiology of Chronic Disease*. Neel, J. V., Shaw, M. W., and Schull, W. J., eds. Washington, DC: U.S. Department of Health, Education, and Welfare, pp. 61–76.

Hawton, K. and van Heeringen, K. 2009. Suicide. *Lancet* 373(9672): 1372–1381.

Heils, A., Teufel, A., Petri, S., Stober, G., Riederer, P., Bengel, D. et al. 1996. Allelic variation of human serotonin transporter gene expression. *J Neurochem* 66(6): 2621–2624.

Heisel, M. J., Duberstein, P. R., Conner, K. R., Franus, N., Beckman, A., and Conwell, Y. 2006. Personality and reports of suicide ideation among depressed adults 50 years of age or older. *J Affect Disord* 90(2–3): 175–180.

Henningsson, S., Borg, J., Lundberg, J., Bah, J., Lindstrom, M., Ryding, E. et al. 2009. Genetic variation in brain-derived neurotrophic factor is associated with serotonin transporter but not serotonin-1A receptor availability in men. *Biol Psychiatry* 66(5): 477–485.

Henriksson, M. M., Aro, H. M., Marttunen, M. J., Heikkinen, M. E., Isometsa, E. T., Kuoppasalmi, K. I. et al. 1993. Mental disorders and comorbidity in suicide. *Am J Psychiatry* 150(6): 935–940.

Hercher, C., Canetti, L., Turecki, G., and Mechawar, N. 2010. Anterior cingulate pyramidal neurons display altered dendritic branching in depressed suicides. *J Psychiatr Res* 44(5): 286–293.

Hercher, C., Parent, M., Flores, C., Canetti, L., Turecki, G., and Mechawar, N. 2009. Alcohol dependence-related increase of glial cell density in the anterior cingulate cortex of suicide completers. *J Psychiatr Neurosci* 34(4): 281–288.

Hesselbrock, V., Dick, D., Hesselbrock, M., Foroud, T., Schuckit, M., Edenberg, H. et al. 2004. The search for genetic risk factors associated with suicidal behavior. *Alcohol Clin Exp Res* 28(5 Suppl): 70S–76S.

Hishimoto, A., Shirakawa, O., Nishiguchi, N., Hashimoto, T., Yanagi, M., Nushida, H. et al. 2006. Association between a functional polymorphism in the renin–angiotensin system and completed suicide. *J Neural Transm* 113(12): 1915–1920.

Ho, L. W., Furlong, R. A., Rubinsztein, J. S., Walsh, C., Paykel, E. S., and Rubinsztein, D. C. 2000. Genetic associations with clinical characteristics in bipolar affective disorder and recurrent unipolar depressive disorder. *Am J Med Genet* 96(1): 36–42.

Hong, C. J., Huo, S. J., Yen, F. C., Tung, C. L., Pan, G. M., and Tsai, S. J. 2003. Association study of a brain-derived neurotrophic-factor genetic polymorphism and mood disorders, age of onset and suicidal behavior. *Neuropsychobiology* 48(4): 186–189.

Hong, C. J., Pan, G. M., and Tsai, S. J. 2004. Association study of onset age, attempted suicide, aggressive behavior, and schizophrenia with a serotonin 1B receptor (A-161T) genetic polymorphism. *Neuropsychobiology* 49(1): 1–4.

Hong, C. J., Wang, Y. C., and Tsai, S. J. 2002. Association study of angiotensin I-converting enzyme polymorphism and symptomatology and antidepressant response in major depressive disorders. *J Neural Transm* 109(9): 1209–1214.

Hotamisligil, G. S. and Breakefield, X. O. 1991. Human monoamine oxidase A gene determines levels of enzyme activity. *Am J Hum Genet* 49(2): 383–392.

Hu, X. Z., Lipsky, R. H., Zhu, G., Akhtar, L. A., Taubman, J., Greenberg, B. D. et al. 2006. Serotonin transporter promoter gain-of-function genotypes are linked to obsessive–compulsive disorder. *Am J Hum Genet* 78(5): 815–826.

Huang, Y. Y., Battistuzzi, C., Oquendo, M. A., Harkavy-Friedman, J., Greenhill, L., Zalsman, G. et al. 2004. Human 5-HT1A receptor C(-1019)G polymorphism and psychopathology. *Int J Neuropsychopharmacol* 7(4): 441–451.

Huang, Y. Y., Grailhe, R., Arango, V., Hen, R., and Mann, J. J. 1999. Relationship of psychopathology to the human serotonin1B genotype and receptor binding kinetics in postmortem brain tissue. *Neuropsychopharmacology* 21(2): 238–246.

Huang, T. L. and Lee, C. T. 2007. Associations between brain-derived neurotrophic factor *G196A* gene polymorphism and clinical phenotypes in schizophrenia patients. *Chang Gung Med J* 30(5): 408–413.

Hwang, J. P., Yang, C. H., Hong, C. J., Lirng, J. F., Yang, Y. M., and Tsai, S. J. 2006. Association of APOE genetic polymorphism with cognitive function and suicide history in geriatric depression. *Dement Geriatr Cogn Disord* 22(4): 334–338.

Iga, J., Ueno, S., Yamauchi, K., Numata, S., Tayoshi-Shibuya, S., Kinouchi, S. et al. 2007. The Val66Met polymorphism of the brain-derived neurotrophic factor gene is associated with psychotic fcature and suicidal behavior in Japanese major depressive patients. *Am J Med Genet B Neuropsychiatr Genet* 144B(8): 1003–1006.

Inada, T., Koga, M., Ishiguro, H., Horiuchi, Y., Syu, A., Yoshio, T. et al. 2008. Pathway-based association analysis of genome-wide screening data suggest that genes associated with the gamma-aminobutyric acid receptor signaling pathway are involved in neuroleptic-induced, treatment-resistant tardive dyskinesia. *Pharmacogenet Genomics* 18(4): 317–323.

Irizarry, R. A., Ladd-Acosta, C., Wen, B., Wu, Z., Montano, C., Onyango, P. et al. 2009. The human colon cancer methylome shows similar hypo- and hypermethylation at conserved tissue-specific CpG island shores. *Nat Genet* 41(2): 178–186.

Johann, M., Putzhammer, A., Eichhammer, P., and Wodarz, N. 2005. Association of the −141C Del variant of the dopamine D2 receptor (DRD2) with positive family history and suicidality in German alcoholics. *Am J Med Genet B Neuropsychiatr Genet* 132B(1): 46–49.

Johnson, B. A., Brent, D. A., Bridge, J., and Connolly, J. 1998. The familial aggregation of adolescent suicide attempts. *Acta Psychiatr Scand* 97(1): 18–24.

Jollant, F., Bellivier, F., Leboyer, M., Astruc, B., Torres, S., Verdier, R. et al. 2005. Impaired decision making in suicide attempters. *Am J Psychiatry* 162(2): 304–310.

Jollant, F., Guillaume, S., Jaussent, I., Castelnau, D., Malafosse, A., and Courtet, P. 2007. Impaired decision-making in suicide attempters may increase the risk of problems in affective relationships. *J Affect Disord* 99(1–3): 59–62.

Jones, J. S., Stanley, B., Mann, J. J., Frances, A. J., Guido, J. R., Traskman-Bendz, L. et al. 1990. CSF 5-HIAA and HVA concentrations in elderly depressed patients who attempted suicide. *Am J Psychiatry* 147(9): 1225–1227.

Jonsson, E. G., Norton, N., Gustavsson, J. P., Oreland, L., Owen, M. J., and Sedvall, G. C. 2000. A promoter polymorphism in the monoamine oxidase A gene and its relationships to monoamine metabolite concentrations in CSF of healthy volunteers. *J Psychiatr Res* 34(3): 239–244.

Kallman, F. J. 1953. *Heredity in Health and Mental Disease: Principles of Psychiatric Genetics in the Light of Comparative Twin Studies*. New York: Norton.

Kaprio, J., Romanov, K., Lonnqvist, J., and Koskenvuo, M. 1995. Suicide in twins: The Finnish twin cohort study. *Psychiatr Genet.* 5:S102.

Karege, F., Vaudan, G., Schwald, M., Perroud, N., and La Harpe, R. 2005. Neurotrophin levels in postmortem brains of suicide victims and the effects of antemortem diagnosis and psychotropic drugs. *Brain Res Mol Brain Res* 136(1–2): 29–37.

Ke, L., Qi, Z. Y., Ping, Y., and Ren, C. Y. 2006. Effect of SNP at position 40237 in exon 7 of the *TPH2* gene on susceptibility to suicide. *Brain Res* 1122(1): 24–26.

Keller, S., Sarchiapone, M., Zarrilli, F., Videtic, A., Ferraro, A., Carli, V. et al. 2010. Increased BDNF promoter methylation in the Wernicke area of suicide subjects. *Arch Gen Psychiatry* 67(3): 258–267.

Khan, A., Khan, S., Kolts, R., and Brown, W. A. 2003. Suicide rates in clinical trials of SSRIs, other antidepressants, and placebo: Analysis of FDA reports. *Am J Psychiatry* 160(4): 790–792.

Kia-Keating, B. M., Glatt, S. J., and Tsuang, M. T. 2007. Meta-analyses suggest association between COMT, but not HTR1B, alleles, and suicidal behavior. *Am J Med Genet B Neuropsychiatr Genet* 144B(8): 1048–1053.

Kim, B., Kim, C. Y., Hong, J. P., Kim, S. Y., Lee, C., and Joo, Y. H. 2008. Brain-derived neurotrophic factor Val/Met polymorphism and bipolar disorder. Association of the Met allele with suicidal behavior of bipolar patients. *Neuropsychobiology* 58(2): 97–103.

Kim, Y. K., Lee, H. P., Won, S. D., Park, E. Y., Lee, H. Y., Lee, B. H. et al. 2007. Low plasma BDNF is associated with suicidal behavior in major depression. *Prog Neuropsychopharmacol Biol Psychiatry* 31(1): 78–85.

Kim, C. D., Seguin, M., Therrien, N., Riopel, G., Chawky, N., Lesage, A. D. et al. 2005. Familial aggregation of suicidal behavior: A family study of male suicide completers from the general population. *Am J Psychiatry* 162(5): 1017–1019.

Kiyohara, C. and Yoshimasu, K. 2010. Association between major depressive disorder and a functional polymorphism of the 5-hydroxytryptamine (serotonin) transporter gene: A meta-analysis. *Psychiatr Genet* 20(2): 49–58.

Klempan, T. A., Rujescu, D., Merette, C., Himmelman, C., Sequeira, A., Canetti, L. et al. 2009. Profiling brain expression of the spermidine/spermine *N*1-acetyltransferase 1 (*SAT1*) gene in suicide. *Am J Med Genet B Neuropsychiatr Genet* 150B(7): 934–943.

Kohli, M. A., Salyakina, D., Pfennig, A., Lucae, S., Horstmann, S., Menke, A. et al. 2010. Association of genetic variants in the neurotrophic receptor-encoding gene NTRK2 and a lifetime history of suicide attempts in depressed patients. *Arch Gen Psychiatry* 67(4): 348–359.

Koller, G., Bondy, B., Preuss, U. W., Bottlender, M., and Soyka, M. 2003. No association between a polymorphism in the promoter region of the *MAOA* gene with antisocial personality traits in alcoholics. *Alcohol Alcohol* 38(1): 31–34.

Kunugi, H., Hashimoto, R., Yoshida, M., Tatsumi, M., and Kamijima, K. 2004. A missense polymorphism (S205L) of the low-affinity neurotrophin receptor *p75NTR* gene is associated with depressive disorder and attempted suicide. *Am J Med Genet B Neuropsychiatr Genet* 129B(1): 44–46.

Labonte, B. and Turecki, G. 2010. The epigenetics of suicide: Explaining the biological effects of early life environmental adversity. *Arch Suicide Res* 14(4): 291–310.

Lachman, H. M., Nolan, K. A., Mohr, P., Saito, T., and Volavka, J. 1998. Association between catechol *O*-methyltransferase genotype and violence in schizophrenia and schizoaffective disorder. *Am J Psychiatry* 155(6): 835–837.

Laje, G., Allen, A. S., Akula, N., Manji, H., John Rush, A., and McMahon, F. J. 2009. Genome-wide association study of suicidal ideation emerging during citalopram treatment of depressed outpatients. *Pharmacogenet Genomics* 19(9): 666–674.

Laje, G., Paddock, S., Manji, H., Rush, A. J., Wilson, A. F., Charney, D. et al. 2007. Genetic markers of suicidal ideation emerging during citalopram treatment of major depression. *Am J Psychiatry* 164(10): 1530–1538.

Lee, R., Petty, F., and Coccaro, E. F. 2009. Cerebrospinal fluid GABA concentration: Relationship with impulsivity and history of suicidal behavior, but not aggression, in human subjects. *J Psychiatr Res* 43(4): 353–359.

Lemonde, S., Turecki, G., Bakish, D., Du, L., Hrdina, P. D., Bown, C. D. et al. 2003. Impaired repression at a 5-hydroxytryptamine 1A receptor gene polymorphism associated with major depression and suicide. *J Neurosci* 23(25): 8788–8799.

Lester, D. 1995. The concentration of neurotransmitter metabolites in the cerebrospinal fluid of suicidal individuals: A meta-analysis. *Pharmacopsychiatry* 28(2): 45–50.

Li, D., Duan, Y., and He, L. 2006. Association study of serotonin 2A receptor (*5-HT2A*) gene with schizophrenia and suicidal behavior using systematic meta-analysis. *Biochem Biophys Res Commun* 340(3): 1006–1015.

Li, D. and He, L. 2007. Meta-analysis supports association between serotonin transporter (5-HTT) and suicidal behavior. *Mol Psychiatry* 12(1): 47–54.

Lieb, R., Bronisch, T., Hofler, M., Schreier, A., and Wittchen, H. U. 2005. Maternal suicidality and risk of suicidality in offspring: Findings from a community study. *Am J Psychiatry* 162(9): 1665–1671.

Lindstrom, M. B., Ryding, E., Bosson, P., Ahnlide, J. A., Rosen, I., and Traskman-Bendz, L. 2004. Impulsivity related to brain serotonin transporter binding capacity in suicide attempters. *Eur Neuropsychopharmacol* 14(4): 295–300.

Linkowski, P., de Maertelaer, V., and Mendlewicz, J. 1985. Suicidal behaviour in major depressive illness. *Acta Psychiatr Scand* 72(3): 233–238.

Liou, Y. J., Tsai, S. J., Hong, C. J., Wang, Y. C., and Lai, I. C. 2001. Association analysis of a functional catechol-*o*-methyltransferase gene polymorphism in schizophrenic patients in Taiwan. *Neuropsychobiology* 43(1): 11–14.

Little, K. Y., McLauglin, D. P., Ranc, J., Gilmore, J., Lopez, J. F., Watson, S. J. et al. 1997. Serotonin transporter binding sites and mRNA levels in depressed persons committing suicide. *Biol Psychiatry* 41(12): 1156–1164.

Lopez de Lara, V. A., Detera-Wadleigh, S., Cardona, I., Kassem, L., and McMahon, F. J. 2007. Nested association between genetic variation in tryptophan hydroxylase II, bipolar affective disorder, and suicide attempts. *Biol Psychiatry* 61(2): 181–186.

Lou, X. Y., Chen, G. B., Yan, L., Ma, J. Z., Zhu, J., Elston, R. C. et al. 2007. A generalized combinatorial approach for detecting gene-by-gene and gene-by-environment interactions with application to nicotine dependence. *Am J Hum Genet* 80(6): 1125–1137.

Lowther, S., De Paermentier, F., Crompton, M. R., Katona, C. L., and Horton, R. W. 1994. Brain 5-HT2 receptors in suicide victims: Violence of death, depression and effects of antidepressant treatment. *Brain Res* 642(1–2): 281–289.

MacKenzie, A. and Quinn, J. 1999. A serotonin transporter gene intron 2 polymorphic region, correlated with affective disorders, has allele-dependent differential enhancer-like properties in the mouse embryo. *Proc Natl Acad Sci U. S. A.* 96(26): 15251–15255.

Magno, L. A., Miranda, D. M., Neves, F. S., Pimenta, G. J., Mello, M. P., De Marco, L. A. et al. 2010. Association between AKT1 but not AKTIP genetic variants and increased risk for suicidal behavior in bipolar patients. *Genes Brain Behav* 9(4): 411–418.

Malloy-Diniz, L. F., Neves, F. S., Abrantes, S. S., Fuentes, D., and Correa, H. 2009. Suicide behavior and neuropsychological assessment of type I bipolar patients. *J Affect Disord* 112(1–3): 231–236.

Malone, K. M., Haas, G. L., Sweeney, J. A., and Mann, J. J. 1995. Major depression and the risk of attempted suicide. *J Affect Disord* 34(3): 173–185.

Mann, J. J., Arango, V. A., Avenevoli, S., Brent, D. A., Champagne, F. A., Clayton, P. et al. 2009. Candidate endophenotypes for genetic studies of suicidal behavior. *Biol Psychiatry* 65(7): 556–563.

Mann, J. J., Bortinger, J., Oquendo, M. A., Currier, D., Li, S., and Brent, D. A. 2005. Family history of suicidal behavior and mood disorders in probands with mood disorders. *Am J Psychiatry* 162(9): 1672–1679.

Mann, J. J., Currier, D., Murphy, L., Huang, Y. Y., Galfalvy, H., Brent, D. et al. 2008. No association between a TPH2 promoter polymorphism and mood disorders or monoamine turnover. *J Affect Disord* 106(1–2): 117–121.

Mann, J. J., Henteleff, R. A., Lagattuta, T. F., Perper, J. A., Li, S., and Arango, V. 1996. Lower 3H-paroxetine binding in cerebral cortex of suicide victims is partly due to fewer high affinity, non-transporter sites. *J Neural Transm* 103(11): 1337–1350.

Manuck, S. B., Flory, J. D., Ferrell, R. E., Mann, J. J., and Muldoon, M. F. 2000. A regulatory polymorphism of the monoamine oxidase-A gene may be associated with variability in aggression, impulsivity, and central nervous system serotonergic responsivity. *Psychiatry Res* 95(1): 9–23.

McGirr, A., Alda, M., Seguin, M., Cabot, S., Lesage, A., and Turecki, G. 2009. Familial aggregation of suicide explained by cluster B traits: A three-group family study of suicide controlling for major depressive disorder. *Am J Psychiatry* 166(10): 1124–1134.

McGirr, A., Diaconu, G., Berlim, M. T., Pruessner, J. C., Sable, R., Cabot, S. et al. 2010. Dysregulation of the sympathetic nervous system, hypothalamic–pituitary–adrenal axis and executive function in individuals at risk for suicide. *J Psychiatry Neurosci* 35(6): 399–408.

McGirr, A., Renaud, J., Bureau, A., Seguin, M., Lesage, A., and Turecki, G. 2008. Impulsive–aggressive behaviours and completed suicide across the life cycle: A predisposition for younger age of suicide. *Psychol Med* 38(3): 407–417.

McGowan, P. O., Sasaki, A., D'Alessio, A. C., Dymov, S., Labonte, B., Szyf, M. et al. 2009. Epigenetic regulation of the glucocorticoid receptor in human brain associates with childhood abuse. *Nat Neurosci* 12(3): 342–348.

McGowan, P. O., Sasaki, A., Huang, T. C., Unterberger, A., Suderman, M., Ernst, C. et al. 2008. Promoter-wide hypermethylation of the ribosomal RNA gene promoter in the suicide brain. *PLoS One* 3(5): e2085.

McGregor, S., Strauss, J., Bulgin, N., De Luca, V., George, C. J., Kovacs, M. et al. 2007. *p75(NTR)* gene and suicide attempts in young adults with a history of childhood-onset mood disorder. *Am J Med Genet B Neuropsychiatr Genet* 144B(5): 696–700.

McGuffin, P., Perroud, N., Uher, R., Butler, A., Aitchison, K. J., Craig, I. et al. 2010. The genetics of affective disorder and suicide. *Eur Psychiatry* 25(5): 275–277.

Melhem, N. M., Brent, D. A., Ziegler, M., Iyengar, S., Kolko, D., Oquendo, M. et al. 2007. Familial pathways to early-onset suicidal behavior: Familial and individual antecedents of suicidal behavior. *Am J Psychiatry* 164(9): 1364–1370.

Merali, Z., Du, L., Hrdina, P., Palkovits, M., Faludi, G., Poulter, M. O. et al. 2004. Dysregulation in the suicide brain: mRNA expression of corticotropin-releasing hormone receptors and GABA(A) receptor subunits in frontal cortical brain region. *J Neurosci* 24(6): 1478–1485.

Mitterauer, B. 1990. A contribution to the discussion of the role of the genetic factor in suicide, based on five studies in an epidemiologically defined area (Province of Salzburg, Austria). *Compr Psychiatry* 31(6): 557–565.

Mouri, K., Hishimoto, A., Fukutake, M., Shiroiwa, K., Asano, M., Nagasaki, Y. et al. 2009. TPH2 is not a susceptibility gene for suicide in Japanese population. *Prog Neuropsychopharmacol Biol Psychiatry* 33(8): 1546–1550.

Mundo, E., Richter, M. A., Sam, F., Macciardi, F., and Kennedy, J. L. 2000. Is the 5-HT(1Dbeta) receptor gene implicated in the pathogenesis of obsessive–compulsive disorder? *Am J Psychiatry* 157(7): 1160–1161.

Mundo, E., Richter, M. A., Zai, G., Sam, F., McBride, J., Macciardi, F. et al. 2002. 5HT1Dbeta receptor gene implicated in the pathogenesis of obsessive–compulsive disorder: Further evidence from a family-based association study. *Mol Psychiatry* 7(7): 805–809.

Mundo, E., Rouillon, F., Figuera, M. L., and Stigler, M. 2001. Fluvoxamine in obsessive–compulsive disorder: Similar efficacy but superior tolerability in comparison with clomipramine. *Hum Psychopharmacol* 16(6): 461–468.

Must, A., Tasa, G., Lang, A., Vasar, E., Koks, S., Maron, E. et al. 2009. Variation in tryptophan hydroxylase-2 gene is not associated to male completed suicide in Estonian population. *Neurosci Lett* 453(2): 112–114.

Nackley, A. G., Shabalina, S. A., Tchivileva, I. E., Satterfield, K., Korchynskyi, O., Makarov, S. S., et al. 2006. Human catechol-*O*-methyltransferase haplotypes modulate protein expression by altering mRNA secondary structure. *Science* 314(5807): 1930–1933.

Nakamura, M., Ueno, S., Sano, A., and Tanabe, H. 2000. The human serotonin transporter gene linked polymorphism (5-HTTLPR) shows ten novel allelic variants. *Mol Psychiatry* 5(1): 32–38.

Neale, B. M. and Sham, P. C. 2004. The future of association studies: Gene-based analysis and replication. *Am J Hum Genet* 75(3): 353–362.

Neves, F. S., Malloy-Diniz, L. F., Romano-Silva, M. A., Aguiar, G. C., de Matos, L. O., and Correa, H. 2010. Is the serotonin transporter polymorphism (5-HTTLPR) a potential marker for suicidal behavior in bipolar disorder patients? *J Affect Disord* 125(1–3): 98–102.

Neves, F. S., Silveira, G., Romano-Silva, M. A., Malloy-Diniz, L., Ferreira, A. A., De Marco, L., et al. 2008. Is the 5-HTTLPR polymorphism associated with bipolar disorder or with suicidal behavior of bipolar disorder patients? *Am J Med Genet B Neuropsychiatr Genet* 147B(1): 114–116.

Nielsen, D. A., Goldman, D., Virkkunen, M., Tokola, R., Rawlings, R., and Linnoila, M. 1994. Suicidality and 5-hydroxyindoleacetic acid concentration associated with a tryptophan hydroxylase polymorphism. *Arch Gen Psychiatry* 51(1): 34–38.

Nishiguchi, N., Shirakawa, O., Ono, H., Nishimura, A., Nushida, H., Ueno, Y. et al. 2002. Lack of an association between 5-HT1A receptor gene structural polymorphisms and suicide victims. *Am J Med Genet* 114(4): 423–425.

Noble, E. P. 2003. D2 dopamine receptor gene in psychiatric and neurologic disorders and its phenotypes. *Am J Med Genet B Neuropsychiatr Genet* 116B(1): 103–125.

Nolan, K. A., Volavka, J., Czobor, P., Cseh, A., Lachman, H., Saito, T. et al. 2000. Suicidal behavior in patients with schizophrenia is related to COMT polymorphism. *Psychiatr Genet* 10(3): 117–124.

Nomura, M., Kusumi, I., Kaneko, M., Masui, T., Daiguji, M., Ueno, T. et al. 2006. Involvement of a polymorphism in the 5-HT2A receptor gene in impulsive behavior. *Psychopharmacology (Berl)* 187(1): 30–35.

Ohara, K., Nagai, M., and Suzuki, Y. 1998a. Low activity allele of catechol-*o*-methyltransferase gene and Japanese unipolar depression. *Neuroreport* 9(7): 1305–1308.

Ohara, K., Nagai, M., Tani, K., Tsukamoto, T., and Suzuki, Y. 1998b. Polymorphism in the promoter region of the alpha 2A adrenergic receptor gene and mood disorders. *Neuroreport* 9(7): 1291–1294.

Ohtani, M., Shindo, S., and Yoshioka, N. 2004. Polymorphisms of the tryptophan hydroxylase gene and serotonin 1A receptor gene in suicide victims among Japanese. *Tohoku J Exp Med* 202(2): 123–133.

Ono, H., Shirakawa, O., Kitamura, N., Hashimoto, T., Nishiguchi, N., Nishimura, A. et al. 2002. Tryptophan hydroxylase immunoreactivity is altered by the genetic variation in postmortem brain samples of both suicide victims and controls. *Mol Psychiatry* 7(10): 1127–1132.

Ono, H., Shirakawa, O., Nushida, H., Ueno, Y., and Maeda, K. 2004. Association between catechol-*O*-methyltransferase functional polymorphism and male suicide completers. *Neuropsychopharmacology* 29(7): 1374–1377.

Oswald, L. M., Wong, D. F., Zhou, Y., Kumar, A., Brasic, J., Alexander, M. et al. 2007. Impulsivity and chronic stress are associated with amphetamine-induced striatal dopamine release. *Neuroimage* 36(1): 153–166.

Pandey, G. N. 1997. Altered serotonin function in suicide. Evidence from platelet and neuroendocrine studies. *Ann N Y Acad Sci* 836: 182–200.

Papiol, S., Arias, B., Gasto, C., Gutierrez, B., Catalan, R., and Fananas, L. 2007. Genetic variability at HPA axis in major depression and clinical response to antidepressant treatment. *J Affect Disord* 104(1–3): 83–90.

Pedroso, I., Lourdusamy, A., Rietschel, M., Nöthen, M. M., Cichon, S., McGuffin, P., Al-Chalabi, A., Barnes, M. R., Breen, G. Common genetic variants and gene-expression changes associated with bipolar disorder are over-represented in brain signaling pathway genes. *Biol Psychiatry* (in press).

Perlis, R. H., Huang, J., Purcell, S., Fava, M., Rush, A. J., Sullivan, P. F. et al. 2010. Genome-wide association study of suicide attempts in mood disorder patients. *Am J Psychiatry* 167(12): 1499–1507.

Perlis, R. H., Purcell, S., Fava, M., Fagerness, J., Rush, A. J., Trivedi, M. H. et al. 2007. Association between treatment-emergent suicidal ideation with citalopram and polymorphisms near cyclic adenosine monophosphate response element binding protein in the STAR*D study. *Arch Gen Psychiatry* 64(6): 689–697.

Perroud, N., Aitchison, K. J., Uher, R., Smith, R., Huezo Diaz, P., Marusic, A. et al. 2009. Genetic predictors of increase in suicidal ideation during antidepressant treatment in the GENDEP project. *Neuropsychopharmacology* 34(12): 2517–2528.

Perroud, N., Neidhart, E., Petit, B., Vessaz, M., Laforge, T., Relecom, C. et al. 2010a. Simultaneous analysis of serotonin transporter, tryptophan hydroxylase 1 and 2 gene expression in the ventral prefrontal cortex of suicide victims. *Am J Med Genet B Neuropsychiatr Genet* 153B(4): 909–918.

Perroud, N., Uher, R., Ng, M. Y., Guipponi, M., Hauser, J., Henigsberg, N. et al. 2010b. Genome-wide association study of increasing suicidal ideation during antidepressant treatment in the GENDEP project. *Pharmacogenomics J.* 12(1): 68–77.

Persson, M. L., Geijer, T., Wasserman, D., Rockah, R., Frisch, A., Michaelovsky, E. et al. 1999a. Lack of association between suicide attempt and a polymorphism at the dopamine receptor D4 locus. *Psychiatr Genet* 9(2): 97–100.

Persson, M. L., Runeson, B. S., and Wasserman, D. 1999b. Diagnoses, psychosocial stressors and adaptive functioning in attempted suicide. *Ann Clin Psychiatry* 11(3): 119–128.

Persson, M. L., Wasserman, D., Geijer, T., Jonsson, E. G., and Terenius, L. 1997. Tyrosine hydroxylase allelic distribution in suicide attempters. *Psychiatry Res* 72(2): 73–80.

Petronis, A. 2010. Epigenetics as a unifying principle in the aetiology of complex traits and diseases. *Nature* 465(7299): 721–727.

Pfeffer, C. R., Normandin, L., and Kakuma, T. 1994. Suicidal children grow up: Suicidal behavior and psychiatric disorders among relatives. *J Am Acad Child Adolesc Psychiatry* 33(8): 1087–1097.

Placidi, G. P., Oquendo, M. A., Malone, K. M., Huang, Y. Y., Ellis, S. P., and Mann, J. J. 2001. Aggressivity, suicide attempts, and depression: Relationship to cerebrospinal fluid monoamine metabolite levels. *Biol Psychiatry* 50(10): 783–791.

Polesskaya, O. O. and Sokolov, B. P. 2002. Differential expression of the "C" and "T" alleles of the 5-HT2A receptor gene in the temporal cortex of normal individuals and schizophrenics. *J Neurosci Res* 67(6): 812–822.

Pompili, M., Serafini, G., Innamorati, M., Moller-Leimkuhler, A. M., Giupponi, G., Girardi, P. et al. 2010. The hypothalamic–pituitary–adrenal axis and serotonin abnormalities: A selective overview for the implications of suicide prevention. *Eur Arch Psychiatry Clin Neurosci* 260(8): 583–600.

Porcelli, S., Drago, A., Fabbri, C., Gibiino, S., Calati, R., and Serretti, A. 2011. Pharmacogenetics of antidepressant response. *J Psychiatry Neurosci* 36(2): 87–113.

Posner, K., Oquendo, M. A., Gould, M., Stanley, B., and Davies, M. 2007. Columbia Classification Algorithm of Suicide Assessment (C-CASA): Classification of suicidal events in the FDA's pediatric suicidal risk analysis of antidepressants. *Am J Psychiatry* 164(7): 1035–1043.

Poulter, M. O., Du, L., Weaver, I. C., Palkovits, M., Faludi, G., Merali, Z. et al. 2008. $GABA_A$ receptor promoter hypermethylation in suicide brain: Implications for the involvement of epigenetic processes. *Biol Psychiatry* 64(8): 645–652.

Poulter, M. O., Du, L., Zhurov, V., Palkovits, M., Faludi, G., Merali, Z. et al. 2010. Altered organization of GABA(A) receptor mRNA expression in the depressed suicide brain. *Front Mol Neurosci* 3: 3.

Powell, J., Geddes, J., Deeks, J., Goldacre, M., and Hawton, K. 2000. Suicide in psychiatric hospital in-patients. Risk factors and their predictive power. *Br J Psychiatry* 176: 266–272.

Preuss, U. W., Koller, G., Soyka, M., and Bondy, B. 2001. Association between suicide attempts and 5-HTTLPR-S-allele in alcohol-dependent and control subjects: Further evidence from a German alcohol-dependent inpatient sample. *Biol Psychiatry* 50(8): 636–639.

Qin, P., Agerbo, E., and Mortensen, P. B. 2002. Suicide risk in relation to family history of completed suicide and psychiatric disorders: A nested case–control study based on longitudinal registers. *Lancet* 360(9340): 1126–1130.

Qin, P., Agerbo, E., and Mortensen, P. B. 2003. Suicide risk in relation to socioeconomic, demographic, psychiatric, and familial factors: A national register-based study of all suicides in Denmark, 1981–1997. *Am J Psychiatry* 160(4): 765–772.

Rigat, B., Hubert, C., Alhenc-Gelas, F., Cambien, F., Corvol, P., and Soubrier, F. 1990. An insertion/deletion polymorphism in the angiotensin I-converting enzyme gene accounting for half the variance of serum enzyme levels. *J Clin Invest* 86(4): 1343–1346.

Risch, N., Herrell, R., Lehner, T., Liang, K. Y., Eaves, L., Hoh, J. et al. 2009. Interaction between the serotonin transporter gene (*5-HTTLPR*), stressful life events, and risk of depression: A meta-analysis. *JAMA* 301(23): 2462–2471.

Rotondo, A., Schuebel, K., Bergen, A., Aragon, R., Virkkunen, M., Linnoila, M. et al. 1999. Identification of four variants in the tryptophan hydroxylase promoter and association to behavior. *Mol Psychiatry* 4(4): 360–368.

Roy, A. 1983. Family history of suicide. *Arch Gen Psychiatry* 40(9): 971–974.

Roy, A. 2001. Characteristics of cocaine-dependent patients who attempt suicide. *Am J Psychiatry* 158(8): 1215–1219.

Roy, A. 2002. Family history of suicide and neuroticism: A preliminary study. *Psychiatry Res* 110(1): 87–90.

Roy, A. 2003. Distal risk factors for suicidal behavior in alcoholics: Replications and new findings. *J Affect Disord* 77(3): 267–271.

Roy, A., Agren, H., Pickar, D., Linnoila, M., Doran, A. R., Cutler, N. R. et al. 1986. Reduced CSF concentrations of homovanillic acid and homovanillic acid to 5-hydroxyindoleacetic acid ratios in depressed patients: Relationship to suicidal behavior and dexamethasone nonsuppression. *Am J Psychiatry* 143(12): 1539–1545.

Roy, A., Hu, X. Z., Janal, M. N., and Goldman, D. 2007. Interaction between childhood trauma and serotonin transporter gene variation in suicide. *Neuropsychopharmacology* 32(9): 2046–2052.

Roy, A., Sarchiopone, M., and Carli, V. 2009. Gene–environment interaction and suicidal behavior. *J Psychiatr Pract* 15(4): 282–288.

Roy, A. and Segal, N. L. 2001. Suicidal behavior in twins: A replication. *J Affect Disord* 66(1): 71–74.

Roy, A., Segal, N. L., Centerwall, B. S., and Robinette, C. D. 1991. Suicide in twins. *Arch Gen Psychiatry* 48(1): 29–32.

Roy, A., Segal, N. L., and Sarchiapone, M. 1995. Attempted suicide among living co-twins of twin suicide victims. *Am J Psychiatry* 152(7): 1075–1076.

Rujescu, D., Giegling, I., Gietl, A., Hartmann, A. M., and Moller, H. J. 2003. A functional single nucleotide polymorphism (V158M) in the *COMT* gene is associated with aggressive personality traits. *Biol Psychiatry* 54(1): 34–39.

Rujescu, D., Giegling, I., Mandelli, L., Schneider, B., Hartmann, A. M., Schnabel, A. et al. 2008. *NOS-I* and *-III* gene variants are differentially associated with facets of suicidal behavior and aggression-related traits. *Am J Med Genet B Neuropsychiatr Genet* 147B(1): 42–48.

Runeson, B. and Asberg, M. 2003. Family history of suicide among suicide victims. *Am J Psychiatry* 160(8): 1525–1526.

Rushton, J. P., Fulker, D. W., Neale, M. C., Nias, D. K., and Eysenck, H. J. 1986. Altruism and aggression: The heritability of individual differences. *J Pers Soc Psychol* 50(6): 1192–1198.

Russ, M. J., Lachman, H. M., Kashdan, T., Saito, T., and Bajmakovic-Kacila, S. 2000. Analysis of catechol-*O*-methyltransferase and 5-hydroxytryptamine transporter polymorphisms in patients at risk for suicide. *Psychiatry Res* 93(1): 73–78.

Sabol, S. Z., Hu, S., and Hamer, D. 1998. A functional polymorphism in the monoamine oxidase A gene promoter. *Hum Genet* 103(3): 273–279.

Saetre, P., Lundmark, P., Wang, A., Hansen, T., Rasmussen, H. B., Djurovic, S. et al. 2010. The tryptophan hydroxylase 1 (*TPH1*) gene, schizophrenia susceptibility, and suicidal behavior: A multi-centre case–control study and meta-analysis. *Am J Med Genet B Neuropsychiatr Genet* 153B(2): 387–396.

Sakai, K., Nakamura, M., Ueno, S., Sano, A., Sakai, N., Shirai, Y. et al. 2002. The silencer activity of the novel human serotonin transporter linked polymorphic regions. *Neurosci Lett* 327(1): 13–16.

Sakinofsky, I., Roberts, R. S., Brown, Y., Cumming, C., and James, P. 1990. Problem resolution and repetition of parasuicide. A prospective study. *Br J Psychiatry* 156: 395–399.

Sarchiapone, M., Carli, V., Roy, A., Iacoviello, L., Cuomo, C., Latella, M. C. et al. 2008. Association of polymorphism (Val66Met) of brain-derived neurotrophic factor with suicide attempts in depressed patients. *Neuropsychobiology* 57(3): 139–145.

Saudou, F., Amara, D. A., Dierich, A., LeMeur, M., Ramboz, S., Segu, L. et al. 1994. Enhanced aggressive behavior in mice lacking 5-HT1B receptor. *Science* 265(5180): 1875–1878.

Schosser, A., Butler, A. W., Ising, M., Perroud, N., Uher, R., Ng, M. Y. et al. 2011. Genome wide association scan of suicidal thoughts and behaviour in major depression. *PLoS One* 6(7): e20690.

Schulsinger, F., Kety, S. S., Rosenthal, D., and Wender, P. H. (1979). A family study of suicide. *Origin, Presentation and Treatment of Affective Disorders*. Schou, M. and Stromgren, S. eds. London, U.K.: Academic Press, pp. 277–287.

Segal, J., Schenkel, L. C., Oliveira, M. H., Salum, G. A., Bau, C. H., Manfro, G. G., et al. 2009. Novel allelic variants in the human serotonin transporter gene linked polymorphism (5-HTTLPR) among depressed patients with suicide attempt. *Neurosci Lett* 451(1): 79–82.

Sen, S., Burmeister, M., and Ghosh, D. 2004. Meta-analysis of the association between a serotonin transporter promoter polymorphism (5-HTTLPR) and anxiety-related personality traits. *Am J Med Genet B Neuropsychiatr Genet* 127B(1): 85–89.

Sequeira, A., Gwadry, F. G., Ffrench-Mullen, J. M., Canetti, L., Gingras, Y., Casero, R. A., Jr. et al. 2006. Implication of SSAT by gene expression and genetic variation in suicide and major depression. *Arch Gen Psychiatry* 63(1): 35–48.

Sequeira, A., Mamdani, F., Ernst, C., Vawter, M. P., Bunney, W. E., Lebel, V. et al. 2009. Global brain gene expression analysis links glutamatergic and GABAergic alterations to suicide and major depression. *PLoS One* 4(8): e6585.

Sequeira, A., Mamdani, F., Lalovic, A., Anguelova, M., Lesage, A., Seguin, M. et al. 2004. Alpha 2A adrenergic receptor gene and suicide. *Psychiatry Res* 125(2): 87–93.

Serretti, A., Calati, R., Giegling, I., Hartmann, A. M., Moller, H. J., Colombo, C. et al. 2007a. 5-HT2A SNPs and the Temperament and Character Inventory. *Prog Neuropsychopharmacol Biol Psychiatry* 31(6): 1275–1281.

Serretti, A., Mandelli, L., Giegling, I., Schneider, B., Hartmann, A. M., Schnabel, A. et al. 2007b. *HTR2C* and *HTR1A* gene variants in German and Italian suicide attempters and completers. *Am J Med Genet B Neuropsychiatr Genet* 144B(3): 291–299.

Sherif, F., Marcusson, J., and Oreland, L. 1991. Brain gamma-aminobutyrate transaminase and monoamine oxidase activities in suicide victims. *Eur Arch Psychiatry Clin Neurosci* 241(3): 139–144.

Shindo, S. and Yoshioka, N. 2005. Polymorphisms of the cholecystokinin gene promoter region in suicide victims in Japan. *Forensic Sci Int* 150(1): 85–90.

Small, K. M., Forbes, S. L., Brown, K. M., and Liggett, S. B. 2000. An asn to lys polymorphism in the third intracellular loop of the human alpha 2A-adrenergic receptor imparts enhanced agonist-promoted Gi coupling. *J Biol Chem* 275(49): 38518–38523.

Smits, K. M., Smits, L. J., Schouten, J. S., Stelma, F. F., Nelemans, P., and Prins, M. H. 2004. Influence of SERTPR and STin2 in the serotonin transporter gene on the effect of selective serotonin reuptake inhibitors in depression: A systematic review. *Mol Psychiatry* 9(5): 433–441.

Soloff, P. H., Lynch, K. G., Kelly, T. M., Malone, K. M., and Mann, J. J. 2000. Characteristics of suicide attempts of patients with major depressive episode and borderline personality disorder: A comparative study. *Am J Psychiatry* 157(4): 601–608.

Soyka, M., Preuss, U. W., Koller, G., Zill, P., and Bondy, B. 2004. Association of 5-HT1B receptor gene and antisocial behavior in alcoholism. *J Neural Transm* 111(1): 101–109.

Spalletta, G., Morris, D. W., Angelucci, F., Rubino, I. A., Spoletini, I., Bria, P., et al. 2010. BDNF Val66Met polymorphism is associated with aggressive behavior in schizophrenia. *Eur Psychiatry* 25(6): 311–313.

Sparks, D. L., Hunsaker, J. C., 3rd, Amouyel, P., Malafosse, A., Bellivier, F., Leboyer, M. et al. 2009. Angiotensin I-converting enzyme I/D polymorphism and suicidal behaviors. *Am J Med Genet B Neuropsychiatr Genet* 150B(2): 290–294.

Stankovic, Z., Saula-Marojevic, B., and Potrebic, A. 2006. Personality profile of depressive patients with a history of suicide attempts. *Psychiatr Danub* 18(3–4): 159–168.

Statham, D. J., Heath, A. C., Madden, P. A., Bucholz, K. K., Bierut, L., Dinwiddie, S. H. et al. 1998. Suicidal behaviour: An epidemiological and genetic study. *Psychol Med* 28(4): 839–855.

Stefulj, J., Buttner, A., Kubat, M., Zill, P., Balija, M., Eisenmenger, W. et al. 2004a. 5HT-2C receptor polymorphism in suicide victims. Association studies in German and Slavic populations. *Eur Arch Psychiatry Clin Neurosci* 254(4): 224–227.

Stefulj, J., Buttner, A., Skavic, J., Zill, P., Balija, M., Eisenmenger, W. et al. 2004b. Serotonin 1B (5HT-1B) receptor polymorphism (G861C) in suicide victims: Association studies in German and Slavic population. *Am J Med Genet B Neuropsychiatr Genet* 127B(1): 48–50.

Suda, A., Kawanishi, C., Kishida, I., Sato, R., Yamada, T., Nakagawa, M. et al. 2009. Dopamine D2 receptor gene polymorphisms are associated with suicide attempt in the Japanese population. *Neuropsychobiology* 59(2): 130–134.

Sun, H. F., Chang, Y. T., Fann, C. S., Chang, C. J., Chen, Y. H., Hsu, Y. P. et al. 2002. Association study of novel human serotonin 5-HT(1B) polymorphisms with alcohol dependence in Taiwanese Han. *Biol Psychiatry* 51(11): 896–901.

Sun, H. F., Fann, C. S., Lane, H. Y., Chang, Y. T., Chang, C. J., Liu, Y. L. et al. 2005. A functional polymorphism in the promoter region of the tryptophan hydroxylase gene is associated with alcohol dependence in one aboriginal group in Taiwan. *Alcohol Clin Exp Res* 29(1): 1–7.

Teicher, M. H., Glod, C., and Cole, J. O. 1990. Emergence of intense suicidal preoccupation during fluoxetine treatment. *Am J Psychiatry* 147(2): 207–210.

Tsai, S. J., Hong, C. J., Yu, Y. W., Chen, T. J., Wang, Y. C., and Lin, W. K. 2004. Association study of serotonin 1B receptor (A-161T) genetic polymorphism and suicidal behaviors and response to fluoxetine in major depressive disorder. *Neuropsychobiology* 50(3): 235–238.

Tsai, S. Y., Kuo, C. J., Chen, C. C., and Lee, H. C. 2002. Risk factors for completed suicide in bipolar disorder. *J Clin Psychiatry* 63(6): 469–476.

Tsankova, N., Renthal, W., Kumar, A., and Nestler, E. J. 2007. Epigenetic regulation in psychiatric disorders. *Nat Rev Neurosci* 8(5): 355–367.

Tsuang, M. T. 1983. Risk of suicide in the relatives of schizophrenics, manics, depressives, and controls. *J Clin Psychiatry* 44(11): 396–397, 398–400.

Tsuang, M. T., Boor, M., and Fleming, J. A. 1985. Psychiatric aspects of traffic accidents. *Am J Psychiatry* 142(5): 538–546.

Turecki, G., Briere, R., Dewar, K., Antonetti, T., Lesage, A. D., Seguin, M. et al. 1999. Prediction of level of serotonin 2A receptor binding by serotonin receptor 2A genetic variation in postmortem brain samples from subjects who did or did not commit suicide. *Am J Psychiatry* 156(9): 1456–1458.

Turecki, G., Sequeira, A., Gingras, Y., Seguin, M., Lesage, A., Tousignant, M. et al. 2003. Suicide and serotonin: Study of variation at seven serotonin receptor genes in suicide completers. *Am J Med Genet B Neuropsychiatr Genet* 118B(1): 36–40.

Uher, R., Caspi, A., Houts, R., Sugden, K., Williams, B., Poulton, R. et al. 2011. Serotonin transporter gene moderates childhood maltreatment's effects on persistent but not single-episode depression: Replications and implications for resolving inconsistent results. *J Affect Disord*. 135(1–3): 56–65.

van Gaalen, M. M., van Koten, R., Schoffelmeer, A. N., and Vanderschuren, L. J. 2006. Critical involvement of dopaminergic neurotransmission in impulsive decision making. *Biol Psychiatry* 60(1): 66–73.

van Heeringen, C., Audenaert, K., Van Laere, K., Dumont, F., Slegers, G., Mertens, J. et al. 2003. Prefrontal 5-HT2a receptor binding index, hopelessness and personality characteristics in attempted suicide. *J Affect Disord* 74(2): 149–158.

Videtic, A., Peternelj, T. T., Zupanc, T., Balazic, J., and Komel, R. 2009a. Promoter and functional polymorphisms of HTR2C and suicide victims. *Genes Brain Behav* 8(5): 541–545.

Videtic, A., Pungercic, G., Pajnic, I. Z., Zupanc, T., Balazic, J., Tomori, M. et al. 2006. Association study of seven polymorphisms in four serotonin receptor genes on suicide victims. *Am J Med Genet B Neuropsychiatr Genet* 141B(6): 669–672.

Videtic, A., Zupanc, T., Pregelj, P., Balazic, J., Tomori, M., and Komel, R. 2009b. Suicide, stress and serotonin receptor 1A promoter polymorphism –1019C>G in Slovenian suicide victims. *Eur Arch Psychiatry Clin Neurosci* 259(4): 234–238.

Vincze, I., Perroud, N., Buresi, C., Baud, P., Bellivier, F., Etain, B. et al. 2008. Association between brain-derived neurotrophic factor gene and a severe form of bipolar disorder, but no interaction with the serotonin transporter gene. *Bipolar Disord* 10(5): 580–587.

Virkkunen, M., De Jong, J., Bartko, J., and Linnoila, M. 1989. Psychobiological concomitants of history of suicide attempts among violent offenders and impulsive fire setters. *Arch Gen Psychiatry* 46(7): 604–606.

Walther, D. J., Peter, J. U., Bashammakh, S., Hortnagl, H., Voits, M., Fink, H. et al. 2003. Synthesis of serotonin by a second tryptophan hydroxylase isoform. *Science* 299(5603): 76.

Wang, S., Zhang, K., Xu, Y., Sun, N., Shen, Y., and Xu, Q. 2009. An association study of the serotonin transporter and receptor genes with the suicidal ideation of major depression in a Chinese Han population. *Psychiatry Res* 170(2–3): 204–207.

Wasserman, D., Geijer, T., Sokolowski, M., Frisch, A., Michaelovsky, E., Weizman, A. et al. 2007. Association of the serotonin transporter promotor polymorphism with suicide attempters with a high medical damage. *Eur Neuropsychopharmacol* 17(3): 230–233.

Wasserman, D., Sokolowski, M., Rozanov, V., and Wasserman, J. 2008. The *CRHR1* gene: A marker for suicidality in depressed males exposed to low stress. *Genes Brain Behav* 7(1): 14–19.

Weaver, I. C., Cervoni, N., Champagne, F. A., D'Alessio, A. C., Sharma, S., Seckl, J. R. et al. 2004. Epigenetic programming by maternal behavior. *Nat Neurosci* 7(8): 847–854.

Weinshilboum, R. M. and Raymond, F. A. 1977. Inheritance of low erythrocyte catechol-*o*-methyltransferase activity in man. *Am J Hum Genet* 29(2): 125–135.

Wender, P. H., Kety, S. S., Rosenthal, D., Schulsinger, F., Ortmann, J., and Lunde, I. 1986. Psychiatric disorders in the biological and adoptive families of adopted individuals with affective disorders. *Arch Gen Psychiatry* 43(10): 923–929.

Williams, R. B., Marchuk, D. A., Gadde, K. M., Barefoot, J. C., Grichnik, K., Helms, M. J. et al. 2003. Serotonin-related gene polymorphisms and central nervous system serotonin function. *Neuropsychopharmacology* 28(3): 533–541.

Willour, V. L., Chen, H., Toolan, J., Belmonte, P., Cutler, D. J., Goes, F. S. et al. 2009. Family-based association of FKBP5 in bipolar disorder. *Mol Psychiatry* 14(3): 261–268.

Willour, V. L., Seifuddin, F., Mahon, P. B., Jancic, D., Pirooznia, M., Steele, J. et al. 2011. A genome-wide association study of attempted suicide. *Mol Psychiatry*. advance online publication: [doi: 10.1038/mp. 2011.4].

Willour, V. L., Zandi, P. P., Badner, J. A., Steele, J., Miao, K., Lopez, V. et al. 2007. Attempted suicide in bipolar disorder pedigrees: Evidence for linkage to 2p12. *Biol Psychiatry* 61(5): 725–727.

Yanagi, M., Shirakawa, O., Kitamura, N., Okamura, K., Sakurai, K., Nishiguchi, N. et al. 2005. Association of 14-3-3 epsilon gene haplotype with completed suicide in Japanese. *J Hum Genet* 50(4): 210–216.

Yen, S., Shea, M. T., Sanislow, C. A., Skodol, A. E., Grilo, C. M., Edelen, M. O. et al. 2009. Personality traits as prospective predictors of suicide attempts. *Acta Psychiatr Scand* 120(3): 222–229.

Yoon, H. K. and Kim, Y. K. 2009. TPH2 –703G/T SNP may have important effect on susceptibility to suicidal behavior in major depression. *Prog Neuropsychopharmacol Biol Psychiatry* 33(3): 403–409.

Zai, C. C., Manchia, M., De Luca, V., Tiwari, A. K., Chowdhury, N. I., Zai, G. C., et al. 2011. The brain-derived neurotrophic factor gene in suicidal behaviour: A meta-analysis. *Int J Neuropsychopharmacol* 30: 1–6. Advance online publication. doi:10.1017/S1461145711001313.

Zai, C. C., Tiwari, A. K., De Luca, V., Muller, D. J., Bulgin, N., Hwang, R. et al. 2009. Genetic study of BDNF, DRD3, and their interaction in tardive dyskinesia. *Eur Neuropsychopharmacol* 19(5): 317–328.

Zalsman, G., Frisch, A., Lewis, R., Michaelovsky, E., Hermesh, H., Sher, L. et al. 2004. DRD4 receptor gene exon III polymorphism in inpatient suicidal adolescents. *J Neural Transm* 111(12): 1593–1603.

Zalsman, G., Huang, Y. Y., Oquendo, M. A., Brent, D. A., Giner, L., Haghighi, F. et al. 2008. No association of COMT Val158Met polymorphism with suicidal behavior or CSF monoamine metabolites in mood disorders. *Arch Suicide Res* 12(4): 327–335.

Zarrilli, F., Angiolillo, A., Castaldo, G., Chiariotti, L., Keller, S., Sacchetti, S. et al. 2009. Brain derived neurotrophic factor (BDNF) genetic polymorphism (Val66Met) in suicide: A study of 512 cases. *Am J Med Genet B Neuropsychiatr Genet* 150B(4): 599–600.

Zhang, Y., Bertolino, A., Fazio, L., Blasi, G., Rampino, A., Romano, R. et al. 2007. Polymorphisms in human dopamine D2 receptor gene affect gene expression, splicing, and neuronal activity during working memory. *Proc Natl Acad Sci USA* 104(51): 20552–20557.

Zhang, B. and Horvath, S. 2005. A general framework for weighted gene co-expression network analysis. *Stat Appl Genet Mol Biol* 4: Article 17.

Zhang, J., Shen, Y., He, G., Li, X., Meng, J., Guo, S. et al. 2008. Lack of association between three serotonin genes and suicidal behavior in Chinese psychiatric patients. *Prog Neuropsychopharmacol Biol Psychiatry* 32(2): 467–471.

Zhuang, X., Gross, C., Santarelli, L., Compan, V., Trillat, A. C., and Hen, R. 1999. Altered emotional states in knockout mice lacking 5-HT1A or 5-HT1B receptors. *Neuropsychopharmacology* 21(2 Suppl): 52S–60S.

Zill, P., Buttner, A., Eisenmenger, W., Moller, H. J., Bondy, B., and Ackenheil, M. 2004. Single nucleotide polymorphism and haplotype analysis of a novel tryptophan hydroxylase isoform (*TPH2*) gene in suicide victims. *Biol Psychiatry* 56(8): 581–586.

Zill, P., Buttner, A., Eisenmenger, W., Muller, J., Moller, H. J., and Bondy, B. 2009. Predominant expression of tryptophan hydroxylase 1 mRNA in the pituitary: A postmortem study in human brain. *Neuroscience* 159(4): 1274–1282.

Zill, P., Preuss, U. W., Koller, G., Bondy, B., and Soyka, M. 2007. SNP- and haplotype analysis of the tryptophan hydroxylase 2 gene in alcohol-dependent patients and alcohol-related suicide. *Neuropsychopharmacology* 32(8): 1687–1694.

Zouk, H., McGirr, A., Lebel, V., Benkelfat, C., Rouleau, G., and Turecki, G. 2007. The effect of genetic variation of the serotonin 1B receptor gene on impulsive aggressive behavior and suicide. *Am J Med Genet B Neuropsychiatr Genet* 144B(8): 996–1002.

Zouk, H., Tousignant, M., Seguin, M., Lesage, A., and Turecki, G. 2006. Characterization of impulsivity in suicide completers: Clinical, behavioral and psychosocial dimensions. *J Affect Disord* 92(2–3): 195–204.

Zubenko, G. S., Maher, B. S., Hughes, H. B., 3rd, Zubenko, W. N., Scott Stiffler, J., and Marazita, M. L. 2004. Genome-wide linkage survey for genetic loci that affect the risk of suicide attempts in families with recurrent, early-onset, major depression. *Am J Med Genet B Neuropsychiatr Genet* 129B(1): 47–54.

12 Approaches and Findings from Gene Expression Profiling Studies of Suicide

Laura M. Fiori and Gustavo Turecki

CONTENTS

12.1 INTRODUCTION

Studies investigating neurological and biological factors associated with suicidal behaviors have been ongoing for decades. Initial molecular genetic studies largely focused on candidate genes and pathways, and while specific neurobiological alterations have been associated with suicide, we are far from reaching a comprehensive understanding of the underlying pathological processes associated with this complex phenotype. In order to expand its focus, suicide research moved toward high-throughput gene expression microarrays in an effort to identify novel biological pathways and molecular mechanisms associated with suicide. By analyzing tissues obtained from suicide completers and examining the expression of a vast number of genes in parallel, these technologies allow researchers to obtain a functional profile of gene expression, thus providing valuable insight into the overall biological processes underlying suicide. This chapter will first discuss the methodologies that have been used to profile gene expression alterations in suicide, and then will examine several of the neurobiological mechanisms that have been implicated. Following this, future uses of gene expression profiling technologies in suicide research will be discussed.

12.2 TECHNOLOGICAL AND METHODOLOGICAL APPROACHES AND CONSIDERATIONS

Microarray technology was first developed almost two decades ago by researchers at Stanford University, who were able to simultaneously quantify the expression of 45 *Arabidopsis* genes using cDNA printed onto glass slides (Schena et al. 1995). In the years since, microarray technology has made huge advancements and can now be used to measure the expression of thousands of gene transcripts in organisms of all classes of life. This technology quickly became an exciting tool, not only for expanding our understanding of processes underlying cellular function, but as a means to identify pathological gene expression changes occurring in disease and in relation to phenotypes such as suicidal behaviors. At the same time, the ability to analyze and interpret microarray findings has faced many challenges and required the development of new statistical methods, including those to address the fact that gene expression can vary over several orders of magnitude, as well as analytical methods to deal with issues related to multiple testing. Furthermore, strategies for proper experimental design have represented important considerations, particularly given the cost of microarray technology. In spite of these challenges, high-throughput gene expression arrays have become an important resource in suicide research, and 15 studies examining the overall transcriptome in suicide completers have been performed to date. In order to better appreciate the findings arising from high-throughput gene expression studies, an understanding of the laboratory methods and technologies, statistical analysis strategies, and experimental design is required. This section will first give a brief overview of the microarray technologies and statistical approaches that have been employed to assess gene expression patterns related to suicide, followed by a discussion of methods that have been used to validate these findings. Following this, aspects related to sample and tissue selection, as well as their inherent limitations, will be explored.

12.2.1 Microarray Technology and Analysis

To date, all microarray studies examining suicide completers have used one-color microarrays, most commonly platforms produced by Affymetrix, which consist of hundreds of thousands of short (typically 25 base pairs) oligonucleotides.

Biotinylated complementary RNA produced from mRNA extracted from biological samples is hybridized with the array, and gene expression is represented by the intensity of fluorescence at each probe following immunochemical treatment. The levels of mRNA carrying each sequence of interest are interrogated by a set of probes (probe set), and data from each probe are combined to generate an overall expression value for that sequence. Illumina BeadChip microarrays, which have also been used in suicide research, use similar methods, with the main difference being the use of larger probes. In both cases, probes have typically been designed to target the 3′ end of mRNA molecules in order to reduce the impact of quality and processing biases. Following hybridization and measurement of fluorescence, several post-processing steps are performed, including quality assessment, normalization of intensity values across chips, and combination of probe intensity values (Miron and Nadon 2006). Although substantial efforts have been put into the development of suitable computational methods for these steps, there remains considerable debate regarding the most appropriate algorithms (Steinhoff and Vingron 2006). To date, the most commonly used analysis methods have been the Microarray Analysis Suite produced by Affymetrix and Robust Multiarray Average (Irizarry et al. 2003). While a variety of statistical approaches have been used to identify gene expression patterns specific to suicide, these have all been required to address the multiple testing issue arising from the vast amount of data generated in these experiments. Although it is generally accepted that standard methods to correct for multiple testing, such as the Bonferroni correction, are excessively stringent for high-throughput studies, it is understood that failing to sufficiently correct will inevitably produce a large number of false positive results. The most typical methods that have been used in suicide research have involved either prespecified fold change and *P*-value cutoffs or false discovery rate corrections.

Technical validation (i.e., confirming that the technology has properly measured mRNA levels) has also generally been performed following the identification of genes displaying differential expression. The current "gold-standard" method of validation is quantitative real-time polymerase chain reaction (qRT-PCR), although in the past less precise methods such as semiquantitative RT-PCR and immunohistochemistry have also been employed. While initially it was believed that a significant correlation of mRNA expression with protein measurements was also an essential aspect of validation, it has now become better recognized that mRNA and protein levels represent distinct aspects of cellular functioning, and that a lack of correlation between these measurements is more often due to the presence of multiple levels of gene regulation. In addition, molecular processes such as alternative splicing or cell type–specific expression can also result in a failure to validate differences in RNA levels if experiments assess different transcripts from those evaluated by a particular probe set, or use RNA extracted from nonidentical tissue samples.

Beyond technical validation, scientific validation is an important consideration in gene expression studies in suicide and other complex phenotypes, where differences in expression may result from confounding factors that may not be directly related to the phenotype being investigated. Greater confidence in the scientific relevance of gene expression findings can be obtained by replication in independent samples. Resources, such as the Stanley Neuropathology Consortium Integrative

Database (SNCID), which holds a large body of gene expression and other information obtained from a well-characterized psychiatric sample, have allowed researchers to readily determine how well their results could be extended to other populations (Kim and Webster 2010b). Moreover, the wealth of information available for these subjects has allowed the combination of expression data with other neurobiological information (Kim and Webster 2010a,b), thus allowing even greater knowledge regarding the nature of dysregulated expression to be obtained. The preselection of genes or probe sets prior to statistical analysis based on knowledge regarding a particular biological pathway (Lalovic et al. 2010; Morita et al. 2005) or chromosomal region (Fiori 2009) has also been used as a means to increase the likelihood that positive results represent biologically relevant findings. In addition to protein measurements, the biological importance of specific genes in suicide has been further investigated through genetic association studies (Sequeira et al. 2006; Yanagi et al. 2005).

12.2.2 Experimental Design

Sample selection represents the most important consideration in microarray studies, as this will determine the ability to detect small magnitude differences in gene expression, as well as allow these results to be properly interpreted in reference to specific research questions. Both the size of the overall sample and the number of samples within each experimental group are essential aspects of experimental design when attempting to identify differential gene expression in general. In order to determine how meaningful and relevant the results are to suicide, both the method by which experimental groups are defined and the ability to identify and address confounding variables play essential roles. Finally, the tissue that is examined will also partially determine which biological pathways are found to be differentially expressed in relation to suicide.

As a result of both the cost of microarray technologies and issues inherent in sample recruitment, sample sizes have generally been small, ranging from 6 (Yanagi et al. 2005) to 90 (Kim et al. 2007) subjects overall, with generally between 10 and 15 subjects within each experimental group. This has posed a significant problem when attempting to identify gene expression changes specific to suicide, as suicide completers represent a heterogeneous group, particularly in terms of psychiatric diagnosis, drug and alcohol use, and gender. Differentiating the effects of suicide from those related to comorbid psychiatric disorders has required careful selection of experimental groups and has typically been approached in one of three ways:

1. Grouping together all suicides irrespective of diagnosis and comparing them to a non-suicide control group
2. Grouping suicide completers within Axis I disorders (particularly major depression, bipolar disorder, or schizophrenia) and then separately comparing each group to the control group
3. Comparing, within a group of subjects with specific Axis I disorders, individuals who died by suicide with those who died of other causes

Confounding factors, including the use of alcohol, medication, or other drugs, are an inherent concern in studies of psychiatric disorders, and have the potential to influence the interpretability of findings in terms of suicide. Unfortunately, these issues have only been addressed in a few studies: approaches have included drug or alcohol use as a covariate in statistical analyses (Kim et al. 2007; Klempan et al. 2009c; Sequeira et al. 2007, 2009), performing a gene expression study in a separate sample of alcohol abusers (Sequeira et al. 2009), and examining the effects of drugs in animals in order to assess their potential influence on gene expression patterns in humans (Ernst et al. 2009; Sequeira et al. 2009). Gender also represents a confounding variable that has not been well addressed. Although gender plays a significant role in the susceptibility and presentation of suicidal behaviors, very few studies have been able to assess the effects of gender, largely as a consequence of difficulties in recruiting female samples due to the much lower rate of suicide completion in females. Accordingly, studies have either typically used exclusively male samples in order to avoid gender-specific effects or examined mixed-gender samples in which the numbers of female subjects have been insufficient to fully examine the influence of gender.

Finally, the selection of the biological sample to examine is also an important factor to consider when designing studies to address specific research questions. To date, all studies examining suicide have used postmortem brain tissues. Given their implication in psychiatric disorders, studies have largely focused on tissues obtained from the prefrontal cortex (Brodmann areas [BA] 8, 9, 10, 11, 44, 45, 46, and 47) or limbic areas (amygdala, hippocampus, BA 24, and BA 29). More recently additional regions, including the motor cortex, temporal cortex, thalamus, hypothalamus, and nucleus accumbens, have also been examined (Ernst et al. 2009; Sequeira et al. 2009). The quality of the brain tissue is also a source of confounding variables. Tissue pH, which can be influenced by both antemortem and postmortem factors, can affect RNA quality, as well as have specific effects on gene expression, including those implicated in psychiatric disorders (Vawter et al. 2006). In addition to being influenced by pH, RNA quality can be influenced by other postmortem variables and laboratory-specific factors, such as sample storage and RNA extraction methods, and must be carefully monitored as the RNA degradation state can directly impact gene expression measurements.

12.3 FINDINGS ARISING FROM MICROARRAY STUDIES

When microarray studies were first undertaken as a means to investigate psychiatric phenotypes, it was believed that they would both confirm and expand on information regarding genes and pathways previously implicated in psychiatry, as well as identify new systems that are involved in their pathology. Interestingly, the majority of microarray studies examining suicide have largely failed to identify altered expression of genes related to the expected pathways. Rather, the microarray studies to date have highlighted many pathways that were previously not suspected to be involved in the neurobiology of suicide, including additional systems related to neurotransmission, stress response, and cellular functioning. Although difficulties in obtaining sufficient numbers of non-suicide psychiatric controls have made it difficult to extricate the gene expression changes associated with suicidal behaviors from those pertaining to underlying psychiatric disorders, the consistency of many findings across samples from different research groups has provided good support for their involvement in suicide.

12.3.1 Neurotransmission

Synaptic transmission in the central nervous system (CNS) underlies much of what makes us who we are and is the main site of the action of the psychopharmaceutical agents currently in use. Consequently, monoaminergic neurotransmission has been the most extensively studied system in suicide research. Interestingly, however, although numerous studies over the last few decades have consistently identified altered functioning of monoaminergic systems in suicide completers, microarray studies have provided only minimal evidence for the dysregulation of genes involved in serotonergic or noradrenergic neurotransmission in suicide completers. Instead, gene expression studies have highlighted the involvement of the glutamatergic and γ-aminobutyric acid (GABA)-ergic neurotransmitter systems, as well as genes that regulate neurotransmission in general. Interestingly, altered expression of genes related to GABA and glutamate signaling are among the strongest findings arising from microarray studies examining suicide, strongly emphasizing the importance of these two pathways in the pathology of suicide.

12.3.1.1 Glutamate

Glutamate is the primary excitatory neurotransmitter in the brain and acts at four classes of receptors: the ionotropic α-amino-3-hydroxy-5-hydroxy-5-methyl-4-isoxazolepropionate (AMPA), kainate, and *N*-methyl-D-aspartate (NMDA) receptors, as well as the metabotropic glutamate receptors (Conn and Pin 1997; Dingledine et al. 1999). Glutamate is synthesized from either glucose obtained from the tricarboxylic acid (TCA) cycle or from glutamine, which is synthesized by glial cells and taken up by neurons (Daikhin and Yudkoff 2000). Glutamate transmission is terminated by reuptake into neurons or astrocytes, which then convert it to glutamine to be resynthesized into glutamate by the enzyme glutaminase (*GLS*) (Daikhin and Yudkoff 2000).

A number of receptor binding studies have been used to examine glutamate transmission in the brains of suicide completers, but have generated largely negative findings. However, alterations in genes related to glutamatergic signaling have consistently emerged from microarray studies, providing strong evidence for a role of this system in suicide. Microarray studies have identified alterations in the levels of several glutamate receptors, including NMDA-like receptor 1A (*GRINL1A*), NMDA receptor 2A (*GRIN2A*), AMPA receptors 1–4 (*GRIA1*, *GRIA2*, *GRIA3*, and *GRIA4*), metabotropic glutamate receptor 3 (*GRM3*), and kainate receptor 1 (*GRIK1*) (Klempan et al. 2009c; Sequeira et al. 2009; Thalmeier et al. 2008). Additionally, downregulated expression has been observed for glutamate–ammonia ligase (glutamine synthetase) (*GLUL*), the enzyme responsible for removing glutamate from synapses, *GLS*, and glial high-affinity glutamate transporters *SLC1A2* and *SLC1A3* (Kim et al. 2007; Klempan et al. 2009c; Sequeira et al. 2009). *GRIA3* is particularly interesting as it has been associated with a number of psychiatric conditions including bipolar disorder, schizophrenia, and citalopram treatment–emergent suicidal ideation (Gécz et al. 1999; Laje et al. 2007; Magri et al. 2008; O'Connor et al. 2007). Interestingly, several of these genes are localized to glia, which lends support for the involvement of dysregulated astroglial functioning in suicide, which will be discussed in a later section.

12.3.1.2 γ-Aminobutyric Acid

GABA is the primary inhibitory neurotransmitter in the CNS and plays many important roles, including cortical development, synaptic plasticity, neurogenesis, and stress responses (Ge et al. 2007; Li and Xu 2008; Nugent and Kauer 2008; Radley et al. 2009). GABA acts upon two classes of receptors: ionotropic $GABA_A$ receptors and metabotropic $GABA_B$ receptors. The metabolism of GABA is intricately tied to that of glutamate, which is the precursor for GABA synthesis by glutamic acid decarboxylase (*GAD*). Additionally, following its release into synapses, GABA is transported into astrocytes and converted to glutamine (Bak et al. 2006).

Although many studies have investigated GABA levels as well as the number and function of GABA receptors, the overall results have been inconsistent. Nonetheless, dysregulated expression of GABAergic genes has been among the most consistent findings arising from microarray studies of suicide completers. Altered expression of numerous GABA receptor subunits have been observed across prefrontal and limbic brain regions, including $GABA_A$, α1 (GABRA1), $GABA_A$, α4 (*GABRA4*), $GABA_A$, α5 (*GABRA5*), $GABA_A$, β1 (*GABRB1*), $GABA_A$, β3 (*GABRB3*), $GABA_A$, δ (*GABRD*), $GABA_A$, γ-1 and γ-2 (*GABRG*1 and *GABRG2*), $GABA_B$, β2 (*GABBR2*), and $GABA_C$, ρ1 (*GABRR1*), as well as $GABA_A$ receptor-associated protein-like 1 (*GABARAPL1*), and a GABA transporter (*SLC6A1*) (Choudary et al. 2005; Kim et al. 2007; Klempan et al. 2009c; Sequeira et al. 2007, 2009). Additionally, one study from our group found that 16% and 36% of the probe sets annotated as involved in GABAergic signaling were significantly differentially expressed in suicide completers in BA 44 and 46, respectively (Klempan et al. 2009c). Microarray findings have been reinforced by candidate gene expression studies, which have identified significant decreases in the expression of several $GABA_A$ receptor subunits in the frontopolar region of suicide completers, and interestingly, there appear to be differences in the interrelations between different $GABA_A$ subunits across the brain of depressed suicide completers relative to controls (Merali et al. 2004; Poulter et al. 2010).

12.3.1.3 Synaptic Structure and Function

In addition to factors related to specific neurotransmitter systems, synaptic transmission is highly regulated by presynaptic factors controlling the vesicle-mediated release of neurotransmitters. Synapse-related proteins have many different roles, including the docking and fusion of synaptic vesicles with the cellular membrane, as well as vesicle packaging and recycling. Considerable evidence is now emerging implicating this system in suicide. Altered expression of numerous synapse-related genes have been identified in suicide completers, including vesicle-associated membrane protein 3 (*VAMP3*), synaptotagmin I (*SYT1*), synaptotagmin IV (*SYT4*), synaptotagmin V (SYT5), synaptotagmin XIII (*SYT13*), synaptophilin (*SNPH*), synaptophysin-like protein (*SYPL*), synapsin II (*SYN2*), synaptosomal-associated protein (23 kDa) (*SNAP23*), synaptosomal-associated protein (25 kDa) (SNAP25), synaptosomal-associated protein (29 kDa) (*SNAP29*), synaptic vesicle glycoprotein 2B (*SV2B*), and synaptopodin 2 (*SYNPO2*) (Klempan et al. 2009c; Sequeira et al. 2006, 2007, 2009). These findings may partially explain how monoaminergic neurotransmission may be altered in suicide completers when the processes involved in monoamine metabolism and reception remain intact.

12.3.2 Stress Systems

One of the most widely used models to conceptualize risk for mental disorders, which has also been adopted in suicide research, is based on the notion of a stress–diathesis interaction. In the case of suicidal behaviors, this model assumes that suicide results from the combination of stressors and predisposing factors. Molecular systems involved in stress response have thus been investigated for their roles in suicide, as they are intricately tied to both stress and predisposition. The two major stress systems in humans include the autonomic nervous system, which is largely influenced by factors influencing the catecholamines, and the hypothalamic–pituitary–adrenal axis system. Although metabolic analyses and candidate gene studies have provided evidence to support a role for these systems in suicide, relatively few related genes have been identified in microarray studies. In contrast, microarray studies highlighted the importance of a third stress pathway, the polyamine system, whose role in suicide had not been previously suspected.

12.3.2.1 Polyamines

The polyamines are ubiquitous aliphatic molecules comprising agmatine, putrescine, spermidine, and spermine. The polyamine system has been identified in all organisms and plays an important role in numerous essential cellular functions, including growth, division, and signaling cascades, as well as stress responses at both the cellular and behavioral levels (Gilad and Gilad 2003; Minguet et al. 2008; Rhee et al. 2007; Seiler and Raul 2005; Tabor and Tabor 1984). Its involvement in schizophrenia and behavioral stress responses has been investigated for several decades (Fiori and Turecki 2008); however, it was not suspected to be involved in suicide until microarray studies identified spermidine/spermine *N*1-acetyltransferase (*SAT1*) as one of the genes displaying the strongest and most consistently altered expression in suicide completers (Sequeira et al. 2006). Evidence for its downregulation in suicide completers has now been extended to additional brain regions and populations, and dysregulated expression of other polyamine-related genes, including spermine synthase (*SMS*), spermine oxidase (*SMOX*), ornithine decarboxylase antizymes 1 and 2 (*OAZ1*, *OAZ2*), *S*-adenosylmethionine decarboxylase (*AMD1*), arginase II (*ARG2*), and ornithine aminotransferase-like 1 (*OATL1*), has also been identified (Fiori et al. 2011; Guipponi et al. 2009; Klempan et al. 2009b,c; Sequeira et al. 2007).

Although the mechanism by which altered polyamine levels can influence risk for suicide or other psychiatric disorders is not yet clear, several mechanisms have been proposed. First, the polyamines can influence neurotransmission through several systems, including the catecholamines (Bastida et al. 2007; Bo et al. 1990; Hirsch et al. 1987; Ritz et al. 1994), glutamate (Williams 1997), GABA (Brackley et al. 1990; Gilad et al. 1992; Morgan and Stone 1983), and nitric oxide (Galea et al. 1996), each of which may be involved in psychiatric disorders. Agmatine itself is believed to act as a neurotransmitter through imidazoline receptors, α2-adrenoceptors, nicotinic acetylcholine receptors, and serotonin 3 receptors: this theory is supported by its storage in synaptic vesicles and capacity to be released upon depolarization (Reis and Regunathan 2000). Interestingly, alterations in imidazoline receptor binding sites have also been implicated in depression and anxiety, and their importance in modulating behavioral stress responses has become accepted (Halaris and Piletz 2003, 2007; Piletz et al. 2000). Second, the importance of neurogenesis in the development and treatment of psychiatric disorders has been recognized (Dranovsky and Hen 2006; Jacobs et al. 2000), and it is therefore of great interest that manipulation of the polyamine system produces significant effects on both CNS development and adult brain neurogenesis (Malaterre et al. 2004; Seiler 1981), and that variations in polyamine levels are associated with both pro- and anti-apoptotic effects in neuronal cells (Harada and Sugimoto 1997; Sparapani et al. 1997). Several studies have found relationships connecting the neurogenic and neuroprotective effects of the polyamines with stress and depression (Li et al. 2006; Zhu et al. 2007, 2008), providing evidence that dysregulation of the polyamine system may act in part through these mechanisms. Finally, both agmatine and putrescine, as well as the polyamine precursor *S*-adenosylmethionine, demonstrate anxiolytic and antidepressant effects in animal and human studies (Bressa 1994; Gong et al. 2006; Lavinsky et al. 2003; Zeidan et al. 2007; Zomkowski et al. 2002, 2004, 2005, 2006), which could be particularly relevant for suicide completers with comorbid mood or anxiety disorders.

12.3.3 Cellular Function

Both synaptic neurotransmission and stress response pathways rely heavily on proper cellular functioning, and both the systems described earlier, as well as many other neurobiological pathways implicated in suicide, can be influenced by pathological alterations to cellular processes. Indeed, microarray studies have demonstrated alterations in several essential components of cellular functioning, including growth factor signaling and energy metabolism. Moreover, these studies have provided strong evidence for cell type–specific alterations in gene expression, pointing toward altered glial functioning as an important pathological component underlying suicidal behavior.

12.3.3.1 Growth Factors

Growth factors are cellular signaling molecules that play essential roles in proliferation, differentiation, and survival (Pawson 1994). Growth factor receptors are found at the cell surface, and following binding of their ligands, they dimerize and activate the tyrosine kinase activity of their intracellular domain, resulting in autophosphorylation that allows the intracellular domains to interact with proteins involved in various second messenger systems (Pawson 1994). Microarray studies have provided evidence for altered expression of genes associated with two classes of growth factors: the neurotrophins and fibroblast growth factor (FGF).

The neurotrophin brain-derived neurotrophic factor (BDNF) plays an important role in neuronal growth and survival through its binding to the receptor neurotrophic tyrosine kinase, type 2 (*NTRK2*). Numerous studies have implicated this pathway in depression (Brunoni et al. 2008; Chen et al. 2001; Sen et al. 2008), and evidence from microarray studies has suggested that it also plays a role in suicide. Although gene expression of BDNF itself has not been found to be altered in suicide completers, decreased expression of *NTRK2* has been observed in several brain regions (Ernst et al. 2009; Kim et al. 2007; Sequeira et al. 2007). Interestingly, this decrease appears to be due to the downregulation of one specific *NTRK2* isoform, TrkB.T1, which is expressed exclusively in astrocytes (Ernst et al. 2009).

Alterations in the expression of components of FGF signaling pathways have also been consistently observed in suicide. There are over 20 FGF ligands in humans, which act upon four different tyrosine kinase receptors (*FGFR1–4*), where they play important roles in cellular proliferation and differentiation during development, as well as in neuronal signal transduction into adulthood (Ornitz and Itoh 2001). Several microarray studies have identified altered expression of genes involved in FGF signaling, including *FGFR2* and *FGFR3*, in suicide completers (Ernst et al. 2008; Kim et al. 2007). Similar to BDNF, many components of the FGF system, including *FGFR2* and *FGFR3*, show glial-specific expression.

12.3.3.2 Energy Metabolism

Processes regulating the production and usage of adenosine triphosphate (ATP) are essential for maintaining adequate energy stores for cellular functions. It is thus of interest that altered expression of several ATP-related genes has been observed in microarray studies of suicide completers. One study found that TCA cycle and ATP-related genes were significantly associated with gene expression differences between depressed and non-depressed suicide completers, and suggested that some defects may be markers for suicidal behavior in the context of depression (Klempan et al. 2009c). Another study identified downregulated expression of the Na^+/K^+-ATPase α3 subunit (*ATP1A3*) in the prefrontal cortex of suicide completers with diagnoses of depression, bipolar disorder, or schizophrenia, indicating that altered energy metabolism in suicide completers is not exclusive to depression (Tochigi et al. 2008). It is also of interest that recent microarray studies by our group examining mood-disordered suicide completers found altered expression of four genes related to creatine metabolism, another important source of high-energy phosphate groups (Fiori et al. 2011).

12.3.3.3 Glial Cells

Glial cells play many distinct roles in the CNS, including the development and maintenance of the nervous system, processing of synaptic transmission, regulation of cerebral blood flow, and immune responses in the brain (Araque 2008; Gehrmann et al. 1995; Koehler et al. 2009; Pfrieger 2009). Although traditional research in suicide and other psychiatric disorders largely ignored glial cells, histopathological studies have demonstrated alterations in glial cell densities and numbers in psychiatric disorders, and microarray studies demonstrating altered expression of glial-specific genes have provided further evidence to support a role for altered glial functioning in suicide.

Astrocytes form the largest group of cells in the CNS and have multiple functions, including roles in the synthesis, release and uptake of neurotransmitters, development of synapses, and formation of the blood–brain barrier (Fiacco et al. 2009; Montana et al. 2006; O'Kusky and Colonnier 1982; Stevens 2008). Many microarray studies have supported the involvement of dysregulated astrocyte functioning in suicide and have identified altered expression of several astrocyte-specific genes including *FGFR2*, *FGFR3*, *GLUL*, S100 calcium-binding protein beta (*S100B*), and *TrkB.T1* (Ernst et al. 2009; Kim et al. 2007; Klempan et al. 2009c; Sequeira et al. 2009). Given the important role of astrocytes in glutamate neurotransmission, it has been proposed that they may also play a role in the alterations of this system in suicide (Sequeira et al. 2009).

While the best-known function of oligodendrocytes is their role in axon myelination, they also participate in neurotransmission, synaptic function, neuronal development, and neuronal survival (Deng and Poretz 2003). Oligodendrocyte-specific genes have demonstrated altered expression in brains of suicide completers, including membrane glycoprotein M6B (*GPM6B*), *S100B*, and quaking homolog, KH domain RNA binding (mouse) (*QKI*) (Fiori et al., 2011; Klempan et al. 2009a,c). Interestingly, *GPM6B* was recently shown to interact with and alter the surface expression of the serotonin transporter, indicating the importance of oligodendrocytes in monoaminergic neurotransmission (Fjorback et al. 2009). Furthermore, given the importance of myelination in neural transmission, cognition, and brain development, it is not surprising that alterations in oligodendrocyte function may be associated with suicide (Fields 2008).

12.4 FUTURE AVENUES FOR GENE EXPRESSION STUDIES

Although our understanding of the neurobiology of suicide has greatly increased within the last few decades, we are still far from developing an integrated picture of how specific biological and neurochemical alterations interact to confer risk for suicidal behaviors and how this knowledge can be used to chemically treat these behaviors. The majority of studies performed to date have focused on genes and proteins of interest, and while these studies continue to be important, they are not sufficient to address these issues. Microarrays have played an essential role in highlighting new molecular pathways involved in suicidal behavior, yet cannot explain either the origin of these alterations or the nature of their short- and long-term effects on brain function. However, continued advances in the fields of both molecular biology and high-throughput technologies can provide the means to answer these questions. By combining gene expression data with information obtained through other measures, such as clinical, epigenetic, genomic, proteomic, or metabolic studies, a more comprehensive view of these processes will be possible. Also, by using more powerful technologies, such as RNA-Seq—next-generation sequencing of the transcriptome—we will be able to gain more precise insight into gene expression profiles associated with the suicide process. In addition, continued sample collection and collaborations between research groups will allow for a greater capacity to investigate the effects of variables such as medication, gender, and the environment, as well as to properly differentiate the effects of suicide from those of Axis I disorders.

12.4.1 Gene Function

Although it is clear that specific alterations in gene expression underlie the suicide process, the functional impact of these alterations remains largely unknown and represents an essential step in understanding the pathological effects of gene expression differences. Several methods are now available to better characterize these effects. As mentioned earlier, gene expression microarray studies of suicide completers have used arrays that are primarily designed to measure the expression of the 3′ end of mRNAs. However, many newer arrays have been developed that allow the expression of each exon within a gene to be quantified. As the functional importance and diversity of alternative splicing has become well recognized over the last decade, the use of exon arrays is an important next step in suicide research. These arrays will allow the identification of genes that show specific splicing differences, thereby improving our understanding of how the function of these genes may be altered in suicide. Similarly, as mentioned earlier, the advance in sequencing methods now allow for the implementation of techniques to directly sequence and quantify the transcriptome, and techniques such as RNA-Seq are powerful alternatives to microarray studies. In addition, although high-throughput methods to examine the proteome have not advanced as quickly as other fields, the technology continues to be improved. In the future, integrating mRNA expression with protein expression data will allow for a better understanding of the functional effects of alterations in gene expression, and how they may ultimately lead to suicide. Finally, while neuroimaging techniques and knowledge regarding the cell type–specific expression of genes have been invaluable in studying the roles of neurons and glia in suicide, they represent somewhat crude approaches for analyzing gene expression differences between different cell types. Protocols for laser capture microdissection have now been optimized for postmortem brain tissues (Pietersen et al. 2009), which will be invaluable in better characterizing the specific roles played by neurons and glia in suicide.

12.4.2 Gene Regulation

Identifying mechanisms involved in gene regulation is also an important step in understanding how pathological changes in gene expression arise; moreover, they can represent potential molecular targets for the development of new treatments. By integrating gene expression data with that obtained from platforms assessing single nucleotide polymorphisms (SNPs) across the genome, essential information regarding the regulation of gene expression can be obtained. Genetic regions that are associated with gene expression differences may represent specific functional genetic variants, or may be due to larger chromosomal alterations, known as copy number variations (CNV), which have been associated with psychiatric disorders including bipolar disorder, schizophrenia, and autism (Cook and Scherer 2008). Epigenetic modifications are also important regulators of gene expression.

These modifications can be environmentally influenced, particularly through physiological and behavioral stressors, and it is believed that they may mediate the interaction between the genome and the environment in conferring risk for suicide (Tsankova et al. 2007). Two important epigenetic modifications that have been examined in the context of suicide are DNA methylation and posttranslational histone modifications. Although the epigenetic studies of suicide to date have only focused on genes of interest, examination at a larger scale is now possible using comparative hybridization arrays, large-scale, genome-based deep sequencing, and chromatin immunoprecipitation-based methods. MicroRNAs represent a third mechanism for the epigenetic modification of gene expression. These short, single-stranded RNA molecules bind to specific mRNA molecules and target them for degradation, and have been implicated in other psychiatric conditions (Abu-Elneel et al. 2008; Chen et al. 2009; Hansen et al. 2007). Examination of each of these three epigenetic mechanisms using high-throughput methods, in conjunction with gene expression microarrays, will allow the systematic investigation of epigenetic effects that may be involved in suicide risk. Furthermore, as epigenetic markings are potentially reversible, this is an exciting area of investigation that creates the opportunity for therapeutic intervention.

14.4.3 Other Considerations

Given the large gender differences in the rates and presentations of suicidal behaviors, gender is an essential consideration in studies of suicide. To address this issue, sample recruitment practices and statistical analyses must be adjusted in order to identify gender-specific factors. Additionally, there is a need for increased sample sizes as well as the availability of well characterized and appropriately selected control groups in order to enable the identification of phenotype-specific gene expression changes—a particularly important challenge in neurobiological studies of suicide. Finally, developing a more comprehensive view of the means by which psychopharmacological agents affect gene expression is an important step in understanding the mechanisms by which they exert their therapeutic effects, which will greatly assist in the development of better treatments directed toward suicidal behaviors.

12.5 CONCLUSIONS

Genomic gene expression profiling has provided an invaluable tool for obtaining a global view of gene expression changes underlying the suicide process, and allowed us to widen our focus beyond the pathways classically studied in suicidal behaviors. While the general lack of evidence supporting alterations in these classical pathways is perplexing, this is a strong indication that gene expression differences represent only one facet of the neurobiological alterations occurring in this complex phenotype. In the future, by integrating gene expression data with that interrogating other molecular and clinical variables, we will be able to obtain a more thorough understanding of the underlying mechanisms involved in the etiology and pathology of suicide, which may ultimately lead toward improved methods to treat and prevent suicidal behaviors.

REFERENCES

Abu-Elneel, K., Liu, T., Gazzaniga, F.S., Nishimura, Y., Wall, D.P., Geschwind, D.H., Lao, K., Kosik, K.S. 2008. Heterogeneous dysregulation of microRNAs across the autism spectrum. *Neurogenetics* 9:153–161.

Araque, A. 2008. Astrocytes process synaptic information. *Neuron Glia Biology* 4:3–10.

Bak, L.K., Schousboe, A., Waagepetersen, H.S. 2006. The glutamate/GABA–glutamine cycle: Aspects of transport, neurotransmitter homeostasis and ammonia transfer. *Journal of Neurochemistry* 98:641–653.

Bastida, C.M., Cremades, A., Castells, M.T., López-Contreras, A.J., López-García, C., Sánchez-Mas, J., Peñafiel, R. 2007. Sexual dimorphism of ornithine decarboxylase in the mouse adrenal: Influence of polyamine deprivation on catecholamine and corticoid levels. *American Journal of Physiology. Endocrinology and Metabolism* 292:E1010–E1017.

Bo, P., Giorgetti, A., Camana, C., Savoldi, F. 1990. EEG and behavioural effects of polyamines (spermine and spermidine) on rabbits. *Pharmacological Research* 22:481–491.

Brackley, P., Goodnow, R., Jr., Nakanishi, K., Sudan, H.L., Usherwood, P.N. 1990. Spermine and philanthotoxin potentiate excitatory amino acid responses of *Xenopus* oocytes injected with rat and chick brain RNA. *Neuroscience Letters* 114:51–56.

Bressa, G.M. 1994. S-Adenosyl-L-methionine (SAMe) as antidepressant: Meta-analysis of clinical studies. *Acta Neurologica Scandinavica Supplement* 154:7–14.

Brunoni, A.R., Lopes, M., Fregni, F. 2008. A systematic review and meta-analysis of clinical studies on major depression and BDNF levels: Implications for the role of neuroplasticity in depression. *International Journal of Neuropsychopharmacology* 11:1169–1180.

Chen, B., Dowlatshahi, D., MacQueen, G.M., Wang, J.F., Young, L.T. 2001. Increased hippocampal BDNF immunoreactivity in subjects treated with antidepressant medication. *Biological Psychiatry* 50:260–265.

Chen, H., Wang, N., Burmeister, M., McInnis, M.G. 2009. MicroRNA expression changes in lymphoblastoid cell lines in response to lithium treatment. *International Journal of Neuropsychopharmacology* 12:975–981.

Choudary, P.V., Molnar, M., Evans, S.J., Tomita, H., Li, J.Z., Vawter, M.P., Myers, R.M., Bunney, W.E., Jr., Akil, H., Watson, S.J., Jones, E.J. 2005. Altered cortical glutamatergic and GABAergic signal transmission with glial involvement in depression. *Proceedings of the National Academy of Sciences of the USA* 102(43):15653–15658.

Conn, P.J., Pin, J.P. 1997. Pharmacology and functions of metabotropic glutamate receptors. *Annual Review of Pharmacology and Toxicology* 37:205–237.

Cook, E.H., Scherer, S.W. 2008. Copy-number variations associated with neuropsychiatric conditions. *Nature* 455:919–923.

Daikhin, Y., Yudkoff, M. 2000. Compartmentation of brain glutamate metabolism in neurons and glia. *Journal of Nutrition* 130:1026S–1031S.

Deng, W., Poretz, R.D. 2003. Oligodendroglia in developmental neurotoxicity. *Neurotoxicology* 24:161–178.

Dingledine, R., Borges, K., Bowie, D., Traynelis, S.F. 1999. The glutamate receptor ion channels. *Pharmacological Reviews* 51:7–61.

Dranovsky, A., Hen, R. 2006. Hippocampal neurogenesis: Regulation by stress and antidepressants. *Biological Psychiatry* 59:1136–1143.

Ernst, C., Bureau, A., Turecki, G. 2008. Application of microarray outlier detection methodology to psychiatric research. *BMC Psychiatry* 8:29.

Ernst, C., Deleva, V., Deng, X., Sequeira, A., Pomarenski, A., Klepman, T., Ernst, N., Quirion, R., Gratton, A., Szyf, M., Turecki, G. 2009. Alternative splicing, methylation state, and expression profile of tropomyosin-related kinase B in the frontal cortex of suicide completers. *Archives of General Psychiatry* 66:22–32.

Fiacco, T.A., Agulhon, C., McCarthy, K.D. 2009. Sorting out astrocyte physiology from pharmacology. *Annual Review of Pharmacology and Toxicology* 49:151–174.

Fields, R.D. 2008. White matter in learning, cognition and psychiatric disorders. *Trends in Neurosciences* 31:361–370.

Fiori, L.M., Turecki, G. 2008. Implication of the polyamine system in mental disorders. *Journal of Psychiatry and Neuroscience* 33:102–110.

Fiori, L.M., Zouk, H., Himmelman, C., Turecki, G. 2011. X Chromosome and suicide. *Molecular Psychiatry* 16:216–226.

Fjorback, A.W., Muller, H.K., Wiborg, O. 2009. Membrane glycoprotein M6B interacts with the human serotonin transporter. *Journal of Molecular Neuroscience* 37:191–200.

Galea, E., Regunathan, S., Eliopoulos, V., Feinstein, D.L., Reis, D.J. 1996. Inhibition of mammalian nitric oxide synthases by agmatine, an endogenous polyamine formed by decarboxylation of arginine. *Biochemical Journal* 316: 247–249.

Ge, S., Pradhan, D.A., Ming, G.L., Song, H. 2007. GABA sets the tempo for activity-dependent adult neurogenesis. *Trends in Neurosciences* 30:1–8.

Gécz, J., Barnett, S., Liu, J., Hollway, G., Donnelly, A., Eyre, H., Eshkevari, H.S., Baltazar, R., Grunn, A., Nagaraja, R., Gilliam, C., Peltonen, L., Sutherland, G.R., Baron, M., Mulley, J.C. 1999. Characterization of the human glutamate receptor subunit 3 gene (*GRIA3*), a candidate for bipolar disorder and nonspecific X-linked mental retardation. *Genomics* 62:356–368.

Gehrmann, J., Matsumoto, Y., Kreutzberg, G.W. 1995. Microglia: Intrinsic immuneffector cell of the brain. *Brain Research Reviews* 20:269–287.

Gilad, G.M., Gilad, V.H. 2003. Overview of the brain polyamine-stress-response: Regulation, development, and modulation by lithium and role in cell survival. *Cellular and Molecular Neurobiology* 23:637–649.

Gilad, G.M., Gilad, V.H., Wyatt, R.J. 1992. Polyamines modulate the binding of $GABA_A$-benzodiazepine receptor ligands in membranes from the rat forebrain. *Neuropharmacology* 31:895–898.

Gong, Z.H., Li, Y.F., Zhao, N., Yang, H.J., Su, R.B., Luo, Z.P., Li, J. 2006. Anxiolytic effect of agmatine in rats and mice. *European Journal of Pharmacology* 550:112–116.

Guipponi, M., Deutsch, S., Kohler, K. 2009. Genetic and epigenetic analysis of *SSAT* gene dysregulation in suicidal behavior. *American Journal of Medical Genetics B. Neuropsychiatric Genetics* 150B:799–807.

Halaris, A., Piletz, J.E. 2003. Relevance of imidazoline receptors and agmatine to psychiatry: A decade of progress. *Annals of the New York Academy of Sciences* 1009:1–20.

Halaris, A., Plietz, J. 2007. Agmatine: Metabolic pathway and spectrum of activity in brain. *CNS Drugs* 21:885–900.

Hansen, T., Olsen, L., Lindow, M., Jakobsen, K.D., Ulum, H., Jonnson, E., Andreassen, O.A., Djurovic, S., Melle, I., Agartz, I., Hall, H., Timm, S., Wang, A.G., Werge, T. 2007. Brain expressed microRNAs implicated in schizophrenia etiology. *PLoS ONE* 2:e873.

Harada, J., Sugimoto, M. 1997. Polyamines prevent apoptotic cell death in cultured cerebellar granule neurons. *Brain Research* 753:251–259.

Hirsch, S.R., Richardson-Andrews, R., Costall, B., Kelly, M.E., de Belleroche, J., Naylor, R.J. 1987. The effects of some polyamines on putative behavioural indices of mesolimbic versus striatal dopaminergic function. *Psychopharmacology (Berlin)* 93:101–104.

Irizarry, R.A., Hobbs, B., Collin, F., Beazer-Barclay, Y.D., Antonellis, K.J., Scherf, U., Speed, T.P. 2003. Exploration, normalization, and summaries of high density oligonucleotide array probe level data. *Biostatistics* 4:249–264.

Jacobs, B.L., Van Praag, H., Gage, F.H. 2000. Adult brain neurogenesis and psychiatry: A novel theory of depression. *Molecular Psychiatry* 5:262–269.

Kim, S., Choi, K.H., Baykiz, A.F., Gershenfeld, H.K. 2007. Suicide candidate genes associated with bipolar disorder and schizophrenia: An exploratory gene expression profiling analysis of post-mortem prefrontal cortex. *BMC Genomics* 8:413.

Kim, S., Webster, M.J. 2010a. Correlation analysis between genome-wide expression profiles and cytoarchitectural abnormalities in the prefrontal cortex of psychiatric disorders. *Molecular Psychiatry* 15:326–336.

Kim, S., Webster, M.J. 2010b. The Stanley Neuropathology Consortium Integrative Database: A novel, web-based tool for exploring neuropathological markers in psychiatric disorders and the biological processes associated with abnormalities of those markers. *Neuropsychopharmacology* 35:473–482.

Klempan, T.A., Ernst, C., Deleva, V., Labonte, B., Turecki, G. 2009a. Characterization of *QKI* gene expression, genetics, and epigenetics in suicide victims with major depressive disorder. *Biological Psychiatry* 66:824–831.

Klempan, T.A., Rujescu, D., Mérette, C., Himmelman, C., Sequeira, A., Canetti, L, Fiori, L.M., Schneider, B., Bureau, A., Turecki, G. 2009b. Profiling brain expression of the spermidine/spermine *N*(1)-acetyltransferase 1 (*SAT1*) gene in suicide. *American Journal of Medical Genetics B. Neuropsychiatric Genetics* 150B:934–943.

Klempan, T.A., Sequeira, A., Canetti, L., Lalovic, A., Ernst, C., ffrench-Mullen, J., Turecki, G. 2009c. Altered expression of genes involved in ATP biosynthesis and GABAergic neurotransmission in the ventral prefrontal cortex of suicides with and without major depression. *Molecular Psychiatry* 14:175–189.

Koehler, R.C., Roman, R.J., Harder, D.R. 2009. Astrocytes and the regulation of cerebral blood flow. *Trends in Neurosciences* 32:160–169.

Laje, G., Paddock, S., Manji, H., Rush, J.A., Wilson, A.F., Charney, D., McMahon, F.J. 2007. Genetic markers of suicidal ideation emerging during citalopram treatment of major depression. *American Journal of Psychiatry* 164:1530–1538.

Lalovic, A., Klempan, T., Sequeira, A., Luheshi, G., Turecki, G. 2010. Altered expression of lipid metabolism and immune response genes in the frontal cortex of suicide completers. *Journal of Affective Disorders* 120:24–31.

Lavinsky, D., Arteni, N.S., Netto, C.A. 2003. Agmatine induces anxiolysis in the elevated plus maze task in adult rats. *Behavioural Brain Research* 141:19–24.

Li, Y.F., Chen, H.X., Liu, Y., Zhang, Y.Z., Liu, Y.Q., Li, J. 2006. Agmatine increases proliferation of cultured hippocampal progenitor cells and hippocampal neurogenesis in chronically stressed mice. *Acta Pharmacologica Sinica* 27:1395–1400.

Li, K., Xu, E. 2008. The role and the mechanism of gamma-aminobutyric acid during central nervous system development. *Neuroscience Bulletin* 24:195–200.

Magri, C., Gardella, R., Valsecchi, P., Barlati, S.D., Guizzetti, L, Imperadori, L, Bonvicini, C., Tura, G.B., Gennarelli, M., Sacchetti, E., Barlati, S. 2008. Study on *GRIA2*, *GRIA3* and *GRIA4* genes highlights a positive association between schizophrenia and GRIA3 in female patients. *American Journal of Medical Genetics B. Neuropsychiatric Genetics* 147B:745–753.

Malaterre, J., Strambi, C., Aouane, A., Strambi, A., Rougon, G., Cayre, M. 2004. A novel role for polyamines in adult neurogenesis in rodent brain. *European Journal of Neuroscience* 20:317–330.

Merali, Z., Du, L., Hrdina, P., Palkovitz, M., Faludi, G., Poulter, M.O., Anisman, H. 2004. Dysregulation in the suicide brain: mRNA expression of corticotropin-releasing hormone receptors and GABA(A) receptor subunits in frontal cortical brain region. *Journal of Neuroscience* 24:1478–1485.

Minguet, E.G., Vera-Sirera, F., Marina, A., Carbonell, J., Blazquez, M.A. 2008. Evolutionary diversification in polyamine biosynthesis. *Molecular Biology and Evolution* 25:2119–2128.

Miron, M., Nadon, R. 2006. Inferential literacy for experimental high-throughput biology. *Trends in Genetics* 22:84–99.

Montana, V., Malarkey, E.B., Verderio, C., Matteoli, M., Parpura, V. 2006. Vesicular transmitter release from astrocytes. *Glia* 54:700–715.

Morgan, P.F., Stone, T.W. 1983. Structure-activity studies on the potentiation of benzodiazepine receptor binding by ethylenediamine analogues and derivatives. *British Journal of Pharmacology* 79:973–977.

Morita, K., Saito, T., Ohta, M., Kawai, K., Teshima-Kondo, S., Rokutan, K. 2005. Expression analysis of psychological stress-associated genes in peripheral blood leukocytes. *Neuroscience Letters* 381:57–62.

Nugent, F.S., Kauer, J.A. 2008. LTP of GABAergic synapses in the ventral tegmental area and beyond. *Journal of Physiology* 586:1487–1493.

O'Connor, J.A., Muly, E.C., Arnold, S.E., Hemby, S.E. 2007. AMPA receptor subunit and splice variant expression in the DLPFC of schizophrenic subjects and rhesus monkeys chronically administered antipsychotic drugs. *Schizophrenia Research* 90:28–40.

O'Kusky, J., Colonnier, M. 1982. A laminar analysis of the number of neurons, glia, and synapses in the adult cortex (area 17) of adult macaque monkeys. *Journal of Comparative Neurology* 210:278–290.

Ornitz, D. M., Itoh, N. 2001. Fibroblast growth factors. *Genome Biology* 2:REVIEWS 3005.

Pawson, T. 1994. Tyrosine kinase signalling pathways. *Princess Takamatsu Symposia* 24:303–322.

Pfrieger, F.W. 2009. Roles of glial cells in synapse development. *Cellular and Molecular Life Sciences* 66:2037–2047.

Pietersen, C.Y., Lim, M.P., Woo, T.U. 2009. Obtaining high quality RNA from single cell populations in human postmortem brain tissue. *Journal of Visualized Experiments*. 30: 1444.

Piletz, J.E., Zhu, H., Ordway, G., Stockmeier, C., Dilly, G., Reis, D., Halaris, A. 2000. Imidazoline receptor proteins are decreased in the hippocampus of individuals with major depression. *Biological Psychiatry* 48:910–919.

Poulter, M.O., Du, L., Zhurov, V., Palkovits, M., Faludi, G., Merali, Z., Anisman, H. 2010. Altered organization of GABA(A) receptor mRNA expression in the depressed suicide brain. *Frontiers in Molecular Neuroscience* 3:3.

Radley, J.J., Gosselink, K.L., Sawchenko, P.E. 2009. A discrete GABAergic relay mediates medial prefrontal cortical inhibition of the neuroendocrine stress response. *Journal of Neuroscience* 29:7330–7340.

Reis, D.J., Regunathan, S. 2000. Is agmatine a novel neurotransmitter in brain? *Trends in Pharmacological Sciences* 21:187–193.

Rhee, H.J., Kim, E.J., Lee, J.K. 2007. Physiological polyamines: Simple primordial stress molecules. *Journal of Cellular and Molecular Medicine* 11:685–703.

Ritz, M.C., Mantione, C.R., London, E.D. 1994. Spermine interacts with cocaine binding sites on dopamine transporters. *Psychopharmacology (Berlin)* 114:47–52.

Schena, M., Shalon, D., Davis, R.W., Brown, P.O. 1995. Quantitative monitoring of gene-expression patterns with a complementary-DNA microarray. *Science* 270:467–470.

Seiler, N. 1981. Polyamine metabolism and function in brain. *Neurochemistry International* 3:95–110.

Seiler, N., Raul, F. 2005. Polyamines and apoptosis. *Journal of Cellular and Molecular Medicine* 9:623–642.

Sen, S., Duman, R., Sanacora, G. 2008. Serum brain-derived neurotrophic factor, depression, and antidepressant medications: Meta-analyses and implications. *Biological Psychiatry* 64:527–532.

Sequeira, A., Gwadry, F.G., Ffrench-Mullen, J.M., Canetti, L., Gingras, Y., Casero, R.A., Jr., Rouleau, G., Benkelfat, C., Turecki, G. 2006. Implication of SSAT by gene expression and genetic variation in suicide and major depression. *Archives of General Psychiatry* 63:35–48.

Sequeira, A., Klempan, T., Canetti, L., ffrench-Mullen, J., Benkelfat, C., Rouleau, G.A., Turecki, G. 2007. Patterns of gene expression in the limbic system of suicides with and without major depression. *Molecular Psychiatry* 12:640–655.

Sequeira, A., Mamdani, F., Ernst, C., Vawter, M.P., Bunney, W.E., Lebel, V., Rehal, S., Klempan, T., Gratton, A., Benkelfat, C., Rouleau G.A., Mechawar, N., Turecki, G. 2009. Global brain gene expression analysis links glutamatergic and GABAergic alterations to suicide and major depression. *PLoS ONE* 4(8):e6585.

Sparapani, M., DallOlio, R., Gandolfi, O., Ciani, E., Contestabile, A. 1997. Neurotoxicity of polyamines and pharmacological neuroprotection in cultures of rat cerebellar granule cells. *Experimental Neurology* 148:157–166.

Steinhoff, C., Vingron, M. 2006. Normalization and quantification of differential expression in gene expression microarrays. *Briefings in Bioinformatics* 7:166–177.

Stevens, B. 2008. Neuron–astrocyte signaling in the development and plasticity of neural circuits. *Neurosignals* 16:278–288.

Tabor, C.W., Tabor, H. 1984. Polyamines. *Annual Review of Biochemistry* 53:749–790.

Thalmeier, A., Dickmann, M., Giegling, I., Schneider, B., Hartmann, A., Maurer, K., Schnabel, A., Kauert, G., Möller, H.J., Rujescu, D. 2008. Gene expression profiling of post-mortem orbitofrontal cortex in violent suicide victims. *International Journal of Neuropsychopharmacology* 11:217–228.

Tochigi, M., Iwamoto, K., Bundo, M., Sasaki, T., Katon, N., Katon, T. 2008. Gene expression profiling of major depression and suicide in the prefrontal cortex of postmortem brains. *Neuroscience Research* 60:184–191.

Tsankova, N., Renthal, W., Kumar, A., Nestler, E.J. 2007. Epigenetic regulation in psychiatric disorders. *Nature Reviews Neuroscience* 8:355–367.

Vawter, M.P., Tomita, H., Meng, F., Bolstad, B., Li, J., Evans, S., Choudary, P., Atz, M., Shao, L., Neal, C., Walsh, D.M., Burmeister, M., Speed, T., Myers, R., Jones, E.G., Watson, S.J., Akil, H., Bunney, W.E. 2006. Mitochondrial-related gene expression changes are sensitive to agonal-pH state: Implications for brain disorders. *Molecular Psychiatry* 11:663–679.

Williams, K. 1997. Modulation and block of ion channels: A new biology of polyamines. *Cellular Signalling* 9:1–13.

Yanagi, M., Shirakawa, O., Kitamura, N., Okamura, K., Sakurai, K., Nishiguchi, N., Hashimoto, T., Nushida, H., Ueno, Y., Kanbe, D., Kawamura, M., Araki, K., Nawa, H., Maeda, K. 2005. Association of 14-3-3 epsilon gene haplotype with completed suicide in Japanese. *Journal of Human Genetics* 50:210–216.

Zeidan, M.P., Zomkowski, A.D., Rosa, A.O., Rodrigues, A.L., Gabilan, N.H. 2007. Evidence for imidazoline receptors involvement in the agmatine antidepressant-like effect in the forced swimming test. *European Journal of Pharmacology* 565:125–131.

Zhu, M.Y., Wang, W.P., Cai, Z.W., Regunathan, S., Ordway, G. 2008. Exogenous agmatine has neuroprotective effects against restraint-induced structural changes in the rat brain. *European Journal of Neuroscience* 27:1320–1332.

Zhu, M.Y., Wang, W.P., Huang, J.J., Regunathan, S. 2007. Chronic treatment with glucocorticoids alters rat hippocampal and prefrontal cortical morphology in parallel with endogenous agmatine and arginine decarboxylase levels. *Journal of Neurochemistry* 103:1811–1820.

Zomkowski, A.D., Hammes, L., Lin, J., Calixto J.B., Santos, A.R., Rodrigues, A.L. 2002. Agmatine produces antidepressant-like effects in two models of depression in mice. *Neuroreport* 13: 387–391.

Zomkowski, A.D., Rosa, A.O., Lin, J., Santos, A.R., Calixto, J.B., Rodrigues, L.S. 2004. Evidence for serotonin receptor subtypes involvement in agmatine antidepressant like-effect in the mouse forced swimming test. *Brain Research* 1023:253–263.

Zomkowski, A.D., Santos, A.R., Rodrigues, A.L. 2005. Evidence for the involvement of the opioid system in the agmatine antidepressant-like effect in the forced swimming test. *Neuroscience Letters* 381:279–283.

Zomkowski, A.D., Santos, A.R., Rodrigues, A.L. 2006. Putrescine produces antidepressant-like effects in the forced swimming test and in the tail suspension test in mice. *Progress in Neuro-Psychopharmacology and Biological Psychiatry* 30:1419–1425.

[illegible]

[illegible]

13 Epigenetic Effects of Childhood Adversity in the Brain and Suicide Risk

Benoit Labonté and Gustavo Turecki

CONTENTS

13.1 DEPRESSION, SUICIDE, AND EARLY-LIFE ADVERSITY

With prevalence estimates ranging between 6.4% and 10.1% [1–5], major depression ranks first among the most significant causes of disability and premature death, thus imposing a continual economic burden on society. For instance, in the United States, the direct and indirect costs are estimated at U.S.$44 billion/year [6]. The greatest loss to our society, however, is the associated mortality by suicide related to major depression. Indeed, it has been estimated that between 50% and 70% of suicide completers will die during an episode of major depression [7,8] and prospective follow-up studies of major depression suggest that between 7% and 15% of these patients will die by suicide [9–12].

Suicide is a complex problem, which is believed to result from the interaction of several different factors [13,14]. Indeed, psychological factors and personality traits such as impulsivity and negative affect [14,15], social factors [16,17], environmental factors such as early-life adversity [18–20], genetic factors [21], and neurobiological factors [22] have been proposed to induce behavioral alterations, which in turn may predispose certain individuals to develop depressive and suicidal behaviors. However, since these factors alone are unlikely to explain suicide and suicide risk, it may be more readily explained when considering the interaction between these different sources of variation [23,24].

Among these risk factors, early-life adversity, particularly childhood sexual abuse (CSA) and childhood physical abuse (CPA), is one of the strongest predictors of mental disorders [25,26] and suicide [18,19]. For example, studies have shown that CSA is associated with early onset of depression, chronic course, and more severe depressive outcome [27–29] but, more importantly, with 12 times higher odds of suicidal behaviors [26,30]. Although less consistently, CPA and neglect have also been associated with suicidal behaviors [19,31]. CSA and CPA have been associated with higher odds of self-harm [20,28,32–34], suicidal ideation [35,36], and suicide attempts [26,35–39]. Moreover, the prevalence of suicidal ideation and suicide attempts has been shown to increase with the severity and intensity of the abuse [35,36,38].

13.2 SUICIDE AND MOLECULAR MECHANISMS

As discussed earlier, substantial theoretical and empirical work supports the relationship between childhood adversity and suicide. On the other hand, if it is generally assumed that childhood adversity affects proper psychological development, the molecular mechanisms that account for these relationships are still poorly understood. Consequently, a critical question remains: "what long-lasting molecular mechanisms take place as a result of the adverse life experience that could be associated with increased risk for suicide?" Despite the complexity of this question, it is now possible to investigate promising hypotheses on the molecular alterations associated with suicide that may be induced by environment. Indeed, a growing body of evidence suggests that the genome may respond to social and environmental stimuli as much as it does to the physical environment and that the basic molecular mechanism of this response is through epigenetic modifications.

Epigenetics refer to the study of the epigenome, chemical and physical modifications taking place in the DNA molecule, and altering the capacity of a gene to produce more or less of its coded mRNA. Given the high complexity of DNA organization, these modifications are expected to follow a defined pattern allowing the underlying molecular mechanisms to be performed correctly and to decode DNA in the context of chromatin. Epigenetic mechanisms refer to DNA methylation [40], histone modifications [41], and, more recently, posttranscriptional mechanisms such as microRNA [42,43]. Generally, DNA methylation has been suggested to direct transcriptional repression [40]. However, recent evidence suggest that this may be particularly true for methylation found within gene's promoter while intra- and intergenic methylation may be associated with the use of alternative promoters and active transcription [44]. At the chromatin level, high histone acetylation and low histone methylation levels have been associated with active transcription [41]. This being said and as mentioned previously, epigenetic mechanisms are thought to be involved in the modification of gene expression induced by environmental factors. As such, it is possible to conceptualize the epigenome as an interface on which environment can act to influence normal genetic processes and to hypothesize that epigenetic mechanisms may account, at least in part, for the regulation of behavior as a response to environmental adversity.

13.3 EARLY-LIFE ADVERSITY INFLUENCES ON BEHAVIOR

Despite the complexity of depressive and suicidal behaviors, common functional and physiological alterations have been consistently reported in the brain of depressed and suicide subjects with a history of childhood abuse. It is accepted that early-life adversity often results in maladaptive patterns of behavioral responses associated with pervasive interpersonal difficulties, enhanced reactivity to stress, and increased risk of psychopathology. In this sense and because of their strong association with depression [45] and childhood abuse [46], genes involved in the stress regulatory systems such as the hypothalamic–pituitary–adrenal (HPA) axis and the polyamine system have been the focus of epigenetic studies performed in the context of early-life adversity in both animals and humans. The epigenetic regulation of cell signaling molecules such as ribosomal RNA (rRNA), brain-derived neurotrophic factor (BDNF) and its receptor the neurotrophic tyrosine kinase receptor (trkB), and quaking (QKI) has also been studied in the brain of animals with a history of early-life stress and in the brain of suicide completers with depression. Finally, neurotransmitters such as the gamma-aminobutyric acid (GABA) and serotonin (5-HT), which have been frequently associated with depression and suicide, have also been the focus of studies assessing epigenetic mechanisms in the brain of suicide completers (summarized in Tables 13.1 and 13.2). In agreement with the alterations in gene and protein expression previously reported, epigenetic modifications have been reported in the promoter of genes involved in the regulation of those systems suggesting that epigenetic alterations could be involved in the maladaptive behavioral responses observed in suicide completers and contribute to an increased risk for psychopathology. The following sections will review these findings.

13.4 HYPOTHALAMIC–PITUITARY–ADRENERGIC AXIS ALTERATIONS INDUCED BY EARLY-LIFE ADVERSITY

The HPA axis is the main stress regulatory system [45]. Under stressful conditions, corticotropin-releasing factor (CRF) and vasopressin (AVP) are released from the hypothalamus. CRF and AVP induce the release of adrenocorticotropic hormone (ACTH) and pro-opiomelanocortin (POMC) from the pituitary gland to the blood, which then travel to the adrenal cortex where they induce the release of glucocorticoids, cortisol in humans and corticosterone in rodents, to the blood. Glucocorticoids then act at each level of the HPA axis to decrease the release of CRF, AVP, POMC,

TABLE 13.1
Summary of Published Studies Assessing Epigenetic Components in Suicide

Studies	Brain Regions	Genes	Findings
McGowan et al. [52]	Hippocampus	*GR*	↑ Methylation in NGFI-A binding site within GR promoter in the hippocampus of suicide completers with history of abuse ↓ Expression of GR in the hippocampus of suicide completers with history of abuse
Alt et al. [63]	Amygdala Hippocampus Inferior frontal gyrus Cingulate gyrus Nucleus accumbens	*GR*	*Amy*: ↓ GRα protein, ↑ expression of 1J, ↓ YY1 transcription factor *HPC*: ↓ Expression of $GR1_F$ and 1C, ↓ NGFI-A transcription factor *IFG*: ↓ GRβ protein, ↓ YY1, and Sp1 transcription factors *CG*: ↓ GRα protein, ↑ expression of 1D, ↓ YY1, NGFIA, and Sp1 transcription factors *NaC*: ↓ Expression of GR1B, ↓ NGFI-A transcription factor No methylation difference in promoters
McGowan et al. [119]	Hippocampus	*rRNA*	Overall hypermethylation of rRNA promoter in the hippocampus of suicide completers with history of abuse ↓ Expression of *rRNA* gene in the hippocampus of suicide completers with history of abuse
Keller et al. [103]	Wernicke area	*BDNF*	Hypermethylation at four CpGs within promoter/exon IV in suicide completers Negative correlation between BDNF promoter methylation levels and expression
Ernst et al. [112]	Frontal cortex	*TrkB-T1*	↑ Methylation in two sites within the promoter of TrkB-T1 in the frontal cortex of suicide completers ↓ Expression of TrkB-T1 in the frontal cortex of suicide completers
Ernst et al. [113]	Frontal cortex	*TrkB-T1*	↑ H3K27 methylation in the frontal cortex of suicide completers

TABLE 13.1 (continued)
Summary of Published Studies Assessing Epigenetic Components in Suicide

Studies	Brain Regions	Genes	Findings
			Negative correlation between H3K27 methylation levels and TrkB-T1 expression in the frontal cortex of suicide completers
Poulter et al. [140]	Frontopolar cortex (FPC) Hippocampus Amygdala Brain stem	*GABA α1* *DNMT1* *DNMT3a* *DNMT3b*	*FPC*: ↓ Expression of DNMT1 mRNA in suicide completers ↑ Expression of DNMT3b mRNA and protein levels in suicide completers ↑ Increased methylation at two sites in the promoter region of GABA receptor subunit α1 in suicide completers *Limbic system*: ↓ Expression of DNMT1 and DNMT3b mRNA levels in suicide completers *Brain stem*: ↓ Expression of DNMT3b mRNA in suicide completers
Grayson et al. [137]	Occipital cortex	*Reelin* *GAD67*	↑ Methylation in CREB-binding sites within reelin promoter in the occipital cortex of schizophrenia subjects
Tamura et al. [138]	Forebrain	*Reelin*	↑ Methylation at three sites within reelin promoter in the forebrain of schizophrenia subjects ↓ Expression of reelin mRNA in the forebrain of schizophrenia subjects
Fiori and Turecki [85]	PFC	*SMOX* *SMS*	No effects of promoter's methylation level in SMOX and SMS on expression levels
Fiori and Turecki [85]	PFC	*SAT1*	Negative correlation between promoter's methylation levels and expression of SAT1
De Luca et al. [127]	PFC	*5-HT$_{2A}$*	↓ Methylation in the promoter region of 5-HT$_{2A}$ receptor associated with a C-allele (trend) in the PFC of suicide completers ↑ Methylation in the promoter region of 5-HT$_{2A}$ receptor associated with a C-allele in leukocytes of suicide attempters
Klempan et al. [147]	Frontal cortex	*QKI*	No difference in methylation pattern between suicide completers and controls

TABLE 13.2
Summary of Published Studies Assessing Epigenetic Components in Animal Models of Stress-Induced Depressive Symptoms

Studies	Animal Model	Brain Regions	Genes	Findings
Weaver et al. [62]	Low/high licking and grooming	Hippocampus	*GR*	*In pups raised by LG mothers* Overall GR17 promoter hypermethylation ↓ H3K9 acetylation in GR17 promoter ↓ Binding of NGFI-A in GR17 promoter
Tsankova et al. [98]	Intruder test	Hippocampus	*BDNF*	*In stressed mice* ↓ Expression of transcripts III and IV ↑ H3K27 dimethylation in transcripts III and IV promoters *Chronic treatment with imipramine* ↑ H3 acetylation and H3K4 dimethylation in transcripts III and IV promoters ↓ HDAC5 in hippocampus of stress mice No DNA methylation difference
Roth et al. [99]	Stress mothers	Prefrontal cortex	*BDNF*	*In maltreated rats* ↓ Expression of BDNF transcript IX from childhood to adulthood ↓ Expression of BDNF transcript IV at adulthood Overall hypermethylation in transcript IV promoter Transgenerational DNA methylation alterations in pups raised by abusive mothers

TABLE 13.2 (continued)
Summary of Published Studies Assessing Epigenetic Components in Animal Models of Stress-Induced Depressive Symptoms

Studies	Animal Model	Brain Regions	Genes	Findings
Murgatroyd et al. [67]	Maternal deprivation	Hypothalamic paraventricular nucleus	*AVP*	*In stressed mice* ↑ Expression of AVP Site-specific hypomethylation in AVP intergenic region Phosphorylation of meCP2
Zhang et al. [139]	Low/high licking and grooming	Hippocampus	*GAD1*	*In pups raised by low LG/ABN mothers* ↓ Expression of GAD1 ↑ Methylation within GAD1 promoter ↑ Expression of DNMT1 ↓ H3K9 acetylation in GAD1 promoter

and ACTH and regulate the stress response. The locus of regulation of the HPA axis lies in the hippocampus where glucocorticoids bind glucocorticoid receptors (GR) and induce an inhibitory feedback on the activation of the HPA axis in order to bring back to basal levels the activity of the stress response.

Hyperactivity of the HPA axis, thought to be related to attenuated glucocorticoid feedback inhibition, is a common feature in depressed patients and childhood abuse victims. High levels of salivary, plasma, and urine glucocorticoids have been reported in depressed patients [47]. Similarly, higher ACTH and cortisol levels have been reported in depressed subjects with a history of childhood abuse following stress and dexamethasone (DEX) challenges [48,49]. Interestingly, in this study, both ACTH and cortisol levels did not differ significantly between depressed subjects without history of childhood abuse and controls [48,49]. In the hippocampus, this impaired inhibitory feedback is thought to be due in part to the binding of glucocorticoids to their receptor (GR). In rats, the development of the HPA axis has been shown to be modulated by maternal behavior. Depression-like behaviors [50] associated with altered HPA axis feedback [51] and low GR mRNA hippocampal levels [51] are common features in rats raised by mothers providing poor maternal care defined by low levels of licking and grooming (LG) and arched-back nursing (ABN). Similarly, low hippocampal GR levels have been reported in suicide completers with a history of childhood abuse [52] but not in non-abused suicide completers. All together, this suggests that the blunted and maladapted responses to stress in depressed patients and suicide completers with a history of abuse may be due to reduced levels of GR in the hippocampus.

Following great efforts, a comprehensive model involving modulations at numerous levels including hormonal, synaptic, and molecular levels has been proposed in an attempt to characterize the molecular pathways involved in the modulatory effects of maternal behavior in rats. High maternal LG/ABN levels, which can also be mimicked by handling pups during early life, induce a physiological response involving the release of thyroid hormone. This was shown to increase 5-HT activity in the raphe and consequently to stimulate serotonin (5-HT) turnover in the hippocampus and frontal cortex [53–55]. Via activation of the G-protein-coupled 5-HT_7 receptor [56], it is believed that 5-HT activates a cAMP/PKA-dependent intracellular cascade that is thought to increase the expression of nerve growth factor 1-A (NGFI-A) and activator protein-2 (AP-2) in the hippocampus [57]. NGFI-A and AP-2 are activating transcription factors with putative binding sites within GR promoter region [58] expected to increase GR mRNA levels in the hippocampus of offspring raised by high LG/ABN mothers. This complex process may be attenuated in rats raised by low LG/ABN mothers in accordance with the molecular and behavioral processes mentioned previously. Interestingly, most of these modifications remain present during adulthood in rats, while one would expect that they would be reinstated after the "mistreatment" disappeared. Moreover, cross-fostering studies report that these behavioral and molecular modifications can be reversed when pups raised by low LG/ABN mothers are transferred to high LG/ABN mothers [50,51] within the first week of life. As such, although providing a mechanism responsible for the molecular consequences of high/poor maternal care, this mechanism does not explain how maternal behavior can induce long-term effects on GR expression, nor explain the reversibility of these effects.

13.5 EFFECTS OF EARLY-LIFE ADVERSITY ON GR EPIGENETIC REGULATION OF EXPRESSION

This led to the hypothesis that the long-term effects of maternal behavior and early-life adversity on GR hippocampal expression could be due to epigenetic modifications. In rats, the *GR* gene is preceded by 10 noncoding exons and by 14 in humans [58,59]. The expression of the noncoding exon 17 in rats and the human homolog 1_F has been shown to be specific to the hippocampus [59]. Multiple transcription factor binding sites, including NGFI-A [60], have been identified in GR promoter, and methylation patterns have been predicted to be highly variable in humans [61]. In rats raised by LG/ABN mothers, CpG methylation levels were significantly increased at almost all CpGs compared to rats raised by high LG/ABN mothers. Of particular interest is the 3′ end of a NGFI-A binding site showing 100% methylation in all rats raised by low LG/ABN mothers [62]. On the other hand, methylation alterations in abused suicide completers were limited to specific CpGs with no clones showing global hypermethylation. However, similarly to what was found in rats, a significant hypermethylation in a NGFI-A binding site was found in abused suicide completers but not in non-abused suicide completers. This epigenetic mark was shown to repress the binding of NGFI-A to its cognate DNA sequence and to decrease GR transcription. Indeed, functional luciferase assays showed that NGFI-A significantly increases $GR1_F$ promoter's transcriptional activity and artificial methylation at its binding site decreases this activity even in the presence of NGFI-A [52]. Altogether, this suggests that early-life adversity could induce specific long-lasting epigenetic alterations affecting gene expression.

In a recent study assessing the expression of numerous GR variants in the limbic system of depressed suicide completers, $GR1_F$ and $GR1_C$ hippocampal expressions were significantly decreased in depressed suicide completers [63]. However, this was not associated with promoter hypermethylation since promoter methylation levels reported were particularly low and did not vary between groups. On the other hand, NGFI-A protein levels in the HPC were significantly decreased in depressed suicide completers suggesting that the decrease in GR expression found in suicide completers may be mediated by different molecular pathways depending on the presence or the absence of early-life adversity.

Beside DNA methylation, chromatin changes were also associated with poor maternal care. For instance, lower H3K9 acetylation levels were found in $hGR1_7$ promoter in low LG/ABN raised rats. Pharmacological challenge with the histone deacetylase inhibitor (TSA) restored methylation levels, increased NGFI-A binding to the promoter, and reinstated H3K9 acetylation and GR hippocampal levels [62]. Treated rats were also less reactive to stressful conditions. H3K9 acetylation has been associated with "opened" euchromatin state [41]. Decreasing H3K9 acetylation would decrease DNA access to transcriptional machinery and DNA binding proteins such as transcription factors and methylated DNA binding proteins. Moreover, the binding of NGFI-A has been shown to be repressed by DNA methylation and has previously been shown to recruit the histone acetylase CREB-binding protein (CBP) [64]. CBP has also been shown to recruit RNA pol II [65,66]. Functionally, these results suggest that early-life adversity in rats and human increases methylation specifically within NGFI-A binding sites and decreases gene expression in part because it represses the binding of transcription factor to DNA, but also by interfering with the recruitment of proteins involved in transcription and associated with euchromatin state.

13.6 *VASOPRESSIN* GENE

Other components of the HPA axis have also been shown to be affected by early-life stress. Recently, early-life infant–maternal separation in mice inducing stress-coping behavioral alterations in pups has been shown to be associated with a long-lasting increase in corticosterone secretion and with an increased expression of POMC and AVP in the paraventricular nucleus (PVN) of the hypothalamus [67]. The *AVP* gene in mice is composed of three coding exons and is oriented tail-to-tail with the oxytocin (*Oxt*) gene. A region within the AVP/Oxt intergenic region has been shown to include an enhancer modulating AVP expression [68]. Interestingly, this region, located ~2.1 kb from the *AVP* gene, has also been shown to be composed of a CpG island [67].

Consistent hypomethylation was measured on multiple CpGs found in AVP's enhancer at 6 weeks, 3 months, and 1 year in the PVN of stressed mice [67]. Consistent with the repressive role of DNA methylation on transcriptional activity, treatment with the demethylating agent aza-5 decreased methylation levels and increased AVP expression in a rat hypothalamic cell line. Deletion of the first part of the enhancer partially reduced transcriptional activity, while removing the entire enhancer almost completely abolished it in transfection assays. Furthermore, methylation of the enhancer also significantly reduced transcriptional activity.

AVP expression was also shown to be significantly increased in stressed mice at 10 days, although no methylation difference was observed in AVP's enhancer. The AVP enhancer can putatively bind the methylated CpG binding protein MeCP2. However, because of the repressive role of MeCP2 on transcription, one would expect the opposite tendency concerning AVP expression. MeCP2 has nevertheless been shown to be susceptible to inactivation by neuronal depolarization-induced phosphorylation leading to its dissociation from putative targets [69,70]. Accordingly, higher neuronal activity-induced CaMKII immunoreactivity and phosphorylated MeCP2 levels have been reported in AVP-expressing neurons in the PVN of 10 day old stressed mice. Altogether, these results suggest that in young stressed mice, methylation patterns in AVP's enhancer allow the binding of MeCP2, which could then repress expression. However, since early-life stress also increases neuronal activity in AVP-expressing neurons, MeCP2 gets phosphorylated and inactivated. Consequently, the repressive effect of MeCP2 on AVP expression is abolished. On the other hand, methylation levels in AVP enhancer decrease with time. This may decrease MeCP2 binding and allow AVP to be expressed at higher levels. Overall, these results suggest that alterations in DNA methylation found outside of the promoter might also be involved in physiological and behavioral modifications induced by environmental factors.

13.7 POLYAMINE SYSTEM

Another system involved in the regulation of stress is the polyamine system [71,72]. Polyamines are released following acute and chronic stress and participate in the cellular responses to stress. In addition, the polyamine stress response has been shown to be dependent on the activation of the HPA axis and on the increased concentrations of circulating glucocorticoids [73]. Furthermore, the emergence of the characteristic adult polyamine stress response is correlated with the cessation of the hyporesponsive period of the HPA axis system [74]. Besides their involvement in stress response, polyamines are important ubiquitous aliphatic molecules also involved in cellular functions including growth, division, and signaling cascades [73,75]. Polyamines include putrescine, spermidine, spermine, and agmatine, as well as polyamine-related proteins including polyamine oxidase (PAO), spermine synthase (SMS), spermidine/spermine *N*1-acetyltransferase (SAT1), spermine oxidase (SMOX), ornithine decarboxylase (ODC), and *S*-adenosylmethionine decarboxylase (AMD1) [76].

Many lines of evidence relate the polyamine system with depressive disorders and suicide. For example, putrescine, spermidine, and spermine hippocampal expression has been shown to be decreased in rats that have been submitted to chronic unpredictable stress [77], while acute stressors have been shown to increase ODC activity and putrescine and agmatine brain levels. In humans, SMS, SAT1, and ornithine aminotransferase-like 1 (OATL1) expressions have been shown to be altered in the limbic system of suicide completers with a history of depressive disorders [78,79]. Pharmacological treatment with bupropion as well as electroconvulsive therapy have been shown to normalize agmatine and PAO plasma levels in depressed patients [80,81]. Moreover, antidepressant and anxiolytic properties have been attributed to agmatine and putrescine in rodents through a NMDA-dependent mechanism [82–84].

Given the alterations in polyamine gene expression reported in the brain of suicide completers with a history of depression, the methylation patterns of particular polyamine genes were recently assessed. Promoter methylation was found to be negatively correlated with the expression of SAT1 although not with SMOX and SMS [85]. Moreover, no association was found between H3K27me3 modification in the promoter regions of SMS, SMOX, or SAT1 and suicide completion or expression of these genes in BA 8/9. These findings suggest that epigenetic alterations in the promoter region of genes involved in the polyamine synthesis do not play a major role in suicidal behavior, although they may partly explain why polyamine gene expression is decreased in the brain of suicide completers.

13.8 *BRAIN-DERIVED NEUROTROPHIC FACTOR* GENE

Because of their involvement in neuronal survival and plasticity, as well as their expression in brain regions from the limbic system, where emotional behaviors are processed, neurotrophic factors have consistently emerged as candidate molecules in the neurobiology of suicide. It is hypothesized that their alteration could partly underlie changes in plasticity observed in the brains of suicides as well as the defective affect observed in depressive patients. While the major neurotrophic factors include NGF, neurotrophin 3 and 4 (NT3/4), fibroblast growth factor (FGF), transforming growth factor (TGF), and BDNF, the latter has received most of the attention concerning the potential implication of neurotrophic factors in depressive disorders and suicide. For instance, low serum and brain BDNF expression has been reported in patients with major depression [86–88]. In addition, antidepressant treatment has been shown to reverse those alterations [89–91]. In mice, BDNF depletion induced depression-like behaviors [92], while, in rats, chronic stress and persistent pain reduced BDNF expression in the hippocampus [93,94]. Here again, those effects have been counteracted by antidepressant treatment [95–97].

BDNF epigenetic regulation has recently been investigated in mice and rat models of stress-induced depressive symptoms [98,99]. In both species, the *BDNF* gene contains nine 5′ noncoding first exons with their own promoter coding for the same protein [100]. The alternative splicing of these exons specifies the tissue in which BDNF is expressed [100]. In both species, epigenetic processes involved in the transcriptional control of BDNF have been shown to be altered by stress. For instance, chronic social stress in mice was shown to decrease the expression of two specific BDNF transcripts (III and IV) in the hippocampus [98], while maternal maltreatment decreased prefrontal cortex (PFC) BDNF mRNA expression in rats [99]. Although similar, these transcriptional alterations were shown to be induced by different epigenetic mechanisms. Indeed, chronic stress in mice raised H3K27 dimethylation levels in transcripts III and IV promoters [98], while site-specific hypermethylation was found in transcripts IV and IX promoters of maltreated rats [99]. In the latter study, site-specific hypermethylation seemed to follow a developmental pattern, exon IX promoter hypermethylation occurred immediately after the maltreatment regimen, while promoter IV methylation increased gradually to reach significantly altered levels only at adulthood. Surprisingly, in a study by Tsankova et al., no DNA methylation difference was found in association with histone modifications, and as reported by Roth et al., no histone modification was reported in association with DNA methylation alterations. These findings illustrate that chronic stress and maternal maltreatment may alter different epigenetic mechanisms with common transcriptional consequences: the first leading to the compaction of chromatin in its heterochromatic state and the second blocking the binding of transcription factors to DNA. On the other hand, those results may also highlight the heterogeneity of stress-induced epigenetic alterations between species.

Pharmacological treatment with the tricyclic antidepressant imipramine was able to reverse the effect of chronic stress on BDNF transcription in mice [98]. However, this reversal does not seem to be due to the reinstatement of altered histone modifications but rather to follow an indirect pathway. Indeed, chronic but not acute imipramine treatment did not reinstate H3K27 basal dimethylation levels, but rather decreased HDAC5 levels in the hippocampus of chronically stressed mice leading to a global hyperacetylation in transcripts III and IV promoter regions. The importance of histone acetylation in the effect of antidepressant treatment has indeed been previously reported in animal models of stress-induced depression [63,101,102]. Additionally, this hyperacetylation was associated with higher hippocampal levels of H3K4 dimethylation in the area of BDNF III and IV promoters with both modifications related to transcriptional activation. Consequently, these results suggest the existence of a compensatory mechanism in the reinstatement of basal BDNF levels by chronic imipramine treatment following chronic stress and emphasize the importance of chromatin hyperacetylation induced by antidepressant treatment.

Recently, the methylation state of BDNF was assessed in postmortem brains from suicide completers [103]. The human *BDNF* gene is also composed of 11 exons preceded by 9 noncoding first exons regulating BDNF expression in different tissues [104]. In a study by Keller et al. [103], three different methods were used to quantify methylation levels in a region encompassing part of noncoding exon IV and its promoter in the Wernicke area. Their results show that methylation in four CpGs located downstream from the promoter IV transcription initiation site was significantly increased in suicide completers compared to controls. These differences were specific to the BDNF promoter since genome-wide methylation assessment did not reveal any significant difference between groups. In addition, BDNF expression in subjects with high methylation levels was significantly lower than in subjects with low and medium methylation levels supporting the repressive effects of methylation within the promoter on transcription.

13.9 TYROSINE KINASE RECEPTOR

The transmembrane receptor TrkB, known as the receptor for BDNF, has consistently been linked to mood disorders and suicide [87,105–107]. Expression microarray studies have reported lower TrkB expression in the PFC of depressed subjects [108,109], and antidepressant treatment has been shown to increase its expression in cultured astrocytes [110].

The *TrkB* gene is found on chromosome 9 at locus q22.1 and has five splice variants. Splice variant T1 or TrkB-T1 is an astrocytic truncated form of TrkB lacking catalytic activity [111]. Recently, analysis of the methylation pattern in the promoter of a subset of suicide completers with low levels of TrkB-T1 expression revealed two sites where methylation levels were higher in suicide completers compared to controls [112]. The methylation pattern at these two sites was negatively correlated with the expression of TrkB-T1 in the low expression subset of suicide completers and was specific to the PFC since no significant difference was found in the cerebellum. Such patterns of expression and methylation are expected to increase predispositions to suicidal behaviors.

In addition, the same subjects showed enrichment of H3K27 methylation in TrkB-T1 promoter [113], suggesting that the astrocytic variant of TrkB may be under the control of epigenetic mechanisms involving histone modifications and DNA methylation. Interestingly, these findings support the involvement of an astrocytic component in suicide.

13.10 *RIBOSOMAL RNA* GENES

The ribosomal RNA (rRNA) is a bottleneck structure for protein synthesis allowing adequate cell function depending on cell needs. Its role is to decode the mRNA into amino acids and to provide enzymatic activity allowing the right amino acid to be properly added to the newly synthesized proteins.

The rRNA promoter is composed of two regulatory regions, namely, the upstream control element (UCE) and the core promoter that binds the upstream binding factor (UBF) [114–116]. The expression of *rRNA* genes has been shown to be epigenetically regulated both in mice [117] and humans [116,118]. In mice, the recruitment of transcription repressors has been suggested to induce chromatin modifications leading to methylation of a single CpG found within UBF binding sites in the UCE. This is thought to prevent UBF binding to its cognate sequence and to decrease rRNA expression [117]. In humans, despite the fact that the CpG density in both promoter regions differs from mice [116,117], rRNA expression has nevertheless also been shown to be epigenetically regulated [118]. Indeed, the active portion of the rRNA promoter associated with pol I has been shown to be completely unmethylated, while the inactive portion is almost fully methylated [118].

The *rRNA* gene expression has recently been shown to be altered in the hippocampus of childhood abuse victims via epigenetic mechanisms [119]. Indeed, methylation in the rRNA core promoter and UCE was shown to be significantly increased in 21 out of 26 CpGs in the hippocampus of abused suicides compared to controls. This was associated with low rRNA hippocampal expression in the abused victims. In other words, abused subjects showed overall promoter hypermethylation compared to controls who exhibited consistently low levels of methylation, which was correlated with lower gene expression. From a mechanistic point of view, it could be hypothesized that promoter hypermethylation represses the interaction of the UBF with the core promoter sequence and consequently decreases both the recruitment of transcriptional cofactors and the transcriptional activity of RNA pol.

These alterations have been suggested to be specific to the HPC. Indeed, no group difference in rRNA methylation pattern was found in the cerebellum. Moreover, genome-wide methylation levels did not reveal any methylation difference between abused suicides and controls further suggesting that this epigenetic alteration was specific to the hippocampus. From a therapeutic point of view, these findings are of major interest since they provide potential tools for the identification of individuals at risk, and, therefore, the possibility of preventive intervention. However, there are major challenges in their potential implementation, not the least of which are access to target tissue in living subjects, modification of epigenetic profiles, and appropriate delivery of such interventions. Consequently, future work should assess whether similar alterations can be found in more accessible tissue and establish correlations between what has been found in the brain and what will be found in other tissues.

13.11 SEROTONIN RECEPTOR TYPE 2A

The serotonergic system has received particular attention in the context of depression and suicide during the past decade. Lower concentration, binding, neurotransmission, and reuptake of serotonin and its metabolites are risk markers for suicidality and major depression [120,121]. Particular attention has been given to the $5\text{-}HT_{2A}$ gene as being a major candidate in association studies of suicidal behavior [122,123]. One of the most extensively studied polymorphism in suicide and depression is the 102 C/T occurring in exon 1 [124,125]. Methylation in the C-allele has previously been associated with higher DNA methyl transferase 1 (DNMT1) expression in the brain and leukocytes of healthy subjects suggesting that allele-specific methylation could alter serotonin receptor type 2A (5-HTR_{2A}) expression [126]. Following this, De Luca et al. [127] investigated the methylation pattern in the C-allele carriers of suicide completers. They reported a nonsignificant hypomethylation in the PFC of C-allele in suicide completers. However, in suicide ideators, methylation appeared to be increased in leukocytes suggesting that methylation levels are different in individuals who committed suicide and those who are planning or attempted suicide. On the other hand, the functional significance of this hypermethylation in leukocytes remains to be explored, and since significance levels were not reached in brain tissue, further research is required.

13.12 GABAERGIC SYSTEM

The GABAergic system has been the focus of much research in postmortem brains of depressed suicide completers [128–130] and in patients with schizophrenia and bipolar disorder who died by suicide [131–134]. For instance, reductions of reelin and GAD1 mRNA [131] and an increase in DNMT1 expression [135,136] were previously reported in postmortem brains of schizophrenia and bipolar subjects who died by suicide. Consistently, promoter hypermethylation was reported for both genes in accordance with the methylating role of DNMT1 [137,138].

More recently, the expression of GAD1 has been shown to be affected by maternal care in rats [139]. Indeed, pups raised by mothers providing poor maternal care characterized by low amount of LG/ABN had lower GAD1 hippocampal expression associated with promoter hypermethylation and lower levels of H3K9ac compared to pups raised by mothers providing high levels of maternal care. Interestingly, this was also associated with higher DNMT1 hippocampal levels. Functional assays revealed that the transcription factor NGFI-A was binding GAD1 promoter in order to increase GAD1 expression. Consequently, these results suggest that similar to the regulation of GR in rat hippocampus, GAD1 expression is modulated by maternal behavior via epigenetic mechanisms involving DNA methylation interfering with the binding of activating transcription factors and by chromatin modifications [139].

Recently, Poulter et al. [140] examined the expression of DNA methyltransferases as well as $GABA_A$ receptor α1 subunit in the frontopolar cortex (FPC), limbic regions, and brain stem of suicide brains. They identified three hypermethylated CpG sites and higher DNMT3b protein expression in the FPC of suicide completers in accordance with DNMT3b role in *de novo* methylation [40]. The expression levels of other DNMTs were also reported to be altered in the limbic system and the brain stem, although these were not significantly associated with any modification in GABAergic gene methylation patterns.

13.13 QUAKING

Quaking (*QKI*) is a brain-specific gene expressed in oligodendrocytes involved in cell development and myelination processes [141,142]. QKI expression was reported to be decreased in schizophrenia subjects, independent of neuroleptic treatment [143,144]. Its expression was also shown to be reduced in cortical regions from suicide completers [128].

QKI gene in human is located on chromosome 6 at locus q26 and undergoes extensive splicing leading to different isoforms: each differing in the 3′ region but all exhibiting a common 5′ region [145,146]. Despite a decreased cortical expression, no DNA methylation alterations have been reported in the promoter of QKI in suicide completers. Indeed, methylation was very low in the promoter region of both suicides and controls, which suggests that other mechanisms may be responsible for the modified expression of QKI in suicide completers [147]. On the other hand, since the promoter region investigated was relatively small, it remains possible that other sites within the promoter control the expression of QKI. It is also possible that the altered expression of this gene is related to posttranslational mechanisms, such as microRNA.

13.14 CONCLUSION

Suicide is a heterogeneous phenomenon involving the interaction of many complex factors, including early-environmental adversity. A growing number of studies using animal models of early environment variation and human samples from extreme phenotypes such as suicide support the notion that early-life environmental adversity may modify behavior via alterations in the epigenetic mechanisms of specific gene sequences. These findings suggest that epigenetics may function as an interface through which environmental factors act and induce long-term behavioral consequences. While these are theoretically appealing concepts, only a handful of studies have been conducted to date and these studies have only investigated specific gene candidates. Additional studies are needed, particularly studies using a more systematic genomic approach.

REFERENCES

1. Kessler, D., D. Sharp, and G. Lewis, Screening for depression in primary care. *Br J Gen Pract*, 2005; 55(518): 659–660.
2. Kessler, R.C. et al., Prevalence, severity, and comorbidity of 12-month DSM-IV disorders in the National Comorbidity Survey Replication. *Arch Gen Psychiatry*, 2005; 62(6): 617–627.
3. Regier, D.A. et al., The NIMH Depression Awareness, Recognition, and Treatment Program: Structure, aims, and scientific basis. *Am J Psychiatry*, 1988; 145(11): 1351–1357.
4. Robins, L.N. and R.K. Price, Adult disorders predicted by childhood conduct problems: Results from the NIMH Epidemiologic Catchment Area project. *Psychiatry*, 1991; 54(2): 116–132.
5. Weissman, M.M., K.K. Kidd, and B.A. Prusoff, Variability in rates of affective disorders in relatives of depressed and normal probands. *Arch Gen Psychiatry*, 1982; 39(12): 1397–1403.
6. Lopez, A.D. and C.C. Murray, The global burden of disease, 1990–2020. *Nat Med*, 1998; 4(11): 1241–1243.
7. Arsenault-Lapierre, G., C. Kim, and G. Turecki, Psychiatric diagnoses in 3275 suicides: A meta-analysis. *BMC Psychiatry*, 2004; 4: 37.
8. Cavanagh, J.T. et al., Psychological autopsy studies of suicide: A systematic review. *Psychol Med*, 2003; 33(3): 395–405.
9. Angst, F. et al., Mortality of patients with mood disorders: Follow-up over 34–38 years. *J Affect Disord*, 2002; 68(2–3): 167–181.
10. Angst, J., F. Angst, and H.H. Stassen, Suicide risk in patients with major depressive disorder. *J Clin Psychiatry*, 1999; 60(Suppl 2): 57–62; discussion 75–76, 113–116.
11. Angst, J., M. Degonda, and C. Ernst, The Zurich Study: XV. Suicide attempts in a cohort from age 20 to 30. *Eur Arch Psychiatry Clin Neurosci*, 1992; 242(2–3): 135–141.
12. Blair-West, G.W. et al., Lifetime suicide risk in major depression: Sex and age determinants. *J Affect Disord*, 1999; 55(2–3): 171–178.
13. Turecki, G., Dissecting the suicide phenotype: The role of impulsive–aggressive behaviours. *J Psychiatry Neurosci*, 2005; 30(6): 398–408.
14. McGirr, A. and G. Turecki, The relationship of impulsive aggressiveness to suicidality and other depression-linked behaviors. *Curr Psychiatry Rep*, 2007; 9(6): 460–466.
15. Yen, S. et al., Personality traits as prospective predictors of suicide attempts. *Acta Psychiatr Scand*, 2009; 120(3): 222–229.
16. Kwok, S.Y. and D.T. Shek, Socio-demographic correlates of suicidal ideation among Chinese adolescents in Hong Kong. *Int J Adolesc Med Health*, 2008; 20(4): 463–472.
17. Ang, R.P. and Y.P. Ooi, Impact of gender and parents' marital status on adolescents' suicidal ideation. *Int J Soc Psychiatry*, 2004; 50(4): 351–360.
18. Santa Mina, E.E. and R.M. Gallop, Childhood sexual and physical abuse and adult self-harm and suicidal behaviour: A literature review. *Can J Psychiatry*, 1998; 43(8): 793–800.
19. Evans, E., K. Hawton, and K. Rodham, Suicidal phenomena and abuse in adolescents: A review of epidemiological studies. *Child Abuse Negl*, 2005; 29(1): 45–58.
20. Fliege, H. et al., Risk factors and correlates of deliberate self-harm behavior: A systematic review. *J Psychosom Res*, 2009; 66(6): 477–493.
21. Brezo, J., T. Klempan, and G. Turecki, The genetics of suicide: A critical review of molecular studies. *Psychiatr Clin North Am*, 2008; 31(2): 179–203.
22. Ernst, C., N. Mechawar, and G. Turecki, Suicide neurobiology. *Prog Neurobiol*, 2009; 89: 315–333.
23. Szyf, M., The early life environment and the epigenome. *Biochim Biophys Acta*, 2009; 1790(9): 878–85. Epub February 3, 2009.

24. McGowan, P.O., M.J. Meaney, and M. Szyf, Diet and the epigenetic (re)programming of phenotypic differences in behavior. *Brain Res*, 2008; 1237: 12–24.
25. Arnow, B.A., Relationships between childhood maltreatment, adult health and psychiatric outcomes, and medical utilization. *J Clin Psychiatry*, 2004; 65(Suppl 12): 10–15.
26. Molnar, B.E., L.F. Berkman, and S.L. Buka, Psychopathology, childhood sexual abuse and other childhood adversities: Relative links to subsequent suicidal behaviour in the US. *Psychol Med*, 2001; 31(6): 965–977.
27. Dinwiddie, S. et al., Early sexual abuse and lifetime psychopathology: A co-twin-control study. *Psychol Med*, 2000; 30(1): 41–52.
28. Gladstone, G.L. et al., Implications of childhood trauma for depressed women: An analysis of pathways from childhood sexual abuse to deliberate self-harm and revictimization. *Am J Psychiatry*, 2004; 161(8): 1417–1425.
29. Jaffee, S.R. et al., Differences in early childhood risk factors for juvenile-onset and adult-onset depression. *Arch Gen Psychiatry*, 2002; 59(3): 215–222.
30. Bensley, L.S. et al., Self-reported abuse history and adolescent problem behaviors. I. Antisocial and suicidal behaviors. *J Adolesc Health*, 1999; 24(3): 163–172.
31. Ystgaard, M. et al., Is there a specific relationship between childhood sexual and physical abuse and repeated suicidal behavior? *Child Abuse Negl*, 2004; 28(8): 863–875.
32. Spinhoven, P. et al., Childhood sexual abuse differentially predicts outcome of cognitive–behavioral therapy for deliberate self-harm. *J Nerv Ment Dis*, 2009; 197(6): 455–457.
33. Akyuz, G. et al., Reported childhood trauma, attempted suicide and self-mutilative behavior among women in the general population. *Eur Psychiatry*, 2005; 20(3): 268–273.
34. Hawton, K. et al., Deliberate self harm in adolescents: Self report survey in schools in England. *Br Med J*, 2002; 325(7374): 1207–1211.
35. Brezo, J. et al., Predicting suicide attempts in young adults with histories of childhood abuse. *Br J Psychiatry*, 2008; 193(2): 134–139.
36. Fergusson, D.M., J.M. Boden, and L.J. Horwood, Exposure to childhood sexual and physical abuse and adjustment in early adulthood. *Child Abuse Negl*, 2008; 32(6): 607–619.
37. Andover, M.S., C. Zlotnick, and I.W. Miller, Childhood physical and sexual abuse in depressed patients with single and multiple suicide attempts. *Suicide Life Threat Behav*, 2007; 37(4): 467–474.
38. Joiner, T.E., Jr. et al., Childhood physical and sexual abuse and lifetime number of suicide attempts: A persistent and theoretically important relationship. *Behav Res Ther*, 2007; 45(3): 539–547.
39. McHolm, A.E., H.L. MacMillan, and E. Jamieson, The relationship between childhood physical abuse and suicidality among depressed women: Results from a community sample. *Am J Psychiatry*, 2003; 160(5): 933–938.
40. Klose, R.J. and A.P. Bird, Genomic DNA methylation: The mark and its mediators. *Trends Biochem Sci*, 2006; 31(2): 89–97.
41. Kouzarides, T., Chromatin modifications and their function. *Cell*, 2007; 128(4): 693–705.
42. Schratt, G., microRNAs at the synapse. *Nat Rev Neurosci*, 2009; 10(12): 842–849.
43. Schratt, G., Fine-tuning neural gene expression with microRNAs. *Curr Opin Neurobiol*, 2009; 19(2): 213–219.
44. Maunakea, A.K. et al., Conserved role of intragenic DNA methylation in regulating alternative promoters. *Nature*, 2010; 466(7303): 253–257.
45. Pariante, C.M. and S.L. Lightman, The HPA axis in major depression: Classical theories and new developments. *Trends Neurosci*, 2008; 31(9): 464–468.

46. Heim, C. et al., The link between childhood trauma and depression: Insights from HPA axis studies in humans. *Psychoneuroendocrinology*, 2008; 33(6): 693–710.
47. Nemeroff, C.B. and W.W. Vale, The neurobiology of depression: Inroads to treatment and new drug discovery. *J Clin Psychiatry*, 2005; 66(Suppl 7): 5–13.
48. Heim, C. et al., Pituitary–adrenal and autonomic responses to stress in women after sexual and physical abuse in childhood. *JAMA*, 2000; 284(5): 592–597.
49. Heim, C. et al., The dexamethasone/corticotropin-releasing factor test in men with major depression: Role of childhood trauma. *Biol Psychiatry*, 2008; 63(4): 398–405.
50. Francis, D. et al., Nongenomic transmission across generations of maternal behavior and stress responses in the rat. *Science*, 1999; 286(5442): 1155–1158.
51. Liu, D. et al., Maternal care, hippocampal glucocorticoid receptors, and hypothalamic–pituitary–adrenal responses to stress. *Science*, 1997; 277(5332): 1659–1662.
52. McGowan, P.O. et al., Epigenetic regulation of the glucocorticoid receptor in human brain associates with childhood abuse. *Nat Neurosci*, 2009; 12(3): 342–348.
53. Mitchell, J.B., L.J. Iny, and M.J. Meaney, The role of serotonin in the development and environmental regulation of type II corticosteroid receptor binding in rat hippocampus. *Brain Res Dev Brain Res*, 1990; 55(2): 231–235.
54. Smythe, J.W., W.B. Rowe, and M.J. Meaney, Neonatal handling alters serotonin (5-HT) turnover and 5-HT_2 receptor binding in selected brain regions: Relationship to the handling effect on glucocorticoid receptor expression. *Brain Res Dev Brain Res*, 1994; 80(1–2): 183–189.
55. Meaney, M.J., D.H. Aitken, and R.M. Sapolsky, Thyroid hormones influence the development of hippocampal glucocorticoid receptors in the rat: A mechanism for the effects of postnatal handling on the development of the adrenocortical stress response. *Neuroendocrinology*, 1987; 45(4): 278–283.
56. Laplante, P., J. Diorio, and M.J. Meaney, Serotonin regulates hippocampal glucocorticoid receptor expression via a 5-HT_7 receptor. *Brain Res Dev Brain Res*, 2002; 139(2): 199–203.
57. Meaney, M.J. et al., Postnatal handling increases the expression of cAMP-inducible transcription factors in the rat hippocampus: The effects of thyroid hormones and serotonin. *J Neurosci*, 2000; 20(10): 3926–3935.
58. McCormick, J.A. et al., 5′-Heterogeneity of glucocorticoid receptor messenger RNA is tissue specific: Differential regulation of variant transcripts by early-life events. *Mol Endocrinol*, 2000; 14(4): 506–517.
59. Turner, J.D. and C.P. Muller, Structure of the glucocorticoid receptor (*NR3C1*) gene 5′ untranslated region: Identification, and tissue distribution of multiple new human exon 1. *J Mol Endocrinol*, 2005; 35(2): 283–292.
60. Meaney, M.J., Maternal care, gene expression, and the transmission of individual differences in stress reactivity across generations. *Annu Rev Neurosci*, 2001; 24: 1161–1192.
61. Turner, J.D. et al., Highly individual methylation patterns of alternative glucocorticoid receptor promoters suggest individualized epigenetic regulatory mechanisms. *Nucleic Acids Res*, 2008; 36(22): 7207–7218.
62. Weaver, I.C. et al., Epigenetic programming by maternal behavior. *Nat Neurosci*, 2004; 7(8): 847–854.
63. Alt, S.R. et al., Differential expression of glucocorticoid receptor transcripts in major depressive disorder is not epigenetically programmed. *Psychoneuroendocrinology*, 2010; 35(4): 544–556.
64. Weaver, I.C. et al., The transcription factor nerve growth factor-inducible protein a mediates epigenetic programming: Altering epigenetic marks by immediate-early genes. *J Neurosci*, 2007; 27(7): 1756–1768.
65. Kee, B.L., J. Arias, and M.R. Montminy, Adaptor-mediated recruitment of RNA polymerase II to a signal-dependent activator. *J Biol Chem*, 1996; 271(5): 2373–2375.

66. Kim, T.K. et al., Widespread transcription at neuronal activity-regulated enhancers. *Nature*, 465(7295): 182–187.
67. Murgatroyd, C. et al., Dynamic DNA methylation programs persistent adverse effects of early-life stress. *Nat Neurosci*, 2009; 12(12): 1559–1566.
68. Gainer, H., R.L. Fields, and S.B. House, Vasopressin gene expression: Experimental models and strategies. *Exp Neurol*, 2001; 171(2): 190–199.
69. Chen, W.G. et al., Derepression of BDNF transcription involves calcium-dependent phosphorylation of MeCP2. *Science*, 2003; 302(5646): 885–889.
70. Zhou, Z. et al., Brain-specific phosphorylation of MeCP2 regulates activity-dependent BDNF transcription, dendritic growth, and spine maturation. *Neuron*, 2006; 52(2): 255–269.
71. Rhee, H.J., E.J. Kim, and J.K. Lee, Physiological polyamines: Simple primordial stress molecules. *J Cell Mol Med*, 2007; 11(4): 685–703.
72. Fiori, L.M. and G. Turecki, Implication of the polyamine system in mental disorders. *J Psychiatry Neurosci*, 2008; 33(2): 102–110.
73. Gilad, G.M. and V.H. Gilad, Overview of the brain polyamine-stress-response: Regulation, development, and modulation by lithium and role in cell survival. *Cell Mol Neurobiol*, 2003; 23(4–5): 637–649.
74. Gilad, G.M. et al., Developmental regulation of the brain polyamine-stress-response. *Int J Dev Neurosci*, 1998; 16(3–4): 271–278.
75. Minguet, E.G. et al., Evolutionary diversification in polyamine biosynthesis. *Mol Biol Evol*, 2008; 25(10): 2119–2128.
76. Moinard, C., L. Cynober, and J.P. de Bandt, Polyamines: Metabolism and implications in human diseases. *Clin Nutr*, 2005; 24(2): 184–197.
77. Genedani, S. et al., Influence of SAMe on the modifications of brain polyamine levels in an animal model of depression. *Neuroreport*, 2001; 12(18): 3939–3942.
78. Sequeira, A. et al., Implication of SSAT by gene expression and genetic variation in suicide and major depression. *Arch Gen Psychiatry*, 2006; 63(1): 35–48.
79. Sequeira, A. et al., Patterns of gene expression in the limbic system of suicides with and without major depression. *Mol Psychiatry*, 2007; 12(7): 640–655.
80. Dahel, K.A., N.M. Al-Saffar, and K.A. Flayeh, Polyamine oxidase activity in sera of depressed and schizophrenic patients after ECT treatment. *Neurochem Res*, 2001; 26(4): 415–418.
81. Halaris, A. et al., Plasma agmatine and platelet imidazoline receptors in depression. *Ann N Y Acad Sci*, 1999; 881: 445–451.
82. Zomkowski, A.D. et al., Agmatine produces antidepressant-like effects in two models of depression in mice. *Neuroreport*, 2002; 13(4): 387–391.
83. Zomkowski, A.D., A.R. Santos, and A.L. Rodrigues, Putrescine produces antidepressant-like effects in the forced swimming test and in the tail suspension test in mice. *Prog Neuropsychopharmacol Biol Psychiatry*, 2006; 30(8): 1419–1425.
84. Gong, Z.H. et al., Anxiolytic effect of agmatine in rats and mice. *Eur J Pharmacol*, 2006; 550(1–3): 112–116.
85. Fiori, L.M. and G. Turecki, Genetic and epigenetic influences on expression of spermine synthase and spermine oxidase in suicide completers. *Int J Neuropsychopharmacol*, 2010; 13(6): 725–736.
86. Brunoni, A.R., M. Lopes, and F. Fregni, A systematic review and meta-analysis of clinical studies on major depression and BDNF levels: Implications for the role of neuroplasticity in depression. *Int J Neuropsychopharmacol*, 2008; 11(8): 1169–1180.
87. Dwivedi, Y. et al., Altered gene expression of brain-derived neurotrophic factor and receptor tyrosine kinase B in postmortem brain of suicide subjects. *Arch Gen Psychiatry*, 2003; 60(8): 804–815.

88. Pandey, G.N. et al., Brain-derived neurotrophic factor and tyrosine kinase B receptor signalling in post-mortem brain of teenage suicide victims. *Int J Neuropsychopharmacol*, 2008; 11(8): 1047–1061.
89. Chen, B. et al., Increased hippocampal BDNF immunoreactivity in subjects treated with antidepressant medication. *Biol Psychiatry*, 2001; 50(4): 260–265.
90. Sen, S., R. Duman, and G. Sanacora, Serum brain-derived neurotrophic factor, depression, and antidepressant medications: Meta-analyses and implications. *Biol Psychiatry*, 2008; 64(6): 527–532.
91. Matrisciano, F. et al., Changes in BDNF serum levels in patients with major depression disorder (MDD) after 6 months treatment with sertraline, escitalopram, or venlafaxine. *J Psychiatr Res*, 2009; 43(3): 247–254.
92. Chan, J.P. et al., Examination of behavioral deficits triggered by targeting BDNF in fetal or postnatal brains of mice. *Neuroscience*, 2006; 142(1): 49–58.
93. Gronli, J. et al., Chronic mild stress inhibits BDNF protein expression and CREB activation in the dentate gyrus but not in the hippocampus proper. *Pharmacol Biochem Behav*, 2006; 85(4): 842–849.
94. Duric, V. and K.E. McCarson, Hippocampal neurokinin-1 receptor and brain-derived neurotrophic factor gene expression is decreased in rat models of pain and stress. *Neuroscience*, 2005; 133(4): 999–1006.
95. Duric, V. and K.E. McCarson, Effects of analgesic or antidepressant drugs on pain- or stress-evoked hippocampal and spinal neurokinin-1 receptor and brain-derived neurotrophic factor gene expression in the rat. *J Pharmacol Exp Ther*, 2006; 319(3): 1235–1243.
96. Rogoz, Z., G. Skuza, and B. Legutko, Repeated treatment with mirtazepine induces brain-derived neurotrophic factor gene expression in rats. *J Physiol Pharmacol*, 2005; 56(4): 661–671.
97. Xu, H. et al., Synergetic effects of quetiapine and venlafaxine in preventing the chronic restraint stress-induced decrease in cell proliferation and BDNF expression in rat hippocampus. *Hippocampus*, 2006; 16(6): 551–559.
98. Tsankova, N.M. et al., Sustained hippocampal chromatin regulation in a mouse model of depression and antidepressant action. *Nat Neurosci*, 2006; 9(4): 519–525.
99. Roth, T.L. et al., Lasting epigenetic influence of early-life adversity on the *BDNF* gene. *Biol Psychiatry*, 2009; 65(9): 760–769.
100. Aid, T. et al., Mouse and rat *BDNF* gene structure and expression revisited. *J Neurosci Res*, 2007; 85(3): 525–535.
101. Schroeder, F.A. et al., Antidepressant-like effects of the histone deacetylase inhibitor, sodium butyrate, in the mouse. *Biol Psychiatry*, 2007; 62(1): 55–64.
102. Covington, H.E., 3rd et al., Antidepressant actions of histone deacetylase inhibitors. *J Neurosci*, 2009; 29(37): 11451–11460.
103. Keller, S. et al., Increased BDNF promoter methylation in the Wernicke area of suicide subjects. *Arch Gen Psychiatry*, 2010; 67(3): 258–267.
104. Pruunsild, P. et al., Dissecting the human BDNF locus: Bidirectional transcription, complex splicing, and multiple promoters. *Genomics*, 2007; 90(3): 397–406.
105. Duman, R.S. and L.M. Monteggia, A neurotrophic model for stress-related mood disorders. *Biol Psychiatry*, 2006; 59(12): 1116–1127.
106. Kim, Y.K. et al., Low plasma BDNF is associated with suicidal behavior in major depression. *Prog Neuropsychopharmacol Biol Psychiatry*, 2007; 31(1): 78–85.
107. Dwivedi, Y. et al., Neurotrophin receptor activation and expression in human postmortem brain: Effect of suicide. *Biol Psychiatry*, 2009; 65(4): 319–328.
108. Aston, C., L. Jiang, and B.P. Sokolov, Transcriptional profiling reveals evidence for signaling and oligodendroglial abnormalities in the temporal cortex from patients with major depressive disorder. *Mol Psychiatry*, 2005; 10(3): 309–322.

109. Nakatani, N. et al., Genome-wide expression analysis detects eight genes with robust alterations specific to bipolar I disorder: Relevance to neuronal network perturbation. *Hum Mol Genet*, 2006; 15(12): 1949–1962.
110. Mercier, G. et al., MAP kinase activation by fluoxetine and its relation to gene expression in cultured rat astrocytes. *J Mol Neurosci*, 2004; 24(2): 207–216.
111. Rose, C.R. et al., Truncated TrkB-T1 mediates neurotrophin-evoked calcium signalling in glia cells. *Nature*, 2003; 426(6962): 74–78.
112. Ernst, C. et al., Alternative splicing, methylation state, and expression profile of tropomyosin-related kinase B in the frontal cortex of suicide completers. *Arch Gen Psychiatry*, 2009; 66(1): 22–32.
113. Ernst, C., E.S. Chen, and G. Turecki, Histone methylation and decreased expression of TrkB.T1 in orbital frontal cortex of suicide completers. *Mol Psychiatry*, 2009; 14(9): 830–832.
114. Haltiner, M.M., S.T. Smale, and R. Tjian, Two distinct promoter elements in the human *rRNA* gene identified by linker scanning mutagenesis. *Mol Cell Biol*, 1986; 6(1): 227–235.
115. Learned, R.M. et al., Human rRNA transcription is modulated by the coordinate binding of two factors to an upstream control element. *Cell*, 1986; 45(6): 847–857.
116. Ghoshal, K. et al., Role of human ribosomal RNA (rRNA) promoter methylation and of methyl-CpG-binding protein MBD2 in the suppression of *rRNA* gene expression. *J Biol Chem*, 2004; 279(8): 6783–6793.
117. Santoro, R. and I. Grummt, Molecular mechanisms mediating methylation-dependent silencing of ribosomal gene transcription. *Mol Cell*, 2001; 8(3): 719–725.
118. Brown, S.E. and M. Szyf, Epigenetic programming of the rRNA promoter by MBD3. *Mol Cell Biol*, 2007; 27(13): 4938–4952.
119. McGowan, P.O. et al., Promoter-wide hypermethylation of the ribosomal RNA gene promoter in the suicide brain. *PLoS One*, 2008; 3(5): e2085.
120. Cronholm, B. et al., Suicidal behaviour syndrome with low CSF 5-HIAA. *Br Med J*, 1977; 1(6063): 776.
121. Bhagwagar, Z. and P.J. Cowen, 'It's not over when it's over': Persistent neurobiological abnormalities in recovered depressed patients. *Psychol Med*, 2008; 38(3): 307–313.
122. Du, L. et al., Serotonergic genes and suicidality. *Crisis*, 2001; 22(2): 54–60.
123. Turecki, G. et al., Prediction of level of serotonin 2A receptor binding by serotonin receptor 2A genetic variation in postmortem brain samples from subjects who did or did not commit suicide. *Am J Psychiatry*, 1999; 156(9): 1456–1458.
124. Du, L. et al., Association of polymorphism of serotonin 2A receptor gene with suicidal ideation in major depressive disorder. *Am J Med Genet*, 2000; 96(1): 56–60.
125. De Luca, V. et al., Differential expression and parent-of-origin effect of the 5-HT_{2A} receptor gene C102T polymorphism: Analysis of suicidality in schizophrenia and bipolar disorder. *Am J Med Genet B Neuropsychiatr Genet*, 2007; 144B(3): 370–374.
126. Polesskaya, O.O., C. Aston, and B.P. Sokolov, Allele C-specific methylation of the 5-HT_{2A} receptor gene: Evidence for correlation with its expression and expression of DNA methylase DNMT1. *J Neurosci Res*, 2006; 83(3): 362–373.
127. De Luca, V. et al., Methylation and QTDT analysis of the 5-HT_{2A} receptor 102C allele: Analysis of suicidality in major psychosis. *J Psychiatr Res*, 2009; 43(5): 532–537.
128. Klempan, T.A. et al., Altered expression of genes involved in ATP biosynthesis and GABAergic neurotransmission in the ventral prefrontal cortex of suicides with and without major depression. *Mol Psychiatry*, 2009; 14(2): 175–189.
129. Merali, Z. et al., Dysregulation in the suicide brain: mRNA expression of corticotropin-releasing hormone receptors and GABA(A) receptor subunits in frontal cortical brain region. *J Neurosci*, 2004; 24(6): 1478–1485.

130. Torrey, E.F. et al., Neurochemical markers for schizophrenia, bipolar disorder, and major depression in postmortem brains. *Biol Psychiatry*, 2005; 57(3): 252–260.
131. Guidotti, A. et al., Decrease in reelin and glutamic acid decarboxylase67 (GAD67) expression in schizophrenia and bipolar disorder: A postmortem brain study. *Arch Gen Psychiatry*, 2000; 57(11): 1061–1069.
132. Akbarian, S. et al., Gene expression for glutamic acid decarboxylase is reduced without loss of neurons in prefrontal cortex of schizophrenics. *Arch Gen Psychiatry*, 1995; 52(4): 258–266.
133. Heckers, S. et al., Differential hippocampal expression of glutamic acid decarboxylase 65 and 67 messenger RNA in bipolar disorder and schizophrenia. *Arch Gen Psychiatry*, 2002; 59(6): 521–529.
134. Volk, D.W. et al., Decreased glutamic acid decarboxylase67 messenger RNA expression in a subset of prefrontal cortical gamma-aminobutyric acid neurons in subjects with schizophrenia. *Arch Gen Psychiatry*, 2000; 57(3): 237–245.
135. Veldic, M. et al., DNA-methyltransferase 1 mRNA is selectively overexpressed in telencephalic GABAergic interneurons of schizophrenia brains. *Proc Natl Acad Sci USA*, 2004; 101(1): 348–353.
136. Kundakovic, M. et al., DNA methyltransferase inhibitors coordinately induce expression of the human reelin and glutamic acid decarboxylase 67 genes. *Mol Pharmacol*, 2007; 71(3): 644–653.
137. Grayson, D.R. et al., Reelin promoter hypermethylation in schizophrenia. *Proc Natl Acad Sci USA*, 2005; 102(26): 9341–9346.
138. Tamura, Y. et al., Epigenetic aberration of the human *REELIN* gene in psychiatric disorders. *Mol Psychiatry*, 2007; 12(6): 519, 593–600.
139. Zhang, T.Y. et al., Maternal care and DNA methylation of a glutamic acid decarboxylase 1 promoter in rat hippocampus. *J Neurosci*, 2010; 30(39): 13130–13137.
140. Poulter, M.O. et al., GABAA receptor promoter hypermethylation in suicide brain: Implications for the involvement of epigenetic processes. *Biol Psychiatry*, 2008; 64(8): 645–652.
141. Ebersole, T.A. et al., The quaking gene product necessary in embryogenesis and myelination combines features of RNA binding and signal transduction proteins. *Nat Genet*, 1996; 12(3): 260–265.
142. Zhao, L. et al., QKI binds MAP1B mRNA and enhances MAP1B expression during oligodendrocyte development. *Mol Biol Cell*, 2006; 17(10): 4179–4186.
143. Aberg, K. et al., Human QKI, a new candidate gene for schizophrenia involved in myelination. *Am J Med Genet B Neuropsychiatr Genet*, 2006; 141B(1): 84–90.
144. Aston, C., L. Jiang, and B.P. Sokolov, Microarray analysis of postmortem temporal cortex from patients with schizophrenia. *J Neurosci Res*, 2004; 77(6): 858–866.
145. Li, Z.Z. et al., Expression of Hqk encoding a KH RNA binding protein is altered in human glioma. *Jpn J Cancer Res*, 2002; 93(2): 167–177.
146. Siomi, H. et al., The pre-mRNA binding K protein contains a novel evolutionarily conserved motif. *Nucleic Acids Res*, 1993; 21(5): 1193–1198.
147. Klempan, T.A. et al., Characterization of QKI gene expression, genetics, and epigenetics in suicide victims with major depressive disorder. *Biol Psychiatry*, 2009; 66(9): 824–831.

14 Genetics of Suicidal Behavior in Children and Adolescents

Gil Zalsman

CONTENTS

14.1 INTRODUCTION

14.1.1 Suicidal Behavior among Children and Adolescents

Suicide is the third leading cause of death in adolescence in the United States. In addition, nonfatal forms of suicidal behavior are the most common reasons for the psychiatric hospitalization of adolescents in many countries (Center for Disease Control 1994). About 1600 youngsters, age 15–19, committed suicide in the United States in 2001; 3.4 million youngsters in this age group seriously considered suicide; 1.7 million made a suicide attempt; and 590,000 made a suicide attempt sufficiently serious to require medical attention (Grunbaum et al. 2004). It is of note that not all suicidal ideation or behavior in pediatric population is directly attributable to depression (Zalsman et al. 2006c). One major survey, the biennial Youth Risk Behavior Survey (YRBS) data for adolescents (Grunbaum et al. 2004), found that during the preceding 12 months, 28.6% of high school students nationwide had felt so sad or hopeless almost every day for ≥2 weeks in a row that they stopped doing some usual activities; 16.9% of students had seriously considered attempting suicide; 16.5% of students nationwide had made a plan to attempt suicide; 8.5% of students had actually attempted suicide one or more times; and 2.9% of students nationwide had made a suicide attempt that had to be treated by a doctor or nurse. Understanding the precursors of suicidal behavior in youths is important for the treatment and prevention of suicidal behavior in this population (Grunbaum et al. 2004).

Despite suicidal behavior being a major public health problem in the youth, relevant genetic studies are still sparse. Adoption studies suggest that there is a genetic susceptibility to suicide that is partially independent of the presence of a psychiatric disorder (Roy 1983). Roy et al. (1983) found suicide rates for monozygotic and dizygotic twins as 11.3% and 1.8%, respectively. When considered together with results from earlier twin studies, even greater differences were noted: 13.2% in monozygotic pairs versus 0.7% in dizygotic pairs.

These findings are supported by the study of Brent et al. (1996) who screened 58 adolescent suicide probands and concluded that suicidal behavior may be transmitted as a familial trait, regardless of the presence of Axis I or II diagnosis.

Suicidal behavior in adolescents has been found to have biochemical, genetic, and psychological correlates (Apter et al. 1990; Brent et al. 2003; Mann 2003). Suicidal behavior refers to the occurrence of suicide attempts and ranges from fatal acts (completed suicide) and high-lethality and failed suicide attempts (where serious intention and careful planning are evident, and survival is fortuitous) to low-lethality attempts.

Usually impulsive attempts that are triggered by social crisis seem to be ambivalent and contain a strong element of appeal for help (Beck et al. 1976; Stengel 1973). Intent and lethality are positively correlated and related to biological abnormalities that mostly involve the serotonergic system. The clinical and neurobiological study of failed suicides can provide information about completed suicide because the two populations are clinically and demographically similar (Mann 2003).

Among adolescents, the annual rate of suicide attempts that require medical attention is 2.6%, while suicide is much less prevalent. Among 15–19 year olds, the rates in 1998 were 14.6 per 100,000 in boys and 2.9 per 100,000 in girls (Brent 2002). There is a strong relationship between attempted suicide in adolescent psychiatric patients and eventual death from suicide (Garrison et al. 1991). Predisposition to suicidal behavior might be genetically transmitted as a trait independent of Axis I or II diagnosis (Brent et al. 1996). Suicide and suicidal behavior of adolescents are linked to a wide variety of psychiatric disorders, including affective illness, alcohol and substance abuse, conduct disorder, and schizophrenia (Shaffer 1998). Over 90% of adolescent and adult suicide victims appear to have at least one Axis I disorder (Brent 1995).

Several neurobiological systems were linked to suicide and suicidal behavior; mainly from neuroendocrine studies of the hypothalamic axes and studies of the serotonergic system (5-HT). Data were collected utilizing several methodologies such as hormonal suppression tests, sleep studies, postmortem studies, and genetic factor analysis. Currently, it is believed that the most plausible biological system related to suicidality, impulsive violence, and anxiety is the serotonergic system (Apter et al. 1990, 1993b; Mann 2003; Zalsman et al. 2006c).

14.2 RISK FACTORS FOR SUICIDE IN PEDIATRIC AGE GROUP

Risk factors for suicidality are discussed elsewhere in this clinical book and are not specific for children and adolescents. Suicide and suicidal behavior of adolescents are linked to a wide variety of psychiatric disorders, including affective illness, alcohol and substance abuse, conduct disorder, and schizophrenia (Shaffer 1998). Over 90% of adolescent and adult suicide victims appear to have at least one Axis I disorder (Brent 1995).

Since the majority of psychiatric patients do not attempt or commit suicide, it appears that psychiatric disorder may be necessary but not a sufficient cause for suicide. In addition to other investigators (for a review, see Mann 2002, 2003), we have found evidence to suggest that certain psychopathological dimensions, namely, a tendency to impulsive aggression and anxiety, may predispose to suicidal behaviors and suicide in this age group (Apter et al. 1991, 1993a,b, 1995) independently from or interactively with psychiatric disorder and that the risk for suicide is substantially increased when psychiatric disorder and certain personality traits occur together.

Currently, the most common biological correlated system related to suicidality, impulsive violence, and anxiety is the serotonergic system (Apter et al. 1993b; Mann 2003; Zalsman et al. 2006b). Suicide attempters have been shown to have lower responsivity and sensitivity of their platelet 5-HT_2 receptors than nonattempters. There appears to be a high correlation between the medical damage resulting from a suicide attempt and the number of 5-HT receptors (Apter et al. 1993b; Mann 2003; Zalsman et al. 2006b). Following the finding of higher B_{max} in the 5-HT_{2A} receptors in suicidal patients independent of diagnosis (Pandey et al. 1995, 2002a, 2004), postmortem studies have demonstrated higher 5-HT_{2A} receptor binding in postmortem brains of adolescent suicide victims (Pandey et al. 2002a), hence replicating the abnormalities observed in adults.

Low platelets, low serum, and low cerebrospinal fluid (CSF) 5-hydroxyindoleacetic acid (5-HIAA) levels were found to be related with suicidal behavior, especially more lethal suicide attempts, regardless of psychiatric diagnosis (Tyano et al. 2006; Zalsman et al. 2005a, 2006b).

14.3 INVOLVEMENT OF THE NEUROENDOCRINE SYSTEM

14.3.1 Hypothalamic–Pituitary–Adrenal Axis

Administration of dexamethasone in healthy adults suppresses the release of cortisol (dexamethasone suppression test [DST]). Nonsuppression or early escape from dexamethasone suppression is considered partly the result of corticosteroid receptor downregulation at the level of the hippocampus, hypothalamus, or pituitary and contributes to hypothalamic–pituitary–adrenal (HPA) axis hyperactivity in depressed adults. Some evidence in adults indicates DST nonsuppression as a predictor of suicide, but not nonfatal suicidal behavior (Coryell and Schlesser 2001). Dahl et al. (1992a) found no difference between suicidal and non-suicidal depressed children on the DST. In another study (Pfeffer et al. 1991), it was found that although there was no association between suicidal behavior and DST nonsuppression in prepubertal children, the patients with persistent suicidal behavior had significantly higher pre-dexamethasone 4 p.m. cortisol levels. In a comparison of depressed adolescents and healthy controls, it was found that both had similar cortisol secretion patterns with the exception of sleep onset when the cortisol secretion in the depressed group was higher. Most of this difference has contributed to suicidal or inpatient depressed adolescents (Dahl et al. 1991). Interestingly, sexually abused girls had significantly lower basal, net, and total corticotropin-releasing-hormone-stimulated ACTH response in comparison to non-abused subjects, but higher rates of suicidal ideation, suicide attempts, and dysthymia (De Bellis et al. 1994). Because the results from adult studies have shown that, in general, the HPA function is altered more frequently in older patients and dexamethasone resistance becomes more common with age, age-related difference may explain some of the discrepancies found in studies of children and adolescents (Zalsman et al. 2006c).

14.3.2 Growth Hormone

Growth hormone (GH) secretion, from the anterior pituitary, is stimulated by GH-releasing hormones (GHRH) and inhibited by somatostatin. Other stimulants of GH release are the α_2-adrenoreceptors, muscarinic cholinergic receptors, and activation of dopamine receptors. In depressed children and adolescents, increased nocturnal secretion of GH was reported (Kutcher et al. 1991; Puig-Antich et al. 1984). It may be that muscarinic hypersensitivity at sleep onset leads to excessive somatostatin inhibition and increased release of GH. Muscarinic hypersensitivity could be the consequence of serotonin deficit, since under normal conditions serotonin modulates the cholinergic surge that accompanies sleep onset (Zalsman et al. 2006c).

Ryan et al. (1988) suggested that lower GH secretion in the response to desmethylimipramine in depressed adolescents may be related to the presence of a suicide plan or attempt during the depressive episode. Dahl et al. (1992b) found no difference between the adolescents with major depression and controls on GH measures. However, the group with depression and suicidal behavior (definite plan or attempt) showed a blunting of sleep GH secretion compared with the non-suicidal group. In a follow-up of the subjects, 10 years after the original study, 13 (38.2%) of the control group adolescents developed some form of depressive disorder, whereas 21 remained depression free. Thirteen subjects (23%) from the depressive group (n = 56) made suicide attempts during the follow-up period. The original data from the sleep studies and GH examination were compared again, reclassifying subjects based on the updated data. During the follow-up period, suicide attempters secreted more GH during the first 4 h of sleep and during the whole 24 h period compared to the other three groups. The depressive non-suicidal group had demonstrated an earlier, steeper GH release curve when compared with depression-free controls, whereas the depressed, suicide-attempting subjects did not differ from controls. The role of GH on adolescents' suicidality remains unclear and needs further investigation.

14.3.3 Serotonergic System

There is accumulating evidence of serotonergic system disruption in suicidality in adults as well as in children and adolescents (Pfeffer et al. 1998; Tyano et al. 2006). There appears to be a high correlation between the medical damage resulting from a suicide attempt and the number of 5-HT receptors in a given individual (Mann et al. 1992). Tyano et al. (2006) compared relationship of measured plasma serotonin and psychometric measures between depressed suicidal adolescents and controls (n = 211). A significant negative correlation was found between plasma serotonin levels and the severity of suicidal behavior among the suicidal inpatients; however, there was no difference in serotonin levels among psychiatric diagnostic categories. The biology of adolescent suicide attempters is understudied and seems to be somewhat different in terms of their response to serotonin-related medication (Brent and Birmaher 2002). Higher levels of B_{max} were found in the 5-HT_{2A} receptors in suicidal patients, independent of diagnosis (Pandey et al. 1995). Postmortem studies have demonstrated higher 5-HT_{2A} receptor binding in postmortem brains of adolescent suicide victims (Pandey et al. 2002b), hence replicating the abnormalities observed in adults.

In adults, low CSF 5-HIAA levels have been found to be related to suicidal behavior, especially more lethal suicide attempts regardless of psychiatric diagnosis (Mann et al. 1992). Moreover, low CSF 5-HIAA has been shown to be predictive of violent suicide attempts and suicide, suggesting that those measures represent trait rather than state measures (Traskman and Asberg 1986). A 2 year follow-up of a group of children and adolescents with disruptive behavior disorders showed that CSF 5-HIAA level predicted severity of physical aggression during follow-up but did not predict suicide attempters (Kruesi et al. 1992). It may be that these results are still preliminary since this group is at risk for suicidality.

14.3.4 Neuroendocrine Challenge Studies

The prolactin response to fenfluramine is an important index of serotonergic function (Malone et al. 1996; Sher et al. 2003). Fenfluramine causes the release of serotonin and inhibits serotonin reuptake. In adults, a history of highly lethal suicide attempts is associated with blunted prolactin response to fenfluramine (Malone et al. 1996). Patients with a history of a very lethal suicide attempts have a blunted prolactin response compared with individuals with major depression but no history of a very lethal suicide attempts, and like low CSF 5-HIAA, a blunted prolactin response to fenfluramine may be a biochemical trait.

In a study (Kaufman et al. 1998) where groups of depressed abused children, depressed non-abused children, and nondepressed, non-abused children were given L-5-hydroxytryptophan challenge in order to examine the measure of prolactin release, it was found that children with a positive family history for suicide attempts, from all three groups, had a more robust prolactin release after the challenge, in comparison to children without a family history of suicide attempts. These results suggested that suicidal behavior in youth is associated with a different pattern of serotonergic abnormality compared with adults, a pattern that may include greater 5-HT_{1A} or 5-HT_{2A} receptor responsiveness (the serotonin affecting prolactin release) (Zalsman et al. 2005b, 2006c).

14.3.5 Platelet Studies

Platelets have been used as a less invasive approach to model brain serotonin neurons, since they share many properties—that is, both are derived from the embryonic neural crest and contain serotonin, 5-HT_{2A} receptors, MAO-B, and serotonin transporters—an approach that is more suitable for the pediatric age group. Yet, this method raised questions regarding its validity, mainly since platelets are not part of the neural circuit that can be modulated by other neurotransmitter systems. A study (Ambrosini et al. 1992) that examined platelet serotonin transporter binding (B_{max}) using ^{3}H-imipramine in depressed children found inverse relationship between ^{3}H-imipramine binding and presence of suicidal behavior. In contrast, Pine et al. (1995) compared ^{3}H-imipramine binding in platelets of adolescents with conduct disorder who had attempted suicide with those of non-suicidal controls with conduct disorder. They found no difference between the two groups; however, the suicidal group showed significant seasonal variations of ^{3}H-imipramine binding with the lowest amount in late winter/early spring.

Serotonin uptake by platelets may represent a number of serotonin transporter binding sites on the membranes of platelets. Suicidality was not found as a factor affecting the lower V_{max} values of depressed adolescents (Modai et al. 1989). The serotonin amplified aggregation of children with mood disorders although lower tryptophan levels were found in the blood of children who had recently attempted suicide as compared to the levels found in the blood of normal controls (Pffefer et al. 1998). Various measures of serotonin receptor contents of precursors in platelets have not yielded a consistent picture (Zalsman et al. 2005a). A significant positive correlation between platelet serotonin transporter density and anger scores and a negative correlation between platelet count and trait anxiety were observed in a suicidal versus a non-suicidal group of adolescent inpatients, although the serotonin transporter promoter (5-HTTLPR) polymorphism was not associated with transporter binding and suicidality (Zalsman et al. 2005).

14.3.6 Postmortem Studies

Pandey et al. (2002b) examined the right prefrontal cortex (Broadmann area 8/9) in 15 teenage suicide victims compared with 15 matched normal subjects. They found that 5-HT_{2A} receptor protein and mRNA levels were higher in the prefrontal cortex and hippocampus but not in the nucleus accumbens of adolescent suicide victims; these higher levels were restricted to the pyramidal cells of layer V. In adults, most autopsy studies of suicide victims reveal various abnormalities in serotonin function in the brain stem and prefrontal cortex, where more 5-HT_{1A} and 5-HT_{2A} binding and lower serotonin transporter bindings are reported, mostly in the ventral prefrontal cortex (Arango et al. 1995). Serotonergic function may be a mechanism whereby genetic factors influence the suicide threshold. Its stability as shown by genetic studies of CSF 5-HIAA over time suggests that serotonergic function is a trait and may contribute to its potential use as a predictor of suicide and suicide attempts (Traskman-Bendz et al. 1992).

14.3.7 Sleep-Related Studies

In addition to disturbances in the quality of sleep reported by adult suicidal patients, some studies have also shown sleep polysomnography abnormalities when comparing to non-suicidal patients, including increased rapid eye movement (REM) time and REM activity (Sabo et al. 1991). In children and adolescents, the findings have not been consistent. Some studies have shown no difference between suicidal and non-suicidal depressed adolescents (Rao et al. 1996), while others show higher sleep latency in suicidal adolescents (Khan and Todd 1990) and possibly shorter REM latency compared with other depressive adolescents (Armitage et al. 2001) as well as significantly higher REM density among suicidal depressives (Goetz et al. 1991). Considering the relationship between sleep and the serotonin system, further studies are needed to clarify the relationship between sleep abnormalitics and suicidal behavior.

14.4 FAMILIAL TRANSMISSION OF SUICIDAL BEHAVIOR

Suicidal behavior runs in families (Brent et al. 1996; Egeland and Sussex 1985; Zalsman et al. 2008). Despite great variability in the methodology of the studies, results consistently demonstrate the familial aggregation of suicidal ideation and behavior. Longitudinal community studies show that family history of suicidal behavior is one among other precursors of youthful suicidal behavior such as depression, suicidal ideation, behavioral symptoms, and child maltreatment (Shaffer 1998).

Family studies cannot distinguish between genetic and environmental causes of transmission. Findings of family-based aggregation of suicidal ideation and behavior can be attributed to unique diathesis, genetic transmission, mediated by transmission of psychiatric disorder, imitation, or shared environmental effect (Brent and Mann 2005). The literature provides clues as to the mechanisms and precursors of familial transmission of suicidal behavior (Zalsman 2010). The transmission of suicidal behavior may be through pure genetic transmission (specific genes and loci), pure environmental transmission (modeling, abuse, etc.), or interaction of these two factors (gene–environment interaction [GXE]) (Brent et al. 1996; Brent and Mann 2005; Mann et al. 2001; Zalsman 2010). The next paragraphs will review the evidence for each of these models.

14.5 GENETIC FACTORS LINKED TO SUICIDALITY

Suicidal behavior runs in families, probably as a factor independent from major depression. Adults who commit suicide or attempt suicide have a higher rate of familial suicidal acts (Roy 1983). Rates for suicide (Roy et al. 1991) and suicide attempts (Roy et al. 1995) are higher in monozygotic versus dizygotic twins. Adoption studies (Schulsinger et al. 1979) have shown a higher reported rate of suicide in the biological parents of adoptees who commit suicide compared with biological relatives of control adoptees, even after controlling for rates of psychosis and mood disorders. Heritability of suicide and suicide attempts is comparable to the heritability of other psychiatric disorders, such as bipolar disorder and schizophrenia (Statham et al. 1998). Several studies have shown a higher rate of suicidal behavior in families of children and adolescents who have attempted or completed suicide, compared with controls, independent of the diagnosis of depression (Pfeffer et al. 1994; Shafii et al. 1985).

These results could be caused by the genetic transmission of underlying psychiatric illnesses or nongenetic family influences, such as imitation or exposure to violence in the family. Brent et al. (1996) conducted a family study of adolescent suicide victims and 55 demographically similar controls. The rate of suicide attempts in first-degree relatives of suicide victims was higher than the controls, even after adjusting for differences in rates of Axis I and II diagnoses in the families, including depression, which suggested that suicidal behavior may be transmitted as an independent factor from Axis I and II disorders. Several studies have attempted to locate genetic markers for suicide, and interest focused on genes for serotonergic systems.

Although genetic factors account for ~45% of the variance in suicidal thoughts and behavior, the specific genes that contribute to vulnerability for suicidal behavior are unknown despite numerous candidate genes association studies, especially relating to the serotonergic system (Mann 2003; Zalsman et al. 2002).

The family-based study design, collecting case-parent triads and other available first-degree relatives, will enrich genetic loading of the sample population and guard against potential population substructure (Zalsman et al. 2001a).

Another strategy that is wildly used in adults is the postmortem studies of suicide victims. This allows a combination of the most serious form of suicidal behavior, genotyping, and use of tissue analyses to determine biological intermediate phenotypes such as the gene transcription and proteins. Postmortem studies in adolescent suicide are still rare (Pandey et al. 2002b). Studies in adolescent population using such combined opportunities may add to our understanding of this fatal behavior in youth.

14.5.1 Candidate Genes

The promoter region of the serotonin transporter gene (*5-HTTLPR*) polymorphism was the center of investigation in relation to depression and suicidality.

Caspi et al. (2003) were the first to demonstrate a gene by environment interaction of this polymorphism with suicidality in young people. The short form of the allele (SL/SS) indicates lower expression of the serotonin transporter, while the opposite is true for the long form (LL). It seems that the long form comes in two varieties: LA express as the short form and LG as the long form (known as "triallelic polymorphism," see Zalsman et al. 2006a). No association was found between the different alleles and depression or suicidality in adolescents. Possible association with aggressive traits was reported (Zalsman et al. 2001a). In the pediatric age group, the candidate genes *5-HT2A*, *COMT*, *TPH1*, and *DRD4III* were also investigated to be associated with depression or suicidality (Zalsman et al. 2001a,b).

14.5.2 Family-Based Studies of Suicidal Behavior

For the child psychiatrist who works with families, the family-based studies are a popular genetic research strategy that have demonstrated advantages over case-control association studies (Schaid and Sommer 1994), particularly its low false-positive and false-negative results in the event of population stratification. The haplotype relative risk (HRR) method (Ott 1989) assesses "trios" of the patient and both biological parents. The alleles transmitted to the patient from the parents are compared with the alleles that were not transmitted. The non-transmitted parental alleles serve as controls. If only one parent is available, parent–child duos may be used in some cases. HRR also enables smaller samples than the case-control method (Ott 1989). The transmission disequilibrium test (TDT) (Spielman and Ewens 1996) analyzes transmitted versus non-transmitted alleles from heterozygote parents as another indicator of linkage disequilibrium. Our group and others have conducted a number of family-based studies on the genetics of suicide in adolescents with contradictory results. No one locus has yet been identified as significantly associated with suicidality, even when family-based methods are used.

14.6 GENE–ENVIRONMENT INTERACTION

Stressful life events (SLE) seem to have an etiologic influence on the vulnerability to depression and suicide. This vulnerability may be caused by genetic factors. The 5-HTTLPR polymorphism was found to moderate the influence of SLE on depression (Caspi et al. 2003); young adults with one or two copies of the short allele exhibited more depressive symptoms and suicidality in response to SLE than individuals who are homozygous for the long allele. This finding was replicated recently in children, adolescents, and young adults (Zalsman et al. 2006a).

14.6.1 Gene–Environment Interactions in Youth Suicidality

A GXE may explain some of the variance in the relationship between SLE and the development and severity of an episode of major depression and suicidal behavior in young people. Caspi et al. (2003) demonstrated that depressive symptoms increased between age 21 and 26 in individuals carrying at least one copy of the S allele of the 5-HTTLPR polymorphism who experienced SLEs after age 21. Furthermore, life events occurring after age 21 predicted depression and suicide ideation or attempt at age 26 among carriers of the S allele who did not have a prior history of depression ($P = 0.02$).

Nonhuman primates having the S allele experience a persistent decline in serotonin activity after maternal separation as measured by CSF 5-hydroxyindoleacetic acid (Suomi 2003). In humans, a decline in serotonin function results in greater aggression, impulsiveness, and risk-taking as an adult. Childhood separation from parents, sexual and physical abuse, and adult losses may precede the onset of major depression and suicidality (Kaufman et al. 2004; Kendler and Prescott 1999; Paykel 1983). Childhood adversity may produce a biological and clinical diathesis for mood disorder that endures into adulthood. SLEs associate with onset of major depression, modulated, at least in part, via an interaction with genetic predisposition (Kaufman et al. 2004; Kendler and Prescott 1999; Moffitt et al. 2005).

Life events predict depression and suicidal ideation or a suicide attempt in children, adolescents, and young adults who carry the S allele of the 5-HTTLPR polymorphism (Caspi et al. 2003; Eley et al. 2004; Kaufman et al. 2004). Analyzing strictly at severe levels of maltreatment requiring separation from parents, Kaufman et al. (2004) found only a potentiating effect of the S allele on the severity of depression. They also showed that therapy ameliorates the effect of the S allele. This relationship needs further study.

We replicated Kaufman et al.'s findings (Zalsman et al. 2006a) using a third functional allele that was originally described by Nakamura et al. (2000). This allele, L_G, is equivalent in expression to the S allele. The allelic frequency of L_G is 0.09–0.14 in Caucasians and 0.24 in African-Americans. Combined with the observation that the three alleles, S, L_A, and L_G, appear to act codominantly, this may partly explain discordant findings in the literature.

The complexity of the GXEs needs to be considered in both directions. For example, Merikangas and Spiker (1982) found that depressed persons tend to select mates with psychiatric illness and therefore increase the genetic risk for depression and suicide in their offspring.

Despite the aforementioned, recent meta-analysis has not yielded evidence that the serotonin transporter genotype alone or in interaction with SLE is associated with an elevated risk of depression in men alone, women alone, or in both sexes combined (Risch et al. 2009). In contrary, a larger meta-analysis has shown a strong interaction and association (Karg et al. 2011), while other replications and nonreplications emphasize the debate (Carli et al. 2011; Lesch 2011; Uher et al. 2011). Thus, the question, whether another interaction is playing a role, especially in the developing brain, arises.

14.7 GENE–ENVIRONMENT AND TIMING

The child psychiatrist who studies behaviors generally uses the *developmental psychopathology* paradigm. One concept in this paradigm, the equifinality approach, conceptualizes behavior as an outcome of many etiologies, including genetics, SLE, nonadaptive cognitions, abnormal affect regulation, low self-esteem, neuroendocrine dysregulation, and defects in brain development.

The recent studies of Giedd et al. (1999) and others show how the adolescent brain develops. Evidence of increased white matter volume with accelerated myelination and synaptic pruning in the cerebral cortex elucidates our understanding of the changes in the developing brain. Studies using magnetic resonance imaging (MRI) and functional MRI (fMRI) demonstrate both *age-related* changes and *gender-related* differences in gray and white matter during adolescents. These findings may explain in part why many psychiatric disorders emerge during adolescence including depression, bipolar disorders, and risk-taking behavior. Suicidal behavior is strongly related to all of these (Giedd et al. 1999; Paus et al. 1999). Such findings may also

TABLE 14.1
Published Studies on Genetics of Adolescent Suicide

Reference	Population	Polymorphisms	Main Findings
Zalsman et al. (2001b)	Family-based study (HRR): 88 inpatient adolescents of Jewish origin who recently attempted suicide and both biological parents of 40 subjects and from one parent of 9 subjects	A218C in intron 7 of tryptophan hydroxylase (*TPH*) gene	HRR method (chi-square = 0.094; $P = 0.76$), the TDT (chi-square = 0.258; $P = 0.61$), or association analysis to known population frequencies (chi-square = 1.667, $P = 0.19$ for Ashkenazi, and chi-square = 0.810, $P = 0.37$ for non-Ashkenazi). Analysis of variance with the Scheffe test demonstrated a significant difference between CC and AA genotypes in suicide risk and depression among the patients ($n = 88$). The findings suggest that polymorphism A218C has no major relevance to the pathogenesis of adolescent suicidal behavior but may have a subtle effect on some related phenotypes
Zalsman et al. (2001a)	Forty-eight Israeli inpatient adolescents who recently attempted suicide using the haplotype relative risk (HRR)	The serotonin transporter-linked promoter region polymorphism (5-HTTLPR)	No significant allelic association of the 5-HTTLPR polymorphism with suicidal behavior was found. Analysis of variance demonstrated a significant difference in violence measures between patients carrying the LL and LS genotypes
Zalsman et al. (2004)	Sixty-nine Israeli inpatient suicidal adolescents who recently attempted suicide and 167 healthy control subjects	Dopamine receptor subtype 4 (*DRD4*) gene exon III 48 bp repeat polymorphism	No significant association between the DRD4 polymorphism and suicidal behavior was found. Analysis of the suicide-related measures demonstrated a significant difference in depression severity between suicidal inpatients homozygote and heterozygote for the DRD4 alleles

(*continued*)

TABLE 14.1 (continued)
Published Studies on Genetics of Adolescent Suicide

Reference	Population	Polymorphisms	Main Findings
Zalsman et al. (2005a)	Thirty-two suicidal and 28 non-suicidal Ashkenazi Israeli adolescent psychiatric inpatients	5-HTTLPR polymorphism and platelet transporter binding	The 5-HTTLPR polymorphism was not associated with transporter binding or with suicidality or other clinical phenotypes. However, in the suicidal group, a significant positive correlation between platelet SERT density and anger scores and a negative correlation between platelet count and trait anxiety were observed
Zalsman et al. (2005b)	A family-based method (HRR): 30 families of inpatient adolescents from Jewish Ashkenazi origin, with a recent suicide attempt	5-HT(2A) receptor gene polymorphism T102C	No difference was found in allelic distribution between transmitted and non-transmitted alleles. There was no significant association of genotype with any of the clinical traits
Cicchetti et al. (2010)	Eight hundred and fifty low-income children (478 maltreated; 372 non-maltreated) with self-reported depressive and suicidal symptoms	5-HTTLPR	Higher suicidal ideation was found among maltreated than non-maltreated children; the groups did not differ in 5-HTTLPR genotype frequencies. Children with one to two maltreatment subtypes and SS or SL genotype had higher suicidal ideation than those with the LL genotype; suicidal ideation did not differ in non-maltreated children or children with three to four maltreatment subtypes based on 5-HTTLPR variation
Zalsman et al. (2010)	Four groups of adolescents were included: suicidal ($N = 35$) and non-suicidal ($N = 30$) psychiatric inpatients, suicide attempters admitted to three psychiatric emergency rooms ($N = 51$), and a community-based control group ($N = 95$)	HTR2A (102T/C), 5-HTTLPR, MAOA, and plasma serotonin	Homozygosity for the T allele of the HTR2A 102T/C polymorphism was associated with lower impulsivity and aggression compared to TC carriers. Low activity MAOA genotypes were associated with suicidality. No association was found with p5HT level

explain the gender differences observed in adolescent suicidal behavior: females tend more to attempt suicide, while males tend more to commit suicide. Gender differences are found in GXE studies in the young. Examining children of both sexes, Eley et al. (2004) found a significant GXE interaction in depression for females only.

Suicidality has been associated with novelty seeking and risk-taking behaviors in adolescents. Recent fMRI studies have demonstrated age-related differences in the nucleus accumbens reaction to sensation and reward during anticipation of monetary gain (Giedd et al. 1999; Paus et al. 1999).

I suggest that limiting the study of suicidal behavior in adolescents to the GXE may miss another factor in the complex interplay, that being timing. One may speculate that only when a *specific genotype* (e.g., S and maybe the L_G allele of the 5-HTTLPR polymorphism) is exposed to a specific *environmental risk* (e.g., child adversity) during a *critical window* of brain development (e.g., specific stage in pruning) will the outcome be suicidality. Surely, this speculation needs evidence-based investigation. Meanwhile, we should keep our depressed adolescents away from means of suicide and keep educating gatekeepers to watch for distress and hopelessness in youngsters in their immediate environment (Table 14.1).

14.8 CONCLUSIONS AND FUTURE DIRECTIONS

It appears that suicidal behavior in children and adolescents correlates with several neurobiological evidences, independent of underlying psychiatric disorders. These include disturbances of the HPA axis as well as GH secretion irregularities. However, the serotonergic system disturbances, as manifested by postmortem findings, serotonin receptor abnormalities on platelets, and metabolite levels as well as genetic studies seem to be the clues for the mechanisms underlying suicidality that are most investigated and considered most critical.

In the future, it is hoped that utilizing a combination of neurobiological tools such as genetic vulnerability, environmental factors, and maybe levels of 5-HIAA in the CSF will help recognizing populations at high risk for suicide.

REFERENCES

Ambrosini PJ, Metz C, Arora RC, Lee JC, Kregel L, Meltzer HY. 1992. Platelet imipramine binding in depressed children and adolescents. *J Am Acad Child Adolesc Psychiatry* 32:298–305.

Apter A, Bleich A, King RA, Kron S, Fluch A, Kotler M, Cohen DJ. 1993a. Death without warning? A clinical postmortem study of suicide in 43 Israeli adolescent males. *Arch Gen Psychiatry* 50:138–142.

Apter A, Gothelf D, Orbach I, Weizman R, Ratzoni G, Har-Even D, Tyano S. 1995. Correlation of suicidal and violent behavior in different diagnostic categories in hospitalized adolescent patients. *J Am Acad Child Adolesc Psychiatry* 34:912–918.

Apter A, Kotler M, Sevy S, Plutchik R, Brown SL, Foster H, Hillbrand M, Korn ML, van Praag HM. 1991. Correlates of risk of suicide in violent and nonviolent psychiatric patients. *Am J Psychiatry* 148:883–887.

Apter A, Plutchik R, van Praag HM. 1993b. Anxiety, impulsivity and depressed mood in relation to suicidal and violent behavior. *Acta Psychiatr Scand* 87:1–5.

Apter A, van Praag HM, Plutchik R, Sevy S, Korn M, Brown SL. 1990. Interrelationships among anxiety, aggression, impulsivity, and mood: A serotonergically linked cluster? *Psychiatry Res* 32:191–199.

Arango V, Underwood MD, Gubbi AV, Mann JJ. 1995. Localized alterations in pre and postsynaptic serotonin binding sites in the ventrolateral prefrontal cortex of suicide victims. *Brain Res* 688:121–133.

Armitage R, Emsile GJ, Hoffmann RF, Rintelmann J, Rush AJ. 2001. Delta sleep EEG in depressed adolescent females and healthy controls. *J Affect Disord* 63:139–148.

Beck AT, Weissman A, Lester D, Trexler L. 1976. Classification of suicidal behaviors. II. Dimensions of suicidal intent. *Arch Gen Psychiatry* 33:835–837.

Brent DA. 1995. Risk factors for adolescent suicide and suicidal behavior. *Suicide Life Threat Behav* 19:52–63.

Brent DA, Birmaher B. 2002. Adolescent depression. *N Engl J Med* 347:667–671.

Brent DA, Bridge J, Johnson BA, Connolly J. 1996. Suicidal behavior runs in families. A controlled family study of adolescent suicide victims. *Arch Gen Psychiatry* 53:1145–1152.

Brent DA, Oquendo M, Birmaher B, Greenhill L, Kolko D, Stanley B, Zelazny J, Brodsky B, Bridge J, Ellis S, Salazar JO, Mann JJ. 2002. Familial pathways to early-onset suicide attempt: risk for suicidal behavior in offspring of mood-disordered suicide attempters. *Arch Gen Psychiatry* 59:801–807.

Brent DA, Oquendo M, Birmaher B, Greenhill L, Kolko D, Stanley B, Zelazny J, Brodsky B, Firinciogullari S, Ellis SP, Mann JJ. 2003. Peripubertal suicide attempts in offspring of suicide attempters with siblings concordant for suicidal behavior. *Am J Psychiatry* 160:1486–1493.

Brent DA, Mann JJ. 2005. Family genetic studies, suicide, and suicidal behavior. *Am J Med Genet C Semin Med Genet* 133:13–24.

Carli V, Mandelli L, Zaninotto L, Roy A, Recchia L, Stoppia L, Gatta V, Sarchiapone M, Serretti A. 2011. A protective genetic variant for adverse environments? The role of childhood traumas and serotonin transporter gene on resilience and depressive severity in a high-risk population. *Eur Psychiatry* 26:471–478.

Caspi A, Sugden K, Moffitt TE, Taylor A, Craig IW, Harrington H, McClay J, Mill J, Martin J, Braithwait A, Poulton R. 2003. Influence of life stress on depression: Moderation by a polymorphism in the *5-HTT* gene. *Science* 301:386–389.

Center for Disease Control. 1994. Deaths resulting from firearm and motor vehicle injuries in the United States, 1968–1991 *MMWR* 43(3):37–42.

Cicchetti D, Rogosch FA, Sturge-Apple M, Toth SL. 2010. Interaction of child maltreatment and 5-HTT polymorphisms: Suicidal ideation among children from low-SES backgrounds. *J Pediatr Psychol* 35:536–546.

Coryell W, Schlesser M. 2001. The dexamethasone suppression test and suicide prediction. *Am J Psychiatry* 158:748–753.

Dahl RE, Kaufman J, Ryan ND, Perel J, al-Shabbout M, Birmaher B, Nelson B, Puig-Antich J. 1992a. The dexamethasone suppression test in children and adolescents: A review and controlled study. *Biol Psychiatry* 32:109–126.

Dahl RE, Ryan ND, Puig-Antich J, Nguyen NA, al-Shabbout M, Meyer VA, Perel J. 1991. 24-Hour cortisol measures in adolescents with major depression: A controlled study. *Biol Psychiatry* 30:25–36.

Dahl RE, Ryan ND, Williamson DE, Ambrosini PJ, Rabinovich H, Novacenko H, Nelson B, Puig-Antich J. 1992b. Regulation of sleep and growth hormone in adolescent depression. *J Am Acad Child Adolesc Psychiatry* 31:615–621.

De Bellis MD, Chrousos GP, Dorn LD, Burke L, Helmers K, Kling MA, Trickett PK, Putnam FW. 1994. Hypothalamic–pituitary–adrenal axis dysregulation in sexually abused girls. *J Clin Endocrinol Metab* 78:249–255.

Egeland JA, Sussex JN. 1985. Suicide and family loading for affective disorders. *JAMA* 254:915–918.

Eley TC, Sugden K, Corsico A, Gregory AM, Sham P, McGuffin P, Plomin R, Craig IW. 2004. Gene–environment interaction analysis of serotonin system markers with adolescent depression. *Mol Psychiatry* 9:908–915.

Garrison CZ, Jackson Kl, Addy CL, McKeowan RE, Waller J. 1991. Suicidal behaviors in young adolescents. *Am J Epidemiol* 133:1005–1014.

Giedd JN, Blumenthal J, Jeffries NO, Castellanos FX, Liu H, Zijdenbos A, Paus T, Evans AC, Rapoport JL. 1999. Brain development during childhood and adolescence: A longitudinal MRI study. *Nat Neurosci* 2:861–863.

Goetz RR, Puig-Antich J, Dahl RE, Ryan ND, Asnis GM, Rabinovich H, Nelson B. 1991. EEG Sleep of young adults with major depression: A controlled study. *J Affect Disord* 22:91–100.

Grunbaum JA, Kann L, Kinchen S, Ross J, Hawkins J, Lowry R, Harris WA, McManus T, Chyen D, Collins J. 2004. Youth risk behavior surveillance—United States, 2003 (Abridged). *J Sch Health* 74:307–324.

Kahn AU, Todd S. 1990. Polysomnographic findings in adolescents with major depression. *Psychiatry Res* 33:313–320.

Karg K, Burmeister M, Shedden K, Sen S. 2011. The serotonin transporter promoter variant (5-HTTLPR), stress, and depression meta-analysis revisited: Evidence of genetic moderation. *Arch Gen Psychiatry* 68:444–454.

Kaufman J, Birmaher B, Perel J, Dahl RE, Stull S, Brent D, Trubnick L, al-Shabbout M, Ryan ND. 1998. Serotoninergic functioning in depressed abused children: Clinical and familial correlates. *Biol Psychiatry* 44:973–981.

Kaufman J, Yang BZ, Douglas-Palumberi H, Houshyar S, Lipschitz D, Krystal JH, Gelernter J. 2004. Social supports and serotonin transporter gene moderate depression in maltreated children. *Proc Natl Acad Sci USA* 101:17316–17321.

Kendler KS, Prescott CA. 1999. A population-based twin study of lifetime major depression in men and women. *Arch Gen Psychiatry* 56:39–44.

Kruesi MJ, Hibbs ED, Zahn TP, Keysor CS, Hamburger SD, Bartko JJ, Rapoport JL. 1992. A 2-year prospective follow-up study of children and adolescents with disruptive behavior disorders: Prediction of cerebrospinal fluid 5-hydroxyindoleacetic acid, homovanillic acid and autonomic measures? *Arch Gen Psychiatry* 49:429–435.

Kutcher S, Malkin D, Silverberg J, Marton P, Williamson P, Malkin A, Szalai J, Katic M. 1991. Nocturnal cortisol, thyroid stimulating hormone and growth hormone secretory profiles in depressed adolescents. *J Am Acad Child Adolesc Psychiatry* 30:407–414.

Lesch KP. 2011. When the serotonin transporter gene meets adversity: The contribution of animal models to understanding epigenetic mechanisms in affective disorders and resilience. *Curr Top Behav Neurosci* 7:251–280.

Malone KM, Corbitt EM, Li S, Mann JJ. 1996. Prolactin response to fenfluramine and suicide attempt lethality in major depression. *Br J Psychiatry* 168:324–329.

Mann JJ, Brent DA, Arango V. 2001. The neurobiology and genetics of suicide and attempted suicide: A focus on the serotonergic system. *Neuropsychopharmacology* 24:467–477.

Mann JJ. 2002. A current perspective of suicide and attempted suicide. *Ann Intern Med* 136:302–311.

Mann JJ. 2003. Neurobiology of suicidal behaviour. *Nat Rev Neurosci* 4:819–828.

Mann JJ, McBride PA, Brown RP, Linnoila M, Leon AC, DeMeo M, Mieczkowski T, Myers J, Stanley M. 1992. Relationship between central and peripheral serotonin indexes in depressed and suicidal patients. *Arch Gen Psychiatry* 49:442–446.

Merikangas KR, Spiker DG. 1982. Assortative mating among in-patients with primary affective disorder. *Psychol Med* 12:753–764.

Modai I, Apter A, Meltzer M, Tyano S, Walevski A, Jerushalmy Z. 1989. Serotonin uptake by platelets of suicidal and aggressive adolescent psychiatric inpatients. *Neuropsychobiology* 21:9–13.

Moffitt TE, Caspi A, Rutter M. 2005. Strategy for investigating interactions between measured genes and measured environment. *Arch Gen Psychiatry* 62:473–481.

Nakamura M, Ueno S, Sano A, Tanabe H. 2000. The human serotonin transporter gene linked polymorphism (5-HTTLPR) shows ten novel allelic variants. *Mol Psychiatry* 5:32–38.

Ott J. 1989. Statistical properties of the haplotype relative risk. *Genet Epidemiol* 6:127–130.

Pandey GN, Dwivedi Y, Rizavi HS, Ren X, Conley RR. 2004. Decreased catalytic activity and expression of protein kinase C isozymes in teenage suicide victims: A postmortem brain study. *Arch Gen Psychiatry* 61:685–693.

Pandey GN, Dwivedi Y, Rizavi HS, Ren X, Pandey SC, Pesold C, Roberts RC, Conley RR, Tamminga CA. 2002a. Higher expression of serotonin 5-HT(2A) receptor in the post-mortem brains of teenage suicide victims. *Am J Psychiatry* 159:419–429.

Pandey GN, Dwivedi Y, Rizavi HS, Ren X, Pandey SC, Pesold C, Roberts RC, Conley RR, Tamminga CA. 2002b. Higher expression of serotonin 5-HT(2A) receptors in the post-mortem brains of teenage suicide victims. *Am J Psychiatry* 159:419–429.

Pandey GN, Pandey SC, Dwivedi Y, Sharma RP, Janicak PG, Davis JM. 1995. Platelet serotonin-2A receptors: A potential biological marker for suicidal behavior. *Am J Psychiatry* 152:850–855.

Paus T, Zijdenbos A, Worsley K, Collins DL, Blumenthal J, Giedd JN, Rapoport JL, Evans AC. 1999. Structural maturation of neural pathways in children and adolescents: In vivo study. *Science* 283:1908–1911.

Paykel ES. 1983. Methodological aspects of life events research. *J Psychosom Res* 27:341–352.

Pfeffer CR, McBridge A, Anderson GM, Kakuma T, Fensterheim L, Khait V. 1998. Peripheral serotonin measures in prepubertal psychiatric inpatients and normal children: Associations with suicidal behavior and its risk factors. *Biol Psychiatry* 44:568–577.

Pfeffer CR, Normandin L, Kakuma T. 1994. Suicidal children grow up: Suicidal behavior and psychiatric disorders among relatives, *J Am Acad Child Adolesc Psychiatry* 33:1087–1097.

Pfeffer CR, Strokes P, Shindledecker R. 1991. Suicidal behavior and hypothalamic–pituitary–adrenocortical axis indices in child psychiatric inpatients. *Biol Psychiatry* 29:909–917.

Pine DS, Trautman PD, Shaffer D, Cohen L, Davies M, Stanley M, Parsons B. 1995. Seasonal rhythm of platelet [^{3}H]imipramine binding in adolescents who attempted suicide. *Am J Psychiatry* 152:923–925.

Puig-Antich J, Goetz R, Davies M, Fein M, Hanlon C, Chambers WJ, Tabrizi MA, Sachar EJ, Weitzman ED. 1984. Growth hormone secretion in prepubertal children with major depression. II. Sleep-related plasma concentrations during a depressive episode. *Arch Gen Psychiatry* 41:463–466.

Rao U, Dahl RE, Ryan ND, Birmaher B, Williamson DE, Giles DE, Rao R, Kaufman J, Nelson B. 1996. The relationship between longitudinal clinical course and sleep and cortisol changes in adolescent depression. *Biol Psychiatry* 40:474–484.

Risch N, Herrell R, Lehner T, Liang KY, Eaves L, Hoh J, Griem A, Kovacs M, Ott J, Merikangas KR. 2009. Interaction between the serotonin transporter gene (*5-HTTLPR*), stressful life events, and risk of depression: A meta-analysis. *JAMA* 301:2462–2471.

Roy A. 1983. Family history of suicide. *Arch Gen Psychiatry* 40:971–974.

Roy A, Segal NL, Centerwall BS, Robinette CD. 1991. Suicide in twins. *Arch Gen Psychiatry* 48:29–32.

Roy A, Segal NL, Sarchiapone M. 1995. Attempted suicide among living co-twins of twin suicide victims. *Am J Psychiatry* 152:1075–1076.

Ryan ND, Puig-Antich J, Rabinovich H, Ambrosini P, Robinson D, Nelson B, Novacenko H. 1988. Growth hormone response to desmethylimipramine in depressed and suicidal adolescents. *J Affect Disord* 15:323–337.

Sabo E, Renolds CF III, Kupfer DJ, Berman SR. 1991. Sleep, depression and suicide. *Psychiatry Res* 36:265–277.

Schaid DJ, Sommer SS. 1994. Comparison of statistics for candidate–gene association studies using cases and parents. *Am J Hum Genet* 55:402–409.

Schulsinger F, Kety SS, Rosenthal D, Wender PH. 1979. A family study of suicide. In Schou M, Stromgren E (Eds.), *Origin, Prevention and Treatment of Affective Disorders*. Academic Press, New York, pp. 277–287.

Shaffer D. 1998. The epidemiology of teen suicide. An examination of risk factors. *J Clin Psychiatry* 49:36–41.

Shafii M, Carrigan S, Whittinghill JR, Derrick A. 1985. Psychological autopsy of completed suicide in children and adolescents. *Am J Psychiatry* 142:1061–1064.

Sher L, Oquendo MA, Li S, Ellis S, Brodsky BS, Malone KM, Cooper TB, Mann JJ. 2003. Prolactin response to fenfluramine administration in patients with unipolar and bipolar depression and healthy controls. *Psychoneuroendocrinology* 28:559–573.

Spielman RS, Ewens WJ. 1996. The TDT and other family-based tests for linkage disequilibrium and association. *Am J Hum Genet* 59:983–989.

Statham DJ, Heath AC, Madden PA, Bucholz KK, Bierut L, Dinwiddie SH, Slutske WS, Dunne MP, Martin NG. 1998. Suicidal behaviour: An epidemiological and genetic study. *Psychol Med* 28:839–855.

Stengel E. 1973. *Suicide and Attempted Suicide*. C. Nicholls & Company Ltd., Middlesex, England.

Suomi SJ. 2003. Gene–environment interactions and the neurobiology of social conflict. *Ann N Y Acad Sci* 1008:132–139.

Traskman L, Asberg M. 1986. Serotonergic function and serotonergic function in personality disorder. In Mann JJ, Stanley M (Eds.), *Psychobiology of Suicidal Behavior*. Annals of American Academy of Sciences, New York, pp. 168–717.

Traskman-Bendz L, Alling C, Oreland L, Regnéll G, Vinge E, Ohman R. 1992. Prediction of suicidal behavior from biologic tests. *J Clin Psychopharmacol* 61:685–693.

Tyano S, Zalsman G, Ofek H, Blum I, Apter A, Wolovik L, Sher L, Sommerfeld E, Harell D, Weizman A. 2006. Plasma serotonin levels and suicidal behavior in adolescents. *Eur Neuropsychopharmacol* 16:49–57.

Uher R, Caspi A, Houts R, Sugden K, Williams B, Poulton R, Moffitt TE. 2011. Serotonin transporter gene moderates childhood maltreatment's effects on persistent but not single-episode depression: Replications and implications for resolving inconsistent results. *J Affect Disord*. 135(1–3): 56–65.

Zalsman G. 2010. Timing is critical: Gene, environment and timing interactions in genetics of suicide in children and adolescents. *Eur Psychiatry* 25:284–286.

Zalsman G, Anderson GM, Peskin M, Frisch A, King RA, Vekslerchik M, Sommerfeld E, Michaelovsky E, Sher L, Weizman A, Apter A. 2005a. Relationships between serotonin transporter promoter polymorphism, platelet serotonin transporter binding and clinical phenotype in suicidal and non-suicidal adolescent inpatients. *J Neural Transm* 112:309–315.

Zalsman G, Brent DA, Weersing VR. 2006a. Depressive disorders in childhood and adolescence: An overview: Epidemiology, clinical manifestation and risk factors. *Child Adolesc Psychiatric Clin N Am* 15:827–841.

Zalsman G, Frisch A, Apter A, Weizman A. 2002. Genetics of suicidal behavior: Candidate association genetic approach. *Isr J Psychiatry Relat Sci* 39:252–261.

Zalsman G, Frisch A, Baruch-Movshovits R, Sher L, Michaelovsky E, King RA, Fischel T, Hermesh H, Goldberg P, Gorlyn M, Misgav S, Apter A, Tyano S, Weizman A. 2005b. Family-based association study of 5-HT(2A) receptor T102C polymorphism and suicidal behavior in Ashkenazi inpatient adolescents. *Int J Adolesc Med Health* 17:231–238.

Zalsman G, Frisch A, Bromberg M, Gelernter J, Michaelovsky E, Campino A, Erlich Z, Tyano S, Apter A, Weizman A. 2001a. A family-based association study of serotonin transporter promoter in suicidal adolescents: Possible role in violence traits. *Am J Med Genet* 105:239–245.

Zalsman G, Frisch A, King RA, Pauls DL, Grice DE, Gelernter J, Alsobrook J, Michaelovsky E, Apter A, Tyano S, Weizman A, Leckman JF. 2001b. Case–control and family based association studies of tryptophan hydroxylase A218C polymorphism and suicidality in adolescents. *Am J Med Genet* 105:451–457.

Zalsman G, Frisch A, Lewis R, Michaelovsky E, Hermesh H, Sher L, Nahshoni E, Wolovik L, Tyano S, Apter A, Weizman R, Weizman A. 2004. DRD4 receptor gene exon III polymorphism in inpatient suicidal adolescents. *J Neural Transm* 111:1593–1603.

Zalsman G, Huang YY, Oquendo MA, Burke AK, Hu XZ, Brent DA, Ellis SP, Goldman D, Mann JJ. 2006b. Association of a triallelic serotonin transporter gene promoter region (5-HTTLPR) polymorphism with stressful life events and severity of depression. *Am J Psychiatry* 163:1588–1593.

Zalsman G, Levy T, Shoval G. 2008. Interaction of child and family psychopathology leading to suicidal behavior. *Psychiatric Clin N Am* 31:237–246.

Zalsman G, Oquendo MA, Greenhill L, Goldberg PH, Kamali M, Martin A, Mann JJ. 2006c. Neurobiology of depression in children and adolescents. *Child Adolesc Psychiatric Clin N Am* 15:843–868.

Zalsman G, Patya M, Frisch A, Ofek H, Schapir L, Blum I, Harell D, Apter A, Weizman A, Tyano S. 2010. Association of polymorphisms of the serotonergic pathways with clinical traits of impulsive-aggression and suicidality in adolescents: A multi-center study. *World J Biol Psychiatry* 12:33–41.

15 Neurobiology of Teenage Suicide

Ghanshyam N. Pandey and Yogesh Dwivedi

CONTENTS

15.1 INTRODUCTION

About 30,000 persons die by suicide each year in the United States alone (Botsis et al., 1997). It is the second or third (depending on the age group and sex) most frequent cause of death for teenagers in the United States (CDC, 2011; Lowy et al., 1984; Moscicki et al., 1988). In 2006, the age-adjusted suicide rate among youth aged 10–19 years in the United States was 4.16 per 100,000. Among this population, the rate of suicide increases with age, and the suicide rate is substantially higher in boys than in girls—in boys between ages 18 and 19 years, the suicide rate is 15–20 per 100,000, and in girls, the rate is 3–4 per 100,000 (Bridge et al., 2006; CDC, 2011). In adults, suicidal behavior is a major symptom of depression and other psychiatric disorders, such as schizophrenia, alcoholism, and personality disorders. Besides psychiatric illnesses, other risk factors include a family history of suicide and a family history of psychiatric disorders and alcoholism, psychosocial stressors, impulsivity, and aggression (Joiner et al., 2005). Abnormalities in neurobiological mechanisms may also be a predisposing or risk factor (Mann et al., 1999; Underwood et al., 2004). Studies conducted on patients with suicidal behavior (Pandey et al., 1995) and on postmortem brain samples from suicide victims (Pandey et al., 2002a) strongly suggest that suicide is associated with neurobiological abnormalities.

Although some progress has been made in elucidating the role of serotonin (5-hydroxytryptamine, 5HT) and other neurobiological mechanisms in adult suicide, the neurobiology of adolescent suicide is understudied.

There is evidence to suggest that some factors associated with adolescent suicide may be different from adult suicide (Brent et al., 1999; Zalsman et al., 2008). Although the impulsive–aggressive behavior is a common risk factor for both adult and teenage suicide, aggression and impulsivity are traits highly related to suicidal behavior in adolescents (Apter et al., 1995). Higher levels of impulsive aggressiveness play a greater role in suicide among younger individuals with decreasing importance with increasing age (Brent et al., 1993). Brent et al. have also shown that adolescents with aggression and conduct disorders may be suicidal even in the absence of depression. Psychosocial factors associated with adolescent suicide, such as stress and contagion, bullying, and peer victimization (Brunstein et al., 2008; Bursztein and Apter, 2009; Klomek et al., 2008), may also be different. Alcohol and drug abuse contribute significantly to the risk of suicide in teenagers (Apter et al., 1990, 1995). Additional potential contributors to suicidal behavior in depressed adolescents are other early defined traits, such as temperament and emotional regulation. One study (Tamas et al., 2007) suggests that suicidal youths are characterized by high maladaptive regulatory responses and low adaptive emotional regulation responses to dysphoria. Since there are both similarities and differences in the risk factors for teenage and adult suicides, it is quite likely that the neurobiology of teenage suicide may be similar in some respects to adult suicide and different in others.

The neurobiology of teenage suicide has been primarily studied by the group of Pandey and colleagues. In this chapter, we summarize these studies and have also discussed the similarities and differences in the findings between teenage and adult suicide victims. Since we also study the neurobiology of adult suicide, we compare these neurobiological findings with particular reference to our own findings and briefly to those reported in the literature.

15.2 $5HT_{2A}$ RECEPTORS IN TEENAGE SUICIDE

The major evidence suggesting that serotonergic abnormalities may be associated with suicide is based on earlier studies of 5-hydroxyindoleacetic acid (5HIAA) in cerebrospinal fluid (CSF) of depressed and suicidal patients. It was reported by Asberg et al. (1976) and several other investigators that 5HIAA was significantly lower in the CSF of depressed patients with suicidal behavior compared with non-suicidal depressed patients or control subjects. Low CSF 5HIAA has also been observed in schizophrenic patients with suicidal behavior as well as in patients with personality disorders (Cooper et al., 1992; Ninan et al., 1984). A meta-analysis of CSF 5HIAA studies by Lester (1995) found significantly lower levels of 5HIAA in patients with a history of suicide and those who subsequently attempted or committed suicide. In addition, one of the serotonin receptors subtypes known as $5HT_{2A}$ receptor was also reported to be higher in the platelets of suicidal patients compared with non-suicidal patients and control subjects (Pandey et al., 1995). Although these studies suggested the role of abnormal serotonin function in suicide, they did not differentiate between suicidal behavior and completed suicide.

The major evidence linking serotonergic abnormality with suicide was obtained from the observation that the binding of $5HT_{2A}$ receptor was significantly increased in the postmortem brain of suicide subjects compared with normal control subjects (Stanley and Mann, 1983). In order to clarify the role of $5HT_{2A}$ receptors in teenage suicide, Pandey et al. (2002a) examined the protein, as well as the mRNA, and histochemical studies of $5HT_{2A}$ receptors in the postmortem brain of teenage suicide victims. This study of $5HT_{2A}$ receptors was performed in the prefrontal cortex (PFC) (Brodmann area 9), hippocampus, and nucleus accumbens (NA), on samples obtained from 15 teenage suicide victims and 15 normal control subjects. The $5HT_{2A}$ binding, using [^{125}I] lysergic acid diethylamide (LSD) as the binding ligand, was used to determine B_{max} and K_D of $5HT_{2A}$ binding sites in the PFC. The binding studies indicated a significant increase in the [^{125}I]LSD binding in the PFC of suicide victims compared with normal control subjects. They also observed that protein expression levels of $5HT_{2A}$ receptors were significantly increased in both PFC and hippocampus of suicide victims compared with normal control subjects; however, no significant differences in protein expression were observed in the NA between suicide victims and normal control subjects. The immunogold labeling technique was used to examine the cellular localization of $5HT_{2A}$ receptors. It was found that the expression of $5HT_{2A}$ receptors was most dense in pyramidal neurons (Layers III, V, and VI) and their atypical dendrites. To localize the changes in expression density of $5HT_{2A}$ receptors at the cellular level, the number of immunogold particles on pyramidal cell somata and in soma-free regions of the neuropil were quantitated. The mean expression levels of $5HT_{2A}$ receptors were significantly greater in the pyramidal cells of Layer V of the teenage suicide victims than in normal control subjects, whereas no changes in the expression levels of $5HT_{2A}$ receptors were found in the pyramidal cells of other cortical layers (Layers III and VI) or in the surrounding neuropil (Layers IV, V, and VI).

In order to examine if the increased $5HT_{2A}$ receptor binding and protein expression were related to increased transcription of the $5HT_{2A}$ receptors, Pandey et al. (2002a) determined the mRNA levels of $5HT_{2A}$ receptors using a quantitative reverse transcription polymerase chain reaction (RT-PCR) technique. It was found that the mean $5HT_{2A}$ mRNA levels were significantly greater in the PFC of 15 teenage suicide victims compared with 15 normal control subjects. It was also found that the mean mRNA levels in the hippocampus of the suicide victims were significantly higher than in the normal control subjects. However, no significant differences in mRNA levels between suicide victims and normal control subjects were found in the NA (Pandey et al., 2002a).

Since an association has been suggested between mental disorders and suicide, the relationship between $5HT_{2A}$ receptors and mental illness in teenage suicide victims was determined. However, no significant differences in either the protein or the mRNA expression levels of $5HT_{2A}$ receptors were observed between subjects who had a history of mental disorders and subjects who did not have such a history. However, levels of both suicide groups were significantly higher compared with those of normal control subjects (Pandey et al., 2002a).

This study thus indicated an increase in $5HT_{2A}$ receptors using a binding technique, in protein expression levels using the Western blot technique, and in mRNA levels using the RT-PCR technique. These results are further supported by the immunohistochemical technique, which showed that the increase in $5HT_{2A}$ receptors observed by the Western blot technique was localized in the pyramidal cells in Layer V of the PFC (Pandey et al., 2002a).

The observation that $5HT_{2A}$ receptor is increased in the postmortem brain of teenage suicide (Pandey et al., 2002a) is similar to that observed in adult suicide victims. However, most of the studies in adult suicide victims have been carried out by ligand binding technique, and the discrepancy in the results of $5HT_{2A}$ receptors in adult suicide victims—half of them finding increased $5HT_{2A}$ receptors and several others not finding any difference—may possibly be related mainly to the type of ligands used for these binding studies. In this study, however, the authors not only performed binding techniques but also determined protein and mRNA expression, thus suggesting that teenage suicide may be related to an increase in $5HT_{2A}$ receptor binding, as well as in protein and gene expression.

15.3 G-PROTEIN STUDIES IN TEENAGE SUICIDE

Dwivedi et al. (2002b) have studied the role of G-proteins in adult as well as in teenage suicide victims. They determined mRNA and protein levels of various α, β, and γ subunits of G-proteins in the postmortem brain. This study included postmortem brain samples of both the adult and the teenage suicide subjects. The study was performed in the PFC (Brodmann areas 8 and 9) obtained from 43 suicide victims and 38 nonpsychiatric normal control subjects. Protein expression levels of G-proteins were determined using the immunolabeling of G-protein subunits by the Western blot technique. The mRNA levels of the various subunits of G-protein were determined using quantitative RT-PCR. They determined the immunolabeling of $G_S\alpha$, $G_{i1}\alpha$, $G_{i2}\alpha$, $G_{q/11}\alpha$, $G\alpha$, and $G\beta\gamma$ subunits. When they compared the protein levels of various G-protein subunits between suicide victims and control subjects, they found that the levels of $G_{i2}\alpha$ and $G_O\alpha$ were significantly decreased and the level of $G_s\alpha_{-S}$ was significantly increased in the PFC of suicide victims compared with normal control subjects. They did not find any significant differences in the levels of $G_S\alpha_{-L}$, $G_{i1}\alpha$, $G_{q/11}\alpha$, $G\beta$, or $G\gamma$ subunits between suicide victims and normal control subjects. As discussed earlier, the characteristics of teenage suicide may differ from those of adult suicide, which suggests that the neurobiology of teenage suicide may, in some respects, be different from that of adult suicide. They then divided the total subjects into two subgroups, teenage suicide victims (age $\leq$19 years) and adult suicide victims (age $\geq$20 years), and analyzed the G-protein subunits of these two populations separately (27 adult suicide victims and 20 adult control subjects, and 16 teenage suicide victims and 16 teenage control subjects). When the suicide population was subdivided on the basis of age, it was observed that the mRNA levels of $G_{i2}\alpha$ and $G_O\alpha$ were significantly decreased and that of $G_s\alpha_{-S}$ were significantly increased in the PFC of adult suicide victims compared with adult control subjects, without any change in mRNA levels of $G_{i1}\alpha$. On the other hand, there were no significant differences in mRNA levels of any of the G-protein subunits between total teenage controls and teenage suicide victims. This observation thus suggests differential abnormalities of G-protein subunits between the adult and the teenage suicide victims.

15.4 SIGNAL TRANSDUCTION STUDIES IN TEENAGE SUICIDE

The functional role of a receptor lies in its ability to activate the signal transduction system causing not only a functional and a behavioral response but also the transcription of some important genes. In fact, it is not surprising that several transduction pathways have been studied in the postmortem brain of suicide victims. These include the studies of phosphoinositide (PI), adenylyl cyclase (AC), MAP kinase, and BDNF signaling systems. Among these signaling systems, the PI and the AC signaling systems appear to have been studied in greater detail and implicated in the pathophysiology of mood disorders and suicide. Some of these receptors, such as $5HT_{2A}$ receptors, which have been found to be altered in the postmortem brain of suicide victims, are linked to the PI signaling system. Agonist-induced activation of these G protein–coupled receptors causes the hydrolysis of phosphatidylinositide-4,5-bisphosphate (PIP_2) by the PI-specific enzyme phospholipase C (PLC), resulting in the formation of two second messengers—diacylglycerol (DAG) and inositol-1,4,5-triphosphate (IP_3). DAG activates the phospholipid- and calcium-dependent enzyme protein kinase C (PKC) and increases its affinity for calcium. PKC subsequently activates several transcription factors, such as cyclic adenosine monophospate (cAMP) response element–binding protein (CREB) and glycogen synthase kinase-3β (GSK-3β), which are also a part of pathway known as the Wnt signaling pathway. Activation of transcription factors by PKC results in the transcription of several important target genes, such as brain-derived neurotrophic factor (BDNF). The other signaling system that has been well studied in mood disorders is the AC signaling system. Several receptors, such as β-adrenergic receptors, are linked to this signaling system. The activation of β-adrenergic receptors, for example, causes the activation of the effector AC, which causes the conversion of ATP to cAMP. cAMP activates the enzyme known as protein kinase A (PKA), which activates several transcription factors, including the CREB family of transcription factors, thus resulting in the transcription of several important genes. The components of the PI and AC signaling systems that had been studied in teenage suicide victims are PKC, PKA and the effector, PLC, the transcription factor CREB, as well as BDNF and TrkB. The results of these studies are described briefly in the following sections.

15.4.1 PHOSPHOLIPASE C IN SUICIDE

Stimulation of G protein–coupled receptors causes the activation of the effector PLC, resulting in the hydrolysis of the substrate phosphoinositide 2 (PIP_2) into DAG and IP_3 (Berridge and Irvine, 1989; Joseph and Williamson, 1989). PLC is quite important in transducting signals from the receptors to the nucleus (Cockcroft and Thomas, 1992; Rhee, 2001). Several neurotransmitter receptors use this pathway. PLC is predominantly present in the cytosol, but its presence in the membrane has also been demonstrated (Banno and Nozawa, 1987; Cockcroft and Thomas, 1992; Lee et al., 1987). PLC isoforms are classified into three major families and further into several subtypes based primarily on sequence homology (Anderson et al., 1990; Exton, 1994; Rhee and Choi, 1992). These three different families include PLCβ, PLCγ, and PLCδ. The PLCβ family consists of three members present in mammalian cells that are found in great abundance in the brain: $PLCβ_1$, $PLCβ_2$, and $PLCβ_3$ (Cockcroft and Thomas, 1992; Rhee, 2001).

PLCγ consists of γ_1 and γ_2 subunits, whereas PLCδ is subtyped into δ_1 and δ_2. Several G-proteins regulate specific PLCs based on the co-expression of G-proteins and specific isoforms of PLC and on pertussis toxin ADP ribosylation. Based on these studies, it appears that PLC is regulated primarily by the G_q or G_O families, specifically G_q and G_{11}. These two widely expressed G-proteins can selectively activate PLCβ_1, but not PLCγ or PLCδ (Rhee and Bae, 1997; Taylor et al., 1991). Although the involvement of receptors coupled to the PI signaling system in the pathophysiology of suicide, depression, and mood disorders has been studied by several investigators, studies of the other components of the PI signaling system, including PLC, have been very limited.

Pacheco et al. (1996) measured the protein expression levels of PLCβ in the PFC region obtained from 13 suicide victims with major depression and 13 normal control comparison subjects. The protein expression of PLCβ was determined in membranes obtained from the PFC using the Western blot technique. When they compared the PLCβ protein expression, they did not find any significant differences between suicide victims and normal control subjects.

Pandey et al. (1999) have reported the only other studies of PLC isozymes in the postmortem brain of suicide victims. They determined the PI–PLC activity in membrane and cytosol fractions of postmortem brain samples obtained from Brodmann areas 8 and 9 from 18 teenage suicide victims and 18 normal control subjects. They also determined the immunolabeling of PLCβ_1, γ_1, and δ_1 isozymes using the Western blot technique. It was observed that the mean PI–PLC activity in Brodmann areas 8 and 9 was significantly decreased in both the membrane as well as the cytosol fractions of the PFC of the teenage suicide victims compared with normal control subjects. In order to examine if the decrease in PI–PLC activity in the postmortem brain of teenage suicide victims is related to changes in any of the isozymes, the protein expression of PLCβ_1, δ_1, and γ_1 was determined in both membrane and cytosol fractions obtained from teenage suicide victims and normal control subjects. The protein expression of PLCβ_1 was found to be significantly decreased in both the membrane and the cytosol fractions of the teenage suicide victims compared with normal control subjects; there were no significant differences observed in the immunolabeling of either PLCδ_1 or PLCγ_1 between the suicide and the normal control groups. These results suggest that adolescents who commit suicide have significantly lower than normal PI–PLC activity in the PFC, and that this low level is due to a selectively lower protein expression of the enzyme PLCβ_1. Further studies of PI–PLC activity and protein expression as well as mRNA expression levels of PLC isozymes are needed in order to examine if this decrease in PLC is specific to the teenage suicide population or if it also occurs in the postmortem brain of adult suicide victims.

15.4.2 Protein Kinase C in Suicide

PKC is a key regulatory enzyme that is present in various tissues (Garcia-Sevilla et al., 1999; Shearman et al., 1987; Tanaka and Nishizuka, 1994). PKC has been shown to be a family of at least 12 structurally related isozymes (Newton, 1995; Nishizuka, 1995; Stabel and Parker, 1991). All PKC isozymes possess two functional domains: an N-terminal regulatory domain and a C-terminal catalytic domain. On the basis of molecular structure and enzymatic characterization, the PKC family has been subgrouped into three classes: conventional (α, βI, βII, and γ) (Hug and Sarre, 1993; Kiley and Jaken, 1994), novel (δ, ε, η, and θ) (Nishizuka, 1992), and atypical (ι, κ, λ, and τ) isozymes (Akimoto et al., 1994; Ono et al., 1989; Tanaka and Nishizuka, 1994). Each isozyme is encoded by a unique gene, except the βI and βII isozymes, which are products of the differential splicing of the same transcript (Nishizuka, 1992). The conventional isozymes are phospholipid- and calcium-dependent (Nishizuka, 1992), whereas the novel PKC isozymes do not require calcium for activation (Nishizuka, 1992).

The atypical isozymes are unresponsive to phorbol esters but can be activated by phosphatidyl serine (Akimoto et al., 1994; Ono et al., 1989). Marked differences occur in the distribution of PKC isozymes. Most PKC isozymes are present in the brain (Nishizuka, 1995; Ono et al., 1988; Shearman et al., 1987), whereas in platelets, only α, β, and δ isozymes have been reported (Baldassare et al., 1992; Grabarek et al., 1992). The biochemical properties of each isozyme have been identified with respect to activation and to phosphorylation, proteolytic activation/degradation, and substrate specificity. PKC has been shown to be present in both the cytosol and the membrane of the cell, and activation of PKC results in its translocation from cytosol to membrane (Kaczmarek, 1987). It has been suggested that an association of PKC with the membrane is required for the subsequent physiological response. The majority of the nonactivated PKC is located in the cytosolic fraction and tends to relocate upon activation (Kaczmarek, 1987).

PKC is involved in the modulation of many neuronal and cellular functions, such as neurotransmitter synthesis and release, regulation of receptors and ion channels, neuronal excitability, gene expression, secretion, and cell proliferation (Nishizuka, 1988). Activation of PKC also causes the activation of transcription factors involved in the transcription of important genes (Nichols et al., 1992; Riabowol et al., 1988; Xie and Rothstein, 1995).

The role of PKC in mental disorders and suicide has not been sufficiently studied; nevertheless, there is both direct and indirect evidence suggesting that PKC may play a crucial role in mental disorders. Friedman et al. (1993) studied PKC activity in the platelets of bipolar patients before and after lithium therapy and observed an increase in the ratio of membrane to cytosolic PKC in manic subjects. Further, 5HT-induced PKC translocation is enhanced in those subjects. Lithium treatment reduced the sensitivity of platelets to PKC translocation by 5HT. Lithium has also been shown to cause a decrease in hippocampal PKC by Manji et al. (1993). Chronic treatment with antidepressant drugs has been reported to cause a decrease in [^{3}H]phorbol dibutyrate (PDBu) binding and in PKC activity in the rat brain (Mann et al., 1995). These data taken together provide some evidence for a role for PKC in mental disorders.

The specific role of PKC in suicide has been studied only in recent years. The function of PKC in the postmortem brain or other tissues in relation to suicide or other mental disorders can be studied in several ways. One can determine the activity of the enzyme PKC, determine PKC binding sites using appropriate radioligands, or determine the protein and mRNA expression levels of different isozymes of PKC in both cytosol and membrane fractions of brain tissues. One can also study the function of PKC by determining PKC-mediated phosphorylation of a substrate, such as MARCKS.

Pandey et al. (2004) examined the role of PKC in the postmortem brain of teenage suicide victims and normal control subjects. In an earlier communication, Pandey et al. (1997) reported decreased PKC binding sites by measuring the binding of [^{3}H]PDBu in membrane and cytosol fractions obtained from the PFC of 17 teenage suicide victims and 17 teenage normal control subjects. We found that the B_{max} of [^{3}H]PDBu binding was significantly decreased in both membrane and cytosol fractions obtained from the PFC of teenage suicide victims compared with control subjects. This study provided some preliminary evidence of abnormalities of PKC in teenage suicide victims. However, [^{3}H]PDBu measures only the regulatory subunits of PKC, and it was not clear from this study whether the decrease in PKC binding sites was associated with either changes in catalytic activity or with changes in the level of any specific isozymes. Because each isozyme is related to specific functions and is region specific, we then determined the catalytic activity of PKC and the protein levels of various isozymes in membrane and cytosol fractions of PFC and hippocampus obtained from postmortem brain samples of teenage suicide victims and nonpsychiatric control subjects. To further examine whether any changes in isozymes are related to altered transcription, the mRNA levels of these isozymes were also determined.

Pandey et al. (2004) found that PKC activity was significantly decreased in the membrane and the cytosol fractions of the PFC and hippocampus of teenage suicide victims compared with control subjects. They also found that the protein expression levels of PKCα, βI, βII, and γ were significantly decreased in membrane and cytosol fractions of the PFC and hippocampus of teenage suicide victims compared with control subjects. They then examined if decreased PKC isozyme levels were related to altered transcription of their respective mRNAs and found that the mRNA levels of PKCα, β, and γ were also significantly decreased in both PFC and hippocampus of suicide victims compared with control subjects.

Although PKC has been implicated in the pathophysiology of mood disorders, schizophrenia (Dean et al., 1997; Friedman et al., 1993; Manji et al., 1999; Pandey et al., 2002b), and Alzheimer's disease (Cole et al., 1988; Masliah et al., 1990; Shimohama et al., 1993; Stokes and Hawthorne, 1987), the role of PKC in suicide has not been extensively studied. Besides the two studies of Pandey et al. (1997, 2004) in teenage suicide victims, there appears to be only one other study in suicide victims (Coull et al., 2000). Coull et al. (2000) determined [^{3}H]PDBu binding in the PFC and hippocampus of antidepressant-treated and antidepressant-free adult depressed suicide victims. They did not find any significant differences in [^{3}H]PDBu binding between antidepressant-treated suicide victims and control subjects. On the other hand, they found a significant increase in the B_{max} of [^{3}H] PDBu binding in the soluble fractions of antidepressant-free suicide victims compared with control subjects. The apparent differences between the Pandey et al.'s (1997) and the Coull et al.'s (2000) studies may be due to the difference in age of the populations studied. Furthermore, these studies of PKC by Coull et al. (2000) examined only [^{3}H]PDBu binding, and it is only the studies of Pandey et al. (2004), which, in addition to the binding, have determined the PKC activity, as well as the protein and mRNA expression of PKC isozymes, in the postmortem brain of teenage suicide victims.

15.4.3 PKA in the Postmortem Brain of Teenage Suicide Victims

PKA, a key component of the AC signaling systems, is activated by cAMP, and the activated PKA phosphorylates several intracellular proteins and activates transcription factors such as CREB. In the absence of cAMP, the PKA holoenzyme exists as an inactive tetramer composed of two catalytic subunits bound to a regulatory subunit dimer. On the basis of elution patterns, two different PKA isozymes, known as PKA I and PKA II, have been identified. These two isozymes have been shown to be composed of two different R subunits, known as RI and RII, which are further composed of subunits known as RIα and RIβ, and RIIα and RIIβ. In addition, three catalytic subunits, known as Cα, Cβ, and Cγ, have also been identified. Each R subunit has two cAMP binding sites, and in activation and binding with cAMP each R subunit dissociates into a dimeric R subunit complex and two monomeric active C subunits (Skalhegg and Tasken, 2000).

In order to examine if abnormalities of PKA are also involved in the pathophysiology of teenage suicide, Pandey et al. (2005) determined the cAMP binding to PKA, PKA activity, and the protein and mRNA expression of different subunits of PKA in cytosol and membrane fractions obtained from the PFC, hippocampus, and NA of postmortem brain samples from 17 teenage suicide victims and 17 nonpsychiatric control subjects. They found that PKA activity was significantly decreased in the PFC but not hippocampus of teenage suicide victims compared with control subjects. However, the protein and mRNA expressions of only two PKA subunits, that is, PKA RIα and PKA RIβ, but not any other subunits, such as Cα, Cβ, RIIα, or RIIβ, were observed to be decreased in the PFC of teenage suicide victims compared with control subjects.

The role of PKA in mood disorders has been studied by many investigators (for review, see Dwivedi and Pandey, 2008). However, the studies of PKA in the postmortem brain of suicide victims are limited. In a study of postmortem brain samples obtained from suicide victims, Dwivedi et al. (2002a) reported that ^{3}[H]-cAMP binding and PKA activity were significantly decreased in the PFC of suicide victims. In a subsequent study, Dwivedi et al. (2004) also observed that the protein and mRNA expressions of PKA subunits, PKA RIIβ and Cβ, were significantly decreased in the PFC of suicide subjects relative to normal controls.

These results in the teenage suicide victims (Pandey et al., 2005), although similar in some respects to those observed in adult suicide victims (Dwivedi et al., 2002a, 2004), were also dissimilar in some other respects. For example, whereas decreased cAMP binding and PKA activity were found in both adult and teenage suicide victims, decreased RIIα and Cβ were found in the adult suicide victims, while the RIα and RIβ subunits were abnormal in the teenage suicide victims. The significance and implications of these observations with regard to the pathophysiology of teenage and adult suicides are unclear at this time.

15.4.4 Transcription Factor CREB in Suicide

Activation of transcription factors is the final step in a signal transduction pathway, which is mediated by the binding of a cell surface receptor with an agonist. Transcription factors can alter the expression of specific genes. The regulation of gene expression by extracellular signals is a fundamental mechanism of development, homeostatics, and adaptation to the environment. One of the mechanisms by which transcription factors are activated is phosphorylation and dephosphorylation. Activation of PKC, as well as of PKA, causes the phosphorylation of several transcription factors, such as the AP-1 family (Jun-B and Jun-D) and CREB (Holian et al., 1991). CREB is one of the important transcription factors and has been recently implicated in the pathophysiology of depression and suicide (Dwivedi et al., 2003a; Pandey et al., 2007).

CREB is a member of the basic leucine zipper family of transcription factors (Borrelli et al., 1992). Phosphorylation of CREB at serine 133 leads to dimerization and activation by binding to cAMP response elements (CREs) at the consensus 5′-TGACGTCA-3′, which is found in the promoters of many neuronally expressed genes (Montminy et al., 1990). In its active phosphorylated form, CREB regulates the transcription of many genes that are involved in several aspects of neuronal functioning, including the excitation of neural cells, CNS development, and long-term synaptic plasticity (Imaki et al., 1994; Moore et al., 1996; Silva et al., 1998). CREB is activated by many kinases, including PKC (Xie et al., 1995) and PKA (Nichols et al., 1992; Riabowol et al., 1988; Xie et al., 1995). That CREB could possibly be involved in such disorders as depression and suicide is evident from studies showing increased expression of CREB in the postmortem brain of depressed patients treated with antidepressants (Dowlatshahi et al., 1998) and from the observation that treatment with almost all antidepressants caused an increase in CREB in the rat brain (Nibuya et al., 1996). Again, studies of CREB in suicide seem to be very limited, although Yamada et al. (2003) determined CREB protein and its phosphorylated form in the orbital frontal cortex of antidepressant-free patients with major depression and found that the immunoreactivity of both CREB and its phosphorylated form was significantly decreased in depressed subjects compared with normal control subjects.

Dwivedi et al. (2003a) studied CREB in the postmortem brain of adult suicide victims and found that the protein expression of CREB was significantly decreased in the nuclear fractions of both the PFC and hippocampus obtained from suicide victims compared with normal control subjects. They also observed that this decrease in protein expression levels was associated with a significant decrease in the mRNA levels of CREB in both PFC and the hippocampus of suicide victims compared with normal control subjects. The CRE–DNA binding activity was significantly decreased in the nuclear fractions of both the PFC as well as hippocampus of suicide victims compared with normal control subjects.

Pandey et al. (2007) determined CREB levels in the postmortem brains obtained from teenage suicide victims and normal control subjects. They found a significant decrease in the protein and mRNA expression levels of CREB in the PFC of teenage suicide victims compared with controls. They also found that the decrease in the protein and mRNA expression of CREB was associated with a significant decrease in the CRE–DNA binding in teenage suicide victims relative to controls. However, they did not find any significant difference in the protein or mRNA expression or in CRE–DNA binding between teenage suicide victims and normal controls in the hippocampus. These observations suggest some differences in the expression of CREB between adult and teenage suicide victims. While CREB expression was found to be decreased in the PFC of both adult and teenage suicide, CREB expression was significantly decreased in the hippocampus of adults but not of teenage suicide victims. These observations indicate another subtle difference in the neurobiology between teenage and adult suicides.

15.4.5 BDNF and TrkB Receptors in Suicide

As described in the previous section, CREB, which is a transcription factor, plays an important role in the regulation of several genes, including *BDNF*. Activation of CREB increases BDNF transcription through the Ca^{2+} and CRE within exon 3 of BDNF (Finkbeiner, 2000). BDNF is a member of the neurotrophin family that includes nerve growth factor and neurotrophins (Huang and Reichardt, 2001). Neurotrophins promote the growth and development of immature neurons and enhance the survival and function of specific neuronal populations, including neuronal growth, plasticity, phenotype maturation, synthesis of proteins, and synaptic functioning (Altar et al., 1997; Bartrup et al., 1997; Thoenen, 1995). The suggestion that BDNF may play a role in the pathophysiology of suicide is derived from studies showing that treatment with antidepressants caused an increase in BDNF in the rat brain (Nibuya et al., 1995).

Dwivedi et al. (2003b) determined the protein and mRNA expression levels of BDNF in the PFC and hippocampus of suicide victims and normal control subjects and found that the protein and mRNA expression level of BDNF was significantly decreased both in the PFC and hippocampus of suicide victims compared with normal control subjects.

Pandey et al. (2008) determined the protein and mRNA expression of BDNF and its receptor, tyrosine kinase B (TrkB) in the PFC and hippocampus of 29 teenage suicide victims and 25 teenage control subjects. The protein expression of BDNF was significantly decreased in the PFC but not the hippocampus of teenage suicide subjects compared with controls. However, mRNA expression of BDNF was significantly decreased in both PFC and hippocampus of teenage suicide subjects compared to controls. The reason for this dissociation between mRNA and protein expression of BDNF in the hippocampus is not clear.

BDNF produces its physiological effects by binding to a receptor known as TrkB, which exists in two forms—a full-length TrkB and a truncated isoform. Pandey et al. (2008) also determined the protein and mRNA expression of TrkB receptors in the PFC and hippocampus of teenage suicide victims. They found that the mRNA and protein expressions of full-length TrkB receptors were significantly decreased in the PFC as well as in the hippocampus of teenage suicide victims compared with controls. No significant changes were observed in the protein and mRNA expressions of truncated TrkB receptors either in the PFC or hippocampus of teenage suicide victims compared with controls.

The results of BDNF and TrkB studies in teenage suicide victims generally appear to be similar to those found in adult suicide victims (Dwivedi et al., 2003b). The main difference was that BDNF protein expression was not significantly different in the hippocampus of teenage suicide victims but was significantly decreased in adult suicide victims. Otherwise, changes in protein and mRNA expression of BDNF and TrkB were similar in both teenage and adult suicides.

15.5 CONCLUSION

Suicide is a major health problem worldwide. Teenage suicide is a particular problem as it is the second or third leading cause of death in this group. Although there are many common risk factors for suicide in adults and teenagers, some characteristics of teenage or adolescent suicide may be different than in adults, as discussed in Section 15.1. There are numerous psychosocial and epidemiological studies of adolescent suicide. On the other hand, neurobiological studies of teenage suicide are scarce. Studies of the neurobiology of teenage suicide are warranted at least for the following reasons. Since some characteristics and risk factors of teenage suicide are different than in adults, there is likely to be a subtle difference in the neurobiology of teenage suicide compared to adults.

In this chapter, we have reviewed the biochemical–biological studies conducted in the postmortem brain of teenage suicide victims, specifically on receptors, signaling systems, and neurotrophins. As discussed, there appear to be several similarities in these findings compared to those reported in adults, but also some differences. There were subtle differences in the abnormalities of PKA subunits and BDNF between teenage and adult suicides. However, the major difference between teenage and adult suicide appears to be abnormalities in the hippocampus. Since stress is a major risk factor for teenage and adult suicides, one would expect this area to be abnormal in both, but more so in teenage suicide. The observation that most of the biological abnormalities were not found in teenage suicide was thus intriguing. Although at this time it is not clear how to explain this anomaly, the imaging and volumetric studies may be able to explain it.

Hippocampal volume changes have been observed in post-traumatic stress disorder (PTSD) (Bremner et al., 1995), schizophrenia (Adriano et al., 2011), and depression (Sheline et al., 2003). Bremmer et al. (1995) also found a reduction in left hippocampal volume in adults with a history of childhood trauma and PTSD. De Bellis et al. (2009) did not find any change in hippocampal volume in children with PTSD. Teicher et al. (1997) did not observe changes in hippocampal volume in severely abused children. One possible explanation of this discrepancy between adolescent and adult suicides may be that the stress of PTSD exerts a very gradual effect on hippocampal morphology, so that stress effects on hippocampus are not observed in children or adolescents but only in adults.

REFERENCES

Adriano F, Caltagirone C, Spalletta G. Hippocampal volume reduction in first-episode and chronic schizophrenia: A review and meta-analysis. *Neuroscientist* 2011 Apr 29. [Epub ahead of print] PMID: 21531988.

Akimoto K, Mizuno K, Osada S, Hirai S, Tanuma S, Suzuki K, Ohno S. A new member of the third class in the protein kinase C family, PKC lambda, expressed dominantly in an undifferentiated mouse embryonal carcinoma cell line and also in many tissues and cells. *J Biol Chem* 1994;269:12677–12683.

Altar CA, Cai N, Bliven T, Juhasz M, Conner JM, Acheson AL, Lindsay RM, Wiegand SJ. Anterograde transport of brain-derived neurotrophic factor and its role in the brain. *Nature* 1997;389:856–860.

Anderson D, Koch CA, Grey L, Ellis C, Moran MF, Pawson T. Binding of SH2 domains of phospholipase C gamma 1, GAP, and Src to activated growth factor receptors. *Science* 1990;250:979–982.

Apter A, Brown S, Korn M, Van Praag HM. Serotonin in childhood psychopathology. In: Brown S, Van Praag HM, eds. *Serotonin in Psychiatry*. New York: Bruner Mazel, 1990. pp 215–238.

Apter A, Gothelf D, Orbach I, Weizman R, Ratzoni G, Har-Even D, Tyano S. Correlation of suicidal and violent behavior in different diagnostic categories in hospitalized adolescent patients. *J Am Acad Child Adolesc Psychiatry* 1995;34:912–918.

Asberg M, Traskman L, Thoren P. 5-HIAA in the cerebrospinal fluid. A biochemical suicide predictor? *Arch Gen Psychiatry* 1976;33:1193–1197.

Baldassare JJ, Henderson PA, Burns D, Loomis C, Fisher GJ. Translocation of protein kinase C isozymes in thrombin-stimulated human platelets. Correlation with 1,2-diacylglycerol levels. *J Biol Chem* 1992;267:15585–15590.

Banno Y, Nozawa Y. Characterization of partially purified phospholipase C from human platelet membranes. *Biochem J* 1987;248:95–101.

Bartrup JT, Moorman JM, Newberry NR. BDNF enhances neuronal growth and synaptic activity in hippocampal cell cultures. *Neuroreport* 1997;8:3791–3794.

Berridge MJ, Irvine RF. Inositol phosphates and cell signalling. *Nature* 1989;341:197–205.

Borrelli E, Montmayeur JP, Foulkes NS, Sassone-Corsi P. Signal transduction and gene control: The cAMP pathway. *Crit Rev Oncog* 1992;3:321–338.

Botsis AF, Soldatos CR, Stefanis, CN. *Suicide: Biopsychosocial Approaches*. Amsterdam, the Netherlands: Elsevier, 1997.

Bremner JD, Randall P, Scott TM, Bronen RA, Seibyl JP, Southwick SM, Delaney RC, McCarthy G, Charney DS, Innis RB. MRI-based measurement of hippocampal volume in patients with combat-related posttraumatic stress disorder. *Am J Psychiatry* 1995;152:973–981.

Brent DA, Baugher M, Bridge J, Chen T, Chiappetta L. Age- and sex-related risk factors for adolescent suicide. *J Am Acad Child Adolesc Psychiatry* 1999;38:1497–1505.

Brent DA, Kolko DJ, Wartella ME, Boylan MB, Moritz G, Baugher M, Zelenak JP. Adolescent psychiatric inpatients' risk of suicide attempt at 6-month follow-up. *J Am Acad Child Adolesc Psychiatry* 1993;32:95–105.

Bridge JA, Goldstein TR, Brent DA. Adolescent suicide and suicidal behavior. *J Child Psychol Psychiatry* 2006;47:372–394.

Brunstein JD, Cline CL, McKinney S, Thomas E. Evidence from multiplex molecular assays for complex multipathogen interactions in acute respiratory infections. *J Clin Microbiol* 2008;46:97–102.

Bursztein C, Apter A. Adolescent suicide. *Curr Opin Psychiatry* 2009;22:1–6.

Centers for Disease Control and Prevention (CDC). National Center for Injury Prevention and Control. Web-based Injury Statistics Query and Reporting System (WISQARS) [online]. Source of data from WISQARS is the National Vital Statistics System from the National Center for Health Statistics. Available from www.cdc.gov/ncipc/wisqars (last accessed 5/2/2011).

Cockcroft S, Thomas GM. Inositol–lipid-specific phospholipase C isoenzymes and their differential regulation by receptors. *Biochem J* 1992;288(Pt 1):1–14.

Cole G, Dobkins KR, Hansen LA, Terry RD, Saitoh T. Decreased levels of protein kinase C in Alzheimer brain. *Brain Res* 1988;452:165–174.

Cooper SJ, Kelly CB, King DJ. 5-Hydroxyindoleacetic acid in cerebrospinal fluid and prediction of suicidal behaviour in schizophrenia. *Lancet* 1992;340:940–941.

Coull MA, Lowther S, Katona CL, Horton RW. Altered brain protein kinase C in depression: A post-mortem study. *Eur Neuropsychopharmacol* 2000;10:283–288.

De Bellis MD, Hooper SR, Spratt EG, Woolley DP. Neuropsychological findings in childhood neglect and their relationships to pediatric PTSD. *J Int Neuropsychol Soc* 2009;15:868–878.

Dean B, Opeskin K, Pavey G, Hill C, Keks N. Changes in protein kinase C and adenylate cyclase in the temporal lobe from subjects with schizophrenia. *J Neural Transm* 1997;104:1371–1381.

Dowlatshahi D, MacQueen GM, Wang JF, Young LT. Increased temporal cortex CREB concentrations and antidepressant treatment in major depression. *Lancet* 1998;352:1754–1755.

Dwivedi Y, Conley RR, Roberts RC, Tamminga CA, Pandey GN. [(3)H]cAMP binding sites and protein kinase A activity in the prefrontal cortex of suicide victims. *Am J Psychiatry* 2002a;159:66–73.

Dwivedi Y, Rao JS, Rizavi HS, Kotowski J, Conley RR, Roberts RC, Tamminga CA, Pandey GN. Abnormal expression and functional characteristics of cyclic adenosine monophosphate response element binding protein in postmortem brain of suicide subjects. *Arch Gen Psychiatry* 2003a;60:273–282.

Dwivedi Y, Rizavi HS, Conley RR, Roberts RC, Tamminga CA, Pandey GN. mRNA and protein expression of selective alpha subunits of G proteins are abnormal in prefrontal cortex of suicide victims. *Neuropsychopharmacology* 2002b;27:499–517.

Dwivedi Y, Rizavi HS, Conley RR, Roberts RC, Tamminga CA, Pandey GN. Altered gene expression of brain-derived neurotrophic factor and receptor tyrosine kinase B in postmortem brain of suicide subjects. *Arch Gen Psychiatry* 2003b;60:804–815.

Dwivedi Y, Rizavi HS, Shukla PK, Lyons J, Faludi G, Palkovits M, Sarosi A, Conley RR, Roberts RC, Tamminga CA, Pandey GN. Protein kinase A in postmortem brain of depressed suicide victims: Altered expression of specific regulatory and catalytic subunits. *Biol Psychiatry* 2004;55:234–243.

Dwivedi Y, Pandey GN. Adenylyl cyclase-cyclicAMP signaling in mood disorders: Role of the crucial phosphorylating enzyme protein kinase A. *Neuropsychiatr Dis Treat* 2008;4:161–176.

Exton JH. Phosphoinositide phospholipases and G proteins in hormone action. *Annu Rev Physiol* 1994;56:349–369.

Finkbeiner S. Calcium regulation of the brain-derived neurotrophic factor gene. *Cell Mol Life Sci* 2000;57:394–401.

Friedman E, Hoau Yan W, Levinson D, Connell TA, Singh H. Altered platelet protein kinase C activity in bipolar affective disorder, manic episode. *Biol Psychiatry* 1993;33:520–525.

Garcia-Sevilla JA, Escriba PV, Guimon J. Imidazoline receptors and human brain disorders. *Ann NY Acad Sci* 1999;881:392–409.

Grabarek J, Raychowdhury M, Ravid K, Kent KC, Newman PJ, Ware JA. Identification and functional characterization of protein kinase C isozymes in platelets and HEL cells. *J Biol Chem* 1992;267:10011–10017.

Holian O, Kumar R, Attar B. Apoprotein A-1 is a cofactor independent substrate of protein kinase C. *Biochem Biophys Res Commun* 1991;179:599–604.

Huang EJ, Reichardt LF. Neurotrophins: Roles in neuronal development and function. *Annu Rev Neurosci* 2001;24:677–736.

Hug H, Sarre TF. Protein kinase C isoenzymes: Divergence in signal transduction? *Biochem J* 1993;291(Pt 2):329–343.

Imaki J, Yoshida K, Yamashita K. A developmental study of cyclic AMP-response element binding protein (CREB) by *in situ* hybridization histochemistry and immunocytochemistry in the rat neocortex. *Brain Res* 1994;651:269–274.

Joiner TE, Jr., Brown JS, Wingate LR. The psychology and neurobiology of suicidal behavior. *Annu Rev Psychol* 2005;56:287–314.

Joseph SK, Williamson JR. Inositol polyphosphates and intracellular calcium release. *Arch Biochem Biophys* 1989;273:1–15.

Kaczmarek LK. The role of protein kinase C in the regulation of ion channels and neurotransmitter release. *Trends Neurosci* 1987;10:30–34.

Kiley SC, Jaken S. Protein kinase C: Interactions and consequences. *Trends Cell Biol* 1994;4:223–227.

Klomek AB, Marrocco F, Kleinman M, Schonfeld IS, Gould MS. Peer victimization, depression, and suicidiality in adolescents. *Suicide Life Threat Behav* 2008;38:166–180.

Lee KY, Ryu SH, Suh PG, Choi WC, Rhee SG. Phospholipase C associated with particulate fractions of bovine brain. *Proc Natl Acad Sci USA* 1987;84:5540–5544.

Lester D. The concentration of neurotransmitter metabolites in the cerebrospinal fluid of suicidal individuals: A meta-analysis. *Pharmacopsychiatry* 1995;28:45–50.

Lowy MT, Reder AT, Antel JP, Meltzer HY. Glucocorticoid resistance in depression: The dexamethasone suppression test and lymphocyte sensitivity to dexamethasone. *Am J Psychiatry* 1984;141:1365–1370.

Manji HK, Bebchuk JM, Moore GJ, Glitz D, Hasanat KA, Chen G. Modulation of CNS signal transduction pathways and gene expression by mood-stabilizing agents: Therapeutic implications. *J Clin Psychiatry* 1999;60(Suppl 2):27–39; discussion 40–41, 113–116.

Manji HK, Etcheberrigaray R, Chen G, Olds JL. Lithium decreases membrane-associated protein kinase C in hippocampus: Selectivity for the alpha isozyme. *J Neurochem* 1993;61:2303–2310.

Mann JJ, Oquendo M, Underwood MD, Arango V. The neurobiology of suicide risk: A review for the clinician. *J Clin Psychiatry* 1999;60(Suppl 2):7–11; discussion 18–20, 113–116.

Mann CD, Vu TB, Hrdina PD. Protein kinase C in rat brain cortex and hippocampus: Effect of repeated administration of fluoxetine and desipramine. *Br J Pharmacol* 1995;115:595–600.

Masliah E, Cole G, Shimohama S, Hansen L, DeTeresa R, Terry RD, Saitoh T. Differential involvement of protein kinase C isozymes in Alzheimer's disease. *J Neurosci* 1990;10:2113–2124.

Montminy MR, Gonzalez GA, Yamamoto KK. Regulation of cAMP-inducible genes by CREB. *Trends Neurosci* 1990;13:184–188.

Moore AN, Waxham MN, Dash PK. Neuronal activity increases the phosphorylation of the transcription factor cAMP response element-binding protein (CREB) in rat hippocampus and cortex. *J Biol Chem* 1996;271:14214–14220.

Moscicki EK, O'Carroll P, Rae DS, Locke BZ, Roy A, Regier DA. Suicide attempts in the Epidemiologic Catchment Area Study. *Yale J Biol Med* 1988;61:259–268.

Newton AC. Protein kinase C: Structure, function, and regulation. *J Biol Chem* 1995;270:28495–28498.

Nibuya M, Morinobu S, Duman RS. Regulation of BDNF and trkB mRNA in rat brain by chronic electroconvulsive seizure and antidepressant drug treatments. *J Neurosci* 1995;15:7539–7547.

Nibuya M, Nestler EJ, Duman RS. Chronic antidepressant administration increases the expression of cAMP response element binding protein (CREB) in rat hippocampus. *J Neurosci* 1996;16:2365–2372.

Nichols M, Weih F, Schmid W, DeVack C, Kowenz-Leutz E, Luckow B, Boshart M, Schutz G. Phosphorylation of CREB affects its binding to high and low affinity sites: Implications for cAMP induced gene transcription. *EMBO J* 1992;11:3337–3346.

Ninan PT, van Kammen DP, Scheinin M, Linnoila M, Bunney WE, Jr., Goodwin FK. CSF 5-hydroxyindoleacetic acid levels in suicidal schizophrenic patients. *Am J Psychiatry* 1984;141:566–569.

Nishizuka Y. The molecular heterogeneity of protein kinase C and its implications for cellular regulation. *Nature* 1988;334:661–665.

Nishizuka Y. Intracellular signaling by hydrolysis of phospholipids and activation of protein kinase C. *Science* 1992;258:607–614.

Nishizuka Y. Protein kinase C and lipid signaling for sustained cellular responses. *FASEB J* 1995;9:484–496.

Ono Y, Fujii T, Ogita K, Kikkawa U, Igarashi K, Nishizuka Y. The structure, expression, and properties of additional members of the protein kinase C family. *J Biol Chem* 1988;263:6927–6932.

Ono Y, Fujii T, Ogita K, Kikkawa U, Igarashi K, Nishizuka Y. Protein kinase C zeta subspecies from rat brain: Its structure, expression, and properties. *Proc Natl Acad Sci USA* 1989;86:3099–3103.

Pacheco MA, Stockmeier C, Meltzer HY, Overholser JC, Dilley GE, Jope RS. Alterations in phosphoinositide signaling and G-protein levels in depressed suicide brain. *Brain Res* 1996;723:37–45.

Pandey GN, Dwivedi Y, Pandey SC, Conley RR, Roberts RC, Tamminga CA. Protein kinase C in the postmortem brain of teenage suicide victims. *Neurosci Lett* 1997;228:111–114.

Pandey GN, Dwivedi Y, Pandey SC, Teas SS, Conley RR, Roberts RC, Tamminga CA. Low phosphoinositide-specific phospholipase C activity and expression of phospholipase C beta1 protein in the prefrontal cortex of teenage suicide subjects. *Am J Psychiatry* 1999;156:1895–1901.

Pandey GN, Dwivedi Y, Ren X, Rizavi HS, Mondal AC, Shukla PK, Conley RR. Brain region specific alterations in the protein and mRNA levels of protein kinase A subunits in the post-mortem brain of teenage suicide victims. *Neuropsychopharmacology* 2005;30:1548–1556.

Pandey GN, Dwivedi Y, Ren X, Rizavi HS, Roberts RC, Conley RR. Cyclic AMP response element-binding protein in post-mortem brain of teenage suicide victims: Specific decrease in the prefrontal cortex but not the hippocampus. *Int J Neuropsychopharmacol* 2007;10:621–629.

Pandey GN, Dwivedi Y, Rizavi HS, Ren X, Conley RR. Decreased catalytic activity and expression of protein kinase C isozymes in teenage suicide victims: A postmortem brain study. *Arch Gen Psychiatry* 2004;61:685–693.

Pandey GN, Dwivedi Y, Rizavi HS, Ren X, Pandey SC, Pesold C, Roberts RC, Conley RR, Tamminga CA. Higher expression of serotonin 5-HT(2A) receptors in the postmortem brains of teenage suicide victims. *Am J Psychiatry* 2002a;159:419–429.

Pandey GN, Dwivedi Y, SridharaRao J, Ren X, Janicak PG, Sharma R. Protein kinase C and phospholipase C activity and expression of their specific isozymes is decreased and expression of MARCKS is increased in platelets of bipolar but not in unipolar patients. *Neuropsychopharmacology* 2002b;26:216–228.

Pandey GN, Pandey SC, Dwivedi Y, Sharma RP, Janicak PG, Davis JM. Platelet serotonin-2A receptors: A potential biological marker for suicidal behavior. *Am J Psychiatry* 1995;152:850–855.

Pandey GN, Ren X, Rizavi HS, Conley RR, Roberts RC, Dwivedi Y. Brain-derived neurotrophic factor and tyrosine kinase B receptor signalling in post-mortem brain of teenage suicide victims. *Int J Neuropsychopharmacol* 2008;11:1047–1061.

Rhee SG. Regulation of phosphoinositide-specific phospholipase C. *Annu Rev Biochem* 2001;70:281–312.

Rhee SG, Bae YS. Regulation of phosphoinositide-specific phospholipase C isozymes. *J Biol Chem* 1997;272:15045–15048.

Rhee SG, Choi KD. Regulation of inositol phospholipid-specific phospholipase C isozymes. *J Biol Chem* 1992;267:12393–12396.

Riabowol KT, Fink JS, Gilman MZ, Walsh DA, Goodman RH, Feramisco JR. The catalytic subunit of cAMP-dependent protein kinase induces expression of genes containing cAMP-responsive enhancer elements. *Nature* 1988;336:83–86.

Shearman MS, Naor Z, Kikkawa U, Nishizuka Y. Differential expression of multiple protein kinase C subspecies in rat central nervous tissue. *Biochem Biophys Res Commun* 1987;147:911–919.

Sheline YI, Gado MH, Kraemer HC. Untreated depression and hippocampal volume loss. *Am J Psychiatry* 2003;160:1516–1518.

Shimohama S, Narita M, Matsushima H, Kimura J, Kameyama M, Hagiwara M, Hidaka H, Taniguchi T. Assessment of protein kinase C isozymes by two-site enzyme immunoassay in human brains and changes in Alzheimer's disease. *Neurology* 1993;43:1407–1413.

Silva AJ, Kogan JH, Frankland PW, Kida S. CREB and memory. *Annu Rev Neurosci* 1998;21:127–148.

Skalhegg BS, Tasken K. Specificity in the cAMP/PKA signaling pathway. Differential expression, regulation, and subcellular localization of subunits of PKA. *Front Biosci* 2000;5:D678–D693.

Stabel S, Parker PJ. Protein kinase C. *Pharmacol Ther* 1991;51:71–95.

Stanley M, Mann JJ. Increased serotonin-2 binding sites in frontal cortex of suicide victims. *Lancet* 1983;1:214–216.

Stokes CE, Hawthorne JN. Reduced phosphoinositide concentrations in anterior temporal cortex of Alzheimer-diseased brains. *J Neurochem* 1987;48:1018–1021.

Tamas Z, Kovacs M, Gentzler AL, Tepper P, Gadoros J, Kiss E, Kapornai K, Vetro A. The relations of temperament and emotion self-regulation with suicidal behaviors in a clinical sample of depressed children in Hungary. *J Abnorm Child Psychol* 2007;35:640–652.

Tanaka C, Nishizuka Y. The protein kinase C family for neuronal signaling. *Annu Rev Neurosci* 1994;17:551–567.

Taylor SJ, Chae HZ, Rhee SG, Exton JH. Activation of the beta 1 isozyme of phospholipase C by alpha subunits of the G_q class of G proteins. *Nature* 1991;350:516–518.

Teicher MH, Ito Y, Glod CA, Andersen SL, Dumont N, Ackerman E. Preliminary evidence for abnormal cortical development in physically and sexually abused children using EEG coherence and MRI. *Ann NY Acad Sci* 1997;821:160–175.

Thoenen H. Neurotrophins and neuronal plasticity. *Science* 1995;270:593–598.

Underwood MD, Mann JJ, Arango V. Serotonergic and noradrenergic neurobiology of alcoholic suicide. *Alcohol Clin Exp Res* 2004;28:57S–69S.

Xie H, Rothstein TL. Protein kinase C mediates activation of nuclear cAMP response element-binding protein (CREB) in B lymphocytes stimulated through surface Ig. *J Immunol* 1995;154:1717–1723.

Yamada S, Yamamoto M, Ozawa H, Riederer P, Saito T. Reduced phosphorylation of cyclic AMP-responsive element binding protein in the postmortem orbitofrontal cortex of patients with major depressive disorder. *J Neural Transm* 2003;110:671–680.

Zalsman G, Levy T, Shoval G. Interaction of child and family psychopathology leading to suicidal behavior. *Psychiatr Clin North Am* 2008;31:237–246.

16 Suicidal Behavior in Pediatric Population

Neurobiology and the Missing Links in Assessing Risk among Patients with Bipolar Disorder

Sonali Nanayakkara, Kiran Pullagurla, and Mani Pavuluri

CONTENTS

Suicidal behavior, in most of the people, is the result of conglomeration of recent negative events in their life. In some there is a planning for the final day, but most of the times, it is a rapid-onset act to relieve oneself from the mental agony (Kessler et al., 1999). *Suicidal ideation* refers to the thoughts of harming or killing oneself, wishing to be dead. *Suicide intent* conveys the seriousness or intensity of the person's desire or wish to die at the time of a suicide attempt. *Suicide attempt* is an act undertaken with the goal of committing suicide. *Parasuicide* is a nonfatal, self-destructive act with the intention of ceasing one's own life. *Self-injury* refers to a range of behaviors that may include cutting, scratching, head banging, self-mutilation with or without specific suicidal ideation or intent. *Suicidality* refers to all the suicidal behavior/acts and suicidal thinking/thoughts referring to an intention to end life (O'Carroll et al., 1996; Posner et al., 2007; Silverman et al., 2007a,b). This term must not be equated with prevalence of completed suicide or imply high correlation between suicidal thoughts, behavior, and death given that there is no adequate scientific data to quantify.

16.1 TRENDS IN SUICIDE IN PEDIATRIC POPULATION IN GENERAL AND BIPOLAR DISORDER IN SPECIFIC

Suicide, the act of deliberately taking one's own life, is the fourth leading cause of death among children between ages 10 and 14 and third leading cause of death among young adults aged 15–24, with an annual loss of life due to suicide at 2000 per year (Degmecić and Filaković, 2008). Nearly 2 million U.S. adolescents attempt suicide and 700,000 receive medical attention as a result of their attempt. According to Centers for Disease Control and Prevention (CDC) youth behavior survey for 2007, 14.5% of high school students reported that they wanted to end their life in the past year (Eaton et al., 2008). Suicidal ideation is often reported in pediatric bipolar disorder (BD), but the number progressing to attempt is around 20% (Klimes-Dougan et al., 1999). Patients with BD who have a past history of suicide attempt are over four times (odds ratio = 4.52, $p < 0.0001$) more prone to repeat the attempt or complete suicide, and 50% of the events (repeat attempt or completed suicide) occurred by 6 months of the first attempt (Marangell et al., 2006). In 2001, after accidents (46%) and homicide (15%), suicide was the third leading cause of death among children aged 10–19 years (CDC, 2004), accounting for 13% of total deaths in this age group.

The three most common methods of suicide, firearms, hanging, and poisoning, account for 92.3% of all completed suicides in the United States, with firearms being the most common method for completed suicide at 55%. Hanging (or suffocation) was reported in one out of five completed suicides, using electric cords, belts, or bed sheets. Overdose, especially along with alcohol, is the most commonly used method in females (Doshi et al., 2005). Pediatric BD has the highest mortality risk (Simpson and Jamison, 1999).

16.2 RISK FACTORS POTENTIALLY APPLICABLE TO SUICIDAL RISK IN BIPOLAR DISORDER

16.2.1 Number of Suicide Attempts

Past history of a suicide attempt is considered the best predictor of a future attempt, regardless of the type of illness. In a prospective naturalistic study on 180 adolescents, the number of prior attempts strongly predicted a suicide attempt in the post-hospitalization period. The highest risk was in the first 6 months to 1 year (Goldston et al., 1999). A longitudinal European multisite follow-up study concluded that among a sample of 1264, 24% had a past history of suicide attempt within the previous year of index attempt. The repeated attempters are at 3.2 times greater risk in future attempt compared to those who attempted once (Hulten et al., 2001). The increased risk conferred by multiple attempts could be due to the persistence of mood disorder (Esposito et al., 2003). Another study on adolescents admitted to an inpatient psychiatric unit concluded that multiple attempters are more likely to be diagnosed with substance abuse and they would have more than one comorbid diagnosis than adolescents with a past history of single attempt or suicidal ideation alone (D'Eramo et al., 2004). Miranda and colleagues studied future suicide attempt outcomes of 228 teenagers who reported recent suicidal ideation or had a past history of suicide attempt (Miranda et al., 2008). With specificity to BD patients, 52 subjects between the ages 21 and 74 years were compared on the basis of single and multiple suicide attempts (Michaelis et al., 2003). Results indicated that single attempters were potentially fatal and serious in their first attempt (odd ratio = 0.65, 95%; confidence interval = 0.43–0.99). Otherwise, there was no other clinical or demographic difference in single and multiple suicide attempters. Two-third of their study sample had multiple suicide attempts indicating that multiple attempts are common in BD.

16.2.2 Age as Moderator

Though many patients with pediatric BD harbor suicidal thoughts, it is the adolescent age group that is more prone for both an attempt and a completion of suicide. Most of the suicides are reported in postpubertal period because of the increased prevalence of psychopathology in adolescents, particularly the combination of mood disorder and substance abuse in them (Brent et al., 1999), or simply because of the increased chronological age (Zubrick et al., 1987). There are various reasons given for this postpubertal increased risk of suicide. The adolescents are cognitively more developed, and plan and execute the plan in a more concrete way (Brent et al., 1999; Shaffer, 1974). A Norwegian study showed that romantic disappointment is one of the factors responsible for increased number of suicide attempts and deaths in older adolescents (Groholt et al., 1998), whereas in children, relatively minor family arguments or disciplinary events have been shown to precipitate a suicide attempt (Beautrais, 2001).

16.2.3 Gender as Moderator

Generally across disorders, girls outnumber boys in having both suicidal ideation and attempt. They are twice as likely to think of suicide and four times as likely to attempt than the boys (Dilsaver et al., 2005). Gender difference for completed suicide does not manifest till mid-adolescence, that is, 15–19 years (Nock et al., 2008). Girls attempt suicide at a much younger age than boys. This gender difference is explained in part by the proneness of female sex to sexual exploitation (Wunderlich et al., 2001).

16.2.4 Hopelessness as Mediator

One of the most important affective state in which young patients commit suicide is hopelessness. Other affective states that are also associated with increased suicide risk are rage, despair, and guilt (Hendin, 1991). The level of hopelessness correlates with the level of suicidal ideation ($r = 0.45, p = 0.004$) and is a key indicator of suicide in patients with BD (Valtonen et al., 2006). Though hopelessness, to some degree, will be present in almost all patients in depressed phase, the severity is more predictive of the suicide risk. Additionally, hopelessness is linked to poor problem-solving abilities (Cannon et al., 1999). In contrast, it has also been shown that hopelessness becomes less of a predictive factor when overall severity of depression is controlled (Goldston et al., 2001).

16.2.5 Comorbid Disorders

Mood, impulse control, alcohol and substance abuse, psychotic, and cluster B personality disorders are found to convey the higher risk for suicide and suicidal behavior, with the rate increasing with multiple disorders (Nock et al., 2008). Comorbid panic disorder was significantly associated with suicide attempts in the pediatric BD (Goldstein et al., 2005).

16.2.6 Impulsive Aggression as Mediator

The gap between suicide ideation and an attempt is explained in part by the personality traits of that patient. Impulsive aggression is noted to be strongly associated with suicide in both pediatric age group and adults (Brent et al., 2003). Presence of impulsivity also contributes for the earlier age of suicide at first attempt. There are studies that have implicated impulsivity as one of the precipitating factors in the biology of suicide attempt and behavior (Mann et al., 2001). Mann et al. (1999) draw a relationship between impulsivity driven by stress where stress leads to suicidal ideation and impulsivity mediates action. Cyclothymic-hypersensitive temperament (CHT) corresponds to irritability, aggressiveness, and explosive anger strongly linked to both suicidal ideations and attempts. CHT increases the risk of suicidal ideation by an odds ratio of 7.4, and attempt by 10.5 (Kochman et al., 2005).

16.2.7 Relationship between Affect Regulation and Impulsivity

A disturbingly high percentage of 20%–47% among BD youth attempt suicide at least once before age 18 (Bhangoo et al., 2003; Goldstein et al., 2005; Lewinsohn et al., 2005; Strober et al., 1995). The high rate of attempts suggests that suicide may be the most common cause of death among adolescent BD patients.

Pediatric BD patients characteristically presents with emotional dysregulation, severe frustration tolerance, rejection sensitivity, aggressive rages, impulsivity, high comorbidity with disruptive behavior disorders, substance abuse and anxiety, mixed episodes with mania and depression, and chronic illness and poor executive functions (Birmaher et al., 2002; Dickstein et al., 2004; Pavuluri et al., 2005, 2006, 2007; Rich et al., 2005). All of these are factors associated with completed suicides (Shaffer et al., 1996; Spirito et al., 2006). Given the prevalence of high risk for suicide in adolescent BD, with several trait characteristics increasing the vulnerability to suicide, understanding the underlying biological mechanisms becomes a top priority.

While suicidal symptoms in Diagnostic and Statistical Manual of Mental Disorders (DSM) IV describe the various gradients of severity (Figure 16.1), we do not yet know whether or if there are unique biological factors that propel distressed patients to complete suicide. The role of self-harm such as "cutting" in suicide is also unknown. Some researchers have suggested that overt suicidal acts may represent a way to reduce "intolerable emotional states" (Zlotnick et al., 1997), with an extreme variant being the completed suicide. In fact, the highly maladaptive phenomenon of repeated suicide attempts has been shown to be highly associated with completed suicide (Brent et al., 2003; Inoue et al., 2006). We depict our clinical model of the spectrum of suicidality in Figure 16.1.

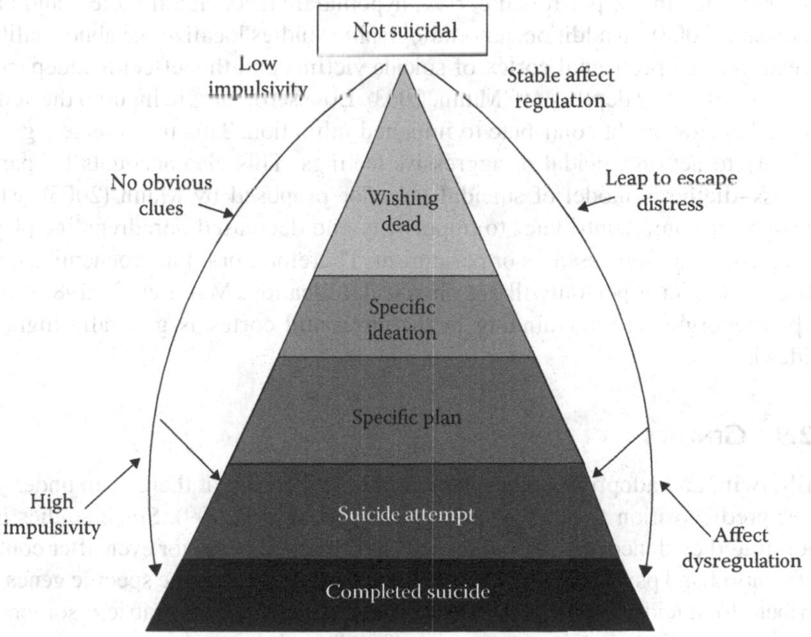

FIGURE 16.1 Model of affect dysregulation and impulsivity.

This model accommodates behaviors ranging from suicidal ideation and attempts to completed suicides. The degree of affect dysregulation and impulsivity may predict why some patients go on to complete suicides without a lag period of rumination and morbidity (Brent et al., 2003).

Therefore, two factors appear to be critical in increasing the risk for suicide in pediatric BD youth. First, severe affect dysregulation, especially in the context of negative emotions, leads to suicide attempts (Davidson et al., 2002; Thompson, 1994). We showed that pediatric BD patients have poor affect regulation, with greater amygdala response to negative emotions, associated with the shutdown of emotional and cognitive control systems, that is, ventrolateral and dorsolateral prefrontal cortex (Pavuluri et al., 2007, 2009). Those who complete suicide are also highly impulsive (Brent et al., 2003). Leibenluft et al. (2007) have also shown poor response inhibition in youth with bipolar diathesis, illustrating the decreased activation in frontostriatal system. There appears to be unexplored relationship between affect dysregulation and impulsivity that drives the youth with BD to be at increased risk for suicidal behavior.

16.2.8 Neurochemistry

Understanding chemical imbalance in neural systems helps encipher the pathophysiology behind increased risk for completed suicide. Many studies have been done on postmortem brain tissue to examine the serotonergic, noradrenergic, and dopaminergic neurotransmitter systems; signal transduction; and cellular morphology in suicide victims. Studies have shown that there are fewer presynaptic serotonin transporter sites in the prefrontal cortex, hypothalamus, occipital cortex, and brain stem (Mann, 2003). In addition, autoradiographic studies localize the abnormality to the ventromedial prefrontal cortex of suicide victims and this effect is independent of a history of major depression (Mann, 2003). Low serotonergic input to the ventral prefrontal cortex might contribute to impaired inhibition. This may create a greater propensity to act on suicidal or aggressive feelings. This also accounts for part of the stress–diathesis model of suicidal behavior proposed by Mann (2003), where decreased serotonin contributes to impulsivity and decreased noradrenaline plays a role in feelings of hopelessness or pessimism. Therefore, brain neurochemistry may modulate whether a person will act on suicidal ideation. Mann et al. (1986) found that β-adrenergic receptor binding in the prefrontal cortex is generally higher in suicide victims.

16.2.9 Genetics

Family, twin, and adoption studies demonstrate evidence that there is an underlying genetic predisposition to suicidal behavior (Ernst et al., 2009). Some studies have demonstrated evidence for familial transition of suicidal behavior even after controlling for mood and psychotic disorders (Nock, 2008). Although the specific genes that contribute to suicide risk independent of other associated psychiatric disorders are unknown, the serotonin transporter gene (*SERT*) and the three serotonin receptors (HTR1A, HTR2A, and HTR1B) and the monoamine oxidase promoter (MAOA) have been studied. The SERT promoter region has long and short allelic variants. Associations have been documented between the short form and violent suicide attempts in mood disorders, alcoholism, and suicide attempts (Bellivier et al., 2000; Bondy et al., 2000; Gorwood et al., 2000). The *HTR1A* gene is implicated because there is altered 5-HT1A binding in the midbrain and ventral prefrontal cortex of depressed suicide victims (Arango et al., 1995; Stockmeier et al., 1998). HTR1B has some involvement as 5-HT1B knockout mice are impulsive, aggressive, and more susceptible to self-administer substances of abuse such as cocaine and possibly alcohol (Ramboz et al., 1996; Rocha et al., 1998). The *MAO* gene has four or more variants based on the number of tandem repeats in the promoter region. Alleles with two to three tandem repeats have been associated with impulsive aggression in males as well as with lower levels of MAO expression (Manuck et al., 2000).

16.2.10 Substance Abuse as Mediator

A strong relation between suicide attempt and alcohol abuse has been shown to exist in children and adolescents (Wu et al., 2004). Cigarette smoking is also linked to suicide attempt. The association between suicidal ideation and substance abuse is insignificant after controlling for depression. The association of alcohol and the risk of suicide attempt is explained by the disinhibition and impaired serotonin system (Audenaert et al., 2001). Chronic use of alcohol can lead to lower levels of serotonin among nondependents (Weiss et al., 1996). Cigarette was worse in BD (Ostacher et al., 2006). Also, smoking was found to be associated with suicidality even after controlling for comorbid conditions and illness severity. This fact can be used to conclude that probably there may be another factor in smokers, like impulsive aggression, that leads to increased suicidality, but this needs further exploration. Somewhat different result was drawn from a case–control, psychological autopsy study, which showed that substance abuse was a risk factor for suicide in males only (Shaffer et al., 1996).

16.2.11 Pharmacotherapy: Implications

16.2.11.1 Selective Serotonin Reuptake Inhibitors: Risk of Suicide

Given the black box warnings from Food and Drug Administration (FDA) to increase awareness of potential suicide due to selective serotonin reuptake inhibitors (SSRIs), a careful appraisal of such risk is warranted. Treating depressed youth with SSRIs may be associated with a small increased risk of suicidality and therefore should only be considered if judicious clinical monitoring is possible. Specific treatment should be based on the individual's needs and mental health treatment guidelines (Williams et al., 2009). The treatment with antidepressants has shown varied results. One study shows inverse relationship of antidepressant use and suicide risk (Olfson et al., 2003), while another one shows increase in the number suicide attempts and suicide deaths in children and adolescents (Olfson et al., 2006). While SSRIs modestly increase the risk of occurrence of suicidal ideation and behavior, several studies also show that their use is associated with a significant decrease in the suicide rates in children and adolescents, probably because of their efficacy, compliance, and low toxicity in overdose (Bailly, 2009). The use of a long-term antidepressant treatment should be adapted to each individual, being cautious of its potential benefits and risks (Sechter, 1995). It is generally agreed that antidepressants are prescribed with caution in the presence of BD or related symptoms, where likelihood of agitation is greater after the first week of taking SSRIs among those that are likely to deteriorate.

Among the treatment options available for BD, lithium has been shown to be most effective in decreasing the risk of suicide attempt as well as suicide death in comparison to other pharmacological agents (Goodwin et al., 2003). Not only this, lithium has been shown to act as an anti-aggressive drug in children and adolescents (Campbell et al., 1984). More studies are needed to show the effectiveness of lithium in suicidal BD. But surprisingly, the discontinuation of lithium in bipolar patients can lead to increase in the number of suicidal acts in the first year of discontinuation in comparison to the later years or the time before the start of lithium treatment (Tondo et al., 1998). The risk of suicide was also affected by a change in the place of treatment. It was noticed that risk of suicide was high immediately after admission to hospital and immediately following discharge from hospital (Hoyer et al., 2004). These findings suggest that the changes introduced with respect to the medical interventions can affect the suicidality in BD.

16.2.12 Heritability

Many studies have been done in the past to demonstrate the link between factors governing suicide in proband and the family. Familial loading of suicidal behavior is directly proportional to the risk of suicide attempt in the offspring (Brent et al., 2003). Risk is further increased with a concordant twin with suicidal behavior. Studies have shown that there may be other factors that regulate the genetics of both suicides and affective disorders (Egeland and Sussex, 1985). It has been hypothesized that the heritability is due to the familial transmission of impulsive aggression. Another study concluded that the relatives of psychiatric patients with suicide attempt had higher risk of suicide attempts, the males being more prone for the risk than the females (Tsuang, 1983).

16.2.13 Phases of Bipolar Disorder

Different phases of the BD and the timing of the disease onset influence suicidality in different ways. Early onset BD increases the risk of suicide in children and adolescents (Jolin et al., 2007). In similar fashion, not all phases of the disorder are associated with suicide. Mixed and depressive phases are shown to increase suicide attempts in pediatric bipolar patients, whereas no suicide attempts were seen in manic/hypomanic phases (Valtonen et al., 2006). The different phases have different factors responsible for the risk of suicide. Hopelessness was related to suicidal behavior during the depressive phase, but during the mixed phase, subjective rating of severity of depression (on Beck Depression Inventory) and younger age were the predictors. In another prospective study, it was shown that the number of days for which the patient was depressed was directly related to the risk of suicide (Marangell et al., 2006). They concluded that an increase of 10% in number of days depressed increased the odds of a suicide attempt by 22%. The type of BD also plays a role in predicting the severity of suicidality. The diagnosis of BD I carries higher risk for suicide attempts over BD not otherwise specified, but no significant difference was noted with respect to the BD II subtype (Goldstein et al., 2005).

16.2.14 Trauma

Physical and sexual abuse in children and adolescents has an escalating effect in suicidal attempts in their future course of illness (Goldstein et al., 2005). It was found that BD has an earlier onset among those who had past history of abuse (physical/sexual), and it is accompanied with other comorbid Axis I, II, and III disorders, including drug and alcohol abuse (Leverich et al., 2002). Sexual abuse alone, from rare to frequent, is shown to increase the risk of suicide attempts. The combination of both physical and sexual abuse has an even greater impact in increasing the risk of suicide attempts as compared to any one alone (Leverich et al., 2003). Family history of abuse has also been shown to act indirectly in increasing suicidality in children. The presence of a history of sexual abuse in a parent increases the likelihood that their children will also be sexually abused, and consequently, the chances of suicide attempt will be increased in them (Brent et al., 2002). Again, it is hypothesized that impulsive–aggressive trait has a role to play in cases of physical and sexual abuse, with consequent increase in suicidality (Brodsky et al., 2001).

16.3 CONCLUSION

Because the causes of suicidal behavior in pediatric BD are multifactorial, the aim to prevent it should be multidimensional. Prevention and intervention must be based on assessing risk factors including further research in neurobiology, affect dysregulation, depression, mixed episode, aggression, and impulsivity. Especially given that adolescents carry higher risk for suicide, patients with BD should be given extra attention and support in transitional phases of their life.

ACKNOWLEDGMENTS

This work is supported by NIMH 1R01MH85639-01A1, NIMH 5R01MH081019, NIMH 1RC1MH088462.

REFERENCES

Arango, V., Underwood, M.D., Gubbi, A.V., Mann, J.J. 1995. Localized alterations in pre- and postsynaptic serotonin binding sites in the ventrolateral prefrontal cortex of suicide victims. *Brain Res* 688(1–2):121–133.

Audenaert, K., Van Laere, K., Dumont, F., Slegers, G., Mertens, J., van Heeringen, C., Dierckx, R.A. 2001. Decreased frontal serotonin 5-HT 2a receptor binding index in deliberate self-harm patients. *Eur J Nucl Med* 28(2):175–182.

Bailly, D. 2009. Antidepressant use in children and adolescents. *Arch Pediatr* 16(10):1415–1418.

Beautrais, A.L. 2001. Child and young adolescent suicide in New Zealand. *Aust N Z J Psychiatry* 35(5):647–653.

Bellivier, F., Szöke, A., Henry, C., Lacoste, J., Bottos, C., Nosten-Bertrand, M., Hardy, P., Rouillon, F., Launay, J.M., Laplanche, J.L., Leboyer, M. 2000. Possible association between serotonin transporter gene polymorphism and violent suicidal behavior in mood disorders. *Biol Psychiatry* 48(4):319–322.

Birmaher, B., Arbelaez, C., Brent, D. 2002. Course and outcome of child and adolescent major depressive disorder. *Child Adolesc Psychiatr Clin N Am* 11(3):619–637.

Bhangoo, R.K., Dell, M.L., Towbin, K., Myers, F.S., Lowe, C.H., Pine, D.S., Leibenluft, E. 2003. Clinical correlates of episodicity in juvenile mania. *J Child Adolesc Psychopharmacol* 13(4):507–514.

Bondy, B., Erfurth, A., de Jonge, S., Krüger, M., Meyer, H. 2000. Possible association of the short allele of the serotonin transporter promoter gene polymorphism (5-HTTLPR) with violent suicide. *Mol Psychiatry* 5(2):193–195.

Brent, D.A., Baugher, M., Bridge, J., Chen, T., Chiappetta, L. 1999. Age- and sex-related risk factors for adolescent suicide. *J Am Acad Child Adolesc Psychiatry* 38(12):1497–1505.

Brent, D.A., Oquendo, M., Birmaher, B., Greenhill, L., Kolko, D., Stanley, B., Zelazny, J., Brodsky, B., Bridge, J., Ellis, S., Salazar, J.O., Mann, J.J. 2002. Familial pathways to early-onset suicide attempt: Risk for suicidal behavior in offspring of mood-disordered suicide attempters. *Arch Gen Psychiatry* 59(9):801–807.

Brent, D.A., Oquendo, M., Birmaher, B., Greenhill, L., Kolko, D., Stanley, B., Zelazny, J., Brodsky, B., Firinciogullari, S., Ellis, S.P., Mann, J.J. 2003. Peripubertal suicide attempts in offspring of suicide attempters with siblings concordant for suicidal behavior. *Am J Psychiatry* 160(8):1486–1493.

Brodsky, B.S., Oquendo, M., Ellis, S.P., Haas, G.L., Malone, K.M., Mann, J.J. 2001. The relationship of childhood abuse to impulsivity and suicidal behavior in adults with major depression. *Am J Psychiatry* 158(11):1871–1877.

Campbell, M., Perry, R., Green, W.H. 1984. Use of lithium in children and adolescents. *Psychosomatics* 25:95–106.

Cannon, B., Mulroy, R., Otto, M.W., Rosenbaum, J.F., Fava, M., Nierenberg, A.A. 1999. Dysfunctional attitudes and poor problem solving skills predict hopelessness in major depression. *J Affect Disord* 55(1):45–49.

Centers for Disease Control (CDC). 2004. *MMWR* 53:471.

Davidson, R.J., Lewis, D.A., Alloy, L.B. et al. 2002. Neural and behavioral substrates of mood and mood regulation. *Biol Psychiatry* 52(6):478–502.

D'Eramo, K.S., Prinstein, M.J., Freeman, J., Grapentine, W.L., Spirito, A. 2004. Psychiatric diagnoses and comorbidity in relation to suicidal behavior among psychiatrically hospitalized adolescents. *Child Psychiatry Hum Dev* 35(1):21–35.

Degmecić, D., Filaković, P. 2008. Depression and suicidality in the adolescents in Osijek, Croatia. *Coll Antropol* 32(1):143–145.

Dickstein, D.P., Treland, J.E., Snow, J., McClure, E.B., Mehta, M.S., Towbin, K.E., Pine, D.S., Leibenluft, E. 2004. Neuropsychological performance in pediatric bipolar disorder. *Biol Psychiatry* 55(1):32–39.

Dilsaver, S.C., Benazzi, F., Rihmer, Z., Akiskal, K.K., Akiskal, H.S. 2005. Gender, suicidality and bipolar mixed states in adolescents. *J Affect Disord* 87(1):11–16.

Doshi A., Boudreaux, E.D., Wang, N., Pelletier, A.J., Camargo, C.A., Jr. National study of US emergency department visits for attempted suicide and self-inflicted injury, 1997–2001. *Ann Emerg Med* 2005;46(4)369–375.

Eaton, D.K., Kann, L., Kinchen, S., Shanklin, S., Ross, J., Hawkins, J., Harris, W.A., Lowry, R., McManus, T., Chyen, D., Lim, C., Brener, N.D., Wechsler, H., Centers for Disease Control and Prevention (CDC). 2008. Youth risk behavior surveillance—United States, 2007. *MMWR* 57(4):1–131.

Egeland, J.A., Sussex, J.N. 1985 Suicide and family loading for affective disorders. *JAMA* 254(7):915–918.

Ernst, C., Mechawar, N., Turecki, G. 2009. Suicide neurobiology. *Prog Neurobiol* 89(4):315–333.

Esposito, C., Spirito, A., Boergers, J., Donaldson, D. 2003. Affective, behavioral, and cognitive functioning in adolescents with multiple suicide attempts. *Suicide Life Threat Behav* 33(4):389–399.

Goldstein, T.R., Birmaher, B., Axelson, D., Ryan, N.D., Strober, M.A., Gill, M.K., Valeri, S., Chiappetta, L., Leonard, H., Hunt, J., Bridge, J.A., Brent, D.A., Keller, M. 2005. History of suicide attempts in pediatric bipolar disorder: Factors associated with increased risk. *Bipolar Disord* 7(6):525–535.

Goldston, D.B., Daniel, S.S., Reboussin, B.A., Reboussin, D.M., Frazier, P.H., Harris, A.E. 2001. Cognitive risk factors and suicide attempts among formerly hospitalized adolescents: A prospective naturalistic study. *J Am Acad Child Adolesc Psychiatry* 40(1):91–99.

Goldston, D.B., Daniel, S.S., Reboussin, D.M., Reboussin, B.A., Frazier, P.H., Kelley, A.E. 1999. Suicide attempts among formerly hospitalized adolescents: A prospective naturalistic study of risk during the first 5 years after discharge. *J Am Acad Child Adolesc Psychiatry* 38(6):660–671.

Goodwin, F.K., Fireman, B., Simon, G.E., Hunkeler, E.M., Lee, J., Revicki, D. 2003. Suicide risk in bipolar disorder during treatment with lithium and divalproex. *JAMA* 290(11):1467–1473.

Gorwood, P., Batel, P., Ades, J., Hamon, M., Boni, C. 2000. Serotonin transporter gene polymorphisms, alcoholism, and suicidal behavior. *Biol Psychiatry* 48(4):259–264.

Groholt, B., Ekeberg, O., Wichstrom, L., Haldorsen, T. 1998. Suicide among children and younger and older adolescents in Norway: A comparative study. *J Am Acad Child Adolesc Psychiatry* 37(5):473–481.

Hendin, H. 1991. Psychodynamics of suicide, with particular reference to the young. *Am J Psychiatry* 148(9):1150–1158.

Hoyer, E.H., Olesen, A.V., Mortensen, P.B. 2004. Suicide risk in patients hospitalised because of an affective disorder: A follow-up study, 1973–1993. *J Affect Disord* 78(3):209–217.

Hulten, A., Jiang, G.X., Wasserman, D., Hawton, K., Hjelmeland, H., De Leo, D., Ostamo, A., Salander-Renberg, E., Scmidtke, A. 2001. Repetition of attempted suicide among teenagers in Europe: Frequency, timing and risk factors. *Eur Child Adolesc Psychiatry* 10(3):161–169.

Inoue, K., Tanii, H., Abe S. et al. 2006. Causative factors as cues for addressing the rapid increase in suicide in Mie Prefecture, Japan: Comparison of trends between 1996–2002 and 1989–1995. *Psychiatry Clin Neurosci* 60(6):736–745.

Jolin, E.M., Weller, E.B., Weller, R.A. 2007. Suicide risk factors in children and adolescents with bipolar disorder. *Curr Psychiatry Rep* 9(2):122–128.

Kessler, R.C., Borges, G., Walters, E.E. 1999. Prevalence of and risk factors for lifetime suicide attempts in the National Comorbidity Survey. *Arch Gen Psychiatry* 56(7):617–626.

Klimes-Dougan, B., Free, K., Ronsaville, D., Stilwell, J., Welsh, C.J., Radke-Yarrow, M. 1999. Suicidal ideation and attempts: A longitudinal investigation of children of depressed and well mothers. *J Am Acad Child Adolesc Psychiatry* 38(6):651–659.

Kochman, F.J., Hantouche, E.G., Ferrari, P., Lancrenon, S., Bayart, D., Akiskal, H.S. 2005. Cyclothymic temperament as a prospective predictor of bipolarity and suicidality in children and adolescents with major depressive disorder. *J Affect Disord* 85(1–2):181–189.

Leibenluft, E., Rich, B.A., Vinton, D.T., Nelson, E.E., Fromm, S.J., Berghorst, L.H., Joshi, P., Robb, A., Schachar, R.J., Dickstein, D.P., McClure, E.B., Pine, D.S. 2007. Neural circuitry engaged during unsuccessful motor inhibition in pediatric bipolar disorder. *Am J Psychiatry* 164(1):52–60.

Leverich, G.S., Altshuler, L.L., Frye, M.A., Suppes, T., Keck, P.E. Jr., McElroy, S.L., Denicoff, K.D., Obrocea, G., Nolen, W.A., Kupka, R., Walden, J., Grunze, H., Perez, S., Luckenbaugh, D.A., Post, R.M. 2003. Factors associated with suicide attempts in 648 patients with bipolar disorder in the Stanley Foundation Bipolar Network. *J Clin Psychiatry* 64(5):506–515.

Leverich, G.S., McElroy, S.L., Suppes, T., Keck, P.E. Jr., Denicoff, K.D., Nolen, W.A., Altshuler, L.L., Rush, A.J., Kupka, R., Frye, M.A., Autio, K.A., Post, R.M. 2002. Early physical and sexual abuse associated with an adverse course of bipolar illness. *Biol Psychiatry* 51(4):288–297.

Lewinsohn, P.M., Olino, T.M., Klein, D.N. 2005. Psychosocial impairment in offspring of depressed parents. *Psychol Med* 35(10):1493–1503.

Mann, J.J. 2003. Neurobiology of suicidal behaviour. *Nat Rev Neurosci* 4(10):819–828.

Mann, J.J., Brent, D.A., Arango, V. 2001. The neurobiology and genetics of suicide and attempted suicide: A focus on the serotonergic system. *Neuropsychopharmacology* 24(5):467–477.

Mann, J.J., Stanley, M., Mcbride, P.A., McEwen, B.S. 1986. Increased serotonin-2 and β-adrenergic receptor binding in the frontal cortices of suicide victims. *Arch Gen Psychiatry* 43:954–959.

Mann, J.J., Waternaux, C., Haas, G.L., Malone, K.M. 1999. Toward a clinical model of suicidal behavior in psychiatric patients. *Am J Psychiatry* 156(2):181–189.

Manuck, S.B., Flory, J.D., Ferrell, R.E., Mann, J.J., Muldoon, M.F. 2000. A regulatory polymorphism of the monoamine oxidase-A gene may be associated with variability in aggression, impulsivity, and central nervous system serotonergic responsivity. *Psychiatry Res* 95(1):9–23.

Marangell, L.B., Bauer, M.S., Dennehy, E.B., Wisniewski, S.R., Allen, M.H., Miklowitz, D.J., Oquendo, M.A., Frank, E., Perlis, R.H., Martinez, J.M., Fagiolini, A., Otto, M.W., Chessick, C.A., Zboyan, H.A., Miyahara, S., Sachs, G., Thase, M.E. 2006. Prospective predictors of suicide and suicide attempts in 1,556 patients with bipolar disorders followed for up to 2 years. *Bipolar Disord* 8(5 Pt 2):566–575.

Michaelis, B.H., Goldberg, J.F., Singer, T.M., Garno, J.L., Ernst, C.L., Davis, G.P. 2003. Characteristics of first suicide attempts in single versus multiple suicide attempters with bipolar disorder. *Compr Psychiatry* 44(1):15–20.

Miranda, R., Scott, M., Hicks, R., Wilcox, H.C., Harris Munfakh, J.L., Shaffer, D. 2008. Suicide attempt characteristics, diagnoses, and future attempts: Comparing multiple attempters to single attempters and ideators. *J Am Acad Child Adolesc Psychiatry* 47(1):32–40.

Nock, M.K., Borges, G., Bromet, E.J. et al. 2008. Cross-national prevalence and risk factors for suicidal ideation, plans and attempts. *Br J Psychiatry* 192(2):98–105.

Nock, M.K., Borges, G., Bromet, E.J., Cha, C.B., Kessler, R.C., Lee, S. 2008. Suicide and suicidal behavior. *Epidemiol Rev* 30:133–154.

O'Carroll, P.W., Berman, A.L., Maris, R.W., Moscicki, E.K., Tanney, B.L., Silverman, M.M. 1996. Beyond the Tower of Babel: A nomenclature for suicidology. *Suicide Life Threat Behav* 26(3):237–252.

Olfson, M., Marcus, S.C., Shaffer, D. 2006. Antidepressant drug therapy and suicide in severely depressed children and adults: A case–control study. *Arch Gen Psychiatry* 63(8):865–872.

Olfson, M., Shaffer, D., Marcus, S.C., Greenberg, T. 2003. Relationship between antidepressant medication treatment and suicide in adolescents. *Arch Gen Psychiatry* 60(10):978–982.

Ostacher, M.J., Nierenberg, A.A., Perlis, R.H., Eidelman, P., Borrelli, D.J., Tran, T.B., Marzilli Ericson, G., Weiss, R.D., Sachs, G.S. 2006. The relationship between smoking and suicidal behavior, comorbidity, and course of illness in bipolar disorder. *J Clin Psychiatry* 67(12):1907–1911.

Pavuluri, M.N., Birmaher, B., Naylor, M.W. 2005. Pediatric bipolar disorder: A review of the past 10 years. *J Am Acad Child Adolesc Psychiatry* 44(9):846–871.

Pavuluri, M.N., Henry, D.B., Nadimpalli, S.S., O'Connor, M.M., Sweeney, J.A. 2006. Biological risk factors in pediatric bipolar disorder. *Biol Psychiatry* 60(9):936–941.

Pavuluri, M.N., O'Connor, M.M., Harral, E., Sweeney, J.A. 2007. Affective neural circuitry during facial emotion processing in pediatric bipolar disorder. *Biol Psychiatry* 62(2):158–167.

Pavuluri, M.N., Passarotti, A.M., Harral, E.M., Sweeney, J.A. 2009. An fMRI study of the neural correlates of incidental versus directed emotion processing in pediatric bipolar disorder. *J Am Acad Child Adolesc Psychiatry* 48(3):308–319.

Posner, K., Oquendo, M.A., Gould, M., Stanley, B., Davies, M. 2007. Columbia Classification Algorithm of Suicide Assessment (C-CASA): Classification of suicidal events in the FDA's pediatric suicidal risk analysis of antidepressants. *Am J Psychiatry* 164(7):1035–1043.

Ramboz, S., Saudou, F., Amara, D.A., Belzung, C., Segu, L., Misslin, R., Buhot, M.C., Hen, R. 1996. 5-HT1B receptor knock out–behavioral consequences. *Behav Brain Res* 73(1–2):305–312.

Rich, B.A., Bhangoo, R.K., Vinton, D.T. et al. 2005. Using affect-modulated startle to study phenotypes of pediatric bipolar disorder. *Bipolar Disord* 7:536–545.

Rocha, B.A., Scearce-Levie, K., Lucas, J.J., Hiroi, N., Castanon, N., Crabbe, J.C., Nestler, E.J., Hen, R. 1998. Increased vulnerability to cocaine in mice lacking the serotonin-1B receptor. *Nature* 393(6681):175–178.

Sechter, D. 1995. Long-term clinical effects of antidepressive agents. *Encephale* 21(Spec No. 2):35–38.

Shaffer, D. 1974. Suicide in childhood and early adolescence. *J Child Psychol Psychiatry* 15(4):275–291.

Shaffer, D., Gould, M.S., Fisher, P., Trautman, P., Moreau, D., Kleinman, M., Flory, M. 1996. Psychiatric diagnosis in child and adolescent suicide. *Arch Gen Psychiatry* 53(4):339–348.

Silverman, M.M., Berman, A.L., Sanddal, N.D., O'Carroll, P.W., Joiner, T.E. 2007a. Rebuilding the tower of Babel: A revised nomenclature for the study of suicide and suicidal behaviors. Part 1: Background, rationale, and methodology. *Suicide Life Threat Behav* 37(3):248–263.

Silverman, M.M., Berman, A.L., Sanddal, N.D., O'Carroll, P.W., Joiner, T.E. 2007b. Rebuilding the tower of Babel: A revised nomenclature for the study of suicide and suicidal behaviors. Part 2: Suicide-related ideations, communications, and behaviors. *Suicide Life Threat Behav* 37(3):264–277.

Simpson, S.G., Jamison, K.R. 1999. The risk of suicide in patients with bipolar disorders. *J Clin Psychiatry* 60(Suppl 2):53–56; discussion 75–76, 113–116.

Spirito, A., Esposito-Smythers, C. 2006. Attempted and completed suicide in adolescence. *Annu Rev Clin Psychol* 2:237–266.

Stockmeier, C.A., Shapiro, L.A., Dilley, G.E., Kolli, T.N., Friedman, L., Rajkowska, G. 1998. Increase in serotonin-1A autoreceptors in the midbrain of suicide victims with major depression-postmortem evidence for decreased serotonin activity. *J Neurosci* 18(18):7394–7401.

Strober, M., Schmidt-Lackner, S., Freeman, R., Bower, S., Lampert, C., DeAntonio, M. 1995. Recovery and relapse in adolescents with bipolar affective illness: A five-year naturalistic, prospective follow-up. *J Am Acad Child Adolesc Psychiatry* 34(6):724–731.

Thompson, E.A., Moody, K.A, Eggert, L.L. 1994. Discriminating suicide ideation among high-risk youth. *J Sch Health* 64(9):361–367.

Tondo, L., Baldessarini, R.J., Hennen, J., Floris, G., Silvetti, F., Tohen, M. 1998. Lithium treatment and risk of suicidal behavior in bipolar disorder patients. *J Clin Psychiatry* 59(8):405–414.

Tsuang, M.T. 1983. Risk of suicide in the relatives of schizophrenics, manics, depressives, and controls. *J Clin Psychiatry* 44(11):396–397, 398–400.

Valtonen, H.M., Suominen, K., Mantere, O., Leppämäki, S., Arvilommi, P., Isometsä, E. 2006. Suicidal behaviour during different phases of bipolar disorder. *J Affect Disord* 97(1–3):101–107.

Weiss, F., Parsons, L.H., Schulteis, G., Hyytiä, P., Lorang, M.T., Bloom, F.E., Koob, G.F. 1996. Ethanol self-administration restores withdrawal-associated deficiencies in accumbal dopamine and 5-hydroxytryptamine release in dependent rats. *J Neurosci* 16(10):3474–3485.

Williams, S.B., O'Connor, E., Eder, M., Whitlock, E. 2009. Screening for child and adolescent depression in primary care settings: A systematic evidence review for the U.S. Preventive Services Task Force. Rockville (MD): Agency for Healthcare Research and Quality (US). Report No. 09-05130-EF-1.

Wu, P., Hoven, C.W., Liu, X., Cohen, P., Fuller, C.J., Shaffer, D. 2004. Substance use, suicidal ideation and attempts in children and adolescents. *Suicide Life Threat Behav* 34(4):408–420.

Wunderlich, U., Bronisch, T., Wittchen, H.U., Carter, R. 2001. Gender differences in adolescents and young adults with suicidal behaviour. *Acta Psychiatr Scand* 104(5):332–339.

Zlotnick, C., Donaldson, D., Spirito, A., Pearlstein, T. 1997. Affect regulation and suicide attempts in adolescent inpatients. *J Am Acad Child Adolesc Psychiatry* 36(6):793–798.

Zubrick, S., Kosky, R., Silburn, S. 1987. Is suicidal ideation associated with puberty? *Aust N Z J Psychiatry* 21(1):54–58.

17 Suicide in Late Life

Olusola A. Ajilore and Anand Kumar

CONTENTS

17.1 INTRODUCTION

Suicide in late life merits special attention for many reasons. Epidemiologically, one of the groups most at-risk for suicide is older adults. It has long been recognized that in the United States, elderly white males have the highest suicide rates. In a review published by Hawton and van Heeringen, it was noted that suicide rates are also highest in the elderly in most countries around the world [1]. The elderly are not only a vulnerable population, but suicide has a more severe impact in this demographic group. For example, one study showed that suicide in late life tends to be associated with less warning and more lethality compared to attempts in younger populations [2]. This has also been shown in a study demonstrating that elderly suicide victims are more likely to have more serious intent with less warning [3]. These epidemiological studies highlight the importance of understanding suicide in late life. Thus, a small but prolific number of researchers have examined why the prevalence of suicide is higher in the elderly. Crucial to this understanding of suicidality in the elderly is the identification of risk factors for proper evaluation and intervention.

17.2 RISK FACTORS

The identification of suicide risk factors in late life has been greatly enhanced by the use of psychological autopsy. These studies attempt to identify differences between populations that have committed suicide or present with suicidality compared to control groups. Additionally, these studies comparing demographic and clinical features of suicide attempters versus age-matched controls have shown great utility in identifying factors that increase the risk of suicide in the elderly. In a Swedish study by Wiktorsson et al., suicide attempters 70 years of age or older tended to be unmarried, have a low education level, live alone, have a history of psychiatric illness, and have a history of a previous attempt [4]. This study highlights both the sociodemographic and clinical risk factors that have identified in a number of studies.

17.2.1 Sociodemographic Factors

Age and gender differences in suicidal ideation, attempts, and attempt lethality have all been identified in the elderly. It has been shown that even among men 50 years of age and older, men 70 years of age and older have higher attempt lethality than those younger than 70 years, while women older than 69 years had lower attempt lethality compared to younger women. The increase in lethality appeared to be driven by increased levels of intent [5]. In a British survey by Harwood et al., the authors reviewed suicides by individuals 60 years of age and older. In their study, men represented 67.7% of cases and the most common means for men was hanging, while it was overdose in women [6]. These studies reflect that males have higher and more lethal means of suicide compared to their female counter parts.

As would be expected, socioeconomic status is an important variable influencing suicidality in the elderly. It has been shown that lower socioeconomic status has been a risk factor for suicide in the elderly [7].

Other sociodemographic factors such as marital status are important risk factors for suicide. Loss of a spouse has been shown to be a risk factor [8]. However, there are particular features of bereavement that correspond to increased suicidality. For example, elderly subjects who rate high on scales of complicated bereavement are more likely to have suicidal ideation compared to lower scoring subjects [9]. The effect of losing a spouse demonstrates the important buffering effect of social relationships. This is also highlighted in a study showing that lower levels of perceived social support are also associated with suicidal ideation [10,11].

17.2.2 Clinical Risk Factors

There has been extensive work looking at mood disorders as a risk factor for suicide in the elderly. A study of completed suicides of patients who had visited a primary care practice within 30 days of suicide compared to control subjects demonstrated that suicide victims had higher levels of depressive illness, physical illness, and functional limitations [12]. In a survey of home health-care utilizers, suicide victims were more likely to have depressed mood, alcohol dependence, chronic pain, lack of social supports, and financial difficulties compared to controls [13]. There is also evidence that even subclinical depressive symptoms can increase the risk of suicide. In a Taiwanese study by Liu and Chiu, elderly patients who attempted suicide had significantly higher scores on the Brief Symptom Rating Scale compared to control subjects [14]. Suicidality also impacts how well mood disorders are treated. In a study by Szanto et al., elderly depressed subjects with suicidality were harder to treat, evidenced by higher relapse rates [15]. In addition to mood disorders, schizophrenia has been associated with increased suicidality in older patients. According to a study by Cohen et al., schizophrenic patients had higher prevalence of suicidal thinking and suicidal attempts compared to control subjects [16].

The presence of Axis II disorders is also a risk factor for suicide. In a study by Heisel et al., narcissistic personality disorder or trait was associated with increased risk of suicidality in later life [17]. Other personality characteristics such as neuroticism have also been associated with increased suicidal ideation [18]. While personality traits associated with suicidality offer limited opportunities for intervention, they are key to identifying at-risk patients.

In addition to the expected association of psychiatric illness and suicidality in the elderly, a number of studies have shown that medical illnesses are key risk factors. In a Finnish study identifying psychosocial stressors as antecedents to suicide in the elderly compared to younger suicide victims, the authors found that elderly suicide victims were more likely to have somatic illness as a stressor compared to loss, financial difficulties, and occupational issues [19]. One of the first studies to examine the impact of medical comorbidities found that a diagnosis of cancer played a large role in the decision to commit suicide [20]. Stroke risk measured by Cerebrovascular Risk Factor (CVRF) is higher in suicides compared to controls [21]. In conjunction with medical comorbidities, polypharmacy and medications can serve as additional risk factors. This was demonstrated by one study that showed the use of sedatives and hypnotics was associated with an increased suicide risk in the elderly [22]. Thus, medical illnesses and its associated complications can be very important factors in suicidality, particularly in the elderly.

17.2.3 Neurobiological Factors

While postmortem studies have been used extensively to understand neuropathological correlates of suicide in younger populations, there has not been extensive study focused on elderly samples. However, there have been attempts to understand the neuroendocrine and neurocognitive correlates of suicidality in the elderly. For example, HPA axis dysregulation has been associated with increased suicide risk. In a longitudinal study by Jokinen and Nordstrom, elderly depressed patients who went on to commit suicide all failed the dexamethasone suppression test [23]. Cognitive impairment has been associated with increased passive and active suicidal ideation in the elderly [24]. More specifically, impairment in executive function, memory, and attention was associated with suicidal ideation [25]. The cognitive impairments have also been linked to poor decision making seen in suicidality. In a paper by Dombrovski et al., elderly suicide attempters had impaired reward/punishment learning characterized by inability to learn from past experience [26].

17.3 EVALUATION

Proper evaluation for suicidality in the elderly depends on effective identification of the aforementioned risk factors. To assess these risk factors, several screening tools have been discussed in the literature. Typically, these screening instruments have been used for the assessment of mood disorders, particularly major depressive disorder. The Hamilton Rating Scale for Depression has a suicidality question with answers ranging from feeling like life is not worth living to previous suicide attempts [27]. Another screening tool that addresses suicidality is the Beck Depression Inventory, a self-administered questionnaire that asks specifically about suicidal thoughts [28]. For the geriatric population, the Geriatric Depression Scale (GDS) has been developed. While it is thought to be limited in that there are no specific questions to address suicidality, a study by Cheng et al. found that the GDS could accurately detect suicidal ideation, particularly in the "old–old," aged 75 and older [29]. Also, it has been demonstrated that the GDS can detect degree of suicidality as well [30].

There have been efforts to develop more specific suicide assessments that can be used in the primary care setting. For example, the "SLAP" interview protocol involves follow-up questions after a patient expresses suicidal ideation. SLAP is an acronym for specificity of the suicide plan, lethality of the means, availability of the means, and proximity of rescuers [31].

17.4 INTERVENTION/PREVENTION

The identification of risk factors facilitates the development of prevention measures to mitigate the likelihood of suicide attempts in the elderly. A number of these measures have been studied extensively in the literature. An important aspect of these intervention/prevention measures is the setting in which they take place. There are many different settings where prevention efforts for suicide should and can occur. Mental health settings are common locations for suicide assessment. Patients are typically assessed for suicidality as part of a standard evaluation. Unfortunately, large number of elderly patients in need of a psychiatric assessment will not utilize these facilities [32]. Instead, primary care settings are an important focal point of contact for a suicidal elderly patient. According to a number of studies cited by Conwell and Duberstein, within 30 days of committing suicide, up to 76% of patients had contact with a primary care provider [33]. In addition, some of the risk factors for suicidality are ideally detected in this setting, such as chronic medical illnesses or functional impairment. This has been highlighted in several studies. One of the ways to prevent late-life suicides that has been best characterized in the literature is effective treatment of late-life depression. The PROSPECT (Prevention of Suicide in Primary Care Elderly: Collaborative Trial) study showed significant reductions in suicidal ideation when tailored treatment guidelines were used in a primary care setting compared to treatment as usual [34]. In another trial in a primary care setting called the Improving Mood: Promoting Access to Collaborative Treatment (IMPACT) study, Unutzer et al. lowered suicidal ideation as early as 6 months and as late as 24 months after an intervention involving Problem Solving Treatment [35].

Another important setting is residential communities, which include independent living facilities, assisted living facilities, and long-term care facilities. In a paper by Podgorski et al., the authors describe approaches to suicide prevention in these communities focusing on at-risk groups and whole populations [36]. "At-risk" approaches focus on targeting individuals who possess risk factors such as mood disorders, functional impairment, and medical illness with programs that include referral for treatment, assessments for suicidality, and reducing stigma. Whole population approaches address issues such as coping, promoting social networks, decreasing access to lethal means, and promoting engagement in positive activities. An example of this last approach was demonstrated in a study by Oyama et al. who used an intervention involving group activities to lower suicide rates in elderly women from a community-based sample [37]. These studies suggest that targeted interventions that address specific risk factors in a variety of settings provide the best outcomes in reducing suicidal ideation and suicide attempts.

17.5 CONCLUSION

Given the changing demographics across the globe, suicide in the elderly is a major public health issue. Suicide in the elderly is characterized by particular challenges in identifying risk factors, evaluating suicidality in the context of these risk factors, and developing targeting interventions designed to meet the specific issues dealt with in late life. In addition to the detecting risk factors and formulating interventions, there is much more work to be done in identifying specific neurobiological substrates of suicidality in the elderly. In particular, there is a dearth of research using postmortem analyses with a focus on elderly samples either from autopsy or from subjects followed prospectively. This is particularly important since the profile of molecular targets identified as altered in suicidal patients may be different in the elderly subjects.

In this chapter, we have also discussed studies using cognitive neuroscience to understand impaired decision making in suicidal elderly subjects. This fascinating work in neurocognitive assessments demonstrates specific cognitive dysfunction associated with suicidal ideation and provides the foundation for future work in neuroimaging and neuropathology.

REFERENCES

1. Hawton K, van Heeringen K. Suicide. *Lancet* 373(9672), 1372–1381, 2009.
2. Salib E, Rahim S, El-Nimr G, Habeeb B. Elderly suicide: An analysis of coroner's inquests into two hundred cases in Cheshire 1. *Med. Sci. Law* 45(1), 71–80, 2005.
3. Conwell Y, Duberstein PR, Cox C, Herrmann J, Forbes N, Caine ED. Age differences in behaviors leading to completed suicide. *Am. J. Geriatr. Psychiatry* 6(2), 122–126, 1998.
4. Wiktorsson S, Runeson B, Skoog I, Ostling S, Waern M. Attempted suicide in the elderly: Characteristics of suicide attempters 70 years and older and a general population comparison group. *Am. J. Geriatr. Psychiatry* 18(1), 57–67, 2010.
5. Dombrovski AY, Szanto K, Duberstein P, Conner KR, Houck PR, Conwell Y. Sex differences in correlates of suicide attempt lethality in late life. *Am. J. Geriatr. Psychiatry* 16(11), 905–913, 2008.
6. Harwood DM, Hawton K, Hope T, Jacoby R. Suicide in older people: Mode of death, demographic factors, and medical contact before death. *Int. J. Geriatr. Psychiatry* 15(8), 736–743, 2000.
7. Cohen A, Gilman SE, Houck PR, Szanto K, Reynolds CF, III. Socioeconomic status and anxiety as predictors of antidepressant treatment response and suicidal ideation in older adults. *Soc. Psychiatry Psychiatr. Epidemiol.* 44(4), 272–277, 2009.
8. Duberstein PR, Conwell Y, Cox C. Suicide in widowed persons. A psychological autopsy comparison of recently and remotely bereaved older subjects. *Am. J. Geriatr. Psychiatry* 6(4), 328–334, 1998.
9. Szanto K, Prigerson H, Houck P, Ehrenpreis L, Reynolds CF, III. Suicidal ideation in elderly bereaved: The role of complicated grief. *Suicide Life Threat. Behav.* 27(2), 194–207, 1997.

10. Harrison KE, Dombrovski AY, Morse JQ, Houck P, Schlernitzauer M, Reynolds CF, III, Szanto K. Alone? Perceived social support and chronic interpersonal difficulties in suicidal elders. *Int. Psychogeriatr*. 22(3), 445–454, 2010.
11. Rowe JL, Conwell Y, Schulberg HC, Bruce ML. Social support and suicidal ideation in older adults using home healthcare services. *Am. J. Geriatr. Psychiatry* 14(9), 758–766, 2006.
12. Conwell Y, Lyness JM, Duberstein P, Cox C, Seidlitz L, DiGiorgio A, Caine ED. Completed suicide among older patients in primary care practices: A controlled study. *J. Am. Geriatr. Soc*. 48(1), 23–29, 2000.
13. Rowe JL, Bruce ML, Conwell Y. Correlates of suicide among home health care utilizers who died by suicide and community controls. *Suicide Life Threat. Behav*. 36(1), 65–75, 2006.
14. Liu IC, Chiu CH. Case–control study of suicide attempts in the elderly. *Int. Psychogeriatr*. 21(5), 896–902, 2009.
15. Szanto K, Mulsant BH, Houck PR, Miller MD, Mazumdar S, Reynolds CF, III. Treatment outcome in suicidal vs. non-suicidal elderly patients. *Am. J. Geriatr. Psychiatry* 9(3), 261–268, 2001.
16. Cohen CI, Abdallah CG, Diwan S. Suicide attempts and associated factors in older adults with schizophrenia. *Schizophr. Res*. 119(1–3), 253–257, 2010.
17. Heisel MJ, Links PS, Conn D, van RR, Flett GL. Narcissistic personality and vulnerability to late-life suicidality. *Am. J. Geriatr. Psychiatry* 15(9), 734–741, 2007.
18. Heisel MJ, Duberstein PR, Conner KR, Franus N, Beckman A, Conwell Y. Personality and reports of suicide ideation among depressed adults 50 years of age or older. *J. Affect. Disord*. 90(2–3), 175–180, 2006.
19. Heikkinen ME, Lonnqvist JK. Recent life events in elderly suicide: A nationwide study in Finland. *Int. Psychogeriatr*. 7(2), 287–300, 1995.
20. Conwell Y, Caine ED, Olsen K. Suicide and cancer in late life. *Hosp. Community Psychiatry* 41(12), 1334–1339, 1990.
21. Chan SS, Lyness JM, Conwell Y. Do cerebrovascular risk factors confer risk for suicide in later life? A case–control study. *Am. J. Geriatr. Psychiatry* 15(6), 541–544, 2007.
22. Carlsten A, Waern M. Are sedatives and hypnotics associated with increased suicide risk of suicide in the elderly? *BMC Geriatr*. 9, 20, 2009.
23. Jokinen J, Nordstrom P. HPA axis hyperactivity as suicide predictor in elderly mood disorder inpatients. *Psychoneuroendocrinology* 33(10), 1387–1393, 2008.
24. Ayalon L, Mackin S, Arean PA, Chen H, McDonel Herr EC. The role of cognitive functioning and distress in suicidal ideation in older adults. *J. Am. Geriatr. Soc*. 55(7), 1090–1094, 2007.
25. Dombrovski AY, Butters MA, Reynolds CF, III, Houck PR, Clark L, Mazumdar S, Szanto K. Cognitive performance in suicidal depressed elderly: Preliminary report. *Am. J. Geriatr. Psychiatry* 16(2), 109–115, 2008.
26. Dombrovski AY, Clark L, Siegle GJ, Butters MA, Ichikawa N, Sahakian BJ, Szanto K. Reward/punishment reversal learning in older suicide attempters. *Am. J. Psychiatry* 167(6), 699–707, 2010.
27. Hamilton M. A rating scale for depression. *J. Neurol. Neurosurg. Psychiatry* 23, 56–62, 1960.
28. Beck AT, Beamesderfer A. Assessment of depression: The depression inventory. *Mod. Probl. Pharmacopsychiatry* 7, 151–169, 1974.
29. Cheng ST, Yu EC, Lee SY, Wong JY, Lau KH, Chan LK, Chan H, Wong MW. The Geriatric Depression Scale as a screening tool for depression and suicide ideation: A replication and extention. *Am. J. Geriatr. Psychiatry* 18(3), 256–265, 2010.

30. Heisel MJ, Flett GL, Duberstein PR, Lyness JM. Does the Geriatric Depression Scale (GDS) distinguish between older adults with high versus low levels of suicidal ideation? *Am. J. Geriatr. Psychiatry* 13(10), 876–883, 2005.
31. Mitty E, Flores S. Suicide in late life. *Geriatr. Nurs.* 29(3), 160–165, 2008.
32. Conwell Y, Thompson C. Suicidal behavior in elders. *Psychiatr. Clin. North Am.* 31(2), 333–356, 2008.
33. Conwell Y, Duberstein PR. Suicide in elders. *Ann. N. Y. Acad. Sci.* 932, 132–147, 2001.
34. Bruce ML, Ten Have TR, Reynolds CF, III, Katz II, Schulberg HC, Mulsant BH, Brown GK, McAvay GJ, Pearson JL, Alexopoulos GS. Reducing suicidal ideation and depressive symptoms in depressed older primary care patients: A randomized controlled trial. *JAMA* 291(9), 1081–1091, 2004.
35. Unutzer J, Tang L, Oishi S, Katon W, Williams JW, Jr., Hunkeler E, Hendrie H, Lin EH, Levine S, Grypma L, Steffens DC, Fields J, Langston C. Reducing suicidal ideation in depressed older primary care patients. *J. Am. Geriatr. Soc.* 54(10), 1550–1556, 2006.
36. Podgorski CA, Langford L, Pearson JL, Conwell Y. Suicide prevention for older adults in residential communities: Implications for policy and practice. *PLoS Med.* 7(5), e1000254, 2010.
37. Oyama H, Watanabe N, Ono Y, Sakashita T, Takenoshita Y, Taguchi M, Takizawa T, Miura R, Kumagai K. Community-based suicide prevention through group activity for the elderly successfully reduced the high suicide rate for females. *Psychiatry Clin. Neurosci.* 59(3), 337–344, 2005.

18 Intermediate Phenotypes in Suicidal Behavior
Focus on Personality

Dan Rujescu and Ina Giegling

CONTENTS

18.1 RISK FACTORS OF SUICIDAL BEHAVIOR

Every year over 1 million people commit and over 10 million people attempt suicide. This is one suicide every 40 s and one suicide attempt every 3 s worldwide. Suicide accounts for almost 2% of the world's death and it has emerged as one of the leading causes of death among individuals aged 15–34 years in most of the countries (Bertolote, 2001; Cantor, 2000; WHO, 2011). Attempted suicide is currently regarded as the most important predictor of a future death from suicide (Gunnell and Lewis, 2005). Almost one-quarter of suicides are preceded by nonfatal suicidal behaviors in the previous year (Owens and House, 1994) and ~2% of suicide attempters end their own life during the 12 months subsequent to the index event (Owens et al., 2002). In the years following an initial suicide attempt, studies indicate a suicide risk ranging from 3.2% (Suokas and Lonnqvist, 1991) to 11.6% (Nielsen et al., 1995) within 5 years, 4.8% (Beck and Steer, 1989) to 12.1% (Nielsen et al., 1995) within 10 years, 6.7% within 18 years (De Moore and Robertson, 1996), and 10%–15% within lifetime (Suominen et al., 2004).

Suicidal behavior is complex and frequently classified by differentiating between suicidal ideation, suicide attempts, and completed suicide. It is not attributable only to one single cause but is a consequence of complex interactions of several risk factors like, for example, medical (e.g., mental disorders and chronic pain), psychological (e.g., hopelessness and aggressiveness), psychosocial (e.g., social isolation), social (e.g., lack of social support), cultural (e.g., religion), socioeconomic (e.g., unemployment), and biological (e.g. genetics and disorders in brain functioning).

Early work especially focusing on demographic, psychiatric, and psychological factors found, among others, that being younger, unmarried, unemployed, having a psychiatric illness, feeling hopeless, and having recently experienced a stressful life event were directly correlated with an increased risk of suicide and suicidal behaviors, whereas religiousness, spirituality, and social support were found to have a protective effect. Furthermore, it was reported that being female was associated with a higher risk of suicidal attempt whereas being male was related to a greater risk of death associated to suicide (Nock et al., 2008). Beside all other risk factors, especially personality seems to have a high impact on suicidal behavior (Brezo et al., 2006).

18.2 PERSONALITY AND SUICIDAL BEHAVIOR

18.2.1 Aggression

Aggression has been associated with suicidal behavior at least since the early findings by Asberg et al. (1976). Since then a high amount of studies could replicate this link (Corbitt et al., 1996; Malone et al., 1995; Oquendo et al., 2000, 2006).

Furthermore, aggression has been related to suicidality in mood disorders. Straub et al. (1992) compared four groups of depressed women with suicide ideation, violent suicide attempt, nonviolent suicide attempt, and depression without suicidality. Aggression scores, together with other psychophysiological features, were shown to differentiate between these groups. In their sample of psychiatric inpatients (51% with mood disorder), Mann et al. (1999) found higher rates of lifetime aggression in suicide attempters compared to non-suicidal patients. Oquendo et al. (2000) and Grunebaum et al. (2006) reported that lifetime aggression traits did correlate with suicide attempts in patients with mood disorder. In a recent study, regression tree analysis applied to a large group of bipolar patients identifying current depression and aggressive traits as indicators of a remote suicide attempt (Mann et al., 2008). A link between aggression and suicidal behavior has also been shown in schizophrenia (Hong et al., 2004; Malone et al., 2003; Mann et al., 2008; McGirr and Turecki, 2008), borderline personality disorder (Brodsky et al., 2006; Horesh et al., 2003), substance use disorders (Sher et al., 2005, 2008; Tremeau et al., 2008), and nonclinical samples (Ille et al., 2001).

18.2.2 Impulsivity

Furthermore, aggressive behaviors have been shown to correlate with impulsivity (Oquendo et al., 2000, 2004; Zouk et al., 2006), suggesting that an impulsive–aggressive dimension may predispose to suicidality (McGirr and Turecki, 2007; Turecki, 2005).

A variety of studies, in particular conducted by the McGill Group for Suicide Studies (Canada), demonstrated an association between impulsive personality traits and lifetime aggression in suicidal subjects (McGirr and Turecki, 2007, 2008; Turecki, 2005; Zouk et al., 2006). Such findings have led researchers to speculate on a common impulsive–aggressive dimension that may predispose to suicidality (McGirr and Turecki, 2007; Turecki, 2005). Conversely, others reported a weak correlation between measures of aggression and impulsivity (Critchfield et al., 2004). Such inconsistent findings may be due to different operationalizing criteria and instruments used to assess aggression and impulsivity in suicide research. Aggression has been measured by self-report questionnaires on aggressive tendencies (Doihara et al., 2008; Giegling et al., 2006, 2007; Pompili et al., 2008) or operationalized as a lifetime history of aggressive behaviors (McGirr and Turecki, 2008; Oquendo et al., 2004, 2007; Renaud et al., 2008; Zouk et al., 2006). Impulsivity may be conceptualized as the inability to resist impulses, which, from the strict phenomenological point of view, refers to explosive and instantaneous, automatic, or semiautomatic psychomotor actions that are characterized by their sudden and incoercible nature (Kempf, 1976). A more behavioral definition considers impulsivity as a predisposition toward rapid, unplanned reactions to internal or external stimuli without regard to the negative consequences of these reactions to the impulsive individual (Moeller et al., 2001). The lack of consensus about a definition of impulsivity leads to difficulties in its measurement. Impulsivity self-report scales exhibit low intercorrelations, are subject to response bias, and incorporate multiple subfactors (Gorlyn, 2005). In clinical evaluation of suicide attempt, premeditation is a critical factor. Indeed, different interventions may be necessary for the subjects who have carefully planned their attempt for a long period or are liable to act out suddenly in response to circumstances. A considerable proportion of attempts are made without premeditation. Although such impulsive suicide attempts are often pointed to being different from planned attempts, few studies have compared the two forms yielding inconsistent results (Brown et al., 1991; Mitrev, 1996; Polewka et al., 2005; Simon et al., 2001; Witte et al., 2008; Wyder and De Leo, 2007). Methods chosen for suicide attempt should be carefully considered in prevention tasks as they vary considerably in terms of violence and lethality. Research suggests that personality traits of anger, aggression, and impulsivity may influence the choice of suicide methods, although there are no consistent results from available studies (Dumais et al., 2005; Held et al., 1998; Seidlitz et al., 2001; Straub et al., 1992).

18.2.3 Anger

Anger has also been indicated as a risk factor for suicidality. Trait anger has been associated with a previous history of attempted suicide in a sample of adolescent inpatients with a variety of psychiatric disorders (Goldston et al., 1996). In adolescent suicide attempters, high levels of anger expression predicted self-mutilative behavior and a self-reported wish to die. In older suicide attempters, higher anger/hostility correlated with a greater number of attempts, while lower levels of anger predicted higher intent to die and higher lethality of methods (Seidlitz et al., 2001). Furthermore, the association between anger and suicidality has been demonstrated in depression (Painuly et al., 2007; Seidlitz et al., 2001; Velting et al., 2000), eating disorders (Nickel et al., 2006; Verkes et al., 1996; Youssef et al., 2004), and alcohol use disorders (Haw et al., 2001). In adolescent samples, studies reveal a strong correlation of anger with self-harm (Hawton et al., 1999) and attempted suicide (Cautin et al., 2001; Esposito et al., 2003; Kirkcaldy et al., 2006; Stein et al., 1998).

18.2.4 Temperament

Temperament and character evaluation has revealed significant differences between individuals with and without suicide attempts. Though there is a relatively wide agreement on the basic definition of temperament (Cloninger et al., 1994; Rettew and McKee, 2005; Thomas and Chess, 1977), a number of different measurement instruments have been used that differ in temperament definitions. However, a large overlap exists among at least three of them: Temperament and Character Inventory (TCI) (Cloninger et al., 1994), NEO Personality Inventory (NEO-PI) (Costa and McCrae, 1992), and Eysenck Personality Questionnaire (EPQ) (Eysenck and Eysenck, 1964, 1975). TCI harm avoidance (HA) and novelty seeking (NS) are, respectively, correlated with neuroticism (N) and extraversion (E) as defined by NEO-PI and EPQ (Costa and McCrae, 1995; De Fruyt et al., 2000).

Available literature has analyzed heterogeneous groups of suicide attempters: patients with mood disorders (Engstrom et al., 2004; Sayin et al., 2007) and eating disorders (Anderson et al., 2002; Bulik et al., 1999; Favaro et al., 2008), substance-dependent patients (Evren and Evren, 2005, 2006), and unselected psychiatric patients (Becerra et al., 2005; Calati et al., 2008; Guillem et al., 2002). Overall these studies report elevation in the temperament dimensions of NS and HA (Becerra et al., 2005; Calati et al., 2008) and decrease in the character dimensions of self-directedness (SD) and cooperativeness (CO) (Becerra et al., 2005; Calati et al., 2008; Evren and Evren, 2006; Favaro et al., 2008) in suicide attempters compared to non-suicidal controls. In addition, a part of the studies indicate temperament and character differences between attempted suicide and self-injurious behaviors (Evren and Evren, 2005; Favaro et al., 2008). Personality traits have been used to identify groups of suicidal individuals such as repeated attempters (Evren and Evren, 1996; Laget et al., 2006), older attempters (Duberstein et al., 2000), and completers (Useda et al., 2007). Overall, these studies report elevation in the temperament dimensions of "NS" and "HA" (Becerra et al., 2005; Calati et al., 2008) and decrease in the character dimensions of "SD" and "CO" (Becerra et al., 2005; Evren and Evren, 2006;

Favaro et al., 2008) in suicide attempters compared to non-suicidal controls. Giegling et al. (2009c) showed that higher aggression scores were predicted by being male, meeting criteria for borderline personality disorder, and having higher angry temperament scores as assessed by STAXI. Furthermore, temperament dimensions associated with self-aggression were high harm avoidance, high impulsivity, and low self-directedness; state anger, inwardly directed anger, and inhibition of aggression were also predictors of self-aggression. Additionally, impulsivity and harm avoidance have emerged as temperament dimensions independently associated with self-aggressive tendencies in personality.

18.2.5 Neuroticism

Neuroticism, a personality trait, reflects a tendency toward negative mood states and has been included in most theories of personality since its introduction. Studies have consistently demonstrated associations between an individual's level of neuroticism and likelihood of having symptoms or syndromes of suicidal behavior, depression, or anxiety (Brandes and Bienvenu, 2006; Widiger and Trull, 1992). Data from a number of studies, including two large, population-based, longitudinal samples (Kendler et al., 1993; Ormel et al., 2004), indicate that neuroticism acts as a premorbid vulnerability trait for major depression. Furthermore, several studies support the hypothesis that neuroticism mediates some of the elevated comorbidity between depressive and anxiety disorders as well as suicidal behavior (Khan et al., 2005).

18.2.6 Personality: An Intermediate Phenotype of Suicidal Behavior?

The identification of personality features related to suicide attempts may be useful to prevent suicide mortality, due to the strong correlation between attempted and complete suicide.

Intermediate phenotypes emerged recently to the most valuable tool in the search for the genetic underprinting of complex traits and diseases. Clinically defined diseases (e.g., stroke) or behaviors (e.g., suicide) can be regarded as the sum of many risk factors, which can be decomposed into intermediate phenotypes (e.g., lipid levels in the case of stroke and trait anger levels in the case of suicide). Genetic factors contributing to intermediate phenotypes will generally be easier to identify because of the improved signal-to-noise ratio in the fraction of variance explained by any single factor. Studies using a single clinical end point (e.g., stroke) have one chance of success, whereas studies that also collect intermediate phenotypes are more likely to help understand the contribution of genetic factors to characteristic components of disease or behavior. Especially suicide-related behavior may represent a heterogeneous trait implying that the current clinically defined phenotype might not be optimal for genetic studies. Therefore, simpler, quantifiable measures of functioning may be more useful in gene discovery. This approach helps to circumvent questions about etiological models.

Intermediates can be regarded as syntheses of subsets of proximal risk factors, both environmental and genetic. The rationale for the use of intermediate phenotypes in gene discovery is that the intermediate phenotypes associated with, for example, a behavioral trait are more elementary compared to clinical phenotypes. This also implies that the number of genes required to produce variations in these traits may be fewer than those involved in producing a diagnostic entity. Intermediate phenotypes are thus likely to bridge the gap between genes and clinical phenotypes.

Some intermediate phenotypes used in psychiatry are state independent, are associated with the illness, and co-segregate in families with the disease. In summary, intermediate phenotypes have emerged as an intriguing concept in the study of complex neuropsychiatric traits. Intermediate phenotypes represent simpler phenotypes than psychiatric diagnoses, which can result in more straightforward and successful genetic analysis (Gottesman and Gould, 2003).

Moreover personality traits, which are partly under genetic control, may be intermediate phenotypes for the genetic component of suicidal behavior (Baud, 2005).

18.3 GENETIC COMPONENT OF SUICIDAL BEHAVIOR IN RELATION TO PERSONALITY

Risk of suicide-related behavior is supposed to be determined by a complex interplay of sociocultural factors, traumatic life experiences, psychiatric history, personality traits, and genetic vulnerability. The latter is supported by adoption and family studies, indicating that suicidal acts have a genetic contribution that is independent of the heritability of Axis I and II psychopathology (Brent and Mann, 2005; Roy and Linnoila, 1986; Roy et al., 1991; Schulsinger et al., 1979). The heritability for serious suicide attempts was estimated to be 55% (Statham et al., 1998), and genetic association studies are valuable bottom-up tools to reveal underlying neurobiological pathways of this complex trait. Most genetic association studies in this context focused on genes that are involved in serotonergic neurotransmission. Serotonergic function was regarded to be crucial for the regulation of impulsive and aggressive behavior, which in turn has been demonstrated to correlate with suicidal behavior in various studies (Mann, 2003).

18.3.1 SEROTONERGIC SYSTEM

Many biological studies investigating the postmortem brains of suicide victims have shown alterations of the serotonergic system, especially in the prefrontal cortex (for review, see Arango et al., 1997). Furthermore, clinical studies found a relationship between lower levels of the serotonin metabolite 5-hydroxyindoleacetic acid (5-HIAA) and suicidal behavior (Asberg, 1997). Low cerebrospinal fluid (CSF) concentrations of the serotonin metabolite 5-HIAA are regarded as a putative indicator for low serotonin turnover (Asberg, 1997; Asberg et al., 1976, 1987). It is a relatively enduring trait that is partially genetically controlled as suggested by studies in humans (Oxenstierna et al., 1986), and demonstrated in rhesus monkeys (Higley et al., 1992; Westergaard et al., 1999). In these animals, low CSF 5-HIAA is associated with aggressiveness, low social affiliation, high-risk behavior, and premature mortality (Higley et al., 1996a,b; Mehlman et al., 1994). In humans, 5-HIAA in the CSF correlates inversely with various aggressive behaviors as demonstrated in healthy and psychiatric samples throughout the life span. Although there are some negative reports, taken together, the vast majority of findings suggest that lowered 5-HIAA is related to the vulnerability for aggressive behavior (Asberg, 1997; Asberg et al., 1987; Lee and Coccaro, 2001), and there seems to be a relationship between lower serotonergic function and aggressive, impulsive, and suicidal behavior. Therefore, genetic studies investigated the serotonergic system in suicidal behavior and personality traits from the very beginning.

18.3.1.1 Tryptophan Hydroxylases 1 and 2

The enzyme tryptophan hydroxylase (TPH) is involved in the biosynthesis of serotonin, converting L-tryptophan in a rate-limiting step into 5-hydroxytryptophan. An early meta-analysis (Lalovic and Turecki, 2002) did not find an association of the commonly studied intron 7 A218C (TPH1) single nucleotide polymorphism (SNP) with suicidal behavior *per se*. A further meta-analysis summarized the results of seven studies investigating thc A218C SNP in Caucasians (Rujescu et al., 2003b) and found a higher frequency of the A218 allele in the patients with suicidal behavior, strongly suggesting that this TPH1 polymorphism is associated with suicidal behavior in Caucasians. A further meta-analysis included nine studies and confirmed the association between the A218C polymorphism and suicidal behavior using both the fixed effect method and the random effect method (Bellivier et al., 2004), and in 2006 Li and He provided a meta-analysis on 22 studies and showed further support for association (Li and He, 2006).

Interestingly, this risk allele "A" was also associated with higher measures of aggression and a tendency to experience unprovoked anger and an expression of anger more outwardly (Manuck et al., 1999). This result was further supported by Rujescu et al. (2002) who showed that A-carrier had higher scores on the Trait Anger Scale of the State Trait Anger Expression Inventory (STAXI) and especially on the subscale "Angry Temperament" in two independent samples (healthy controls and suicide attempters). An involvement of TPH1 in anger and aggression phenotypes could furthermore recently been presented by Baud et al. (2009) who showed that suicide attempters carrying the AA genotype scored significantly lower on the Anger Control subscale than C-allele carrier, which confirms the hypothesis that the *TPH1* gene could confer a vulnerability to suicidal behavior through a reduced capacity to control anger, which in turn may represent a common psychopathological and behavioral pathway to suicidal behavior in an important subgroup of clinical subjects.

Walther et al. (2003) discovered a second TPH isoform in mice, termed TPH2, that seems to control the brain serotonin synthesis (Zhang et al., 2004). Breidenthal et al. (2004) screened the coding and exon-flanking intronic sequence of the *TPH2* gene and identified several genetic variants that might serve as markers for association studies. The investigation of 10 SNPs in the *TPH2* gene in a sample of 263 suicide victims and 266 ethnically matched healthy controls showed an association of one SNP with completed suicide (Zill et al., 2004). Since then several case-control studies were provided showing positive and negative results (for review, see Tsai et al., 2011). Interestingly, polymorphisms in the *TPH2* gene were also associated with personality traits like aggression or impulsivity, for example, Kulikov et al. (2005a,b) reported that the C1473G polymorphism in the *TPH2* gene is associated with intermale aggression in mice (see also Osipova et al., 2009, 2010).

18.3.1.2 Monoamine Oxidase A

The gene coding for the monoamine oxidase A (MAOA) contains a variable number of tandem repeats (VNTR) polymorphism in the promoter region. The 30 bp repeated sequence is present in 3, 3.5, 4, or 5 copies, and alleles with 3.5 or 4 copies are transcribed 2–10 times more efficiently than those with 3 or 5 copies (Sabol et al., 1998). MAOA is one of the key enzymes in the metabolism of serotonin as well as noradrenalin. Studies investigating the possible association of this MAOA-uVNTR polymorphism and suicidal behavior yielded inconsistent results. Several groups found no significant differences in genotype or allele distribution between subjects with suicidal behavior and comparison groups (Huang et al., 2004b; Kunugi et al., 1999; Ono et al., 2002), whereas one investigation delivered a significant association (Ho et al., 2000). Ho et al. examined this VNTR and the Fnu4H1 polymorphism in a sample of patients suffering from bipolar affective disorder. They found the VNTR variant to be associated with a history of suicide attempts, especially in females. The Fnu4H1 RFLP only showed significant differences in allele frequencies for female subjects, but not in the total sample. Another study showed a strong association of the high activity-related EcoRV allele and depressed suicide in male subjects, but not in females or the total sample (Du et al., 2002).

Additionally, the *MAOA* gene seems to be implicated in the control of aggressive or impulsive behavior. Brunner et al. (1993a,b) reported a large Dutch family with a new form of X-linked mild mental retardation, with all affected males showing aggressive, impulsive, and sometimes violent behavior and one case of attempted suicide. A second investigation showed that each of five affected males had a point mutation in the *MAOA* gene, which changes a glutamine to a termination codon (Brunner et al., 1993a,b). Interestingly, a recent study by Zalsman et al. (2011) could report an association with low activity MAOA genotypes and clinical traits of impulsive aggression and suicidality in adolescents.

18.3.1.3 Serotonin Transporter

The serotonin transporter (5-HTT) is located on the presynaptic membrane of serotonergic neurons and is responsible for the reuptake of released serotonin from the synaptic cleft. The 5′-promotor region of the *5-HTT* gene contains a functional insertion/deletion variant (5-HTTLPR) with former two and latter three common alleles that were designated as "short" (s) and "long" (l_A and l_G). The 5-HTT transcription and serotonin reuptake are higher in cells containing the homozygous ll genotype, compared with cells having the ls or ss forms (Lesch et al., 1996).

A meta-analysis conducted by Anguelova et al. (2003) included 12 studies investigating the 5-HTT promoter polymorphism. The study samples compromised 10 Caucasian populations, one U.S. population (80% Caucasians, 20% others) and one Chinese sample. They pooled a total number of 1168 cases (suicide completers and suicide attempters) and 1371 controls and found a significant association of the s-allele with suicidal behavior as a whole. They furthermore observed an overall association if only studies that investigated suicide attempters and control subjects were considered, but no association was found by comparing only suicide completers versus controls.

A second meta-analysis, including 18 studies with 1521 suicide attempters or completers and 2429 controls, delivered different results (Lin and Tsai, 2004). In contrast to the investigation of Anguelova et al. (2003), Lin and Tsai (2004) found no overall association of 5-HTTLPR alleles with suicidal behavior. This was also true if only the 15 studies with subjects of Caucasian origin were examined. The authors also compared the allelic and genotype distribution between 190 violent suicide attempters or completers and 733 normal control subjects. They observed a significant association of the s-allele with violent suicidal behavior, which is mainly characterized by the use of highly lethal and violent methods like hanging or shooting, but not with nonviolent suicide. Lin and Tsai (2004) concluded that violent suicidal subjects might be a relatively homogeneous group and that patients carrying the s-allele are likely to act more impulsive and aggressive. Additionally, the genotypes containing the s-allele were more frequent in suicide attempters than in nonattempters with the same psychiatric diagnoses. Li and He (2007) reviewed 39 studies that examined the association between functional polymorphism of 5-HTTLPR and suicide attempts in groups with psychiatric diagnoses. Their report suggested a significant association with the s-allele. In conclusion, there is evidence that the 5-HTTLPR polymorphism is involved in the predisposition to suicidal behavior. More detailed studies are required, as it seems that this functional variant is associated with particular intermediate phenotypes of suicidal behavior (for review, see Canli and Lesch, 2007; Lesch, 2007; Murphy and Lesch, 2008).

18.3.1.4 5-HT1A Receptor

Lemonde et al. (2003) examined the common C-1019G SNP in the promoter region of the *5-HT1A* gene in a sample of 102 suicide victims and 116 healthy controls, all of French-Canadian origin. They found the G-allele to be significantly overrepresented in the suicide group, and the homozygous G/G genotype was four times more frequent among suicide completers. This association could not be replicated by Huang et al. (2004a). The investigation of the structural polymorphisms Pro16Leu and Gly272Asp revealed no association with suicidal behavior in Japanese subjects (Nishiguchi et al., 2002), and the result for the Pro16Leu SNP was replicated by a second Japanese group (Ohtani et al., 2004). But so far no positive association could be found for anger- or aggression-related phenotypes and genetic variants in this gene (Serretti et al., 2007a,b, 2009).

18.3.1.5 5-HT1B Receptor

Suicidality and impulsive aggression are partially heritable, and postmortem brain studies suggest that abnormalities in serotonin 1B may be associated with suicide. Studies of serotonin 1B "knockout" mice show an increase in aggressive behavior relative to wild-type mice (Saudou et al., 1994). Interestingly, Jensen et al. (2009) could show that a common polymorphism in serotonin receptor 1B mRNA moderates regulation by miR-96 and associates with aggressive human behaviors. Furthermore, Conner et al. (2010) show that functional polymorphisms in the serotonin 1B receptor gene (*HTR1B*) predict self-reported anger and hostility among young men. Whereas there seems some support for the involvement of the *HTR1B* gene in aggression, no association could be found with suicidal behavior *per se* so far (Arango et al., 2003; Hong et al., 2004; Huang et al., 1999, 2003; New et al., 2001; Nishiguchi et al., 2001; Pooley et al., 2003; Rujescu et al., 2003c; Stefulj et al., 2004; Tsai et al., 2004b; Turecki et al., 2003), and also a recent meta-analysis showed no significance (Kia-Keating et al., 2007).

18.3.1.6 5-HT2A Receptor

The increased density of brain and platelet serotonin 2A (5-HT2A) receptors in subjects with suicidal behavior is evidenced by several studies (Bachus et al., 1997). Thus, the 5-HT2A receptor gene has been regarded as a major candidate for the genetic susceptibility to this behavior (Mann, 1998). However, a meta-analysis pooling nine studies with altogether 596 suicide completers or attempters and 1003 healthy controls could not find any association with the T102C SNP (Anguelova et al., 2003). A later meta-analysis of 25 studies could furthermore not find any association with this SNP but interestingly, the meta-analysis of a promoter SNP in this gene (A-1438G; rs20070040) could show association with suicidal behavior (Li et al., 2006). To test for intermediate phenotypes, Giegling et al. (2006) investigated the association of the three SNPs in the *5-HT2A* gene with anger- and aggression-related personality traits. CC homozygotes for the functional SNP rs6311 reported more anger-related traits in general and had higher scores on the subscales "Trait Anger" with its "Angry Reaction" component and lower scores on the subscale "Anger Out." Accordingly, the C-allele of the functional SNP was also associated with aggression-related traits and more specifically with less aggression inhibition. The SNPs predominantly analyzed in association with personality traits to date were rs6311 and rs6313. These SNPs are in strong linkage disequilibrium (LD) being in the promoter and at the beginning of the gene, respectively. A large quantity of studies was published. In particular, platelet serotonin 2A receptor sites have been found associated with impulsivity and aggression in a positron emission tomography (PET) study (Coccaro et al., 1997). Moreover, rs6311 was found to be associated with impulsive traits in alcohol dependents, measured with the Baratt Impulsiveness Scale (BIS) (Preuss et al., 2000). This was the first report on an association of a 5-HT2A promoter SNP with personality dimensions, which is partially confirmed by more recent reports on the same variant (Ni et al., 2006; Nomura, 2006). Subsequently, 5-HT2A SNPs have been found to be associated with low anxiety-related traits, which are negatively related to impulsive traits (Golimbet et al., 2004; Rybakowski et al., 2006). In addition, TCI self-transcendence was related to the rs6311 SNP (Ham et al., 2004). On the other hand, a considerable number of negative results have also been reported regarding the involvement of 5-HT2A in personality traits (Berggard et al., 2003; Blairy et al., 2000; Jonsson et al., 2001; Kusumi et al., 2002; Lochner et al., 2007; Tochigi et al., 2005). Serretti et al. (2007a,b) investigated a few SNPs with personality traits as measured with the TCI in three independent samples including healthy subjects and patients. Although the SNP rs594242 showed an association with SD ($p = 0.003$) in the German sample, and rs6313 was marginally associated with NS ($p = 0.01$) in the Italian sample, this study does not support a major effect of these SNPs on temperament so far.

18.3.2 Dopaminergic System and Catecholamines

CSF studies provided support for an involvement of the dopaminergic system in suicidal behavior showing correlations between low levels of the dopamine metabolite homovanillic acid and suicidal behavior (Ryding et al., 2008).

18.3.2.1 Dopamine Receptors

The first study investigated a SNP in the 3′-UTR of exon 8 in the dopamine D2 receptor gene and showed that the A/A genotype was associated with an increased number of suicide attempts (Finckh et al., 1997). Furthermore, the functionally relevant −141C insertion/deletion polymorphism was investigated by Ho et al. (2000) who found no differences in allele frequencies in subgroups of unipolar and bipolar patients with or without suicidal behavior. Johann et al. (2005) showed an association for this variant to suicidality in patients with alcohol dependence, although this association did not remain significant after Bonferroni correction, and Suda et al. (2009) reported an association of the −141C Ins/Del and TaqIA polymorphisms with suicide attempters. Two studies examined the association of a 48 bp repeat polymorphism in the dopamine receptor D4 gene with suicide attempts. Both groups did not find any evidence of an implication of this polymorphism in suicidal behavior (Persson et al., 1999; Zalsman et al., 2004).

The first study on genetic variants in the dopamine receptor genes by Sweet et al. (1998) showed no association with the DRD2 S311C polymorphism nor the presence of long alleles for the DRD4 exon III repeat sequence and also not with the *DRD3* gene, while there was a weak association with one SNP in the *DRD1* gene an aggression. Furthermore, Chen et al. (2005) provided evidence for a positive correlation of both the dopamine D2 receptor gene (*DRD2*) and the dopamine transporter gene (*DAT1*) polymorphisms with pathological violence in adolescents. As the efficacy of dopamine D2 receptor (DRD2) blocking antipsychotic drugs in borderline personality disorder treatment also suggests involvement of the dopamine system in the neurobiology of that disease, Nemoda et al. (2010) tested the dopamine dysfunction hypothesis of impulsive self- and other damaging behaviors. The DRD2 TaqI B1-allele and A1-allele were associated with borderline traits, also, the DRD4 −616 CC genotype appeared as a risk factor, but only the DRD4 promoter finding was replicated in the independent sample of psychiatric inpatients. No association was found with the COMT and DAT1 polymorphisms. Further support for the involvement of genetic polymorphisms in aggression and impulsivity comes from Retz et al. (2003) whose results suggest that variations of the *DRD3* gene are likely involved in the regulation of impulsivity and some psychopathological aspects of ADHD related to violent behavior. There are many other studies on this matter (for review, see Craig and Halton 2009).

18.3.2.2 Dopa Decarboxylase

Components of the dopaminergic system have been previously associated to suicidal behaviors (Oquendo and Mann, 2000; Rujescu et al., 2003a; Ryding et al., 2008) and aggression regulation (Pitchot et al., 1992, 2001; Rujescu et al., 2003a), hence making the pathway a good candidate to be studied in this condition. Dopamine is synthesized from the amino acid tyrosine undergoing two steps: the first producing DOPA by tyrosine hydroxylase (TH) and the second from DOPA to dopamine by the DOPA decarboxylase (DDC). Both steps could therefore modulate the system. Low concentrations of homovanillic acid, a metabolite produced by the catabolism of dopamine, have been found in CSF of depressed patients who attempted suicide compared to controls (Engstrom et al., 1999), supporting the hypothesis of a diminished dopaminergic neurotransmission in suicidal behavior (Roy et al., 1992), though not unequivocally (Asberg, 1997). DDC is located on chromosome 7p12.3-p12.1, and consists of 15 exons spanning >85 kb (Sumi-Ichinose et al., 1992). *DDC* gene variants have been investigated in relation to suicide-related behavior or phenotypes by Giegling et al. (2008a, 2009b) who showed some marginal associations with suicide, violence, anger, aggression, and temperament but no major effect was seen for this gene.

18.3.2.3 Tyrosine Hydroxylase

The TH is the rate-limiting enzyme in the biosynthesis of noradrenalin, and different studies reported alterations of TH levels in the locus coeruleus of suicide victims (Pandey and Dwivedi, 2007). A Swedish study (Persson et al., 1999) examined a penta-allelic short tandem repeat in the first intron of the *TH* gene. The sample consisted of 118 adult suicide attempters and 78 control subjects. This study reported a tendency for a low incidence of the TH-K1 allele among all suicide attempters compared to the controls. Furthermore, a significant association between the TH-K3 allele in a subgroup of patients with adjustment disorders and attempted suicide was found (Persson et al., 1999). Giegling et al. (2008a, 2009b) investigated SNPs in the *TH* gene in 571 suicide attempters and controls and found only marginal associations with suicidal behavior, which did not hold after correction for multiple testing (rs3842727 uncorrected $p = 0.023$). As other SNPs showed no association, this study does not support a major involvement of *TH* gene variants in suicidal behavior and related traits. Interestingly, De Luca et al. (2008) found a trend for a TH polymorphism and a higher severity of suicidal behavior ($p = 0.060$) but the power was too low and further studies are needed to make any conclusions regarding the involvement of TH SNPs in suicidal behavior. Furthermore, Giegling et al. (2009b) presented interaction effects of genotype and diagnosis with TH SNPs and temperament having a greater effect on the respective personality dimensions in the group of suicide attempters.

18.3.2.4 Catechol-*O*-Methyltransferase

The catechol-*O*-methyltransferase (COMT) is a major enzyme involved in the inactivation of the catecholamines, dopamine and noradrenalin. One functional polymorphism in the *COMT* gene, in which valine (Val) at codon 158 is replaced with methionine (Met), was studied intensively in psychiatric diseases as well as in suicidal behavior and related traits including aggression.

Lachman et al. (1996) showed that this polymorphism is due to a G-to-A transition at codon 158 of the *COMT* gene, resulting in a valine-to-methionine substitution. This SNP causes differences in the functional ability of the enzyme to catabolize dopamine (Lachman et al., 1996).

Several studies investigated the association of COMT SNPs with a wide range of phenotypes such as psychiatric disorders and personality traits. COMT has been hypothesized as a risk factor, for example, for schizophrenia (Williams et al., 2007), attention deficit hyperactivity disorder (Kebir et al., 2009), alcoholism (Serý et al., 2006), bipolar disorder, and others (Dickinson and Elvevåg, 2009).

For suicidal behavior, Kia-Keating et al. (2007) provided a meta-analysis based on six related studies including 519 cases and 933 control subjects. There was evidence of a significant association between the COMT 158Met polymorphism and the suicidal behavior (odds ratio [OR] = 1.25, 95% confidence interval [CI] = 1.01–1.56, $z = 2.03$, $p = 0.04$). Although the results for COMT were not influenced by publication bias, the significance of the combined results was related to the gender of the case and control subjects. The results for COMT support past studies that have found a relationship between suicidal behavior and COMT, and have also found that the relationship differs for males and females. They speculate that due to a higher number of suicide attempts among females and a higher number of committed suicides among males, it could be possible that COMT is related to the lethality of suicide attempts.

The latest meta-analysis on COMT and suicidal behavior was provided by Calati et al. (2011) who included 10 studies with 1324 patients and could not find an association.

Interestingly, Rujescu et al. (2003a) found that Val-carriers expressed their anger more outwardly, whereas Met-carriers expressed it more inwardly and reported more state anger, as assessed by the self-report questionnaire STAXI. Furthermore, COMT was found related to HA and neuroticism (Eley et al., 2003; Enoch et al., 2003; Stein et al., 2005), NS and extraversion (Reuter and Henning, 2005; Tsai et al., 2004a), and anger-related traits connected with violent suicide (Rujescu et al., 2003a). A detailed summary on 46 studies of the association of COMT and personality traits can be found in Calati et al. (2011).

18.3.3 Outlook

There is evidence for the involvement of genetic variants in other than serotonergic and dopaminergic genes in suicidal behavior and related phenotypes. Further studies will show to which proportion they will play a role and which findings will be replicated in the future. There are also several studies on aggression or impulsivity coming directly from animal studies showing some evidence for involvement (e.g., NCAM1, neuronal cell adhesion molecule gene, Giegling et al., 2010; estrogen receptor genes, Giegling et al., 2008b, 2009a; *NOS-I* and *NOS-III* genes, Rujescu et al., 2008; tachykinin receptor 1 gene, Giegling et al., 2007; ABCG1, Giegling et al., 2006; Gietl et al., 2007; Rujescu et al., 2000; and many more, Rujescu et al., 2007).

Beside candidate gene approaches, new hypothesis-free approaches arised due to the new technical possibilities in genotyping. Recent progress in the field of parallel SNPs typing of up to 1 million variants at once has provided proof of principle and yielded several genes showing a strong association with complex diseases. For example, van den Oord et al. (2008) published a genome-wide study on neuroticism as an intermediate phenotype of suicidal behavior. More than 420,000 genetic markers were tested for their association with neuroticism in a genome-wide association study (GWAS). The GWAS sample consisted of 1227 healthy individuals ascertained from a U.S. national sampling frame and available from the National Institute of Mental Health genetics repository. The most promising markers were subsequently tested in a German replication sample comprising 1880 healthy individuals. The most promising results were SNPs in the gene *MAMDC1*, which is proposed to be involved in regulating neuronal migration and axonal guidance. Further genome-wide studies on personality were provided by Terracciano et al. (2010) and Calboli et al. (2010). The future will show if these results can be replicated.

REFERENCES

Anderson CB, Carter FA, McIntosh VV, Joyce PR, Bulik CM. Self-harm and suicide attempts in individuals with bulimia nervosa. *Eating Disord* 2002;10:227–243.

Anguelova M, Benkelfat C, Turecki G. A systematic review of association studies investigating genes coding for serotonin receptors and the serotonin transporter: II. Suicidal behavior. *Mol Psychiatry* 2003;8(7):646–653.

Arango V, Huang YY, Underwood MD, Mann JJ. Genetics of the serotonergic system in suicidal behavior. *J Psychiatr Res* 2003;37(5):375–386.

Arango V, Underwood MD, Mann JJ. Postmortem findings in suicide victims. Implications for in vivo imaging studies. *Ann NY Acad Sci* 1997;836:269–287.

Asberg M. Neurotransmitters and suicidal behavior. The evidence from cerebrospinal fluid studies. *Ann NY Acad Sci* 1997;836:158–181.

Asberg M, Schalling D, Traeskman-Bendz L, Waegner A. Psychobiology of suicide, impulsivity, and related phenomena. In: Meltzer HY (ed.), *Psychopharmacology. The Third Generation of Progress*. Raven Press, New York, 1987, pp. 655–668.

Asberg M, Traskman L, Thoren P. 5-HIAA in the cerebrospinal fluid. A biochemical suicide predictor? *Arch Gen Psychiatry* 1976;33(10):1193–1197.

Bachus SE, Hyde TM, Akil M, Weickert CS, Vawter MP, Kleinman JE. Neuropathology of suicide. A review and an approach. *Ann NY Acad Sci* 1997;836:201–219.

Baud P. Personality traits as intermediary phenotypes in suicidal behavior: Genetic issues. *Am J Med Genet Part C Semin Med Genet* 2005;133:34–42.

Baud P, Perroud N, Courtet P, Jaussent I, Relecom C, Jollant F, Malafosse A. Modulation of anger control in suicide attempters by TPH-1. *Genes Brain Behav* 2009;8(1):97–100.

Becerra B, Paez F, Robles-Garcia R, Vela GE. Temperament and character profile of persons with suicide attempt. *Actas Esp Psiquiatr* 2005;33:117–122.

Beck AT, Steer RA. Clinical predictors of eventual suicide: A 5- to 10-year prospective study of suicide attempters. *J Affect Disord* 1989;17:203–209.

Bellivier F, Chaste P, Malafosse A. Association between the *TPH* gene A218C polymorphism and suicidal behavior: A meta-analysis. *Am J Med Genet B Neuropsychiatr Genet* 2004;124B(1):87–91.

Berggard C, Damberg M, Longato-Stadler E, Hallman J, Oreland L, Garpenstrand H. The serotonin 2A −1438 G/A receptor polymorphism in a group of Swedish male criminals. *Neurosci Lett* 2003;347:196–198.

Bertolote JM. Suicide in the world: An epidemiological overview 1959–2000. In: Wasserman D (ed.), *Suicide—An Unnecessary Death*. Martin Dunitz, London, U.K., 2001, pp. 3–10.

Blairy S, Massat I, Staner L, Le Bon O, Van Gestel S, Van Broeckhoven C, Hilger C, Hentges F, Souery D, Mendlewicz J. 5-HT2a receptor polymorphism gene in bipolar disorder and harm avoidance personality trait. *Am J Med Genet* 2000;96:360–364.

Brandes M, Bienvenu OJ. Personality and anxiety disorders. *Curr Psychiatry Rep* 2006;8:263–269.

Breidenthal SE, White DJ, Glatt CE. Identification of genetic variants in the neuronal form of tryptophan hydroxylase (TPH2). *Psychiatr Genet* 2004;14(2):69–72.

Brent DA, Mann JJ. Family genetic studies, suicide, and suicidal behavior. *Am J Med Genet C Semin Med Genet* 2005;133C(1):13–24.

Brezo J, Paris J, Turecki G. Personality traits as correlates of suicidal ideation, suicide attempts, and suicide completions: A systematic review. *Acta Psychiatr Scand* 2006;113:180–206.

Brodsky BS, Groves SA, Oquendo MA, Mann JJ, Stanley B. Interpersonal precipitants and suicide attempts in borderline personality disorder. *Suicide Life Threat Behav* 2006;36:313–322.

Brown LK, Overholser J, Spirito A, Fritz GK. The correlates of planning in adolescent suicide attempts. *J Am Acad Child Adolesc Psychiatry* 1991;30:95–99.

Brunner HG, Nelen M, Breakefield XO, Ropers HH, van Oost BA. Abnormal behavior associated with a point mutation in the structural gene for monoamine oxidase A. *Science* 1993a;262(5133):578–580.

Brunner HG, Nelen MR, van Zandvoort P, Abeling NG, van Gennip AH, Wolters EC, Kuiper MA, Ropers HH, van Oost BA. X-linked borderline mental retardation with prominent behavioral disturbance: Phenotype, genetic localization, and evidence for disturbed monoamine metabolism. *Am J Hum Genet* 1993b;52(6):1032–1039.

Bulik CM, Sullivan PF, Joyce PR. Temperament, character and suicide attempts in anorexia nervosa, bulimia nervosa and major depression. *Acta Psychiatr Scand* 1999;100:27–32.

Calati R, Giegling I, Rujescu D, Hartmann AM, Möller HJ, De Ronchi D, Serretti A. Temperament and character of suicide attempters. *J Psychiatr Res* 2008;42(11):938–945.

Calati R, Porcelli S, Giegling I, Hartmann AM, Möller HJ, De Ronchi D, Serretti A, Rujescu D. Catechol-*o*-methyltransferase gene modulation on suicidal behavior and personality traits: Review, meta-analysis and association study. *J Psychiatr Res* 2011;45(3):309–321.

Calboli FC, Tozzi F, Galwey NW, Antoniades A, Mooser V, Preisig M, Vollenweider P, Waterworth D, Waeber G, Johnson MR, Muglia P, Balding DJ. A genome-wide association study of neuroticism in a population-based sample. *PLoS One* 2010;5(7):e11504.

Canli T, Lesch KP. Long story short: The serotonin transporter in emotion regulation and social cognition. *Nat Neurosci* 2007;10(9):1103–1109 [Review].

Cantor CH. Suicide in the Western World. In: Hawton K, van Heeringen K (eds.), *Suicide and Attempted Suicide*. Wiley, Chichester, U.K., 2000, pp. 9–28.

Cautin RL, Overholser JC, Goetz P. Assessment of mode of anger expression in adolescent psychiatric inpatients. *Adolescence* 2001;36:163–170.

Chen TJ, Blum K, Mathews D, Fisher L, Schnautz N, Braverman ER, Schoolfield J, Downs BW, Comings DE. Are dopaminergic genes involved in a predisposition to pathological aggression? Hypothesizing the importance of "super normal controls" in psychiatric genetic research of complex behavioral disorders. *Med Hypotheses* 2005;65(4):703–707.

Cloninger CR, Przybeck TR, Svrakic DM, Wetzel RD. *The Temperament and Character Inventory (TCI): A Guide to its Development and Use*. Center for Psychobiology of Personality Washington University, St. Louis, MO, 1994.

Coccaro EF, Kavoussi RJ, Sheline YI, Berman ME, Csernansky JG. Impulsive aggression in personality disorder correlates with platelet 5-HT2A receptor binding. *Neuropsychopharmacology* 1997;16(3):211–216.

Conner TS, Jensen KP, Tennen H, Furneaux HM, Kranzler HR, Covault J. Functional polymorphisms in the serotonin 1B receptor gene (HTR1B) predict self-reported anger and hostility among young men. *Am J Med Genet B Neuropsychiatr Genet* 2010;153B(1):67–78.

Corbitt EM, Malone KM, Haas GL, Mann JJ. Suicidal behavior in patients with major depression and comorbid personality disorders. *J Affect Disord* 1996;39(1):61–72.

Costa P, McCrae R. *Revised NEO Personality Inventory and NEO Five Factor Inventory Professional Manual*, Odessa, FL: Psychological Assessment Resources, 1992.

Costa PT Jr, McCrae RR. Primary traits of Eysenck's P-E-N system: Three- and five-factor solutions. *J Pers Soc Psychol* 1995;69:308–317.

Craig IW, Halton KE. Genetics of human aggressive behaviour. *Hum Genet* 2009;126(1):101–113.

Critchfield KL, Levy KN, Clarkin JF. The relationship between impulsivity, aggression, and impulsive-aggression in borderline personality disorder: An empirical analysis of self-report measures. *J Pers Disord* 2004;18:555–570.

De Fruyt F, Van De Wiele L, Van Heeringen C. Cloninger's psychobiological model of temperament and character and the five-factor model of personality. *Pers Individual Diff* 2000;29:441–452.

De Luca V, Strauss J, Kennedy JL. Power based association analysis (PBAT) of serotonergic and noradrenergic polymorphisms in bipolar patients with suicidal behaviour. *Prog Neuropsychopharmacol Biol Psychiatry* 2008;32:197–203.

De Moore GM, Robertson AR. Suicide in the 18 years after deliberate self-harm a prospective study. *Br J Psychiatry* 1996;169:489–494.

Dickinson D, Elvevåg B. Genes, cognition and brain through a COMT lens. *Neuroscience* 2009;164(1):72–87.

Doihara C, Kawanishi C, Yamada T, Sato R, Hasegawa H, Furuno T, Nakagawa M, Hirayasu Y. Trait aggression in suicide attempters: A pilot study. *Psychiatry Clin Neurosci* 2008;62:352–354.

Du L, Faludi G, Palkovits M, Sotonyi P, Bakish D, Hrdina PD. High activity-related allele of *MAO-A* gene associated with depressed suicide in males. *Neuroreport* 2002;13(9):1195–1198.

Duberstein PR, Conwell Y, Seidlitz L, Denning DG, Cox C, Caine ED. Personality traits and suicidal behavior and ideation in depressed inpatients 50 years of age and older. *J Gerontol B Psychol Sci Soc Sci* 2000;55(1):P18–P26.

Dumais A, Lesage AD, Lalovic A, Seguin M, Tousignant M, Chawky N, Turecki G. Is violent method of suicide a behavioral marker of lifetime aggression? *Am J Psychiatry* 2005;162:1375–1378.

Eley TC, Tahir E, Angleitner A, Harriss K, McClay J, Plomin R, Riemann R, Spinath F, Craig I. Association analysis of MAOA and COMT with neuroticism assessed by peers. *Am J Med Genet B Neuropsychiatr Genet* 2003;120:90–96.

Engstrom G, Alling C, Blennow K, Regnell G, Traskman-Bendz L. Reduced cerebrospinal HVA concentrations and HVA/5-HIAA ratios in suicide attempters. Monoamine metabolites in 120 suicide attempters and 47 controls. *Eur Neuropsychopharmacol* 1999;9:399–405.

Engstrom C, Brandstrom S, Sigvardsson S, Cloninger CR, Nylander PO. Bipolar disorder. III: Harm avoidance a risk factor for suicide attempts. *Bipolar Disord* 2004;6:130–138.

Enoch MA, Xu K, Ferro E, Harris CR, Goldman D. Genetic origins of anxiety in women: A role for a functional catechol-*O*-methyltransferase polymorphism. *Psychiatr Genet* 2003;13:33–41.

Esposito C, Spirito A, Boergers J, Donaldson D. Affective, behavioral, and cognitive functioning in adolescents with multiple suicide attempts. *Suicide Life Threat Behav* 2003;33:389–399.

Evren C, Evren B. Self-mutilation in substance-dependent patients and relationship with childhood abuse and neglect, alexithymia and temperament and character dimensions of personality. *Drug Alcohol Depend* 2005;80:15–22.

Evren C, Evren B. The relationship of suicide attempt history with childhood abuse and neglect, alexithymia and temperament and character dimensions of personality in substance dependents. *Nord J Psychiatry* 2006;60:263–269.

Eysenck HJ, Eysenck SBG. *Manual of the Eysenck Personality Inventory*. University Press, London, U.K., 1964.

Eysenck HJ, Eysenck SBG. *Manual of the Eysenck Questionnaire*. EdITS, San Diego, CA, 1975.

Favaro A, Santonastaso P, Monteleone P, Bellodi L, Mauri M, Rotondo A, Erzegovesi S, Maj M. Self injurious behavior and attempted suicide in purging bulimia nervosa: Associations with psychiatric comorbidity. *J Affect Disord* 2008;105:285–289.

Finckh U, Rommelspacher H, Kuhn S, Dufeu P, Otto G, Heinz A, Dettling M, Giraldo-Velasquez M, Pelz J, Gräf KJ, Harms H, Sander T, Schmidt LG, Rolfs A. Influence of the dopamine D2 receptor (DRD2) genotype on neuroadaptive effects of alcohol and the clinical outcome of alcoholism. *Pharmacogenetics* 1997;7:271–281.

Giegling I, Chiesa A, Calati R, Hartmann AM, Möller HJ, De Ronchi D, Rujescu D, Serretti A. Do the estrogen receptors 1 gene variants influence the Temperament and Character Inventory scores in suicidal attempters and healthy subjects? *Am J Med Genet B Neuropsychiatr Genet* 2009a;150B(3):434–438.

Giegling I, Chiesa A, Mandelli L, Gibiino S, Hartmann AM, Möller HJ, Schneider B, Schnabel A, Maurer K, De Ronchi D, Rujescu D, Serretti A. Influence of neuronal cell adhesion molecule (NCAM1) variants on suicidal behaviour and correlated traits. *Psychiatry Res* 2010;179(2):222–225.

Giegling I, Hartmann AM, Moller HJ, Rujescu D. Anger- and aggression-related traits are associated with polymorphisms in the *5-HT-2A* gene. *J Affect Disord* 2006;96(1–2):75–81.

Giegling I, Moreno-De-Luca D, Calati R, Hartmann AM, Möller HJ, De Ronchi D, Rujescu D, Serretti A. Tyrosine hydroxylase and DOPA decarboxylase gene variants in personality traits. *Neuropsychobiology* 2009b;59(1):23–27.

Giegling I, Moreno-De-Luca D, Rujescu D, Schneider B, Hartmann AM, Schnabel A, Maurer K, Möller HJ, Serretti A. Dopa decarboxylase and tyrosine hydroxylase gene variants in suicidal behavior. *Am J Med Genet B Neuropsychiatr Genet* 2008a;147(3):308–315.

Giegling I, Olgiati P, Hartmann AM, Calati R, Möller HJ, Rujescu D, Serretti A. Personality and attempted suicide. Analysis of anger, aggression and impulsivity. *J Psychiatr Res* 2009c;43(16):1262–1271.

Giegling I, Rujescu D, Mandelli L, Schneider B, Hartmann AM, Schnabel A, Maurer K, De Ronchi D, Möller HJ, Serretti A. Tachykinin receptor 1 variants associated with aggression in suicidal behavior. *Am J Med Genet B Neuropsychiatr Genet* 2007;144B(6):757–761.

Giegling I, Rujescu D, Mandelli L, Schneider B, Hartmann AM, Schnabel A, Maurer K, De Ronchi D, Möller HJ, Serretti A. Estrogen receptor gene 1 variants are not associated with suicidal behavior. *Psychiatry Res* 2008b;160(1):1–7.

Gietl A, Giegling I, Hartmann AM, Schneider B, Schnabel A, Maurer K, Möller HJ, Rujescu D. *ABCG1* gene variants in suicidal behavior and aggression-related traits. *Eur Neuropsychopharmacol* 2007;17(6–7):410–416.

Goldston DB, Daniel S, Reboussin DM, Kelley A, Ievers C, Brunstetter R. First-time suicide attempters, repeat attempters, and previous attempters on an adolescent inpatient psychiatry unit. *J Am Acad Child Adolesc Psychiatry* 1996;35:631–639.

Golimbet VE, Alfimova MV, Mitiushina NG. Polymorphism of the serotonin 2A receptor gene (*5HTR2A*) and personality traits. *Mol Biol (Mosk)* 2004;38:404–412.

Gorlyn M. Impulsivity in the prediction of suicidal behavior in adolescent populations. *Int J Adolesc Med Health* 2005;17:205–209.

Gottesman II, Gould TD. The endophenotype concept in psychiatry: Etymology and strategic intentions. *Am J Psychiatry* 2003;160(4):636–645 [Review].

Grunebaum MF, Ramsay SR, Galfalvy HC, Ellis SP, Burke AK, Sher L, Printz DJ, Kahn DA, Mann JJ, Oquendo MA. Correlates of suicide attempt history in bipolar disorder: A stress-diathesis perspective. *Bipolar Disord* 2006;8:551–557.

Guillem E, Pelissolo A, Notides C, Lepine JP. Relationship between attempted suicide, serum cholesterol level and novelty seeking in psychiatric in-patients. *Psychiatry Res* 2002;112:83–88.

Gunnell D, Lewis G. Studying suicide from the life course perspective: Implications for prevention. *Br J Psychiatry* 2005;187:206–208.

Ham BJ, Kim YH, Choi MJ, Cha JH, Choi YK, Lee MS. Serotonergic genes and personality traits in the Korean population. *Neurosci Lett* 2004;354:2–5.

Haw C, Houston K, Townsend E, Hawton K. Deliberate self-harm patients with alcohol disorders: Characteristics, treatment, and outcome. *Crisis* 2001;22:93–101.

Hawton K, Kingsbury S, Steinhardt K, James A, Fagg J. Repetition of deliberate selfharm by adolescents: The role of psychological factors. *J Adolescence* 1999;22:369–378.

Held T, Hawellek B, Dickopf-Kaschenbach K, Schneider-Axmann T, Schmidtke A, Moller HJ. Violent and non-violent methods of parasuicide: What determines the choice? *Fortschr Neurol Psychiatr* 1998;66:505–511.

Higley JD, Mehlman PT, Higley SB, Fernald B, Vickers J, Lindell SG, Linnoila M. Excessive mortality in young free-ranging male nonhuman primates with low cerebrospinal fluid 5-hydroxyindoleacetic acid concentrations. *Arch Gen Psychiatry* 1996a;53(6):537–543.

Higley JD, Mehlman PT, Poland RE, Taub DM, Vickers J, Suomi SJ, Linnoila M. CSF testosterone and 5-HIAA correlate with different types of aggressive behaviors. *Biol Psychiatry* 1996b;40(11):1067–1082.

Higley JD, Mehlman PT, Taub DM, Higley SB, Suomi SJ, Vickers JH, Linnoila M. Cerebrospinal fluid monoamine and adrenal correlates of aggression in free-ranging rhesus monkeys. *Arch Gen Psychiatry* 1992;49(6):436–441.

Ho LW, Furlong RA, Rubinsztein JS, Walsh C, Paykel ES, Rubinsztein DC. Genetic associations with clinical characteristics in bipolar affective disorder and recurrent unipolar depressive disorder. *Am J Med Genet* 2000;96(1):36–42.

Hong CJ, Pan GM, Tsai SJ. Association study of onset age, attempted suicide, aggressive behavior, and schizophrenia with a serotonin 1B receptor (A-161T) genetic polymorphism. *Neuropsychobiology* 2004;49(1):1–4.

Horesh N, Orbach I, Gothelf D, Efrati M, Apter A. Comparison of the suicidal behavior of adolescent inpatients with borderline personality disorder and major depression. *J Nerv Mental Dis* 2003;191:582–588.

Huang YY, Battistuzzi C, Oquendo MA, Harkavy-Friedman J, Greenhill L, Zalsman G, Brodsky B, Arango V, Brent DA, Mann JJ. Human 5-HT1A receptor C(−1019)G polymorphism and psychopathology. *Int J Neuropsychopharmacol* 2004a;7(4):441–451.

Huang YY, Cate SP, Battistuzzi C, Oquendo MA, Brent D, Mann JJ. An association between a functional polymorphism in the monoamine oxidase a gene promoter, impulsive traits and early abuse experiences. *Neuropsychopharmacology* 2004b;29(8):1498–1505.

Huang YY, Grailhe R, Arango V, Hen R, Mann JJ. Relationship of psychopathology to the human serotonin1B genotype and receptor binding kinetics in postmortem brain tissue. *Neuropsychopharmacology* 1999;21(2):238–246.

Huang YY, Oquendo MA, Friedman JM, Greenhill LL, Brodsky B, Malone KM, Khait V, Mann JJ. Substance abuse disorder and major depression are associated with the human 5-HT1B receptor gene (*HTR1B*) G861C polymorphism. *Neuropsychopharmacology* 2003;28(1):163–169.

Ille R, Huber HP, Zapotoczky HG. Aggressive and suicidal behavior. A cluster analysis study of suicidal and nonclinical subjects. *Psychiatr Prax* 2001;28:24–28.

Jensen KP, Covault J, Conner TS, Tennen H, Kranzler HR, Furneaux HM. A common polymorphism in serotonin receptor 1B mRNA moderates regulation by miR-96 and associates with aggressive human behaviors. *Mol Psychiatry* 2009;14(4):381–389.

Johann M, Putzhammer A, Eichhammer P, Wodarz N. Association of the −141C Del variant of the dopamine D2 receptor (DRD2) with positive family history and suicidality in German alcoholics. *Am J Med Genet B Neuropsychiatr Genet* 2005;132B(1):46–49.

Jonsson EG, Nothen MM, Gustavsson JP, Berggard C, Bunzel R, Forslund K, Rylander G, Mattila-Evenden M, Propping P, Asberg M, Sedvall G. No association between serotonin 2A receptor gene variants and personality traits. *Psychiatr Genet* 2001;11:11–17.

Kebir O, Tabbane K, Sengupta S, Joober R. Candidate genes and neuropsychological phenotypes in children with ADHD: Review of association studies. *J Psychiatry Neurosci* 2009;34(2):88–101.

Kempf E. *Psychopathology*. Arno Press, New York, 1976.

Kendler KS, Neale MC, Kessler RC, Heath AC, Eaves LJ. A longitudinal twin study of personality and major depression in women. *Arch Gen Psychiatry* 1993;50:853–862.

Khan AA, Jacobson KC, Gardner CO, Prescott CA, Kendler KS. Personality and comorbidity of common psychiatric disorders. *Br J Psychiatry* 2005;186:190–196.

Kia-Keating BM, Glatt SJ, Tsuang MT. Meta-analyses suggest association between COMT, but not HTR1B, alleles, and suicidal behavior. *Am J Med Genet B Neuropsychiatr Genet* 2007;144B(8):1048–1053.

Kirkcaldy BD, Siefen GR, Urkin J, Merrick J. Risk factors for suicidal behavior in adolescents. *Minerva Pediatr* 2006;58:443–450.

Kulikov AV, Osipova DV, Naumenko VS, Popova NK. The C1473G polymorphism in the tryptophan hydroxylase-2 gene and intermale aggression in mice. *Dokl Biol Sci* 2005a;402:208–210.

Kulikov AV, Osipova DV, Naumenko VS, Popova NK. Association between *Tph2* gene polymorphism, brain tryptophan hydroxylase activity and aggressiveness in mouse strains. *Genes Brain Behav* 2005b;4(8):482–485.

Kunugi H, Ishida S, Kato T, Tatsumi M, Sakai T, Hattori M, Hirose T, Nanko S. A functional polymorphism in the promoter region of monoamine oxidase-A gene and mood disorders. *Mol Psychiatry* 1999;4(4):393–395.

Kusumi I, Suzuki K, Sasaki Y, Kameda K, Sasaki T, Koyama T. Serotonin 5-HT (2A) receptor gene polymorphism, 5-HT(2A) receptor function and personality traits in healthy subjects: A negative study. *J Affect Disord* 2002;68:235–241.

Lachman HM, Papolos DF, Saito T, Yu YM, Szumlanski CL, Weinshilboum RM. Human catechol-*O*-methyltransferase pharmacogenetics: Description of a functional polymorphism and its potential application to neuropsychiatric disorders. *Pharmacogenetics* 1996;6(3):243–250.

Laget J, Plancherel B, Stephan P, Bolognini M, Corcos M, Jeammet P, Halfon O. Personality and repeated suicide attempts in dependent adolescents and young adults. *Crisis* 2006;27:164–711.

Lalovic A, Turecki G. Meta-analysis of the association between tryptophan hydroxylase and suicidal behavior. *Am J Med Genet* 2002;114(5):533–540.

Lee R, Coccaro E. The neuropsychopharmacology of criminality and aggression. *Can J Psychiatry* 2001;46(1):35–44.

Lemonde S, Turecki G, Bakish D, Du L, Hrdina PD, Bown CD, Sequeira A, Kushwaha N, Morris SJ, Basak A, Ou XM, Albert PR. Impaired repression at a 5-hydroxytryptamine 1A receptor gene polymorphism associated with major depression and suicide. *J Neurosci* 2003;23(25):8788–8799.

Lesch KP. Linking emotion to the social brain. The role of the serotonin transporter in human social behaviour. *EMBO Rep* 2007;8(Spec No.):S24–S29 [Review].

Lesch KP, Bengel D, Heils A, Sabol SZ, Greenberg BD, Petri S, Benjamin J, Müller CR, Hamer DH, Murphy DL. Association of anxiety-related traits with a polymorphism in the serotonin transporter gene regulatory region. *Science* 1996;274(5292):1527–1531.

Li D, Duan Y, He L. Association study of serotonin 2A receptor (*5-HT2A*) gene with schizophrenia and suicidal behavior using systematic meta-analysis. *Biochem Biophys Res Commun* 2006;340(3):1006–1015.

Li D, He L. Meta-analysis supports association between serotonin transporter (5-HTT) and suicidal behavior. *Mol Psychiatry* 2007;12(1):47–54.

Li D, He L. Meta-analysis shows association between the tryptophan hydroxylase (*TPH*) gene and schizophrenia. *Hum Genet* 2006;120(1):22–30.

Lin PY, Tsai G. Association between serotonin transporter gene promoter polymorphism and suicide: Results of a meta-analysis. *Biol Psychiatry* 2004;55(10):1023–1030.

Lochner C, Hemmings S, Seedat S, Kinnear C, Schoeman R, Annerbrink K, Olsson M, Eriksson E, Moolman-Smook J, Allgulander C, Stein DJ. Genetics and personality traits in patients with social anxiety disorder: A case–control study in South Africa. *Eur Neuropsychopharmacol* 2007;17(5):321–327.

Malone KM, Haas GL, Sweeney JA, Mann JJ. Major depression and the risk of attempted suicide. *J Affect Disord* 1995;34:173–185.

Malone KM, Waternaux C, Haas GL, Cooper TB, Li S, Mann JJ. Cigarette smoking, suicidal behavior, and serotonin function in major psychiatric disorders. *Am J Psychiatry* 2003;160:773–779.

Mann JJ. The neurobiology of suicide. *Nat Med* 1998;4(1):25–30.

Mann JJ. Neurobiology of suicidal behaviour. *Nat Rev Neurosci* 2003;4(10):819–828.

Mann JJ, Ellis SP, Waternaux CM, Liu X, Oquendo MA, Malone KM, Brodsky BS, Haas GL, Currier D. Classification trees distinguish suicide attempters in major psychiatric disorders: A model of clinical decision making. *J Clin Psychiatry* 2008;69:23–31.

Mann JJ, Waternaux C, Haas GL, Malone KM. Toward a clinical model of suicidal behavior in psychiatric patients. *Am J Psychiatry* 1999;156:181–189.

Manuck SB, Flory JD, Ferrell RE, Dent KM, Mann JJ, Muldoon MF. Aggression and anger-related traits associated with a polymorphism of the tryptophan hydroxylase gene. *Biol Psychiatry* 1999;45(5):603–614.

McGirr A, Turecki G. The relationship of impulsive aggressiveness to suicidality and other depression-linked behaviors. *Curr Psychiatry Rep* 2007;9:460–466.

McGirr A, Turecki G. What is specific to suicide in schizophrenia disorder? Demographic, clinical and behavioural dimensions. *Schizophr Res* 2008;98:217–224.

Mehlman PT, Higley JD, Faucher I, Lilly AA, Taub DM, Vickers J, Suomi SJ, Linnoila M. Low CSF 5-HIAA concentrations and severe aggression and impaired impulse control in nonhuman primates. *Am J Psychiatry* 1994;151(10):1485–1491.

Mitrev I. A study of deliberate self-poisoning in patients with adjustment disorders. *Folia Med* 1996;38:11–16.

Moeller FG, Barratt ES, Dougherty DM, Schmitz JM, Swann AC. Psychiatric aspects of impulsivity. *Am J Psychiatry* 2001;158:1783–1793.

Murphy DL, Lesch KP. Targeting the murine serotonin transporter: Insights into human neurobiology. *Nat Rev Neurosci* 2008;9(2):85–96.

Nemoda Z, Lyons-Ruth K, Szekely A, Bertha E, Faludi G, Sasvari-Szekely M. Association between dopaminergic polymorphisms and borderline personality traits among at-risk young adults and psychiatric inpatients. *Behav Brain Funct* 2010;6:4.

New AS, Gelernter J, Goodman M, Mitropoulou V, Koenigsberg H, Silverman J, Siever LJ. Suicide, impulsive aggression, and HTR1B genotype. *Biol Psychiatry* 2001;50(1):62–65.

Ni X, Bismil R, Chan K, Sicard T, Bulgin N, McMain S, Kennedy JL. Serotonin 2A receptor gene is associated with personality traits, but not to disorder, in patients with borderline personality disorder. *Neurosci Lett* 2006;408:214–219.

Nickel MK, Simek M, Lojewski N, Muehlbacher M, Fartacek R, Kettler C, Bachler E, Egger C, Rother N, Buschmann W, Pedrosa Gil F, Kaplan P, Mitterlehner FO, Anvar J, Rother WK, Loew TH, Nickel C. Familial and sociopsychopathological risk factors for suicide attempt in bulimic and in depressed women: Prospective study. *Int J Eating Disord* 2006;39:410–417.

Nielsen B, Petersen P, Rask PH, Krarup G. Suicide and other causes of death in patients admitted for attempted suicide. 10-year follow-up. *Ugeskrift Laeger* 1995;157:2149–2153.

Nishiguchi N, Shirakawa O, Ono H, Nishimura A, Nushida H, Ueno Y, Maeda K. No evidence of an association between 5HT1B receptor gene polymorphism and suicide victims in a Japanese population. *Am J Med Genet* 2001;105(4):343–345.

Nishiguchi N, Shirakawa O, Ono H, Nishimura A, Nushida H, Ueno Y, Maeda K. Lack of an association between 5-HT1A receptor gene structural polymorphisms and suicide victims. *Am J Med Genet* 2002;114(4):423–425.

Nock MK, Borges G, Bromet EJ, Cha CB, Kessler RC, Lee S. Suicide and suicidal behavior. *Epidemiol Rev* 2008;30:133–154 [Review].

Nomura M. Involvement of a polymorphism in the 5-HT2A receptor gene in impulsive behavior. *Nippon Yakurigaku Zasshi* 2006;127:9–13.

Ohtani M, Shindo S, Yoshioka N. Polymorphisms of the tryptophan hydroxylase gene and serotonin 1A receptor gene in suicide victims among Japanese. *Tohoku J Exp Med* 2004;202(2):123–133.

Ono H, Shirakawa O, Nishiguchi N, Nishimura A, Nushida H, Ueno Y, Maeda K. No evidence of an association between a functional monoamine oxidase a gene polymorphism and completed suicides. *Am J Med Genet* 2002;114(3):340–342.

Oquendo MA, Bongiovi-Garcia ME, Galfalvy H, Goldberg PH, Grunebaum MF, Burke AK, Mann JJ. Sex differences in clinical predictors of suicidal acts after major depression: A prospective study. *Am J Psychiatry* 2007;164:134–141.

Oquendo MA, Galfalvy H, Russo S, Ellis SP, Grunebaum MF, Burke A, Mann JJ. Prospective study of clinical predictors of suicidal acts after a major depressive episode in patients with major depressive disorder or bipolar disorder. *Am J Psychiatry* 2004;161:1433–1441.

Oquendo MA, Mann JJ. The biology of impulsivity and suicidality. *Psychiatr Clin North Am* 2000;23(1):11–25.

Oquendo MA, Russo SA, Underwood MD, Kassir SA, Ellis SP, Mann JJ, Arango V. Higher postmortem prefrontal 5-HT2A receptor binding correlates with lifetime aggression in suicide. *Biol Psychiatry* 2006;59(3):235–243.

Oquendo MA, Waternaux C, Brodsky B, Parsons B, Haas GL, Malone KM, Mann JJ. Suicidal behavior in bipolar mood disorder: Clinical characteristics of attempters and nonattempters. *J Affect Disord* 2000;59(2):107–117.

Ormel J, Oldehinkel AJ, Vollebergh W. Vulnerability before, during, and after a major depressive episode: A 3-wave population-based study. *Arch Gen Psychiatry* 2004;61:990–996.

Osipova DV, Kulikov AV, Mekada K, Yoshiki A, Moshkin MP, Kotenkova EV, Popova NK. Distribution of the C1473G polymorphism in tryptophan hydroxylase 2 gene in laboratory and wild mice. *Genes Brain Behav* 2010;9(5):537–543.

Osipova DV, Kulikov AV, Popova NK. C1473G polymorphism in mouse *tph2* gene is linked to tryptophan hydroxylase-2 activity in the brain, intermale aggression, and depressive-like behavior in the forced swim test. *J Neurosci Res* 2009;87(5):1168–1174.

Owens D, Horrocks J, House A. Fatal and non-fatal repetition of self-harm. Systematic review. *Br J Psychiatry* 2002;181:193–199.

Owens D, House A. General hospital services for deliberate self-harm. Haphazard clinical provision, little research, no central strategy. *J Roy Coll Phys Lond* 1994;28:370–371.

Oxenstierna G, Edman G, Iselius L, Oreland L, Ross SB, Sedvall G. Concentrations of monoamine metabolites in the cerebrospinal fluid of twins and unrelated individuals—A genetic study. *J Psychiatr Res* 1986;20(1):19–29.

Painuly N, Sharan P, Mattoo SK. Antecedents, concomitants and consequences of anger attacks in depression. *Psychiatry Res* 2007;153:39–45.

Pandey GN, Dwivedi Y. Noradrenergic function in suicide. *Arch Suicide Res* 2007;11:235–246.

Persson ML, Geijer T, Wasserman D, Rockah R, Frisch A, Michaelovsky E, Jönsson EG, Apter A, Weizman A. Lack of association between suicide attempt and a polymorphism at the dopamine receptor D4 locus. *Psychiatr Genet* 1999;9(2):97–100.

Pitchot W, Hansenne M, Ansseau M. Role of dopamine in non-depressed patients with a history of suicide attempts. *Eur Psychiatry* 2001;16(7):424–427.

Pitchot W, Hansenne M, Moreno AG, Ansseau M. Suicidal behavior and growth hormone response to apomorphine test. *Biol Psychiatry* 1992;31(12):1213–1219.

Polewka A, Mikolaszek-Boba M, Chrostek Maj J, Groszek B. The characteristics of suicide attempts based on the suicidal intent scale scores. *Przeglad Lekarski* 2005;62:415–418.

Pompili M, Innamorati M, Raja M, Falcone I, Ducci G, Angeletti G, Lester D, Girardi P, Tatarelli R, De Pisa E. Suicide risk in depression and bipolar disorder: Do impulsiveness–aggressiveness and pharmacotherapy predict suicidal intent? *Neuropsychiatr Dis Treat* 2008;4:247–255.

Pooley EC, Houston K, Hawton K, Harrison PJ. Deliberate self-harm is associated with allelic variation in the tryptophan hydroxylase gene (TPH A779C), but not with polymorphisms in five other serotonergic genes. *Psychol Med* 2003;33(5):775–783.

Preuss UW, Koller G, Bahlmann M, Soyka M, Bondy B. No association between suicidal behavior and 5-HT2A-T102C polymorphism in alcohol dependents. *Am J Med Genet* 2000;96(6):877–878.

Renaud J, Berlim MT, McGirr A, Tousignant M, Turecki G. Current psychiatric morbidity, aggression/impulsivity, and personality dimensions in child and adolescent suicide: A case–control study. *J Affect Disord* 2008;105:221–228.

Rettew DC, McKee L. Temperament and its role in developmental psychopathology. *Harv Rev Psychiatry* 2005;13:14–27.

Retz W, Rösler M, Supprian T, Retz-Junginger P, Thome J. Dopamine D3 receptor gene polymorphism and violent behavior: Relation to impulsiveness and ADHD-related psychopathology. *J Neural Transm* 2003;110(5):561–572.

Reuter M, Hennig J. Association of the functional catechol-*O*-methyltransferase VAL158MET polymorphism with the personality trait of extraversion. *Neuroreport* 2005;16:1135–1138.

Roy A, Karoum F, Pollack S. Marked reduction in indexes of dopamine metabolism among patients with depression who attempt suicide. *Arch Gen Psychiatry* 1992;49:447–450.

Roy A, Linnoila M. Alcoholism and suicide. *Suicide Life Threat Behav* 1986;16(2):244–273.

Roy A, Segal NL, Centerwall BS, Robinette CD. Suicide in twins. *Arch Gen Psychiatry* 1991;48(1):29–32.

Rujescu D, Giegling I, Bondy B, Gietl A, Zill P, Moller HJ. Association of anger-related traits with SNPs in the *TPH* gene. *Mol Psychiatry* 2002;7:1023–1029.

Rujescu D, Giegling I, Dahmen N, Szegedi A, Anghelescu I, Gietl A, Schäfer M, Müller-Siecheneder F, Bondy B, Möller HJ. Association study of suicidal behavior and affective disorders with a genetic polymorphism in ABCG1, a positional candidate on chromosome 21q22.3. *Neuropsychobiology* 2000;42(Suppl 1):22–25.

Rujescu D, Giegling I, Gietl A, Hartmann AM, Moller HJ. A functional single nucleotide polymorphism (V158M) in the *COMT* gene is associated with aggressive personality traits. *Biol Psychiatry* 2003a;54(1):34–39.

Rujescu D, Giegling I, Mandelli L, Schneider B, Hartmann AM, Schnabel A, Maurer K, Möller HJ, Serretti A. *NOS-I* and *-III* gene variants are differentially associated with facets of suicidal behavior and aggression-related traits. *Am J Med Genet B Neuropsychiatr Genet* 2008;147B(1):42–48.

Rujescu D, Giegling I, Sato T, Hartmann AM, Moller HJ. Genetic variations in tryptophan hydroxylase in suicidal behavior: Analysis and meta-analysis. *Biol Psychiatry* 2003b;54(4):465–473.

Rujescu D, Giegling I, Sato T, Moller HJ. Lack of association between serotonin 5-HT1B receptor gene polymorphism and suicidal behavior. *Am J Med Genet B Neuropsychiatr Genet* 2003c;116B(1):69–71.

Rujescu D, Thalmeier A, Moller HJ, Bronisch T, Giegling I. Molecular genetic findings in suicidal behavior: What is beyond the serotonergic system? *Arch Suicide Res* 2007;11(1):17–40.

Rybakowski F, Slopien A, Dmitrzak-Weglarz M, Czerski P, Rajewski A, Hauser J. The 5-HT2A −1438 A/G and 5-HTTLPR polymorphisms and personality dimensions in adolescent anorexia nervosa: Association study. *Neuropsychobiology* 2006;53:33–39.

Ryding E, Lindström M, Träskman-Bendz L. The role of dopamine and serotonin in suicidal behaviour and aggression. *Prog Brain Res* 2008;172:307–315.

Sabol SZ, Hu S, Hamer D. A functional polymorphism in the monoamine oxidase A gene promoter. *Hum Genet* 1998;103(3):273–279.

Saudou F, Amara DA, Dierich A, LeMeur M, Ramboz S, Segu L, Buhot MC, Hen R. Enhanced aggressive behavior in mice lacking 5-HT1B receptor. *Science* 1994;265(5180):1875–1878.

Sayin A, Kuruoglu AC, Yazici Gulec M, Aslan S. Relation of temperament and character properties with clinical presentation of bipolar disorder. *Compre Psychiatry* 2007;48:446–451.

Schulsinger F. A family study of suicide. In: Schov M, Stromgren G (eds.), *Origin, Prevention and Treatment of Affective Disorders*. Academic Press, London, U.K., 1979, pp. 277–287.

Seidlitz L, Conwell Y, Duberstein P, Cox C, Denning D. Emotion traits in older suicide attempters and non-attempters. *J Affect Disord* 2001;66:123–131.

Serretti A, Calati R, Giegling I, Hartmann AM, Möller HJ, Colombo C, Rujescu D. 5-HT2A SNPs and the Temperament and Character Inventory. *Prog Neuropsychopharmacol Biol Psychiatry* 2007a;31(6):1275–1281.

Serretti A, Calati R, Giegling I, Hartmann AM, Möller HJ, Rujescu D. Serotonin receptor HTR1A and HTR2C variants and personality traits in suicide attempters and controls. *J Psychiatr Res* 2009;43(5):519–525.

Serretti A, Mandelli L, Giegling I, Schneider B, Hartmann AM, Schnabel A, Maurer K, Möller HJ, Rujescu D. *HTR2C* and *HTR1A* gene variants in German and Italian suicide attempters and completers. *Am J Med Genet B Neuropsychiatr Genet* 2007b;144B(3):291–299.

Serý O, Didden W, Mikes V, Pitelová R, Znojil V, Zvolský P. The association between high-activity COMT allele and alcoholism. *Neuro Endocrinol Lett* 2006;27(1–2):231–235.

Sher L, Oquendo M, Galfalvy H, Grunebaum M, Burke A, Zalsman G, Mann JJ. The relationship of aggression to suicidal behavior in depressed patients with a history of alcoholism. *Addict Behav* 2005;30:1144–1153.

Sher L, Stanley BH, Harkavy-Friedman JM, Carballo JJ, Arendt M, Brent DA, Sperling D, Lizardi D, Mann JJ, Oquendo MA. Depressed patients with co-occurring alcohol use disorders: A unique patient population. *J Clin Psychiatry* 2008;69(6):907–915.

Simon OR, Swann AC, Powell KE, Potter LB, Kresnow MJ, O'Carroll PW. Characteristics of impulsive suicide attempts and attempters. *Suicide Life Threat Behav* 2001;32:49–59.

Statham DJ, Heath AC, Madden PA, Bucholz KK, Bierut L, Dinwiddie SH, Slutske WS, Dunne MP, Martin NG. Suicidal behaviour: An epidemiological and genetic study. *Psychol Med* 1998;28(4):839–855.

Stefulj J, Buttner A, Skavic J, Zill P, Balija M, Eisenmenger W, Bondy B, Jernej B. Serotonin 1B (5HT-1B) receptor polymorphism (G861C) in suicide victims: Association studies in German and Slavic population. *Am J Med Genet B Neuropsychiatr Genet* 2004;127B(1):48–50.

Stein D, Apter A, Ratzoni G, Har-Even D, Avidan G. Association between multiple suicide attempts and negative affects in adolescents. *J Am Acad Child Adolesc Psychiatry* 1998;37:488–494.

Stein MB, Fallin MD, Schork NJ, Gelernter J. COMT polymorphisms and anxiety-related personality traits. *Neuropsychopharmacology* 2005;30:2092–2102.

Straub R, Wolfersdorf M, Keller F, Hole G. Personality, motivation and affect as modulating factors in suicidal behavior of depressed women. *Fortschr Neurol Psychiatr* 1992;60:45–53.

Suda A, Kawanishi C, Kishida I, Sato R, Yamada T, Nakagawa M, Hasegawa H, Kato D, Furuno T, Hirayasu Y. Dopamine D2 receptor gene polymorphisms are associated with suicide attempt in the Japanese population. *Neuropsychobiology* 2009;59:130–134.

Sumi-Ichinose C, Ichinose H, Takahashi E, Hori T, Nagatsu T. Molecular cloning of genomic DNA and chromosomal assignment of the gene for human aromatic L-amino acid decarboxylase, the enzyme for catecholamine and serotonin biosynthesis. *Biochemistry* 1992;31:2229–2238.

Suokas J, Lonnqvist J. Outcome of attempted suicide and psychiatric consultation: Risk factors and suicide mortality during a five-year follow-up. *Acta Psychiatr Scand* 1991;84:545–549.

Suominen K, Isometsa E, Suokas J, Haukka J, Achte K, Lonnqvist J. Completed suicide after a suicide attempt: A 37-year follow-up study. *Am J Psychiatry* 2004;161:562–563.

Sweet RA, Nimgaonkar VL, Kamboh MI, Lopez OL, Zhang F, DeKosky ST. Dopamine receptor genetic variation, psychosis, and aggression in Alzheimer disease. *Arch Neurol* 1998;55(10):1335–1340.

Terracciano A, Sanna S, Uda M, Deiana B, Usala G, Busonero F, Maschio A, Scally M, Patriciu N, Chen WM, Distel MA, Slagboom EP, Boomsma DI, Villafuerte S, Sliwerska E, Burmeister M, Amin N, Janssens AC, van Duijn CM, Schlessinger D, Abecasis GR, Costa PT Jr. Genome-wide association scan for five major dimensions of personality. *Mol Psychiatry* 2010;15(6):647–656.

Thomas A, Chess S. *Temperament and Development*. Bruner/Mazel, New York, 1977, pp. 1–270.

Tochigi M, Umekage T, Kato C, Marui T, Otowa T, Hibino H, Otani T, Kohda K, Kato N, Sasaki T. Serotonin 2A receptor gene polymorphism and personality traits: No evidence for significant association. *Psychiatr Genet* 2005;15:67–69.

Tremeau F, Darreye A, Staner L, Correa H, Weibel H, Khidichian F, Macher JP. Suicidality in opioid-dependent subjects. *Am J Addict* 2008;17:187–194.

Tsai SJ, Hong CJ, Liou YJ. Recent molecular genetic studies and methodological issues in suicide research. *Prog Neuropsychopharmacol Biol Psychiatry* 2011;35(4):809–817.

Tsai SJ, Hong CJ, Yu YW, Chen TJ. Association study of catechol-*O*-methyltransferase gene and dopamine D4 receptor gene polymorphisms and personality traits in healthy young Chinese females. *Neuropsychobiology* 2004a;50:153–156.

Tsai SJ, Hong CJ, Yu YW, Chen TJ, Wang YC, Lin WK. Association study of serotonin 1B receptor (A-161T) genetic polymorphism and suicidal behaviors and response to fluoxetine in major depressive disorder. *Neuropsychobiology* 2004b;50(3):235–238.

Turecki G. Dissecting the suicide phenotype: The role of impulsive–aggressive behaviours. *J Psychiatry Neurosci* 2005;30:398–408.

Turecki G, Sequeira A, Gingras Y, Seguin M, Lesage A, Tousignant M, Chawky N, Vanier C, Lipp O, Benkelfat C, Rouleau GA. Suicide and serotonin: Study of variation at seven serotonin receptor genes in suicide completers. *Am J Med Genet B Neuropsychiatr Genet* 2003;118B(1):36–40.

Useda JD, Duberstein PR, Conner KR, Beckman A, Franus N, Tu X, Conwell Y. Personality differences in attempted suicide versus suicide in adults 50 years of age or older. *J Consult Clin Psychol* 2007;75:126–133.

van den Oord EJ, Kuo PH, Hartmann AM, Webb BT, Möller HJ, Hettema JM, Giegling I, Bukszár J, Rujescu D. Genome wide association analysis followed by a replication study implicates a novel candidate gene for neuroticism. *Arch Gen Psychiatry* 2008;65(9):1062–1071.

Velting DM, Rathus JH, Miller AL. MACI personality scale profiles of depressed adolescent suicide attempters: A pilot study. Million adolescent clinical inventory. *J Clin Psychol* 2000;56:1381–1385.

Verkes RJ, Pijl H, Meinders AE, Van Kempen GM. Borderline personality, impulsiveness, and platelet monoamine measures in bulimia nervosa and recurrent suicidal behavior. *Biol Psychiatry* 1996;40:173–180.

Walther DJ, Peter JU, Bashammakh S, Hortnagl H, Voits M, Fink H, Bader M. Synthesis of serotonin by a second tryptophan hydroxylase isoform. *Science* 2003;299(5603):76.

Westergaard GC, Suomi SJ, Higley JD, Mehlman PT. CSF 5-HIAA and aggression in female macaque monkeys: Species and interindividual differences. *Psychopharmacology (Berl)* 1999;146(4):440–446.

WHO. *World Health Report. Health Systems*. WHO, Geneva, Switzerland. 2011. http://www.who.int/mental_health/prevention/suicide/country-reports/en/index.html

Widiger TA, Trull TJ. Personality and psychopathology: An application of the five-factor model. *J Pers* 1992;60:363–393.

Williams HJ, Owen MJ, O'Donovan MC. Is *COMT* a susceptibility gene for schizophrenia? *Schizophr Bull* 2007;33(3):635–641.

Witte TK, Merrill KA, Stellrecht NE, Bernert RA, Hollar DL, Schatschneider C, Joiner TE Jr. "Impulsive" youth suicide attempters are not necessarily all that impulsive. *J Affect Disord* 2008;107:107–116.

Wyder M, De Leo D. Behind impulsive suicide attempts: Indications from a community study. *J Affect Disord* 2007;104:167–173.

Youssef G, Plancherel B, Laget J, Corcos M, Flament MF, Halfon O. Personality trait risk factors for attempted suicide among young women with eating disorders. *Eur Psychiatry* 2004;19:131–139.

Zalsman G, Frisch A, Lewis R, Michaelovsky E, Hermesh H, Sher L, Nahshoni E, Wolovik L, Tyano S, Apter A, Weizman R, Weizman A. DRD4 receptor gene exon III polymorphism in inpatient suicidal adolescents. *J Neural Transm* 2004;111(12):1593–1603.

Zalsman G, Patya M, Frisch A, Ofek H, Schapir L, Blum I, Harell D, Apter A, Weizman A, Tyano S. Association of polymorphisms of the serotonergic pathways with clinical traits of impulsive-aggression and suicidality in adolescents: A multi-center study. *World J Biol Psychiatry* 2011;12(1):33–41.

Zhang X, Beaulieu JM, Sotnikova TD, Gainetdinov RR, Caron MG. Tryptophan hydroxylase-2 controls brain serotonin synthesis. *Science* 2004;305(5681):217.

Zill P, Buttner A, Eisenmenger W, Moller HJ, Bondy B, Ackenheil M. Single nucleotide polymorphism and haplotype analysis of a novel tryptophan hydroxylase isoform (*TPH2*) gene in suicide victims. *Biol Psychiatry* 2004;56(8):581–586.

Zouk H, Tousignant M, Seguin M, Lesage A, Turecki G. Characterization of impulsivity in suicide completers: Clinical, behavioral and psychosocial dimensions. *J Affect Disord* 2006;92:195–204.

19 *Toxoplasma gondii*, the Immune System, and Suicidal Behavior

Olaoluwa Okusaga and Teodor T. Postolache

CONTENTS

19.1 INTRODUCTION

Each year suicide leads to the tragic and premature deaths of over 1 million individuals around the world with an estimated annual mortality of 14.5 per 100,000 people. This translates to one death occurring every 40 s. Suicide is the 10th leading cause of death, making up 11.5% of all deaths (Hawton and van Heeringen 2009), though this burden is probably underestimated considering many third world countries appear to underreport suicide 9–10 times the actual amount (Hawton and van Heeringen 2009). While suicide rates have remained constant for the last decade, the three greatest causes of death (heart disease, cancer, and cerebrovascular disease) have all seen a decrease in death rates in this time period. Two of the most important risk factors for suicide are history of past suicide attempt (Harris and Barraclough 1997; Mann 2003) and a history of mood disorder. Every suicide is preceded by an estimated 8–25 suicide attempts, and 4% of depressed individuals die from suicide (Hawton and van Heeringen 2009). Additionally, more than half of individuals who attempt suicide had a major depressive episode at the time of the attempt.

For the past 7 years, our team at the University of Maryland School of Medicine Mood and Anxiety Program has been focused on studying triggers and vulnerabilities for suicide originating in the *natural* environment, that is, physical, chemical, and biological. In particular, we have been interested in the highly consistent peaks of suicide (Postolache et al. 2010) during certain seasons and their possible triggers. Specifically, we have identified (1) a relationship between atmospheric peaks of aeroallergens and suicide attempts in women (Postolache et al. 2005), confirmed in Denmark (Qin et al. 2011), (2) a relationship between suicide and allergy (Qin et al. 2011), and (3) an increased expression of allergy-related cytokines in the prefrontal cortex of suicide victims (Tonelli et al. 2008b). We have also reported that intranasal administration of allergens induces animal behaviors that are analogous to certain suicide risk factors such as aggression (Tonelli et al. 2008a) and anxiety (Tonelli et al. 2009). Our intermediate conclusion is that molecular and cellular mechanisms involved in the allergic immune response might attenuate functional capabilities of areas of the prefrontal cortex to act as behavioral breaks via multisynaptic inhibition of infralimbic centers. Following this line of thought, if allergy (a misdirected immune response against innocuous substances that were "misperceived" by the immune system as invasive pathogens) is associated with suicidal behavior, one would expect real neurotropic parasites to also be associated with suicide behavior. This led us to investigate *Toxoplasma gondii* and the anti-*T. gondii* immune response. A possible connection between *T. gondii* and suicidal behavior was suggested by the relatively high seroprevalence, its neurotropism (Flegr 2007), the immune activation involved in the defense against the parasite leading to elevation of cytokines previously found related to suicidal behavior (see Section 19.3.2), the occurrence of induced self-destructive behavior in rodent models (Lamberton et al. 2008; Vyas et al. 2007; Webster 2007), behavioral changes in humans (Flegr et al. 2002), and the parasite's association with mental illness (Niebuhr et al. 2008; Torrey et al. 2007). We will first briefly review the immune system and the evidence connecting immune activation with suicidal behavior, and then we will describe the immune response to *T. gondii*, followed by a description of the parasite and the evidence associating *T. gondii* infection with suicidal behavior.

19.2 IMMUNE SYSTEM

The immune system is charged with distinguishing self from nonself and attacking and eliminating the nonself. In particular, the immune system protects the individual from pathogenic microbes, their products, and neoplastic cells, while avoiding responses that could produce excessive damage of self-tissues and eliminate beneficial commensal microbes. The microbial defense system is comprised of two domains that include innate immunity and adaptive (or acquired) immunity (Turvey and Broide 2010). These two aspects interact to provide an integrated immune system (Figure 19.1).

19.2.1 Innate Immunity

Innate immunity (Chaplin 2010) includes anatomical and physiological barriers such as epithelial cell layers (e.g., intact skin), mucociliary clearance mechanisms (as seen in the respiratory tract), acidic pH of the stomach, bacteriolytic lysozyme in

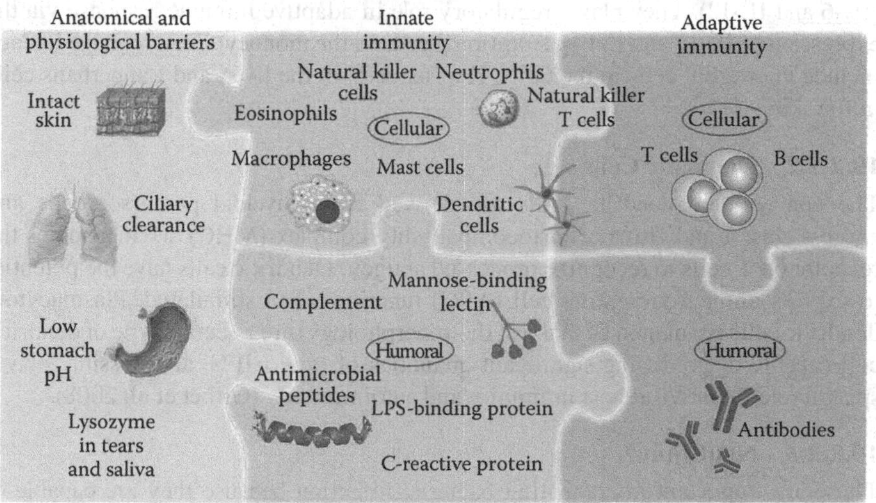

FIGURE 19.1 **(See color insert.)** Various subdivisions of the human immune system. (From Turvey, S.E. and Broide, D.H., *J. Allergy*, 25, S24, 2010. With permission.)

tears, saliva, and other secretions. Innate immunity also includes a number of soluble proteins and small bioactive molecules that are present in some body fluids or released from activated cells. Examples of soluble proteins and bioactive molecules involved in innate immunity include complement proteins, defensins, chemokines, free radical species, and lipid mediators of inflammation. The remaining aspect of innate immunity involves membrane-bound receptors and cytoplasmic proteins that target conserved microbial components shared by large groups of pathogens (Chaplin 2010).

Innate immunity (like adaptive immunity) has both cellular and humoral components. The various hematopoietic cells (i.e., macrophages, dendritic cells, mast cells, neutrophils, eosinophils, natural killer [NK] cells, and NK T cells) make up the cellular component, while complement proteins, LPS-binding protein, C-reactive protein, and other acute-phase reactants, antimicrobial peptides, and mannose-binding lectin make up the humoral component.

19.2.2 Cellular Components of Innate Immunity

A synoptic review of the principal cellular components of the innate immune system is presented below.

19.2.2.1 Monocytes and Macrophages

Both originate from myeloid stem cells and ingest microbes and particles bound to immunoglobulin, complement, or both (i.e., microbes and particles marked for clearance from the body). They produce nitric oxide, which is very effective in killing microbial pathogens, and are involved in both acute inflammatory responses and granulomatous processes. Activated macrophages produce large amounts of tumor necrosis factor (TNF), interferon gamma (IFN-γ), and interleukins 6 and 12 (IL-6 and IL-12). They play a regulatory role in adaptive immune response via the expression of IL-12 and INF-γ. Additional cells in the monocyte/macrophage lineage include microglial cells in the CNS, Kupffer cells in the liver, and Langerhans cells in the skin.

19.2.2.2 Dendritic Cells

The conventional dendritic cells are derived from myeloid precursor cells and express class I and II major histocompatibility complex (MHC). MHC enables the receptor on T cells to recognize processed antigen. Dendritic cells have the potential to express antigen-presenting cell (APC) function when stimulated. Plasmacytoid dendritic cells (so named because of their morphology) are a second type of dendritic cell capable of producing significant quantities of type I IFN, and possibly play a special role in antiviral host immunity and autoimmunity (Gilliet et al. 2008).

19.2.2.3 Neutrophils

These are important in combating bacterial infection because they are capable of engulfing bacterial pathogens, and produce reactive oxygen species that are cytotoxic to bacteria. They also engulf particulate matter, and have recently been observed to express significant quantities of IL-12, TNF, and certain chemokines (Chaplin 2010).

19.2.2.4 Natural Killer Cells

NK cells are related to both B and T lymphocytes because they are derived from the common lymphoid progenitor cell. They play an important and individual role in antiviral responses, and they also attack tumor cells (Paraskevas 2004).

19.2.2.5 Eosinophils, Basophils, and Mast Cells

Eosinophils are prominent in allergic responses. They possess prominent cytoplasmic granules containing toxic molecules and enzymes active against helminthes and other parasites. Basophils and mast cells are important in immediate hypersensitivity responses, and upon contact with selected antigens they release preformed and newly synthetized inflammatory mediators, including histamine, prostaglandins, and cytokines (Stone et al. 2010).

19.2.2.6 Antigen-Presenting Cells

Phagocytic cells of the monocyte/macrophage lineage and dendritic cells are referred to as Antigen-Presenting Cells (APC). They have the ability to "ingest" microbial antigen, process the antigen by proteolysis, and present them in forms (usually 9–11 amino acid peptides) that can activate T cells (see Section 19.2.3.2). Epithelial and endothelial cells may also be stimulated to produce MHC class II in the presence of INF-γ, and therefore act as APC to CD4+ T cells at sites of inflammation (Chaplin 2010). In addition, other host cells that have been invaded by viruses or those that have become tumorous may also produce peptides (bound to MHC) capable of activating T cells.

19.2.2.7 Major Histocompatibility Molecules

MHC molecules are glycoprotein molecules present on the surface of cells and bind peptide fragments of proteins that have either been synthesized locally in the cell (self) or proteolytically processed from protein antigens ingested by phagocytosis (nonself proteins). MHC molecules are involved in the presentation of antigens to T cells (Raposo et al. 1997). If the T cell recognizes the protein as nonself, then a cascade of reactions is triggered, which ultimately results in the killing of the cell expressing the nonself protein (antigen). MHC molecules are also called HLA (human leukocyte antigen) antigens. There are two classes: MHC I (binds endogenous or "self" peptides) and MHC II (binds peptides derived from exogenous or "nonself" antigens).

19.2.3 Adaptive Immunity

The cellular component of the adaptive immune response includes T and B lymphocytes, while immunoglobulins secreted by B cells constitute the humoral component. Both T and B lymphocytes (T: thymus and B: bone marrow) are derived from the common lymphoid progenitor. After a period of development in the primary lymphoid organs (bone marrow and thymus), the lymphocytes migrate to the secondary lymphoid organs (spleen, lymph nodes, Peyer's patches of the gut, tonsils, and adenoids). The secondary lymphoid organs are the site of origin of adaptive immune response, often under the influence of innate immune system signals, resulting either directly from circulating pathogens or indirectly by pathogen-activated APC that have migrated to the secondary lymphoid organs. Lymphocytes that have developed in the spleen and lymph nodes move out and spread to the various parts of the body to initiate immune effector functions (Bonilla and Oettgen 2010).

19.2.3.1 T-Lymphocyte Families

T lymphocytes differ in terms of their effector functions. Those T lymphocytes involved in "helping" other T and B cells by enhancing immunologic cell responses are traditionally referred to as T-helper (T_H) cells. T_H cells express CD4 (cell surface molecules). Based on the predominant cytokines produced by the cells, T_H cells are further subdivided into T_H1, T_H2, and, recently, T_H17. T_H1 cells mainly produce IFN-γ and TNF-β, but not IL-4 and IL-5. T_H1 cytokines promote cell-mediated responses for killing of intracellular microbes. T_H2 cells primarily produce IL-4, IL-5, IL-9, and IL-13, but not IFN-γ. T_H2 cytokines enhance antibody production and allergic immune responses. T_H17 cells express IL-17, a family of six cytokines (IL-17A–F) some of which are proinflammatory and important in immune response to extracellular pathogens (Chaplin 2010).

19.2.3.2 Activation of T and B Cells

T-cell activation is initiated when receptors (T-cell receptors, abbreviated TCRs) on the T cells interact with antigenic peptides (9–11 amino acids) complexed with MHC molecules. The TCRs on $CD8^+$ T cells interact with peptides complexed with MHC class I, while TCRs on $CD4^+$ T cells interact with peptides complexed with MHC class II. Peptides bearing MHC class I are generally derived from proteins produced inside host cells and such proteins are encoded either by the host genome (e.g., cancerous cells) or intracellular pathogens (e.g., viruses). MHC class II peptides reside on APC and can be induced through stimulation of innate immune system (Bonilla and Oettgen 2010). The interaction of the TCRs with the respective MHC/peptide complex triggers a sequence of biochemical events that mobilize the molecular killing machinery in $CD8^+$ T cells (Paraskevas 2004).

Most of the B-cell immune activation and response to antigens involves T cells, but a few antigens are capable of eliciting an antibody response in the absence of T cells. Such antigens are called T-independent antigens (TI antigens). Proteins and glycoproteins that require participation of T cells to generate B-cell antibody response are referred to as T-dependent antigens (Bonilla and Oettgen 2010). T-dependent antigens initiate B-cell activation by first interacting with immunoglobulin receptors on the cells, and making them cross-link. Cross-linking of the receptors triggers intracellular signaling that makes the B cell capable of interacting with a T cell. The peptides can be formed from antigens that were internalized by interaction with the receptors on the B cell. Interaction of the peptide/MHC II with a $CD4^+$ T cell renders the $CD4^+$ T cell capable of aiding the development and differentiation of the B cells to antibody-producing plasma cells or memory cells (that rapidly produce antibodies when rechallenged with the same antigen). The first immune response to an antigen challenge is called the primary response and IgM antibodies of low affinity predominate during this phase. A second presentation of the same antigen results in a stronger and faster response by the immune system, and IgG of high affinity (or IgD and IgE) predominate at this stage (Schroeder and Cavacini 2010).

19.2.3.3 Cytokines

Cytokines are proteins representing molecular signals that the immune cells use to communicate with each other. They help modulate the immune system through their effects on numerous aspects of cell growth, differentiation, and activation (Commins et al. 2010). Cytokines are grossly classified as proinflammatory (Th1) and anti-inflammatory (Th2).

19.3 NEUROIMMUNE INTERACTIONS

Interactions between the nervous and immune systems are increasingly being recognized, and neuroimmunology—the synthesis of neurobiology and immunology—is a rapidly evolving discipline (Bhat and Steinman 2009). Until recently, the central nervous system (CNS) was generally believed to be an immunologically privileged anatomical site (Tansey 2010), and this dogma was based on the observation that immune responses are blunted in the CNS. Furthermore, the blood–brain barrier (BBB) does not readily allow the passage of immune cells and molecules into the CNS. In addition, immune cells and other diffusible molecules from the periphery were thought to permeate the BBB only during infection, CNS trauma, and other significant insults. However, recent findings from various biological sciences indicate that regular cellular and molecular cross talk between the immune and nervous systems might occur even in the absence of any CNS pathology. It has also been suggested recently that the BBB becomes relatively more permeable with healthy aging (McAllister and van de Water 2009), as well as in systemic and distant inflammation.

A number of molecules involved in innate and adaptive immunity have been found to have dual roles in the immune system and normal brain physiology. Conversely, some molecules involved in the physiology of the CNS or its neurosecretory appendages have also been found to function as immune modulators. For example, complement molecules that function as opsonins in the immune system also eliminate synapses in the CNS; MHC, involved in antigen presentation, modulates CNS plasticity; and gamma amino butyric acid (GABA), an inhibitory neurotransmitter, also inhibits inflammation triggered by the immune system.

19.3.1 Neuroimmune Interactions in Mood Disorders

Through their modulation of neuronal anatomy and function, cytokines and other immune molecules have been found to impact neuropsychiatric functions such as mood and cognition. For example TNF-α and IL-1β, when present at pathophysiologically elevated levels, have the tendency to inhibit long-term potentiation and impair neuronal plasticity (Loftis et al. 2010). Neuronal synaptic plasticity describes the phenomenon whereby postsynaptic neurons are able to vary their response to presynaptic stimulation and long-term potentiation is a long-term increase in synaptic strength. Both synaptic plasticity and long-term potentiation are generally believed to be the basis for learning and memory (Berretta et al. 2008).

An accumulating body of evidence now supports the notion that neuroimmune interactions play a significant role in the pathogenesis of depressive disorders and other psychiatric conditions. An immune (inflammatory) activation consequence is "sickness behavior," which is characterized by fatigue, sleep disturbance, appetite disturbance, decreased social interaction, and loss of interest in usual activities—all of which are also seen in major depressive disorder (MDD) (Dantzer 2009). Sickness behavior is mediated by proinflammatory cytokines such as IL-1, IL-6, TNF-α, and IFN-γ. Several studies have reported elevated levels of proinflammatory cytokines in patients with MDD even without any apparent infection or inflammation (Raison et al. 2006). A corollary to this is the finding that most antidepressant medication as well as electroconvulsive therapy (ECT) inhibit the production of proinflammatory cytokines (Müller et al. 2009).

In MDD, cellular, molecular, and morphological studies in animals and human subjects have demonstrated an imbalance between neuroprotection and neurotoxicity in favor of the latter (Duman 2009). Neuroimmune interactions are involved in the neurotoxic mechanisms of depressive disorder. Proinflammatory cytokines such as IL-2, IFN-γ, and TNF-α increase the activity of indoleamine 2,3-dioxygenase (IDO) and kynurenine monooxygenase (KMO), two enzymes involved in the metabolism of tryptophan. Tryptophan is an amino acid that serves as "raw material" for the synthesis of serotonin (a neurotransmitter). IDO catalyzes the breakdown of tryptophan to kynurenine, thus resulting in a relative tryptophan depletion. The shunting of tryptophan toward production of kynurenine makes tryptophan unavailable for serotonin synthesis ultimately resulting in low serotonin levels in the brain. Low serotonin has been implicated in the pathogenesis of depression (Dursun et al. 2001). In addition, suicide attempters (especially those with violent attempts) have been found to have significantly lower cerebrospinal fluid (CSF) levels of 5-hydroxyindoleacetic acid (5-HIAA), a key metabolite of serotonin, relative to healthy volunteers (Träskman et al. 1981). Kynurenine crosses freely from the periphery to the brain and from the brain to the periphery. It has recently been implicated in depression and depressive-like behaviors (Dantzer et al. 2011; Raison et al. 2010). Kynurenine metabolites are potent immunomodulators (Schwarcz and Pellicciari 2002); specifically under the influence of KMO, kynurenine is catalyzed to 3-hydroxykynurenine (3-OH-kynurenine) and quinolinic (QUIN) acid in a two-step process. Both 3-OH-kynurenine and QUIN can induce neurodegeneration through the induction of excitotoxicity and generation of neurotoxic radicals (Müller et al. 2009). These pathways have specific cellular substrate. For instance, the microglia are the cells responsible for the rate-limiting pathway of transformation of kynurenine via kynurenine 3-monooxygenase (KMO) to QUIN. Astrocytes are responsible for the transformation of kynurenine via kynurenine aminotransferases (KAT) I and II to kynurenine acid (KA) (Wonodi and Schwarcz 2010).

19.3.2 Immune Activation and Suicidal Behavior

In contrast to the number of published studies on immune dysregulation and mood disorders, only a few studies have identified a possible link between immune mechanisms and suicidal behavior. In one study (Nassberger and Traskman-Bendz 1993), the plasma concentrations of soluble interleukin-2 receptor (S-IL-2R) in medication-free suicide attempters were significantly higher than those found in healthy controls. Most recently, Janelidze et al. (2010) evaluated blood cytokine levels in 47 suicide attempters, 17 non-suicidal depressed patients, and 16 healthy controls, and found increased levels of IL-6 and TNF-α in suicide attempters relative to non-suicidal depressed patients and healthy controls. While this cytokine activation was found in the "periphery," IL-6 levels have also been reported to be elevated in the CSF of suicide attempters relative to controls (Lindqvist et al. 2009). Another study found elevated levels of Th2 cytokine mRNAs in postmortem brain tissue samples within the orbitofrontal cortex of suicide victims (Tonelli et al. 2008b). Microglia cells in the brain are capable of expressing cytokines, and significant microgliosis has been observed in the brains of patients who committed suicide (Steiner et al. 2008).

19.4 *Toxoplasma gondii* AND SUICIDE

T. gondii, a widespread neurotropic protozoan parasite (Ajioka and Soldati 2007), affects approximately one-third of all humans worldwide. Symptoms of infection range from none to minimal depending on the host's immune response adequacy. Congenital infection, occurring if a mother has a primary infection during pregnancy and passes it to the fetus, is relatively rare. Within the animal world, felids have been identified as the definitive host to *T. gondii*. It is within the cat's gut that the parasite can sexually reproduce and spread via oocysts. Humans may be infected by *T. gondii* via ingestion of the parasite's oocysts, which can spread from the feces of infected cats. Other routes of transmission include consumption of undercooked meat that has been infected with *T. gondii* cysts or ingestion of contaminated water. When ingested by an intermediate host, the parasite spreads from the intestine to other organs, finally localizing in muscle and brain. In the brain, the parasite will hide within neurons and glial cells, intracellularly, ultimately in cystic structures. These structures have minimal exposure to cellular and molecular mediators of the immune system that contain the infection successfully, but fail to eradicate it. Previous research in rodents has revealed that *T. gondii* localizes in multiple structures of the brain, including the prefrontal cortex and predominantly the amygdala (Vyas et al. 2007). These areas have a primary role in emotional and behavioral regulation, and they show major histopathological changes in suicide victims (Mann 2003). It is possible that because *T. gondii* occupies these areas, *T. gondii* infection may disrupt the balance of affective and behavioral modulation and in turn elevate risk of suicide.

19.4.1 Immune Responses to *T. gondii*: Innate Resistance

Production of IFN-γ by T cells and NK cells is critical to resistance to *T. gondii*. In addition, IL-12 production is also critical as the absence of IL-12 incapacitates replication, probably via a reduced production of IFN-γ. Both IL-12-deficient as well as IFN-γ-deficient animals succumb to infection. The major source of IL-12 *in vitro* appeared to be the macrophages (Gazzinelli et al. 1993a,b), but more recent work *in vivo* points toward dendritic cells (which synthesize it) and neutrophils (which have it prestored in vacuoles to be released upon contact); thus, multiple cells types are involved in resistance (Pepper and Hunter 2007).

Two proximal signaling events are involved in the production of IL-12—first (Luangsay et al. 2003) is via the chemokines receptor CCR5 (and G protein–coupled signaling) and the second (Medzhitov 2001) is via toll-like receptors (TLRs). Downstream from

TLRs, signaling involves MyD88 and TRAF-6, and from there it branches to either MAPK or IKK and NF-κB both signaling production if IL-12.

NK cells-mediated resistance of *T. gondii* has been clearly demonstrated by the successful resistance to *T. gondii* of SCID mice that have normal NK cells but lack T and B cells. IFN- γ derived from NK or T cells plays the major role in controlling *T. gondii*, with multiple downstream pathways including (a) priming phagocytes to produce leukotrienes and reactive oxygen intermediaries that kill the parasite and (b) depriving the parasite of tryptophan via activating IDO. There is evidence that TNF-α also plays a role, and the two cytokines act synergistically to increase activity of enzyme inducible nitric oxide synthase (iNOS) and nitric oxide production (Pepper and Hunter 2007).

In addition to the innate immunity mechanisms described earlier, adaptive immunity mechanisms including nongermline-encoded clonal receptors BCR and TCR are essential for long-term resistance. A major expansion of B cells results in high levels of IgM (acute infection) and IgG (chronic infection; Montoya and Liesenfeld 2004). These antibodies opsonize extracellular parasites and induce complement activation and lysis, as well as phagocytosis by macrophages. μMT mice, that lack B cells, are able to survive the acute phase of infection to *T. gondii* only to die from fatal toxoplasma-caused encephalitis <1 month postinfection. The role of T cells in *T. gondii* resistance is illustrated by the reactivation of infection in AIDS patients, as well as immunosuppressive therapies and cancers that lead to T cells deficiency in number or function (Israelski and Remington 1993; Israelski et al. 1993).

Initiation of the adaptive immune response is dependent upon presentation by accessory cells, of peptides derived from the parasite in the context of MHC molecules. The parasite-derived MHC class I molecules could be presented to $CD8^+$ T cells by any cells but most commonly by infected cells, while MHC class II molecules can be presented to $CD4^+$ cells by dendritic cells (predominantly), macrophages, and B cells (Pepper and Hunter 2007).

After priming, T cells acquire cytolytic properties against infected cells. However, in toxoplasmosis the main role of T cells is to produce IFN-γ for which controls parasite replication. In addition to Major Histocompatibility Complex/T-cell Receptor (MHC/TCR) interactions, co-stimulatory molecules such as CD28, ICOS (Inducible T-cell co-stimulator) and IL-2 production by $CD4^+$ cells all appear to contribute to adequate IFN-γ production.

19.4.2 Immune Manipulation of the Parasite

T. gondii uses the immune system to facilitate its spread. The manipulation of the immune system by the parasite starts in the small intestine, where *T. gondii* starts replicating in the lamina propria (Speer and Dubey 2005). Immature dendritic cells are recruited by the chemokine-like activities of the parasite (Diana et al. 2005) in the small intestine (Luangsay et al. 2003), further amplified by inflammatory cells attracted at the site of infection. Dendritic cells and monocytes that cross the lamina propria in the intestine act as Trojan horses to safely disseminate *T. gondii* to its final tissue destination, including the brain (Courret et al. 2006). In order to survive, *T. gondii* must ensure not only its own survival, but also the survival of its host (it is unlikely that cats will eat corpses). Thus, *T. gondii* plays a sophisticated balancing game between signaling, allowing itself to be kept in check by the immune system, and also avoiding complete elimination by the immune system. From a molecular standpoint, one of the most important mechanisms is interference of the parasite with NF-κB and STAT1 signaling and activates STAT3, resulting in reduced production of IL-12 and inhibition of proinflammatory events (Pepper and Hunter 2007).

19.4.3 Pathological Immune Responses

Systemic inflammation, high levels of inflammatory cytokines, and flu-like symptoms occur during the acute infection (Liesenfeld et al. 1996). While replicating parasites may have a role, the immune-mediated (particularly CD4$^+$ cells) bystander effect is to the greatest extent the culprit (Mordue et al. 2001). Anti-inflammatory mechanisms involving TGF-β, IL-4, IL-10, and IL-5 are necessary to tune down or prevent hyperinflammation (Buzoni-Gatel et al. 2001; Gazzinelli et al. 1996; Nickdel et al. 2004). Other more recently discovered breaking mechanisms are MHC CD1 and the γδ T cells (Egan et al. 2005; Smiley et al. 2005). In fact, elevations of self-targeting antibodies (for instance, antithyroid antibodies) are also observed in chronic "latent" toxoplasmosis in contrast to other chronic or latent infections (Wasserman et al. 2009).

19.4.4 Toxoplasma in the Brain

T. gondii transforms from tachyzoite to bradyzoite forms under the pressure of the adaptive immune response. The bradyzoite live in cystic structures that persist for the life of the individual, occasionally release bradyzoite that are able to infect other cells and transform into tachyzoites for a brief time, considering the immediate pressure from the adaptive immune system. Immunodeficiency may result in toxoplasmic encephalitis. The ability to control *T. gondii* in the brain is a paradox for those who still believe in the brain as an immunoprivileged organ, considering the BBB that limits access to molecular and cellular mediators of inflammation, the small constitutive expression of class I and II molecules, and the absence of a lymphatic system. In mice models, dendritic cells, macrophages, CD4$^+$, and CD8$^+$ cells migrate to the brain and produce Th1 cytokines (Fischer et al. 2000; Hunter and Remington 1994). Prior to T-cell infiltration, astrocyte activation has been observed (Hunter et al. 1993). Astrocytes infected with *T. gondii* produce the proinflammatory cytokines IL-1, IL-6, and TNF-α (Fischer et al. 1997) as well as additional cytokines with a possible role in resistance, IL-12 and IL-10.

Essentially, the production of IFN-γ by T cells, in combination with TNF-α, is probably the most important factor for the control of *T. gondii* in the CNS, via activation of microglia and infiltrating macrophages and their production of NO controlling replication (Chao et al. 1993). Mice that lack TNF-α develop toxoplasma encephalitis, consistent with a recent report in a patient treated with anti-TNF who developed toxoplasma encephalitis (Young and McGwire 2005).

Thus, the immune response in the brain involves activation of immune mechanisms previously implicated in suicide and suicide risk factors—some involved in protection against *T. gondii* invasion, such as TNF-α (Janelidze et al. 2010), IL-6 (Janelidze et al. 2010), and microglia activation (Steiner et al. 2008), and others to minimize immune pathology, such as IL-4 and IL-5 (Tonelli et al. 2008b).

19.4.5 Fatal Attraction to Cats in Rodents

Normally, rodents display an instinctual avoidance of feline odors. However, infection with *T. gondii* induces not only a reduction in avoidance of, but also attraction to feline odors (Berdoy et al. 2000; Vyas et al. 2007). The attraction appears rather specific for felids and is not present for odors of other predators. The infected rodents therefore become more vulnerable to predation and correspondingly contribute, albeit unwittingly, to *T. gondii* reproduction. This striking phenomenon appears to be consistent with many well-documented examples of "behavioral manipulation" of the host by parasite. In addition to the specific changes in behavior related to attraction to cats, nonspecific effects of decreasing neophobia (Hay et al. 1984; Webster 2007) have also been reported, though unconfirmed by other studies (Vyas et al. 2007). A decrease in neophobia might be a consequence of the functional impairment of fear circuits in the amygdala, where *T. gondii* predominantly localize (Vyas et al. 2007).

19.4.6 Toxoplasma and Suicide Attempts: First Reported Association in Patients with Mood Disorders

The first report of a relationship between *T. gondii* infection and suicidal behavior was reported by our group at the Mood and Anxiety Program at the University of Maryland School of Medicine (Arling et al. 2009).

Two hundred and eighteen individuals classified as having MDD or bipolar disorder by the Structured Clinical Interview for DSM-IV Disorders (SCID) (First 2003) along with 39 healthy controls had *T. gondii* IgG antibodies measured through solid-phase enzyme immunoassay as was previously described (Leweke et al. 2004). Titers of anti-*T. gondii* antibodies greater than 10 international units per individual sample were considered to be positive for seropositivity analysis purposes. Quantitative analysis for antibody levels (serointensity) was also performed using a ratio of the sample's optical density (OD) and a standard with 10 international units of anti-*T. gondii* antibody.

If patients failed to meet the criteria for either bipolar disorder or MDD or if they met criteria for cognitive disorders, psychotic disorders, or substance dependence, they were excluded from the study. A semistructured questionnaire, The Columbia Suicide History Form, was used to collect the history of suicide attempt (Oquendo et al. 2003). An individual's diagnosis or suicide attempt status was unknown to the laboratory staff.

Chi-square tests and one-way analysis of variance were used to compare demographic characteristics and *T. gondii* serointensity values of the three groups (i.e., mood disorder patients with and without history of suicide attempt and controls), which, because of their positive skew, were log transformed. Log-transformed *T. gondii* antibody titers were compared among groups using linear regression analyses, with adjustments for race (white or other), gender, and age. Geometric means with 95% confidence intervals were calculated by exponentiation of the mean log-transformed titers of *T. gondii* antibody, and are given for the comparison of adjusted and non-adjusted values of the three groups. Finally, logistic regressions were carried out with attempt status as dependent variables. SAS 9.1 (SAS Institute Inc., Cary, NC) was used to perform all statistical analyses, with $p < 0.05$ as a level of significance.

Patients with recurrent depression who previously attempted suicide had mean values for antibodies to *T. gondii* higher than either the patients with recurrent depression and no history of suicide attempts ($p = 0.04$), or the normal control group ($p = 0.12$). When adjusted for race, age, and gender, the suicidal patients with recurrent depression versus non-suicidal patients with recurrent depression had a greater mean IgG titer of 0.51 versus 0.37 ($p = 0.017$) (Figure 19.2). Logistic regression models revealed that serointensity predicted suicide attempts with OR of 1.55 (1.14–2.12), $p = 0.006$. However, there was a nonsignificant relation of seropositivity with suicide attempt, OR = 1.62 (0.72–3.65).

No significant difference was found between *T. gondii* positive and *T. gondii* negative patients when compared using Pearson's *chi-square* test in relation to proportion with versus without recurrent mood disorder ($p = 0.62$), proportions having MDD versus having bipolar ($p = 0.82$), or proportion with compared to without psychotic symptoms ($p = 0.66$). *T. gondii* antibody levels showed no difference in regard to subjects with versus without a mood disorder ($p = 0.22$), having MDD versus having bipolar disorder ($p = 0.55$), or with in comparison to without psychotic symptoms ($p = 0.34$).

Limitations of this study included the cross-sectional design, as well as an unmeasured difference among socioeconomic groups, which could explain the variations in suicidality and titers of *T. gondii* antibody.

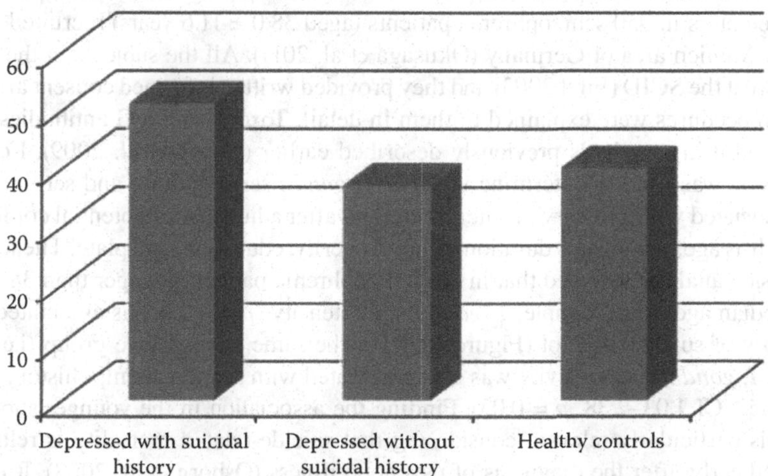

FIGURE 19.2 Increased antibody titers in patients with history of depression and suicide attempt than in patients with history of depression without suicide attempts and healthy controls (geometric means × 100). (From Arling et al. 2009. With permission.)

19.4.7 Toxoplasma and Suicide Attempts: A Replication

An independent group in Turkey (Yagmur et al. 2010), 1 year after the Arling et al.'s report, published a study confirming the association between *T. gondii* IgG (but not IgM) antibodies and suicide attempts. They compared 200 psychiatric inpatients who had attempted suicide and 200 controls recruited from "case workers" and family members/visitors of patients, matched for age, gender, and urban versus rural residence. The IgG seroprevalence was 41% in the patient group and this was significantly higher than in the control group with a seroprevalence of 28% ($p = 0.004$ by chi-square test). This study, performed in a country with a higher seropositivity for *T. gondii*, identified not only serointensity but also seropositivity as related to suicidal behavior. All limitations of the Arling's (2009) study are also present in the Yagmur et al.'s (2010) study. In addition, the Yagmur et al.'s study does not have diagnostic information, and the patient group is likely more heterogeneous. If one would look solely at the Yagmur et al.'s study, it would be impossible to rule out the possibility that mental illness is in fact associated with *T. gondii* seropositivity. It has already been established that there is an increased rate of *T. gondii* seropositivity in psychotic disorders (see Torrey et al. 2007 for meta-analysis), and thus it might be possible that the individuals with psychotic illness present only in the suicide group "drove" the observed association. Teasing apart suicidality from psychosis requires a study comparing psychotic attempters and psychotic nonattempters. Some of these limitations are addressed in the next study.

19.4.8 *Toxoplasma gondii* IgG Antibodies and Suicide Attempts in Patients with Psychotic Disorders

Supported by a grant from the American Foundation for Suicide Prevention (PI Postolache, CoPI Rujescu), the project involved comparing suicide attempters with nonattempters in 950 schizophrenia patients (aged 38.0 ± 11.6 years) recruited in the greater Munich area of Germany (Okusaga et al. 2011). All the subjects in the study underwent the SCID (First 2003) and they provided written informed consent after the study procedures were explained to them in detail. Toxoplasma IgG antibodies were measured using methods previously described earlier (Arling et al. 2009). Logistic regression was used to determine whether *T. gondii* seropositivity and serointensity are associated with a history of suicide attempt, after adjusting for potential confounders such as age, sex, illness duration, illness severity, education, and plate. The logistic regression analysis revealed that in the schizophrenia patients younger than 38 years, the median age of the sample, *T. gondii* serointensity ($p = 0.02$) was associated with a history of suicide attempt (Figure 19.3). In the same younger age group (i.e., <38 years), *T. gondii* seropositivity was also associated with suicide attempt history (odds ratio 1.57; CI 1.03–2.38, $p = 0.03$). Finding the association in the younger subpopulation is particularly relevant considering that suicide-related mortality is relatively elevated early after the diagnosis of psychotic illness (Osborn et al. 2008). It is also important to note that the association of *T. gondii* with history of suicidal behavior is unlikely to have been mediated through symptom severity since *T. gondii* was not associated with symptom severity in this sample of schizophrenia patients.

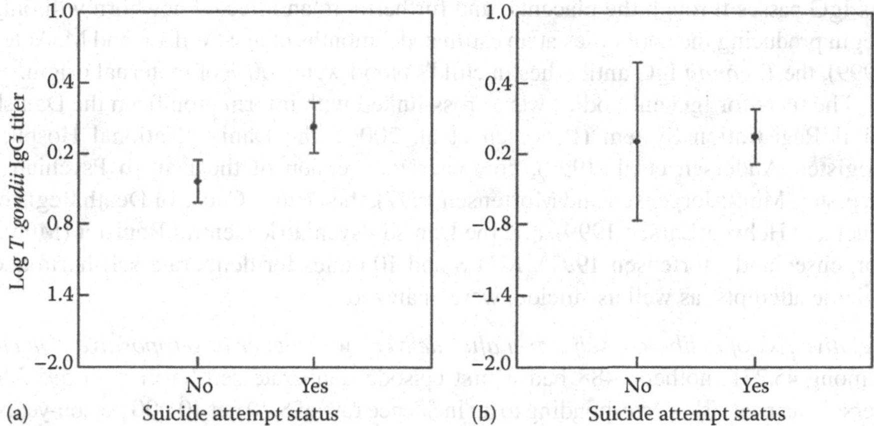

FIGURE 19.3 Comparison of *T. gondii* IgG antibody titers in schizophrenia patients with and without history of suicide attempt. (a) Difference is significant in the subgroup younger than 38 years (the median of the sample). $F = 10.607$, $p = 0.001$. (b) Difference is not significant in patients 38 years and older. $F = 1.141$, $p = 0.258$. (From Okusaga, O. et al., *Schizophr. Res.*, 133, 150, 2011. With permission.)

A question arises: Could this relationship be an artifact, the result of a general immune activation, or antibody elevation in patients at risk for suicide? We tested this hypothesis by analyzing antibodies to a number of neurotropic viruses as well as a food antigen, gliadin, and found no differences, using identical methods with those used for *T. gondii* antibodies analysis, for the cytomegalovirus ($p = 0.22$), herpes 1 virus ($p = 0.36$), and gliadin ($p = 0.92$). Thus, the increase in *T. gondii* IgG antibodies in patients who attempted suicide is unlikely to be attributed to a general nonspecific increase in antibody production.

All these results presented so far are cross-sectional associations. The following study will have a model of predictive association, as the determination of *T. gondii* antibodies will precede (often by many years) suicidal behavior.

19.4.9 Predictive Association between *T. gondii* Seropositivity and Suicidal Behavior in Women

Participants in this study (Pedersen et al. 2012) were women living in five counties in Denmark from 1992 to 1995 and were originally recruited for a project on *T. gondii* neonatal screening. In order to test for phenyl ketonuria and other metabolic abnormalities, a heel stick blood sample was obtained 5–10 days after birth. Only the first birth was included in the study. Two 3.2 mm disks were obtained from the children whose mothers agreed to participate in the *T. gondii* study and analyzed using an enzyme immunosorbant assay (EIA) for *T. gondii* IgG antibodies (Lebech and Petersen 1995). The level of those antibodies was then given as a percentage of the OD obtained for the World Health Organization (WHO 1994) international standard serum. The IgG titer was calculated as the mean of the titers for the two disks. Mothers with children that had an IgG titer above 24 were defined *T. gondii* positive at the time of delivery.

As IgG passes through the placenta, and furthermore an infected newborn will only begin producing the antibodies at an estimated 3 months of age (Wilson and McAuley 1999), the *T. gondii* IgG antibodies in child's blood were 100% of maternal origin.

The titers for IgG antibodies were cross-linked with information from the Danish Civil Registration System (Pedersen et al. 2006), the Danish National Hospital Register (Andersen et al. 1999), computerized version of the Danish Psychiatric Register (Munk-Jorgensen and Mortensen 1997), the Danish Cause of Death Register (Juel and Helweg-Larsen 1999), and the Danish Psychiatric Central Register (Munk-Jorgensen and Mortensen 1997). ICD 8 and 10 codes for deliberate self-harm and suicide attempts, as well as suicide, were analyzed.

Relative risk of deliberate self-harm after delivery was higher in seropositive women: Among 45,271 mothers, 488 had a first episode deliberate self-harm with 595,306 person-years at risk, corresponding to an incidence rate of 8.20 per 10,000 person-years. *T. gondii* seropositive women had a 1.53-fold (95% CI, 1.27–1.85, $p = 0.002$) significant relative risk of deliberate self-harm compared to *T. gondii* seronegative mothers. The risk of deliberate self-harm increased with increasing IgG level—stratified as quartiles (women with an IgG level above 71 had a relative risk of 1.89 [95% CI 1.42–2.46] compared to seronegative women). Contrary to our expectations, the association was stronger in women without a diagnosis of a psychiatric illness than in those with a psychiatric illness. There was no significant increase in repeated attempts in women who attempted suicide prior to testing in *T. gondii* positive women.

Relative risk of violent suicide attempt was higher in seropositive women: Seventy eight of 45,745 women had a violent suicide attempt during 603,876 person-years at risk (incidence rate of 1.29 per 10,000 person-years). The risk was 1.81 (95% CI, 1.13–2.84, $p = 0.014$) higher in *T. gondii* positive women when compared to *T. gondii* seronegative women.

Suicides (total 18) were too few to be analyzed with satisfactory statistical power. Nevertheless, the relative risk appeared to be nonsignificantly higher in women who were *T. gondii* positive, as compared to women who were *T. gondii* negative 2.05 (95% CI, 0.78–5.20, $p = 0.142$).

The major strength of this study is that, for the first time, it connects *predictively T. gondii seropositivity* with suicide attempts. Limitations include insufficient coverage for medical illness, not including women who did not give birth, and the potential-limited generalizability from Denmark to the rest of the world. Finally, this study was not powered to study suicides, but only attempts. It is unclear if associations with attempts would translate to associations with suicide. Our next study, even if preliminary, is a first connection of *T. gondii* with suicide.

19.4.10 National Suicide Rates Positively Correlate with Seroprevalence Rates for *T. gondii* in Women

Very recently, an abstract reported that seroprevalence for 20 European countries was significantly correlated with increased suicide rates (Lester 2010). Due to the fact that blood samples were taken from women only, we retested this association

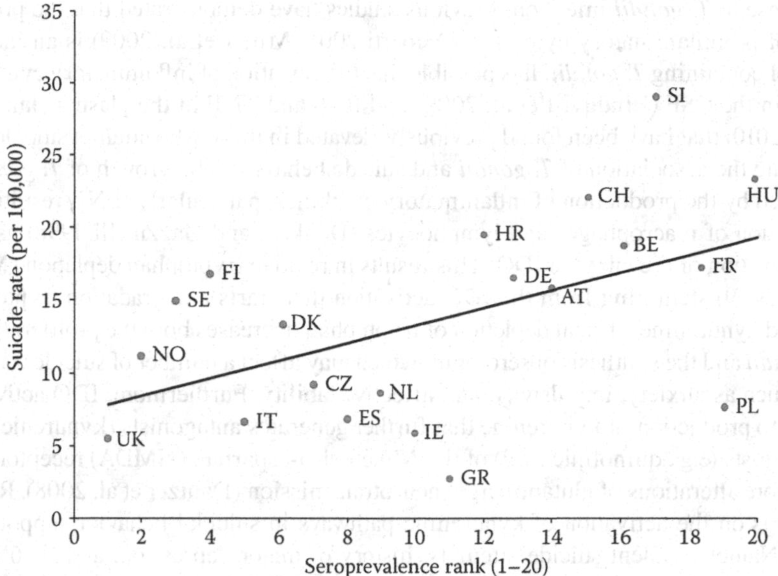

FIGURE 19.4 Relationship between suicide rates and *T. gondii* seroprevalence rank across 20 European countries in women aged 60–74 years old. Linear regression ($t = 2.54$, standardized coefficient = 0.51, $p = 0.020$; after adjustment for GDP rank: $t = 3.02$, standardized coefficient = 0.33, $p = 0.008$). Rank order is reversed. AT, Austria; BE, Belgium; HR, Croatia; CZ, Czech Republic; DK, Denmark; FI, Finland; FR, France; DE, Germany; GR, Greece; HU, Hungary; IE, Ireland; IT, Italy; NL, the Netherlands; NO, Norway; PL, Poland; SI, Slovenia; ES, Spain; SE, Sweden; CH, Switzerland; UK, United Kingdom. (From Ling, V.J. et al., *J. Nerv. Ment. Dis.*, 199, 440, 2011. With permission.)

in women only, stratified by age (Ling et al. 2011, Figure 19.4). In women 60 years and older, simple correlations between ranked *T. gondii* seropositivity and suicide rate identified were significant ($p < 0.05$); adjusting for GDP, positive relationships also occurred in the 45 and above age group. After adjustment for GDP, the relationship ($p = 0.007$) resisted Bonferroni adjustment for multiple comparisons only in the 60–74 age-group (Figure 19.4). In conclusion, the results confirm a relationship between suicide and *T. gondii* seropositivity, especially in the postmenopausal age group. As the methodology is ecological, and thus prone to inherent fallacies, individually linked methodologies and prospective studies are necessary to further confirm the association.

In conclusion, cross-sectional studies in mood disorders (Arling et al. 2009), psychiatric inpatients (Yagmur et al. 2010), schizophrenic patients (Okusaga et al. 2011), a prospective cohort study in mothers (Pedersen et al. 2012), and an ecological study in Europe (Ling et al. 2011) strongly support an association between *T. gondii* and suicidal behavior.

What are possible mechanisms mediating the relationship between T. gondii *and suicidal behavior?* In addition to reactivation of the latent parasite (i.e., a direct effect), one of the important potential mechanisms is the host's immune system activation in response to *T. gondii* infection. Previous studies have demonstrated that the production of proinflammatory cytokines (Aliberti 2005; Miller et al. 2009) is an integral part of containing *T. gondii*. It is possible that the elevation of inflammatory cytokines IL-6 in the CSF (Lindqvist et al. 2009) and IL-6 and TNF in the plasma (Janelidze et al. 2010) that have been found previously elevated in those who attempt suicide may mediate the association of *T. gondii* and suicide behavior. The growth of *T. gondii* is blocked by the production of inflammatory cytokines, particularly IFN-γ resulting in activation of macrophages and lymphocytes (Denkers and Gazzinelli 1998) as well as activation of the enzyme IDO. This results in relative tryptophan depletion (Miller et al. 2009) stemming from the IDO activation that starts a degradation tryptophan toward kynurenines. Local depletion of tryptophan decreases both the proliferation of *T. gondii* and the synthesis of serotonin, which may affect a number of suicide risk factors such as anxiety, impulsivity, and affective lability. Furthermore, IDO activation leads to production of kynurenine that further generates antagonists (kynurenic acid) or agonists (e.g., quinolinic acid) of the *N*-methyl-D-aspartate (NMDA) receptors and therefore alterations of glutaminergic neurotransmission (Dantzer et al. 2008). Recent findings on the activation of kynurenine pathways in suicidal behavior support this idea. Namely, violent suicide attempts, history of major depression, and IL-6 levels (Lindqvist 2010) have been found to be associated with kynurenic acid concentrations in the CSF. Our collaborative study with Dr. Mann's group at Columbia University has found that patients with mood disorders who have a history of suicide attempt relative to those without a history of attempts had an elevated level of kynurenine in plasma (Sublette et al. 2011) (Figure 19.5).

In addition, previous research has suggested that infection with *T. gondii* may elevate testosterone levels (Flegr 2007), and in turn that elevation in testosterone may lead to an increase in aggression, which has been identified as an intermediate phenotype in suicide (Kovacsics et al. 2009; Mann et al. 2009). Testosterone has been linked to the suppression of neural circuitry that is related to both impulse control and emotional regulation (Mehta and Beer 2010). This could help explain why the association between *T. gondii* and suicide is observed in older women, but not in younger women, whose androgens are balanced by endogenous estrogen and progesterone.

T. gondii infection may have the potential to heighten the risk factors that lead to attempting suicide. Joiner et al. (2009) presents a two-factor theory that states that there are two components that lead to attempting suicide when occurring simultaneously. The first domain is psychological, and it is expressed as a desire to die. This commonly results from a lack of the feeling of belongingness and a perception that one is a burden. The second domain is behavioral and is expressed as an acquired capability to attempt suicide through the habituation to the fear of death, dying, and the beyond; it is sometimes a result of witnessing or experiencing violence, or having painful and fearful occurrences.

T. gondii infection may contribute to the *capability* to engage in self-injurious behavior rather than just the increased *wish* to commit suicide. In experimentally infected immunocompetent rodents, *T. gondii* cysts are predominantly found in the amygdala, an area implicated in the expression of fear, of its host (Vyas et al. 2007), leading to a degree of atrophy of the dendritic tree and deafferentation. Furthermore, *T. gondii* contains two genes encoding tyrosine hydroxylase producing L-DOPA

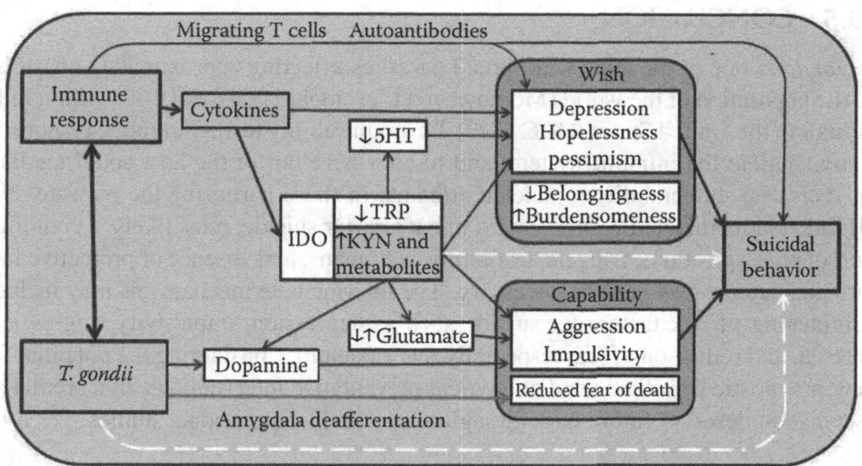

FIGURE 19.5 Hypothesized relationship between *T. gondii* and suicidal behavior and possible mechanisms. White dashed lines mark available preliminary evidence for the relationships between *T. gondii*, kynurenine, and suicidal behavior. The model integrates hypothesized molecular impact of *T. gondii* infection in an immunocompetent host with elements of Joiner's Interpersonal (Joiner 2009) and Mann's Stress Diathesis (Mann 2003) theories of suicide. The immune response to *T. gondii* releases cytokines that may directly or indirectly affect the brain and induce depression, elevating hopelessness and pessimism (diathesis) or desire to die (in Joiner's theory). In addition to cytokines, migrating T cells and autoantibodies could affect the centers involved with emotional regulation and dysregulation and lead to suicidal behavior. *T. gondii* has the enzymatic capacity to produce its own dopamine, and thus lead to potential increased aggression, arousal, and changes in impulsivity. In addition, activation by interferon-γ and TNF-α of the enzyme idolomine-2, 3-deoxygenase (IDO) results in the stealing of tryptophan from the synthesis of serotonin toward kynurenine and its metabolites. Kynurenine's metabolites, kynurenic acid and quinolinic acid, modulate glutaminergic and dopaminergic neurotransmission, further leading to changes in impulsivity and aggression. Decreased tryptophan may lead to decreased serotonin turnover and depression, hopelessness, and increased pessimism. Neuroanatomically, *T. gondii* parasitizes brain structures involved in behavioral regulation, the predominant site being the amygdala. The *T. gondii* infestation of the amygdala may result in deafferentation and the shrinkage of neurotic processes (Vyas et al. 2007). This may be contributory, hypothetically, to a reduced fear in general and potentially a reduced fear of death, an important factor in Joiner's theory for *acquired capability* to commit suicide. Important factors such as availability of means and deterrents are omitted. Also omitted is the reported elevation of testosterone in individuals harboring *T. gondii*, with potential for increased aggression.

(Gaskell et al. 2009), which in turn may lead to an increase in dopamine and the ability to act on suicidal impulses and overcoming an innate fear of death. In additional to increased localization in the amygdala and olfactory bulb, *T. gondii* is localized in the prefrontal cortex (Vyas et al. 2007). Histopathological changes in certain areas of the prefrontal cortex, namely, the ventrolateral prefrontal cortex, have been implicated in suicidal behavior (Mann 2003). It is also possible that the ability of the prefrontal cortex to act as a behavioral "braking mechanism" on impulses and emotions produced in the subcortical structures of the limbic system is inhibited.

19.5 CONCLUSION

T. gondii is one of the most widespread parasites affecting approximately one-third of the population of the world (Montoya and Liesenfeld 2004), and ~60 million individuals in the United States (CDC 2010). Its unique ability to alter immune responses, to manipulate the immune system, and to alter behavior of the host could mediate an increased vulnerability to suicide attempts in those harboring the parasite. Not all individuals infected with *T. gondii* are at risk for suicide; most likely, a combination of predispositions, triggers, availability of means, and absence of protective factors and deterrents would be necessary. The intermediate mechanisms may include heightening of risk factors for suicide such as depression, impulsivity, aggression, arousal, and reduction of fear (especially fear of death). Considering the potential for new prognostic paradigms and etiological preventative interventions, this predictive association deserves future larger, longitudinal, and interventional studies.

ACKNOWLEDGMENTS

This work was supported by the American Foundation for Suicide Prevention (PI Postolache). The authors thank Dr. Gagan Virk for her overall support during the early stage, Dr. Dipika Vaswani during the final stages of this project, and to Vinita J. Ling for editorial and bibliographic assistance.

REFERENCES

Ajioka, J.W., Soldati, D. 2007. *Toxoplasma Molecular and Cellular Biology*. Norfolk, U.K.: Horizon Bioscience.

Aliberti, J. 2005. Host persistence: Exploitation of anti-inflammatory pathways by *T. gondii*. *Nat Rev Immunol*. 5: 162–170.

Andersen, T.F., Madsen, M., Jorgensen, J. et al. 1999. The Danish National Hospital Register. A valuable source of data for modern health sciences. *Dan Med Bull*. 46:263–268.

Arling, T.A., Yolken, R.H., Lapidus, M. et al. 2009. *Toxoplasma gondii* antibody titers and history of suicide attempts in patients with recurrent mood disorders. *J Nerv Ment Dis*. 197: 905–908.

Berdoy, M., Webster, J.P., Macdonald, D.W. 2000. Fatal attraction in rats infected with *T. gondii*. *Proc Biol Sci*. 267: 1591–1594.

Berretta, N., Nistico, R., Bernadi, G. et al. 2008. Synaptic plasticity in the basal ganglia: A similar code for physiological and pathological conditions. *Prog Neurobiol*. 84: 343–362.

Bhat, R., Steinman, L. 2009. Innate and adaptive autoimmunity directed to the central nervous system. *Neuron*. 64: 123–132.

Bonilla, F.A., Oettgen, H.C. 2010. Adaptive immunity. *J Allergy Clin Immunol*. 125: S33–S40.

Buzoni-Gatel, D., Debbabi, H., Mennechet, F.J. et al. 2001. Murine ileitis after intracellular parasite infection is controlled by TGF-beta-producing intraepithelial lymphocytes. *Gastroenterology*. 120: 914–924.

Centers for disease control and prevention. 2010. Toxoplasmosis. http://www.cdc.gov/parasites/toxoplasmosis/; last reviewed November 2, 2010. Accessed February 1, 2011.

Chao, C.C., Hu, S., Gekker, G. et al. 1993. Effects of cytokines on multiplication of *Toxoplasma gondii* in microglial cells. *J Immunol*. 150: 3404–3410.

Chaplin, D. 2010. Overview of the immune response. *J Allergy Clin Immunol*. 125: S3–S23.

Commins, S.P., Borish, L., Steinke, J.W. 2010. Immunologic messenger molecules: Cytokines, interferons, and chemokines. *J Allergy Clin Immunol*. 125: S53–S72.

Courret, N., Darche, S., Sonigo, P. et al. 2006. CD11c- and CD11b-expressing mouse leukocytes transport single *Toxoplasma gondii* tachyzoites to the brain. *Blood*. 107: 309–316.

Dantzer, R. 2009. Cytokine, sickness behavior, and depression. *Immunol Allergy Clin North Am*. 29: 247–264.

Dantzer, R., O'Connor J.C., Freund, G.G. et al. 2008. From inflammation to sickness and depression: When the immune system subjugates the brain. *Nat Rev Neurosci*. 9: 46–56.

Dantzer, R., O'Connor, J.C., Lawson, M.A. et al. 2011. Inflammation-associated depression: From serotonin to kynurenine. *Psychoneuroendocrinology*. 36: 426–436.

Denkers, E.Y., Gazzinelli, R.T. 1998. Regulation and function of T-cell-mediated immunity during *T. gondii* infection. *Clin Microbiol Rev*. 11: 569–588.

Diana, J., Vincent, C., Peyron, F. et al. 2005. *Toxoplasma gondii* regulates recruitment and migration of human dendritic cells via different soluble secreted factors. *Clin Exp Immunol*. 141: 475–484.

Duman, R.S. 2009. Neuronal damage and protection in the pathophysiology and treatment of psychiatric illness: Stress and depression. *Dialogues Clin Neurosci*. 11: 239–255.

Dursun, S.M., Blackburn, J.R., Kutcher, S.P. 2001. An exploratory approach to the serotonergic hypothesis of depression: Bridging the synaptic gap. *Med Hypotheses*. 56: 235–243.

Egan, C.E., Dalton, J.E., Andrew, E.M. et al. 2005. Requirement for the Vgamma1+ subset of peripheral gamma delta T cells in the control of the systemic growth of *Toxoplasma gondii* and infection-induced pathology. *J Immunol*. 175: 8191–8199.

First, M.B. 2003. Standardized evaluation in clinical practice. In *Review of Psychiatry*, J.M. Oldham, M.B. Riba, eds., p. 167. Washington, DC: American Psychiatric Publishing.

Fischer, H.G., Bonifas, U., Reichmann, G. 2000. Phenotype and functions of brain dendritic cells emerging during chronic infection of mice with *Toxoplasma gondii*. *J Immunol*. 164: 4826–4834.

Fischer, H.G., Nitzgen, B., Reichmann, G. et al. 1997. Cytokine responses induced by *Toxoplasma gondii* in astrocytes and microglial cells. *Eur J Immunol*. 27: 1539–1548.

Flegr, J. 2007. Effects of *T. gondii* on human behavior. *Schizophr Bull*. 33: 757–760.

Flegr, J., Havlicek, J., Kodym, P. et al. 2002. Increased risk of traffic accidents in subjects with latent toxoplasmosis: A retrospective case–control study. *BMC Infect Dis*. 2: 11.

Gaskell, E.A., Smith, J.E., Pinney, J.W. 2009. A unique dual activity amino acid hydroxylase in *Toxoplasma gondii*. *PLoS One*. 4: e4801.

Gazzinelli, R.T., Eltoum, I., Wynn, T.A. et al. 1993a. Acute cerebral toxoplasmosis is induced by in vivo neutralization of TNF-alpha and correlates with the down-regulated expression of inducible nitric oxide synthase and other markers of macrophage activation. *J Immunol*. 151: 3672–3681.

Gazzinelli, R.T., Hieny, S., Wynn, T.A. et al. 1993b. Interleukin 12 is required for the T lymphocyte-independent induction of interferon gamma by an intracellular parasite and induces resistance in T-cell-deficient hosts. *Proc Natl Acad Sci U. S. A*. 90: 6115–6119.

Gazzinelli, R.T., Wysocka, M., Hieny, S. et al. 1996. In the absence of endogenous IL-10, mice acutely infected with *Toxoplasma gondii* succumb to a lethal immune response dependent on CD4+ T cells and accompanied by overproduction of IL-12, IFN-gamma and TNF-alpha. *J Immunol*. 157: 798–805.

Gilliet, M., Cao, W., Liu, Y.J. 2008. Plasmacytoid dendritic cells: Sensing nucleic acids in viral infection and autoimmune diseases. *Nat Rev Immunol*. 8: 594–606.

Harris, E.C., Barraclough, B. 1997. Suicide as an outcome for mental disorders. A meta-analysis. *Br J Psychiatry*. 170: 205–228.

Hawton, K., van Heering, K. 2009. Suicide. *Lancet*. 373: 1372–1381.

Hay, J., Aitken, P.P., Hair, D.M. et al. 1984. The effect of congenital *T. gondii* infection on mouse activity and relative preference for exposed areas over a series of trials. *Ann Trop Med Parasitol.* 78: 611–618.

Hunter, C.A., Abrams, J.S., Beaman, M.H. et al. 1993. Cytokine mRNA in the central nervous system of SCID mice infected with *Toxoplasma gondii*: Importance of T-cell-independent regulation of resistance to *T. gondii. Infect Immun.* 61: 4038–4044.

Hunter, C.A., Remington, J.S. 1994. Immunopathogenesis of toxoplasmic encephalitis. *J Infect Dis.* 170: 1057–1067.

Israelski, D.M., Remington, J.S. 1993. Toxoplasmosis in patients with cancer. *Clin Infect Dis.* 17(Suppl 2): S423–S435.

Israelski, D.M., Chmiel, J.S., Poggensee, L. et al. 1993. Prevalence of toxoplasma infection in a cohort of homosexual men at risk of AIDS and toxoplasmic encephalitis. *J Acquir Immune Defic Syndr.* 6: 414–418.

Janelidze, S., Mattei, D., Westrin, Å. et al. 2010. Cytokine levels in the blood may distinguish suicide attempters from depressed patients. *Brain Behav Immun.* 25: 335–339.

Joiner, T.E. Jr, Van Orden, K.A., Witte, T.K. 2009. Main predictions of the interpersonal–psychological theory of suicidal behavior: Empirical tests in two samples of young adults. *J Abnorm Psychol.* 118: 634–646.

Juel, K., Helweg-Larsen, K. 1999. The Danish registers of causes of death. *Dan Med Bull.* 46: 354–357.

Kovacsics, C.E., Gottesman, I.I., Gould, T.D. 2009. Lithium's antisuicidal efficacy: Elucidation of neurobiological targets using endophenotype strategies. *Annu Rev Pharmacol Toxicol.* 49: 175–198.

Lamberton, P.H., Donnelly, C.A., Webster, J.P. 2008. Specificity of the *T. gondii* altered behavior to definitive versus non-definitive host predation risk. *Parasitology.* 135: 1143–1150.

Lebech, M., Petersen, E. 1995. Detection by enzyme immunosorbent assay of *Toxoplasma gondii* IgG antibodies in dried blood spots on PKU-filter paper from newborns. *Scand J Infect Dis.* 27: 259–263.

Lester, D. 2010. Brain parasites and suicide. *Psychol Rep.* 107: 424.

Leweke, F.M., Gerth, C.W., Koethe, D. et al. 2004. Antibodies to infectious agents in individuals with recent onset schizophrenia. *Eur Arch Psychiatry Clin Neurosci.* 254: 4–8.

Liesenfeld, O., Kosek, J., Remington, J.S. et al. 1996. Association of CD4+ T cell-dependent, interferon-gamma-mediated necrosis of the small intestine with genetic susceptibility of mice to peroral infection with *Toxoplasma gondii. J Exp Med.* 184: 597–607.

Lindqvist, D. 2010. *Refining Suicidal Behavior: Rating Scales and Biomarkers.* Lund, Sweden: University Press.

Lindqvist, D., Janelidze, S., Hagell, P. et al. 2009. Interleukin-6 is elevated in the cerebrospinal fluid of suicide attempters and related to symptom severity. *Biol Psychiatry.* 66: 287–292.

Ling, V.J., Lester, D., Mortensen, P.B. et al. 2011. *Toxoplasma gondii* seropositivity and suicide rates in women. *J Nerv Ment Dis.* 199:440–444.

Loftis, J.M., Huckans, M., Morasco, B.J. 2010. Neuroimmune mechanisms of cytokine-induced depression: Current theories and novel treatment strategies. *Neurobiol Dis.* 37: 519–533.

Luangsay, S., Kasper, L.H., Rachinel, N. et al. 2003. CCR5 mediates specific migration of *Toxoplasma gondii*-primed CD8 lymphocytes to inflammatory intestinal epithelial cells. *Gastroenterology.* 125: 491–500.

Mann, J.J. 2003. Neurobiology of suicidal behavior. *Nat Rev Neurosci.* 4: 819–828.

Mann, J.J., Arango, V.A., Avenevoli, S. et al. 2009. Candidate endophenotypes for genetic studies of suicidal behavior. *Biol Psychiatry.* 65: 556–563.

McAllister, K.A., van de Water, J. 2009. Breaking boundaries in neural–immune interactions. *Neuron*. 64: 9–12.

Medzhitov, R. 2001. Toll-like receptors and innate immunity. *Nat Rev Immunol*. 1: 135–145.

Mehta, P.H., Beer, J. 2010. Neural mechanisms of the testosterone-aggression relation: The role of orbitofrontal cortex. *J Cogn Neurosci*. 22: 2357–2368.

Miller, C.M., Boulter, N.R., Ikin, R.J. et al. 2009. The immunobiology of the innate response to *T. gondii*. *Int J Parasitol*. 39: 23–39.

Montoya, J.G., Liesenfeld, O. 2004. Toxoplasmosis. *Lancet*. 363: 1965–1976.

Mordue, D.G., Monroy, F., La Regina, M. et al. 2001. Acute toxoplasmosis leads to lethal overproduction of Th1 cytokines. *J Immunol*. 167: 4574–4584.

Müller, N., Myint, A.-M, Schwarz, M.J. 2009. The impact of neuroimmune dysregulation on neuroprotection and neurotoxicity in psychiatric disorders—Relation to drug treatment. *Dialogues Clin Neurosci*. 11: 319–332.

Munk-Jorgensen, P., Mortensen, P.B. 1997. The Danish Psychiatric Central Register. *Dan Med Bull*. 44: 82–84.

Nassberger, L., Traskman-Bendz, L. 1993. Increased soluble interleukin-2 receptor concentrations in suicide attempters. *Acta Psychiatr Scand*. 88: 48–52.

Nickdel, M.B., Lyons, R.E., Roberts, F., Brombacher, F., Hunter, C.A., Alexander, J., Roberts, C.W. 2004. Intestinal pathology during acute toxoplasmosis is IL-4 dependent and unrelated to parasite burden. *Parasite Immunol*. 26: 75–82.

Niebuhr, D.W., Millikan, A.M., Cowan, D.N. et al. 2008. Selected infectious agents and risk of schizophrenia among U.S. military personnel. *Am J Psychiatry*. 165: 99–106.

Okusaga, O., Langenberg, P., Sleemi, A. et al. 2011. *Toxoplasma gondii* antibody titers and history of suicide attempts in patients with schizophrenia. *Schizophr Res*. 133: 150–155.

Oquendo, M.A., Halberstam, B., Mann, J.J. 2003. Risk factors for suicidal behavior: The utility and limitations of research instruments. In *Standardized Evaluation in Clinical Practice*, M.B. First, ed., pp. 103–130. Arlington, VA: American Psychiatric Publishing.

Osborn, D., Levy, G., Nazareth, I., King, M. 2008. Suicide and severe mental illnesses. Cohort study within the UK general practice research database. *Schizophr Res*. 99: 134–138.

Paraskevas, F. 2004. T Lymphocytes and natural killer cells. In *Wintrobe's Clinical Hematology*, J.P. Greer, J. Foerster, J.N. Lukens, G.M. Rodgers, F. Paraskevas, B. Glader, eds., p. 491. Philadelphia, PA: Lippincott Williams & Wilkins.

Pedersen, C.B., Gotzsche, H., Moller, J.O. et al. 2006. The Danish Civil Registration System. A cohort of eight million persons. *Dan Med Bull*. 53: 441–449.

Pedersen, M.G, Mortensen, P.B., Nørgaard-Pedersen, B. et al. 2012. *Toxoplasma gondii* infection and deliberate self-harm in mothers (submitted).

Pepper, M., Hunter, C.A. 2007. Innate recognition and the regulation of protective immunity to *Toxoplasma gondii*. In *Toxoplasma: Molecular and Cellular Biology*, J.W. Ajioka, D. Soldati, eds. Norfolk, U.K.: Horizon Bioscience.

Postolache, T.T., Mortensen, P.B., Tonelli, L.H. et al. 2010. Seasonal spring peaks of suicide in victims with and without prior history of hospitalization for mood disorders. *J Affect Disord*. 121: 88–93.

Postolache, T.T., Stiller, J.W., Herrell, R. et al. 2005. Tree pollen peaks are associated with increased nonviolent suicide in women. *Mol Psychiatry*. 10: 232–235.

Qin, P., Mortensen, P.B., Waltoft, B.L. et al. 2011. Allergy is associated with suicide completion with a possible mediating role of mood disorder—A population-based study. *Allergy*. 66: 658–664.

Qin, P., Waltoft, B.L., Mortensen, P.B. et al. 2012. Incidents of suicide death are associated with pollen counts in the environment—A study based on Danish longitudinal data (under review).

Raison, C.L., Capuron, L., Miller, A.H. 2006. Cytokines sing the blues: Inflammation and the pathogenesis of depression. *Trends Immunol.* 27: 24–31.

Raison, C.L., Dantzer, R., Kelley, K.W. et al. 2010. CSF concentrations of brain tryptophan and kynurenines during immune stimulation with IFN-alpha: Relationship to CNS immune responses and depression. *Mol Psychiatry.* 15: 393–403.

Raposo, G., Tenza, D., Mecheri, S. et al. 1997. Accumulation of major histocompatibility complex class II molecules in mast cell secretory granules and their release upon degranulation. *Mol Biol Cell.* 8: 2631–2645.

Schroeder, H.W., Cavacini, L. 2010. Structure and function of immunoglobulins. *J Allergy Clin Immunol.* 125: S41–S52.

Schwarcz, R., Pellicciari, R. 2002. Manipulation of brain kynurenines: Glial targets, neuronal effects, and clinical opportunities. *J Pharmacol Exp Ther.* 303: 1–10.

Smiley, S.T., Lanthier, P.A., Couper, K.N., Szaba, F.M., Boyson, J.E., Chen, W., Johnson, L.L. 2005. Exacerbated susceptibility to infection-stimulated immunopathology in CD1d-deficient mice. *J Immunol.* 174: 7904–7911.

Speer, C.A., Dubey, J.P. 2005. Ultrastructural differentiation of *Toxoplasma gondii* schizonts (types B to E) and gamonts in the intestines of cats fed bradyzoites. *Int J Parasitol.* 35: 193–206.

Steiner, J., Bielau, H., Brisch, R. et al. 2008. Immunological aspects in the neurobiology of suicide: Elevated microglial density in schizophrenia and depression is associated with suicide. *J Psychiatr Res.* 42: 151–157.

Stone, K.D., Prussin, C., Metcalfe, D.D. 2010. IgE, mast cells, basophils, and eosinophils. *J Allergy Clin Immunol.* 125: S73–S80.

Sublette, M.E., Galfalvy, H.C., Fuchs, D., Lapidus, M., Grunebaum, M.F., Oquendo, M.A., Mann, J.J., Postolache, T.T. 2011. Plasma kynurenine levels are elevated in suicide attempters with major depressive disorder. *Brain Behav Immun.* 25:1272–1278.

Tansey, M.G. 2010. Inflammation in neuropsychiatric disease (Editorial). *Neurobiol Dis.* 37: 491–492.

Tonelli, L.H., Hoshino, A., Katz, M. et al. 2008a. Acute stress promotes aggressive-like behavior in rats made allergic to tree pollen. *Int J Child Health Hum Dev.* 1: 305–312.

Tonelli, L.H., Katz, M., Kovacsics, C.E. et al. 2009. Allergic rhinitis induces anxiety-like behavior and altered social interaction in rodents. *Brain Behav Immun.* 23: 784–793.

Tonelli, L.H., Stiller, J., Rujescu, D. et al. 2008b. Elevated cytokine expression in the orbitofrontal cortex of victims of suicide. *Acta Psychiatr Scand.* 117: 198–206.

Torrey, E.F., Bartko, J.J., Lun, Z.R. et al. 2007. Antibodies to *T. gondii* in patients with schizophrenia: A meta-analysis. *Schizophr Bull.* 33: 729–736.

Träskman, L., Åsberg, M., Bertilsson, L. et al. 1981. Monoamine metabolites in CSF and suicidal behavior. *Arch Gen Psychiatry.* 38: 631–634.

Turvey, S.E., Broide, D.H. 2010. Innate immunity. *J Allergy.* 25: S24–S32.

Vyas, A., Kim, S.K., Giacomini, N. et al. 2007. Behavioral changes induced by *T. gondii* infection of rodents are highly specific to aversion of cat odors. *Proc Natl Acad Sci U. S. A.* 104: 6442–6447.

Wasserman, E.E., Nelson, K., Rose, N.R. et al. 2009. Infection and thyroid autoimmunity: A seroepidemiologic study of TPOaAb. *Autoimmunity.* 42(5): 439–446.

Webster, J.P. 2007. The effect of *T. gondii* on animal behavior: Playing cat and mouse. *Schizophr Bull.* 33:752–756.

Wilson, M., McAuley, J.B. 1999. Toxoplasma. In *Manual of Clinical Microbiology.* P.R. Murray, E.J. Baron, M.A. Pfaller et al., eds., pp. 1374–1382. Washington, DC: ASM Press.

Wonodi, I., Schwarcz, R. 2010. Cortical kynurenine pathway metabolism: A novel target for cognitive enhancement in schizophrenia. *Schizophr Bull.* 36: 211–218.

World Health Organization (WHO). 1994. *WHO ICD-10: Mental and Behavioural Disorders. Classification and Diagnostic Criteria.* Copenhagen, Denmark: WHO.

Yagmur, F., Yazar, S., Temel, H.O. et al. 2010. May *Toxoplasma gondii* increase suicide attempt-preliminary results in Turkish subjects? *Forensic Sci Int.* 199: 15–17.

Young, J.D., McGwire, B.S. 2005. Infliximab and reactivation of cerebral toxoplasmosis. *N Engl J Med.* 353: 1530–1531.

20 Peripheral Biomarkers for Suicide

Ghanshyam N. Pandey and Yogesh Dwivedi

CONTENTS

20.1 INTRODUCTION

Suicide is a major public health concern. Almost 30,000 people die of suicide each year in the United States, and 1 million people die worldwide. In the teenage population, it is the second leading cause of death. Early identification of suicidal behavior is crucial for its treatment and the prevention of completed suicide. Chemical, behavioral, and psychological risk factors are important in identifying these patients; however, many times they produce a false positive or a false negative diagnosis. Several studies suggest that abnormal biology may be a risk factor for suicidal behavior. Therefore, a combination of biological factors with psychosocial factors might more accurately predict suicidal behavior and identify suicidal patients. It is thus important to develop biomarkers for suicidal behavior. A useful biomarker should not only reflect the psychopathology, in this case suicidal behavior, but it should also be measured in a noninvasive manner.

Because of the inaccessibility of the human brain, initial studies of the biology of suicidal behavior and development of biomarkers focused on peripheral tissues such as cerebrospinal fluid (CSF), urine, platelets, serum, etc. In this chapter, we discuss studies pertaining to the use of peripheral tissues for the studies of pathophysiology of suicidal behavior and their potential role as biomarkers.

Initial studies of peripheral tissues focused on the levels of neurotransmitters and their metabolites in suicidal behavior. Most recent studies have found functional abnormalities of these neurotransmitters—for example, in the receptors and receptor-linked signaling systems for the serotonin (5-hydroxytryptamine [5HT]) and norepinephrine (NE) neurotransmitters.

The intent of this chapter is not to exhaustively review all the studies pertaining to these two monoamines (5HT and NE), their metabolites, their receptors, and the related signal transduction mechanisms in all peripheral tissues. Instead, this chapter focuses on those studies of peripheral tissues that examined and found potential peripheral markers primarily for suicidal behavior.

20.2 SEROTONERGIC FUNCTION IN SUICIDE

Two approaches have been used to examine the role of the serotonergic system in suicidal patients. One approach is to determine 5HT, its metabolite, 5-hydroxyindoleacetic acid (5HIAA), its receptors, $5HT_{2A}$, and its transporter, 5HTT, in platelets. The other approach is to use a neuroendocrine challenge strategy. These strategies are described briefly in the following sections.

20.2.1 Serotonin in Suicidal Behavior

There are several studies that determined the levels of 5HT in whole blood or platelets in patients with mood disorders. However, the results appear to be inconsistent as some investigations find low, no change, or increased levels of 5HT in patients with mood disorders. On the other hand, although the studies of 5HT levels in suicidal patients are few, they are more consistent.

Rao et al. (1998) reported significantly lower blood 5HT concentrations in suicidal patients compared with controls. Spreux-Varoquaux et al. (2001) reported significantly lower plasma 5HT in suicide attempters compared with controls. Almeida-Montes et al. (2000) reported significantly lower serum levels of 5HT in depressed patients who attempted suicide compared with those depressed patients who never attempted suicide. Roggenbach et al. (2007) and Mann et al. (1992a) reported lower levels of 5HT in platelets of suicidal patients who made suicidal attempts compared to nonattempters or nonsuicidal patients. Muck-Seler et al. (1996) and Tyano et al. (2006) found a negative relationship between platelet or plasma 5HT levels and severity of suicidal behavior in patients. In summary, the studies of 5HT in blood and/or platelets in patients with suicidal behavior appear to be more consistent than studies in patients with mood disorders, suggesting a relationship between lower blood 5HT levels and suicidal behavior.

20.2.2 5HIAA in Suicidal Behavior

Earlier evidence suggesting a role for 5HT in the pathophysiology of suicide is derived from studies of 5HIAA, a major metabolite of 5HT, in the CSF of suicidal patients. It was based on the assumption that studies of CSF 5HIAA may reflect similar changes in the 5HT system in the brain. This assumption was strengthened by the observations of Stanley et al. (1985), who measured simultaneously 5HIAA in the CSF and prefrontal cortex (PFC) of autopsied subjects and found a positive correlation between the two. Asberg et al. (1976), in a study of suicidal patients, observed a bimodal distribution in CSF 5HIAA in depressed patients. They also observed that depressed patients with a suicide attempt were found significantly more often in the group with low CSF 5HIAA levels, suggesting a relationship between low 5HIAA and suicidal behavior. Several subsequent studies observed low 5HIAA in the CSF of suicidal patients (Banki et al. 1984; Malone 1997; Mann et al. 1996; Traskman-Bendz et al. 1984).

Several other investigations have failed to find a correlation between suicidal behavior and low CSF 5HIAA (Gardner et al. 1990; Mann et al. 1996; Roy et al. 1985; Roy-Byrne et al. 1983). In a review, Roggenbach et al. (2002) questioned a link between low CSF 5HIAA and suicidality. Also, Asberg (1997) and Kamali et al. (2001) have reviewed the CSF studies in suicidal patients. A meta-analysis of CSF metabolite studies by Lester (1995) found significantly lower levels of CSF 5HIAA in subjects who made prior suicide attempts and those who subsequently committed suicide or made an attempt.

20.2.3 Serotonin Transporter in Suicidal Behavior

The earlier hypothesis of abnormal serotonergic function in depression and suicide was primarily based on the observation of decreased 5HIAA in the CSF in depressed suicidal patients. One other major line of evidence to support the involvement of serotonin in suicide and depression was derived from the observation that serotonin uptake in platelets of depressed patients was decreased compared with that of normal control subjects (Maes and Meltzer 1995; Mann 1999; Owens and Nemeroff 1994; Paul et al. 1981a,b). Since an abnormal serotonin uptake mechanism and decreased serotonin transporter levels have been implicated in the pathophysiology of depression, and since depression is a risk factor for suicide, it is not surprising to find that the serotonin transporter has been quite intensively studied in the postmortem brain of suicide victims.

Serotonin uptake in platelets or in synaptosomes occurs through a protein known as the serotonin transporter, and it was found that this transporter could be labeled by the antidepressant imipramine (Langer et al. 1981). As mentioned earlier, serotonin uptake and serotonin transporter levels are decreased in the platelets of patients with suicidal behavior (Arora and Meltzer 1989b; Crow et al. 1984; Meltzer et al. 1981; Perry et al. 1983; Rausch et al. 1986; Stanley et al. 1982; Tuomisto and Tukiainen 1976). Various other studies then determined the binding of imipramine in the postmortem brain of suicide victims (Gross-Isseroff et al. 1998; Stockmeier 2003). Those studies reported an increase (Meyerson et al. 1982), a decrease (Perry et al. 1983; Stanley et al. 1982), or no change (Arora and Meltzer 1989a; Crow et al. 1984; Gross-Isseroff et al. 1990; Lawrence et al. 1998; Owen et al. 1986) of [^{3}H]imipramine binding in the frontal cortex of suicide victims. Subsequently, other ligands were developed that were deemed to be more specific and selective for the labeling of serotonin transporters, and these include the ligands [^{3}H]paroxetine, [^{3}H]citalopram, and [^{125}I]cyanoimipramine (Arranz et al. 1994; Gurevich and Joyce 1996). Again, as in the case with imipramine binding, the results of studies of the serotonin transporter in suicide victims using other ligands have also been mixed. Although most investigators, including Leake et al. (1991), Laruelle et al. (1993), Joyce et al. (1993), and Arango et al. (1995), found a significant decrease in binding using these ligands, several investigators, including Lawrence et al. (1990, 1998), Hrdina et al. (1993), and Little et al. (1997), did not find any significant difference. Most recently, Bligh-Glover et al. (2000) determined [^{3}H]paroxetine binding in the dorsal raphe (DR) nucleus or its subnuclei in suicide subjects and control subjects. They did not find any significant difference in the binding in the DR between suicide victims with major depression (MD) and normal control subjects.

Receptors-coupled phosphoinositide (PI) signaling system, such as $5HT_{2A}$ receptor, cause the hydrolysis of phosphatidylinositide-4,5-bisphosphate (PIP_2) by the PI-specific enzyme phospholipase C (PLC), resulting in the formation of two second messengers—diacylglycerol (DAG) and inositol-1,4,5-triphosphate (IP_3). DAG activates the phospholipid- and the calcium-dependent enzyme protein kinase C (PKC) and increases its affinity for calcium. PKC subsequently activates several transcription factors, such as cAMP response element-binding protein (CREB) and glycogen synthase kinase (GSK) 3-β, that are part of another pathway known as WNT signaling pathway. PKC also phosphorylates important substrates including myristoylated alanine-rich protein kinase C substrate (MARCKS), and this activation also results in transcription of several important genes such as brain-derived neurotrophic factor (*BDNF*).

Another signaling pathway implicated in mood disorders and suicide is the adenylyl cyclase (AC) signaling system whose components are similar to those of the PI signaling system. Activation of certain receptors, such as β-adrenergic, causes the conversion of ATP to cyclic AMP with the second messenger and activates a phosphorylating enzyme known as protein kinase A (PKA). PKA, like PKC, activates several transcription factors including CREB, finally resulting in the transcription of several important genes. The role of several of these components of the PI and AC signaling system has been studied in the postmortem brain of suicide subjects, and they are briefly discussed later.

20.2.4 $5HT_{2A}$ Receptors in Mood Disorders and Suicide

As discussed earlier, initial studies of serotonin function in MD and suicide were primarily carried out by investigating 5HIAA in the CSF of depressed patients with and without suicidal behavior. Since a decrease in 5HIAA was found in the CSF of depressed and/or suicidal patients compared with control subjects, this implied that the function and the turnover of 5HT may be altered in suicidal subjects. The alterations in 5HT turnover could result not only in an altered level of 5HT in the synaptic cleft but also in changes in the postsynaptic receptors for 5HT. Therefore, it is not surprising that 5HT receptor subtypes have been studied in the platelets and the postmortem brain of depressed and suicidal subjects (Pandey et al. 1995; Stanley and Mann 1983).

Determining 5HT receptors is a complex process because several types of 5HT receptors have been identified (Barnes and Sharp 1999; Hoyer and Martin 1997;

Humphrey et al. 1993; Teitler and Herrick-Davis 1994). Serotonin receptors have been classified into several subtypes, known as $5HT_1$, $5HT_2$, $5HT_3$, $5HT_4$, $5HT_5$, $5HT_6$, and $5HT_7$. These receptors have been further subdivided: $5HT_1$ into $5HT_{1A}$ and $5HT_{1B}$ and $5HT_2$ into $5HT_{2A}$, $5HT_{2B}$, and $5HT_{2C}$ receptors (for a review, see Humphrey et al. 1993). Of these receptors, only the $5HT_{2A}$ receptor subtype has been unequivocally shown to be present in peripheral blood cells (specifically the platelets) of human subjects (Andres et al. 1993; Conn and Sanders-Bush 1986; Elliott and Kent 1989; Kusumi et al. 1991a,b). Therefore, it is not surprising that platelet $5HT_{2A}$ receptors have been extensively studied in patients with MD and in suicide.

$5HT_{2A}$ receptors in platelets have generally been studied by radioligand binding techniques using various ligands, such as [^{125}I]- or [^{3}H]LSD and [^{3}H]ketanserin. The presence of $5HT_{2A}$ receptors in human platelets has been reported by several investigators (Biegon et al. 1987; Hrdina et al. 1995; Pandey et al. 1990, 1995). The pharmacological profile of the $5HT_{2A}$ receptors in platelets appears to be very similar to that observed in the brain (Conn and Sanders-Bush 1986; Kusumi et al. 1991a,b). For example, Elliott and Kent (1989) studied rabbit platelet and brain $5HT_{2A}$ receptors and showed that agonist and antagonist displacement appeared to be very similar in both tissues. Similar observations were made by Ostrowitzki et al. (1993), who observed that the binding characteristics and the expression of $5HT_{2A}$ receptors in the platelets and the brains of pigs appear to be very similar. Additional investigators have studied $5HT_{2A}$ receptors in the platelets of depressed or bipolar patients with and/or without suicidal behavior (Arora and Meltzer 1989b; Biegon et al. 1990a; Pandey and Dwivedi 2006; Pandey et al. 1990). The studies of platelet $5HT_{2A}$ receptors in depression have been reviewed by Mendelson (2000).

In our initial studies of $5HT_{2A}$ receptors in platelets, we studied 23 depressed patients during a drug-free period and 20 normal control subjects using [^{125}I]LSD as the ligand (Pandey et al. 1990). We found that the [^{125}I]LSD binding B_{max} was higher in depressed patients compared with normal control subjects. However, when we divided the depressed patients into those with suicidal behavior and those with no suicidal behavior, we found that the $5HT_{2A}$ receptors were even more elevated in the depressed patients with suicidal behavior compared with normal control subjects and nonsuicidal depressed patients. To examine if the increase in $5HT_{2A}$ receptors in platelets of depressed suicidal patients is independent of diagnosis, we studied $5HT_{2A}$ receptors in patients with different diagnoses—for example, depression, bipolar depression, mania, schizoaffective disorder, or schizophrenia (Pandey et al. 1995). We observed that the mean B_{max} of the total group of suicidal patients was significantly higher than that of nonsuicidal patients or normal control subjects. We also found that B_{max} was significantly higher in 23 suicidal depressed patients, 5 suicidal patients with bipolar illness, 5 suicidal patients with schizophrenia, and 9 suicidal patients with schizoaffective disorders, compared with normal control subjects. This study thus suggested a significant increase in platelet $5HT_{2A}$ receptors in suicidal patients independent of diagnosis. To examine the predictive value of platelet $5HT_{2A}$ receptors, we determined the sensitivity and the specificity of platelet $5HT_{2A}$ receptors. Since the B_{max} of $5HT_{2A}$ receptors for 90% of the comparison subjects was <70 fmol/mg protein, this value was arbitrarily assigned as a cutoff point for the upper limit of normal. Hence, the specificity of normal comparison subjects was 90% according to this cutoff point. This cutoff point has a sensitivity of 55% in predicting suicidal behavior (that is, it correctly identifies 55% of patients as having suicidal behavior) with a specificity of 76%, that is 76% of patients whose B_{max} was below 70 fmol/mg protein were correctly identified as not having suicidal behavior.

Platelet $5HT_{2A}$ receptors in depression have been studied by several investigators. At least nine investigators (Arora and Meltzer 1989b; Biegon et al. 1987, 1990a; Butler and Leonard 1988; Hrdina et al. 1995, 1997; Pandey et al. 1990; Rao et al. 1998; Sheline et al. 1995a) found an increase in platelet $5HT_{2A}$ receptors in depressed patients. There are, however, several studies that did not find differences in the $5HT_{2A}$ receptors between depressed patients and normal control subjects (Bakish et al. 1997; Cowen et al. 1987; Mann et al. 1992b; Pandey et al. 1995; Rosel et al. 1999; Sheline et al. 1995b). A relationship between an increase in $5HT_{2A}$ receptors and suicide was observed by Biegon et al. (1990b) who found that $5HT_{2A}$ receptors were significantly increased in suicidal men. Others have generally found an increase in $5HT_{2A}$ receptors associated with depression (Bakish et al. 1997; Mann et al. 1992b; McBride et al. 1994).

There have been several studies of $5HT_{2A}$ receptors in the postmortem brain of suicide victims. Half of these studies indicated no change (Arranz et al. 1994; Crow et al. 1984; Joyce et al. 1993; Kafka et al. 1985; Lowther et al. 1994; Owen et al. 1986; Rosel et al. 2000; Stockmeier et al. 1997), while the other half indicated an increase in $5HT_{2A}$ receptors very similar to that in the platelets (Arango et al. 1990; Arora and Meltzer 1989a; Hrdina et al. 1993; Laruelle et al. 1993; Mann et al. 1986; Pandey et al. 2002; Stanley and Mann 1983). Most of these studies of $5HT_{2A}$ receptors were carried out using radioligands, such as $[^{125}I]$LSD or $[^{3}H]$ketanserin, none of which were specific in labeling the $5HT_{2A}$ receptors. More recently, we investigated $5HT_{2A}$ receptors in the postmortem brains of teenage suicide victims. We measured the LSD binding, as well as the protein and the mRNA expression of $5HT_{2A}$ receptors in the PFC and hippocampus and found an increase, not only in the binding, but also in the protein and the mRNA expression of $5HT_{2A}$ receptors in teenage suicide subjects compared with normal control subjects.

The increase in $5HT_{2A}$ receptors in depression and/or suicide has also been substantiated by functional studies of $5HT_{2A}$ receptors. $5HT_{2A}$ receptors are linked to the PI system. Mikuni et al. (1991) determined the 5HT-stimulated IP_3 formation in platelets of depressed patients and found that $5HT_{2A}$ receptor responsiveness was increased, which suggests an increase in $5HT_{2A}$ receptor number, as well as in $5HT_{2A}$ receptor function in depression.

In summary, it appears that there is a similarity in the binding, as well as in other pharmacological characteristics, between $5HT_{2A}$ receptors in the platelets and in the human postmortem brain. In general, there appears to be an increase in $5HT_{2A}$ receptors in the platelets of suicidal patients compared with normal control subjects, whereas there may or may not be an increase in $5HT_{2A}$ receptors in depressed patients without suicidal behavior. Increased $5HT_{2A}$ receptors in bipolar patients have also been reported by our lab but appear to be related more to suicidal behavior (Pandey et al. 1995). The specificity and predictive value of the $5HT_{2A}$ receptor is high in predicting suicidal behavior and it may be a useful marker for suicide rather than depression.

20.2.5 Neuroendocrine Studies of Serotonin Function

Neuroendocrine studies, often called the window to the brain, provide another useful method for studying central serotonergic function using peripheral sources. The procedure involves the administration of serotonergic probes, such as 5HT precursors like 5-hydroxytryptophan, or agents that cause the release of 5HT, or antagonists, such as fenfluramine, or a 5HT agonist. A 5HT agonist/antagonist, such as metachlorophenyl piperzine (mCPP), buspirone, or ipsapirone, acts on and stimulates 5HT receptor subtypes. Certain hormones such as prolactin, adrenocorticotropic hormone (ACTH), or cortisol that are released as a result of 5HT acting on the serotonergic system can then be measured. This strategy has been used extensively to study 5HT function in depressed patients. Using 5HT precursor studies, Meltzer and colleagues (Meltzer 1984; Meltzer et al. 1984a–c) measured the 5HT-induced cortisol levels of 40 patients with MD compared with control subjects and found that the cortisol response was significantly increased in patients who had made a suicide attempt or who had a history of suicidal behavior.

The measurement of fenfluramine-induced prolactin or cortisol release in patients with depression is another method used to study central serotonergic function. The administration of fenfluramine causes a release of 5HT from presynaptic neurons and the blockade of 5HT uptake into the presynaptic terminal. The released 5HT acts on the postsynaptic 5HT receptors and causes the release of anterior pituitary hormones, such as prolactin and ACTH. ACTH in turn causes the release of cortisol from the adrenals. Using this method of study, several investigators have observed a decrease in prolactin release following the administration of DL-fenfluramine in depressed patients, although some studies have not observed such differences. Coccaro et al. (1996) found that the prolactin response to DL-fenfluramine was significantly decreased in depressed patients and personality disorder patients who had a history of suicide attempts. Similar results were reported by Malone et al. (1996) who also observed decreased prolactin response to fenfluramine in patients with a history of suicide attempts. O'Keane and Dinan (1991) studied the fenfluramine-induced prolactin response in depressed patients and control subjects and found reduced prolactin and cortisol response in patients with depression compared with control subjects. More recently, Cleare et al. (1996) found a decreased prolactin and cortisol response in patients with depression compared with normal control subjects. However, they also observed that patients with a history of suicide attempts had a lower cortisol response than normal control subjects and patients without a history of suicide attempts. Both mCPP and ipsaperone have also been studied in children with depression with mixed results.

20.3 NORADRENERGIC FUNCTION IN SUICIDE

Initial studies of NE function associated with suicidal behavior focused on determining the levels of NE or its metabolite, 3-methoxy-4-hydroxyphenylglycol (MHPG), in the CSF or urine of suicidal patients. Since the major rate-limiting factor in the synthesis of NE from phenylalanine is the enzyme tyrosine hydroxylase (TH), the role of TH has also been studied in suicide. On the other hand, NE produces its functional effects by interacting with its various receptors and initiating a signaling cascade primarily through the AC system, and hence, the receptors for NE, β- and α-adrenoceptor (AD) as well as components of the AC, have also been studied to some extent either in the postmortem brain of suicide victims or in the peripheral cells obtained from suicidal patients. In this chapter, we briefly review each of these studies of NE function in suicide. We first discuss human patient studies followed by postmortem studies.

The noradrenergic (NA) function in suicide has been primarily studied using two different strategies. Strategy 1 involves the studies of NA function in suicidal patients by determining either the levels of NE or its metabolite MHPG in the urine, plasma, or CSF of suicidal patients or, in some cases, using a neuroendocrine strategy. Strategy 2 involves determining NA receptors, primarily α- and β-AD in the postmortem brain obtained from suicidal subjects, along with the enzyme TH, which is a rate-limiting step in the synthesis of NE from phenylalanine. Functional consequences of NA receptor activation have also been studied by determining the various components of the AC signaling system to which these ADs are linked.

20.3.1 Norepinephrine and MHPG in Suicide

MHPG is a major metabolite of NE, and it has been suggested by some investigators that the level of MHPG in urine is a major index of NE function in the brain. MHPG has therefore been studied in the urine of depressed patients and hence MHPG, as well as NE, has also been studied in the urine of suicidal patients. Ostroff et al. (1985) observed that three depressed patients who had made serious suicide attempts had a significantly lower 24 h urine NE to epinephrine (EPI) ratio than 19 depressed patients who had made no suicide attempts. A comprehensive study of NE and MHPG was conducted by the National Institute of Mental Health (NIMH) collaborative study, and it was reported that depressed patients with suicidal behavior showed decreased urinary secretion of MHPG and lower plasma level of MHPG compared with those patients with no suicidal behavior (Secunda et al. 1986). Brown et al. (1979) observed that suicidal patients with personality disorders have significantly higher CSF levels of NE and higher levels of MHPG compared with nonsuicidal patients with personality disorders. They also observed a significant positive correlation between aggression scores and CSF MHPG levels. A similar significant correlation between CSF MHPG levels and several components of suicidality was also reported in other studies (Agren 1980b; Agren et al. 1983; Roy et al. 1985, 1989). Roy et al. (1985) examined the relationship between suicidal behavior and NE function and compared depressed patients, who either had or had never attempted suicide, and controls. They found that unipolar patients who had attempted suicide had higher plasma NE levels than control subjects, although there were no significant differences in the plasma MHPG levels. They also found that the patients who never attempted suicide had significantly higher plasma levels of NE than patients who had attempted suicide. However, they did not find a significant relationship of NE abnormalities with suicidal behavior as far as MHPG and NE levels were concerned.

Some earlier studies also determine the levels of NE in the postmortem brain of suicide victims, but these have been inconclusive (Beskow et al. 1976).

Some investigators have studied the relationship of NE function with the lethality of suicide attempts. Sher et al. (2006) found that the MHPG levels were negatively correlated with the maximum lethality of suicide attempts in bipolar patients. Roy et al. (1985) also studied the relationship between the lethality of suicide attempts and MHPG in suicidal patients; however, they did not find any significant relationship between the levels of MHPG in the CSF of 27 chronic schizophrenic patients, and they found no significant differences between patients with violent or nonviolent suicide attempts and CSF concentrations of MHPG. On the other hand, Garvey et al. (1994) found a weak trend for current but not past suicide lethality to be associated with decreased MHPG levels. In a more recent study, Tripodianakis et al. (2002) determined the levels of urinary MHPG in 54 subjects with adjustment disorder, 25 with depression, 16 with schizophrenia, and 16 with personality disorder. They found that urinary MHPG was significantly higher in all diagnostic groups compared with controls. Urinary MHPG was found to be higher in all diagnostic groups who had attempted suicide; however, no difference in MHPG level was found between those subjects who attempted violent suicide and those with nonviolent attempts.

20.4 ABNORMAL HPA AXIS AS A BIOMARKER FOR SUICIDE

An abnormal HPA axis in depression is one of the most consistent findings in biological psychiatry. Most patients with depression have been shown to have increased concentrations of cortisol in their plasma and CSF, increased cortisol response to ACTH, and a deficient feedback mechanism, as evidenced by an abnormal dexamethasone suppression test (DST), and enlarged pituitary and adrenal glands. The release of CRF from the paraventricular nucleus (PVN) of the hypothalamus causes the release of ACTH from the pituitary, which stimulates the production of glucocorticoids (cortisol in humans, corticosterone in animals) from the adrenals. Glucocorticoids regulate the HPA axis through a negative feedback mechanism while binding to soluble glucocorticoid receptors in the pituitary and the hypothalamus and inhibiting the release of CRF and ACTH.

Since depression and stress are important risk factors for suicide and both regulate HPA function, the role of HPA axis in suicide has also been investigated. In a study of DST nonsuppressors, Yerevanian et al. (2004) found that DST nonsuppressors were significantly more likely to commit and complete suicide than DST suppressors. Coryell and Schlesser (1981) have studied the relationship between DST and suicidal behavior. In a study of 243 depressed patients, they found that 4 patients who eventually committed suicide were nonsuppressors, as opposed to one suppressor who committed suicide. In a subsequent study, Coryell and Schlesser (2001) followed 78 inpatients over a 15 year period. Based on this follow-up study, they concluded that the estimated risk for suicide DST in nonsuppressors was 26.8% compared to only 2.9% among patients who had normal DST results.

Other investigators have also found an association between DST nonsuppression and suicide (Coccaro et al. 1989; Coryell and Schlesser 1981; Mann 1995; Mann et al. 1989; Targum et al. 1983). In a meta-analysis, van Heeringen et al. (2000) found that suicide completions, but not attempts, were associated with DST nonsuppression. They found that DST nonsuppression was a risk factor for suicide. The findings on DST nonsuppression and suicide attempts are inconsistent. Some find an association between suicide attempts and DST nonsuppression (Coryell and Schlesser 2001; Meltzer and Lowy 1987), while others do not (Agren 1980a; Asberg et al. 1984, 1976). It thus appears that HPA axis abnormally may be more strongly related to suicide completers than attempters, suggesting the importance of studying HPA axis abnormalities in the brain of suicide victims. HPA axis abnormalities therefore appear to be potential markers for suicide, specifically the DST studies.

20.5 OTHER POTENTIAL BIOMARKERS IN SUICIDE: BDNF

Although several peripheral biomarkers were studied in depression, the studies of biomarkers in suicide are limited. BDNF has been shown to be altered in the serum and platelets of patients with mood disorders (Karege et al. 2002; Pandey et al. 2010). The role of peripheral BDNF levels in suicide has been reported by one investigator (Deveci et al. 2007). Kim et al. (2007) determined plasma BDNF levels in suicidal and nonsuicidal depressed patients and normal controls and found that plasma BDNF levels were significantly lower in suicidal depressed patients compared with nonsuicidal depressed patients and normal controls. The low plasma BDNF was found to have a high degree of sensitivity and specificity in predicting suicidal behavior in these depressed patients. It has also been reported that brain BDNF levels are altered in the postmortem brain of suicide victims (Dwivedi et al. 2003; Pandey et al. 2008).

20.6 PERIPHERAL CELLS AS NEUROPROBES AND BIOMARKERS: USEFULNESS FOR PSYCHIATRIC DISORDER STUDIES

Access to the human brain for studying either the pathophysiology or clinical response in psychiatric disorders is limited. Studies of postmortem brain samples, functional neuroimaging, as well as neuroimaging techniques provide important information with regard to the cellular biochemistry, either of receptors or the signaling mechanism. Biochemical studies have also been performed in peripheral tissues, including blood cells, CSF, plasma, and urine. Although limited, they have been useful in providing information with regard to pathophysiological abnormalities. Such studies also present possible prognostic and diagnostic markers in psychiatric disorders.

Leukocytes, platelets, and to a certain degree red cells are highly useful sources for studying many cellular mechanisms. The lymphocytes (white cells) have become particularly important and interesting because of the role they play in immunological function, which has been implicated in MD, and in their communication with the CNS. Their role as a central probe has been elegantly reviewed by Gladkevich et al. (2004). The usefulness of lymphocytes as a neuroprobing marker in psychiatric disorders has especially become evident concerning (1) the role they play in immune response (cytokine production) alterations in production of different lymphocytes and (2) their role in HPA axis dysfunction and neuroendocrine regulation. Because of the page limitation, these two aspects cannot be discussed in detail. In brief, abnormalities of cytokines in depression have been reported by many investigators. Lymphocytes have also been suggested to be important in studying gene expression in various psychiatric disorders (Hamalainen et al. 2001). Abnormalities in mRNA of many receptors and signal transduction molecules have been demonstrated by us and other investigators in depression and suicide. Many of these genes are also expressed in lymphocytes, and several of these genes have similar characteristics in both brain and lymphocytes. This presents another use of lymphocytes in studying gene expression in psychiatric disorders.

The other peripheral marker used extensively in studies of psychiatric disorders are the platelets, as reviewed by Plein and Berk (2001). Platelets have been used for studies of neurotransmitter function in patients, including monoamine oxidase, adrenergic receptors, $5HT_{2A}$ receptors, and BDNF. Although these studies provide important information, the significance of these studies and of the use of platelets as models of CNS function in psychiatric patients is less clear. Blood platelets exhibit various components that are similar to those in the CNS neurotransmitter system—for example, intracellular levels for biogenic amines, metabolizing enzymes, such as monoamine oxidase, and several other membrane receptors. However, the organization of the CNS 5HT system is much more complex and is modulated by other neurotransmitter systems. One of the most compelling similarities between the receptors, especially the 5HT receptors, and serotonin uptake is the observation that the proteins for the human platelet serotonin uptake site and the brain serotonin transporter are identical in structure and are encoded by the same single-copy gene assigned to chromosome-17 (Lesch et al. 1993). In summary, although there may be some dissimilarity between the characteristics of the receptor systems and signaling systems between platelets and the brain, there are many similarities, and platelets have the potential to be very useful for studying these systems in psychiatric disorders and may possibly result in diagnostic and prognostic biomarkers.

20.7 CONCLUSION AND FUTURE DIRECTIONS

The diagnosis as well as the therapeutic response in patients with psychiatric disorders is primarily based on clinical assessments. This often provides false-positive or false-negative results. A simple, easily accessible biomarker is necessary to complement the clinical assessments for both diagnostic and prognostic purposes. A biomarker is an indicator of a normal or pathological process or pharmacological response to a therapeutic intervention. Biomarkers may be highly useful in early identification of illness (e.g., depression or bipolar illness) or in identification of symptoms (e.g., suicidal behavior) for the purpose of early intervention and prevention of suicide. Biomarkers can therefore be used as a diagnostic tool for the identification of individuals with certain illnesses, such as bipolar illness or depression. They can also be used as indicators of disease progression and/or prognosis, and for monitoring and predicting clinical response to a particular therapeutic intervention.

Because of the inaccessibility of the human brain, peripheral biomarkers (e.g., in serum, blood cells, saliva, or urine) are more accessible than the brain or CSF. Abnormalities of the serotonergic and NA system in peripheral tissues have been studied in great detail, as reviewed in this chapter. Many of these results have been quite encouraging, but many of the studies were not only conflicting but produced inconsistent results. Nonetheless, these studies do indicate some potentially useful biomarkers for these illnesses. Some of the peripheral biomarkers discussed in this chapter, such as platelet $5HT_{2A}$ receptors and CSF 5HIAA, have great potential as biomarkers for suicidal behavior. Other useful biomarkers, such as HPA axis components and cytokines, have great potential as biomarkers for suicide, but need to be studied further.

REFERENCES

Agren, H. 1980a. Symptom patterns in unipolar and bipolar depression correlating with monoamine metabolites in the cerebrospinal fluid: I. General patterns. *Psychiatry Res* 3:211–223.

Agren, H. 1980b. Symptom patterns in unipolar and bipolar depression correlating with monoamine metabolites in the cerebrospinal fluid: II. Suicide. *Psychiatry Res* 3:225–236.

Agren, H., Osterberg, B., Niklasson, F., Franzen, O. 1983. Depression and somatosensory evoked potentials: I. Correlations between SEP and monoamine and purine metabolites in CSF. *Biol Psychiatry* 18:635–649.

Almeida-Montes, L.G., Valles-Sanchez, V., Moreno-Aguilar, J., Chavez-Balderas, R.A., Garcia-Marin, J.A., Cortes Sotres, J.F., Hheinze-Martin, G. 2000. Relation of serum cholesterol, lipid, serotonin and tryptophan levels to severity of depression and to suicide attempts. *J Psychiatry Neurosci* 25:371–377.

Andres, A.H., Rao, M.L., Ostrowitzki, S., Entzian, W. 1993. Human brain cortex and platelet serotonin2 receptor binding properties and their regulation by endogenous serotonin. *Life Sci* 52:313–321.

Arango, V., Ernsberger, P., Marzuk, P.M., Chen, J.S., Tierney, H., Stanley, M., Reis, D.J., Mann, J.J. 1990. Autoradiographic demonstration of increased serotonin $5\text{-}HT_2$ and beta-adrenergic receptor binding sites in the brain of suicide victims. *Arch Gen Psychiatry* 47:1038–1047.

Arango, V., Underwood, M.D., Gubbi, A.V., Mann, J.J. 1995. Localized alterations in pre- and postsynaptic serotonin binding sites in the ventrolateral prefrontal cortex of suicide victims. *Brain Res* 688:121–133.

Arora, R.C., Meltzer, H.Y. 1989a. ^{3}H-imipramine binding in the frontal cortex of suicides. *Psychiatry Res* 30:125–135.

Arora, R.C., Meltzer, H.Y. 1989b. Increased serotonin2 ($5\text{-}HT_2$) receptor binding as measured by 3H-lysergic acid diethylamide (^{3}H-LSD) in the blood platelets of depressed patients. *Life Sci* 44:725–734.

Arranz, B., Eriksson, A., Mellerup, E., Plenge, P., Marcusson, J. 1994. Brain $5\text{-}HT_{1A}$, $5\text{-}HT_{1D}$, and $5\text{-}HT_2$ receptors in suicide victims. *Biol Psychiatry* 35:457–463.

Asberg, M. 1997. Neurotransmitters and suicidal behavior. The evidence from cerebrospinal fluid studies. *Ann N Y Acad Sci* 836:158–181.

Asberg, M., Bertilsson, L., Martensson, B., Scalia-Tomba, G.P., Thoren, P., Traskman-Bendz, L. 1984. CSF monoamine metabolites in melancholia. *Acta Psychiatr Scand* 69:201–219.

Asberg, M., Traskman, L., Thoren, P. 1976. 5-HIAA in the cerebrospinal fluid. A biochemical suicide predictor? *Arch Gen Psychiatry* 33:1193–1197.

Bakish, D., Cavazzoni, P., Chudzik, J., Ravindran, A., Hrdina, P.D. 1997. Effects of selective serotonin reuptake inhibitors on platelet serotonin parameters in major depressive disorder. *Biol Psychiatry* 41:184–190.

Banki, C.M., Arato, M., Papp, Z., Kurcz, M. 1984. Biochemical markers in suicidal patients. Investigations with cerebrospinal fluid amine metabolites and neuroendocrine tests. *J Affect Disord* 6:341–350.

Barnes, N.M., Sharp, T. 1999. A review of central 5-HT receptors and their function. *Neuropharmacology* 38:1083–1152.

Beskow, J., Gottfries, C.G., Roos, B.E., Winblad, B. 1976. Determination of monoamine and monoamine metabolites in the human brain: Post mortem studies in a group of suicides and in a control group. *Acta Psychiatr Scand* 53:7–20.

Biegon, A., Essar, N., Israeli, M., Elizur, A., Bruch, S., Bar-Nathan, A.A. 1990a. Serotonin 5-HT_2 receptor binding on blood platelets as a state dependent marker in major affective disorder. *Psychopharmacology (Berl)* 102:73–75.

Biegon, A., Grinspoon, A., Blumenfeld, B., Bleich, A., Apter, A., Mester, R. 1990b. Increased serotonin 5-HT_2 receptor binding on blood platelets of suicidal men. *Psychopharmacology (Berl)* 100:165–167.

Biegon, A., Weizman, A., Karp, L., Ram, A., Tiano, S., Wolff, M. 1987. Serotonin 5-HT_2 receptor binding on blood platelets—A peripheral marker for depression? *Life Sci* 41:2485–2492.

Bligh-Glover, W., Kolli, T.N., Shapiro-Kulnane, L., Dilley, G.E., Friedman, L., Balraj, E., Rajkowska, G., Stockmeier, C.A. 2000. The serotonin transporter in the midbrain of suicide victims with major depression. *Biol Psychiatry* 47:1015–1024.

Brown, G.L., Goodwin, F.K., Ballenger, J.C., Goyer, P.F., Major, L.F. 1979. Aggression in humans correlates with cerebrospinal fluid amine metabolites. *Psychiatry Res* 1:131–139.

Butler, J., Leonard, B.E. 1988. The platelet serotonergic system in depression and following sertraline treatment. *Int Clin Psychopharmacol* 3:343–347.

Cleare, A.J., Murray, R.M., O'Keane, V. 1996. Reduced prolactin and cortisol responses to D-fenfluramine in depressed compared to healthy matched control subjects. *Neuropsychopharmacology* 14:349–354.

Coccaro, E.F., Berman, M.E., Kavoussi, R.J., Hauger, R.L. 1996. Relationship of prolactin response to D-fenfluramine to behavioral and questionnaire assessments of aggression in personality-disordered men. *Biol Psychiatry* 40:157–164.

Coccaro, E.F., Siever, L.J., Klar, H.M., Maurer, G., Cochrane, K., Cooper, T.B., Mohs, R.C., Davis, K.L. 1989. Serotonergic studies in patients with affective and personality disorders. Correlates with suicidal and impulsive aggressive behavior. *Arch Gen Psychiatry* 46:587–599.

Conn, P.J., Sanders-Bush, E. 1986. Regulation of serotonin-stimulated phosphoinositide hydrolysis: Relation to the serotonin 5-HT-2 binding site. *J Neurosci* 6:3669–3675.

Coryell, W., Schlesser, M.A. 1981. Suicide and the dexamethasone suppression test in unipolar depression. *Am J Psychiatry* 138:1120–1121.

Coryell, W., Schlesser, M. 2001. The dexamethasone suppression test and suicide prediction. *Am J Psychiatry* 158:748–753.

Cowen, P.J., Charig, E.M., Fraser, S., Elliott, J.M. 1987. Platelet 5-HT receptor binding during depressive illness and tricyclic antidepressant treatment. *J Affect Disord* 13:45–50.

Crow, T.J., Cross, A.J., Cooper, S.J., Deakin, J.F., Ferrier, I.N., Johnson, J.A., Joseph, M.H., Owen, F., Poulter, M., Lofthouse, R. 1984. Neurotransmitter receptors and monoamine metabolites in the brains of patients with Alzheimer-type dementia and depression, and suicides. *Neuropharmacology* 23:1561–1569.

Deveci, A., Aydemir, O., Taskin, O., Taneli, F., Esen-Danaci, A. 2007. Serum BDNF levels in suicide attempters related to psychosocial stressors: A comparative study with depression. *Neuropsychobiology* 56:93–97.

Dwivedi, Y., Rizavi, H.S., Conley, R.R., Roberts, R.C., Tamminga, C.A., Pandey, G.N. 2003. Altered gene expression of brain-derived neurotrophic factor and receptor tyrosine kinase B in postmortem brain of suicide subjects. *Arch Gen Psychiatry* 60:804–815.

Elliott, J.M., Kent, A. 1989. Comparison of [^{125}I]iodolysergic acid diethylamide binding in human frontal cortex and platelet tissue. *J Neurochem* 53:191–196.

Gardner, D.L., Lucas, P.B., Cowdry, R.W. 1990. CSF metabolites in borderline personality disorder compared with normal controls. *Biol Psychiatry* 28:247–254.

Garvey, M.J., Hollon, S.D., Tuason, V.B. 1994. Relationship between 3-methoxy-4-hydroxyphenylglycol and suicide. *Neuropsychobiology* 29:112–116.

Gladkevich, A., Kauffman, H.F., Korf, J. 2004. Lymphocytes as a neural probe: Potential for studying psychiatric disorders. *Prog Neuropsychopharmacol Biol Psychiatry* 28:559–576.

Gross-Isseroff, R., Biegon, A., Voet, H., Weizman, A. 1998. The suicide brain: A review of postmortem receptor/transporter binding studies. *Neurosci Biobehav Rev* 22:653–661.

Gross-Isseroff, R., Dillon, K.A., Fieldust, S.J., Biegon, A. 1990. Autoradiographic analysis of alpha 1-noradrenergic receptors in the human brain postmortem. Effect of suicide. *Arch Gen Psychiatry* 47:1049–1053.

Gurevich, E.V., Joyce, J.N. 1996. Comparison of [^{3}H]paroxetine and [^{3}H]cyanoimipramine for quantitative measurement of serotonin transporter sites in human brain. *Neuropsychopharmacology* 14:309–323.

Hamalainen, H., Zhou, H., Chou, W., Hashizume, H., Heller, R., Lahesmaa, R. 2001. Distinct gene expression profiles of human type 1 and type 2 T helper cells. *Genome Biol* 2(7):RESEARCH0022.

Hoyer, D., Martin, G. 1997. 5-HT receptor classification and nomenclature: Towards a harmonization with the human genome. *Neuropharmacology* 36:419–428.

Hrdina, P.D., Bakish, D., Chudzik, J., Ravindran, A., Lapierre, Y.D. 1995. Serotonergic markers in platelets of patients with major depression: Upregulation of 5-HT$_2$ receptors. *J Psychiatry Neurosci* 20:11–19.

Hrdina, P.D., Bakish, D., Ravindran, A., Chudzik, J., Cavazzoni, P., Lapierre, Y.D. 1997. Platelet serotonergic indices in major depression: Up-regulation of 5-HT$_{2A}$ receptors unchanged by antidepressant treatment. *Psychiatry Res* 66:73–85.

Hrdina, P.D., Demeter, E., Vu, T.B., Sotonyi, P., Palkovits, M. 1993. 5-HT uptake sites and 5-HT$_2$ receptors in brain of antidepressant-free suicide victims/depressives: Increase in 5-HT$_2$ sites in cortex and amygdala. *Brain Res* 614:37–44.

Humphrey, P.P., Hartig, P., Hoyer, D. 1993. A proposed new nomenclature for 5-HT receptors. *Trends Pharmacol Sci* 14:233–236.

Joyce, J.N., Shane, A., Lexow, N., Winokur, A., Casanova, M.F., Kleinman, J.E. 1993. Serotonin uptake sites and serotonin receptors are altered in the limbic system of schizophrenics. *Neuropsychopharmacology* 8:315–336.

Kafka, M.S., Siever, L.J., Nurnberger, J.I., Uhde, T.W., Targum, S., Cooper, D.M., van Kammen, D.P., Tokola, N.S. 1985. Platelet alpha-adrenergic receptor function in affective disorders and schizophrenia. *Psychopharmacol Bull* 21:599–602.

Kamali, M., Oquendo, M.A., Mann, J.J. 2001. Understanding the neurobiology of suicidal behavior. *Depress Anxiety* 14:164–176.

Karege, F., Perret, G., Bondolfi, G., Schwald, M., Bertschy, G., Aubry, J.M. 2002. Decreased serum brain-derived neurotrophic factor levels in major depressed patients. *Psychiatry Res* 109:143–148.

Kim, Y.K., Lee, H.P., Won, S.D., Park, E.Y., Lee, H.Y., Lee, B.H., Lee, S.W., Yoon, D., Han, C., Kim, D.J., Choi, S.H. 2007. Low plasma BDNF is associated with suicidal behavior in major depression. *Prog Neuropsychopharmacol Biol Psychiatry* 31:78–85.

Kusumi, I., Koyama, T., Yamashita, I. 1991a. Effect of various factors on serotonin-induced Ca2+ response in human platelets. *Life Sci* 48:2405–412.

Kusumi, I., Koyama, T., Yamashita, I. 1991b. Serotonin-stimulated Ca^{2+} response is increased in the blood platelets of depressed patients. *Biol Psychiatry* 30:310–312.

Langer, S.Z., Javoy-Agid, F., Raisman, R., Briley, M., Agid, Y. 1981. Distribution of specific high-affinity binding sites for [^{3}H]imipramine in human brain. *J Neurochem* 37:267–271.

Laruelle, M., Abi-Dargham, A., Casanova, M.F., Toti, R., Weinberger, D.R., Kleinman, J.E. 1993. Selective abnormalities of prefrontal serotonergic receptors in schizophrenia. A postmortem study. *Arch Gen Psychiatry* 50:810–818.

Lawrence, K.M., De Paermentier, F., Cheetham, S.C., Crompton, M.R., Katona, C.L., Horton, R.W. 1990. Brain 5-HT uptake sites, labelled with [^{3}H]paroxetine, in antidepressant-free depressed suicides. *Brain Res* 526:17–22.

Lawrence, K.M., Kanagasundaram, M., Lowther, S., Katona, C.L., Crompton, M.R., Horton, R.W. 1998. [^{3}H]Imipramine binding in brain samples from depressed suicides and controls: 5-HT uptake sites compared with sites defined by desmethylimipramine. *J Affect Disord* 47:105–112.

Leake, A., Fairbairn, A.F., McKeith, I.G., Ferrier, I.N. 1991. Studies on the serotonin uptake binding site in major depressive disorder and control post-mortem brain: Neurochemical and clinical correlates. *Psychiatry Res* 39:155–165.

Lesch, K.P., Wolozin, B.L., Murphy, D.L., Reiderer, P. 1993. Primary structure of the human platelet serotonin uptake site: Identity with the brain serotonin transporter. *J Neurochem* 60:2319–2322.

Lester, D. 1995. The concentration of neurotransmitter metabolites in the cerebrospinal fluid of suicidal individuals: A meta-analysis. *Pharmacopsychiatry* 28:45–50.

Little, K.Y., McLauglin, D.P., Ranc, J., Gilmore, J., Lopez, J.F., Watson, S.J., Carroll, F.I., Butts, J.D. 1997. Serotonin transporter binding sites and mRNA levels in depressed persons committing suicide. *Biol Psychiatry* 41:1156–1164.

Lowther, S., De Paermentier, F., Crompton, M.R., Katona, C.L., Horton, R.W. 1994. Brain 5-HT_2 receptors in suicide victims: Violence of death, depression and effects of antidepressant treatment. *Brain Res* 642:281–289.

Maes, M., Meltzer, H.Y. 1995. *The Serotonin Hypothesis of Major Depression*. New York: Raven Press.

Malone, K.M. 1997. Pharmacotherapy of affectively ill suicidal patients. *Psychiatr Clin North Am* 20:613–624.

Malone, K.M., Corbitt, E.M., Li, S., Mann, J.J. 1996. Prolactin response to fenfluramine and suicide attempt lethality in major depression. *Br J Psychiatry* 168:324–329.

Mann, J.J. 1995. Violence and aggression. In: Bloom, F.E., Kupfer, D., (eds.) *Psychopharmacology: The Fourth Generation of Progress*. New York: Raven Press, pp. 1919–1928.

Mann, J.J. 1999. Role of the serotonergic system in the pathogenesis of major depression and suicidal behavior. *Neuropsychopharmacology* 21:99S–105S.

Mann, J.J., Malone, K.M., Psych, M.R., Sweeney, J.A., Brown, R.P., Linnoila, M., Stanley, B., Stanley, M. 1996. Attempted suicide characteristics and cerebrospinal fluid amine metabolites in depressed inpatients. *Neuropsychopharmacology* 15:576–586.

Mann, J.J., Marzuk, P.M., Arango, V., McBride, P.A., Leon, A.C., Tierney, H. 1989. Neurochemical studies of violent and nonviolent suicide. *Psychopharmacol Bull* 25:407–413.

Mann, J.J., McBride, P.A., Anderson, G.M., Mieczkowski, T.A. 1992a. Platelet and whole blood serotonin content in depressed inpatients: Correlations with acute and life-time psychopathology. *Biol Psychiatry* 32:243–257.

Mann, J.J., McBride, P.A., Brown, R.P., Linnoila, M., Leon, A.C., DeMeo, M., Mieczkowski, T., Myers, J.E., Stanley, M. 1992b. Relationship between central and peripheral serotonin indexes in depressed and suicidal psychiatric inpatients. *Arch Gen Psychiatry* 49:442–446.

Mann, J.J., Stanley, M., McBride, P.A., McEwen, B.S. 1986. Increased serotonin2 and beta-adrenergic receptor binding in the frontal cortices of suicide victims. *Arch Gen Psychiatry* 43:954–959.

McBride, P.A., Brown, R.P., DeMeo, M., Keilp, J., Mieczkowski, T., Mann, J.J. 1994. The relationship of platelet 5-HT_2 receptor indices to major depressive disorder, personality traits, and suicidal behavior. *Biol Psychiatry* 35:295–308.

Meltzer, H.Y. 1984. Serotonergic function in the affective disorders: The effect of antidepressants and lithium on the 5-hydroxytryptophan-induced increase in serum cortisol. *Ann N Y Acad Sci* 430:115–137.

Meltzer, H.Y., Arora, R.C., Baber, R., Tricou, B.J. 1981. Serotonin uptake in blood platelets of psychiatric patients. *Arch Gen Psychiatry* 38:1322–1326.

Meltzer, H.Y., Lowy, M.T. 1987. The serotonin hypothesis of depression. In: Meltzer, H., (ed.) *Psychopharmacology: The Third Generation of Progress*. New York: Raven Press, pp. 609–615.

Meltzer, H.Y., Lowy, M., Robertson, A., Goodnick, P., Perline, R. 1984a. Effect of 5-hydroxytryptophan on serum cortisol levels in major affective disorders. III. Effect of antidepressants and lithium carbonate. *Arch Gen Psychiatry* 41:391–397.

Meltzer, H.Y., Perline, R., Tricou, B.J., Lowy, M., Robertson, A. 1984b. Effect of 5-hydroxytryptophan on serum cortisol levels in major affective disorders. II. Relation to suicide, psychosis, and depressive symptoms. *Arch Gen Psychiatry* 41:379–387.

Meltzer, H.Y., Umberkoman-Wiita, B., Robertson, A., Tricou, B.J., Lowy, M., Perline, R. 1984c. Effect of 5-hydroxytryptophan on serum cortisol levels in major affective disorders. I. Enhanced response in depression and mania. *Arch Gen Psychiatry* 41:366–374.

Mendelson, S.D. 2000. The current status of the platelet 5-HT(2A) receptor in depression. *J Affect Disord* 57:13–24.

Meyerson, L.R., Wennogle, L.P., Abel, M.S., Coupet, J., Lippa, A.S., Rauh, C.E., Beer, B. 1982. Human brain receptor alterations in suicide victims. *Pharmacol Biochem Behav* 17:159–163.

Mikuni, M., Kusumi, I., Kagaya, A., Kuroda, Y., Mori, H., Takahashi, K. 1991. Increased 5-HT-2 receptor function as measured by serotonin-stimulated phosphoinositide hydrolysis in platelets of depressed patients. *Prog Neuropsychopharmacol Biol Psychiatry* 15:49–61.

Muck-Seler, D., Jakovljevic, M., Pivac, N. 1996. Platelet 5-HT concentrations and suicidal behaviour in recurrent major depression. *J Affect Disord* 39:73–80.

O'Keane, V., Dinan, T.G. 1991. Prolactin and cortisol responses to D-fenfluramine in major depression: Evidence for diminished responsivity of central serotonergic function. *Am J Psychiatry* 148:1009–1015.

Ostroff, R.B., Giller, E., Harkness, L., Mason, J. 1985. The norepinephrine-to-epinephrine ratio in patients with a history of suicide attempts. *Am J Psychiatry* 142:224–227.

Ostrowitzki, S., Rao, M.L., Redei, J., Andres, A.H. 1993. Concurrence of cortex and platelet serotonin2 receptor binding characteristics in the individual and the putative regulation by serotonin. *J Neural Transm Gen Sect* 93:27–35.

Owen, F., Chambers, D.R., Cooper, S.J., Crow, T.J., Johnson, J.A., Lofthouse, R., Poulter, M. 1986. Serotonergic mechanisms in brains of suicide victims. *Brain Res* 362:185–188.

Owens, M.J., Nemeroff, C.B. 1994. Role of serotonin in the pathophysiology of depression: Focus on the serotonin transporter. *Clin Chem* 40:288–295.

Pandey, G.N., Dwivedi, Y. 2006. Monoamine receptors and signal transduction mechanisms in suicide. *Curr Psychiatr Rev* 2:51–75.

Pandey, G.N., Dwivedi, Y., Rizavi, H.S., Ren, X., Pandey, S.C., Pesold, C., Roberts, R.C., Conley, R.R., Tamminga, C.A. 2002. Higher expression of serotonin 5-HT(2A) receptors in the postmortem brains of teenage suicide victims. *Am J Psychiatry* 159:419–429.

Pandey, G.N., Dwivedi, Y., Rizavi, H.S., Ren, X., Zhang, H., Pavuluri, M.N. 2010. Brain-derived neurotrophic factor gene and protein expression in pediatric and adult depressed subjects. *Prog Neuropsychopharmacol Biol Psychiatry* 34:645–651.

Pandey, G.N., Pandey, S.C., Dwivedi, Y., Sharma, R.P., Janicak, P.G., Davis, J.M. 1995. Platelet serotonin-2A receptors: A potential biological marker for suicidal behavior. *Am J Psychiatry* 152:850–855.

Pandey, G.N., Pandey, S.C., Janicak, P.G., Marks, R.C., Davis, J.M. 1990. Platelet serotonin-2 receptor binding sites in depression and suicide. *Biol Psychiatry* 28:215–222.

Pandey, G.N., Ren, X., Rizavi, H.S., Conley, R.R., Roberts, R.C., Dwivedi, Y. 2008. Brain-derived neurotrophic factor and tyrosine kinase B receptor signalling in post-mortem brain of teenage suicide victims. *Int J Neuropsychopharmacol* 11:1047–1061.

Paul, S.M., Rehavi, M., Rice, K.C., Ittah, Y., Skolnick, P. 1981a. Does high affinity [^{3}H]imipramine binding label serotonin reuptake sites in brain and platelet? *Life Sci* 28:2753–2760.

Paul, S.M., Rehavi, M., Skolnick, P., Ballenger, J.C., Goodwin, F.K. 1981b. Depressed patients have decreased binding of tritiated imipramine to platelet serotonin "transporter." *Arch Gen Psychiatry* 38:1315–1317.

Perry, E.K., Marshall, E.F., Blessed, G., Tomlinson, B.E., Perry, R.H. 1983. Decreased imipramine binding in the brains of patients with depressive illness. *Br J Psychiatry* 142:188–192.

Plein, H., Berk, M. 2001. The platelet as a peripheral marker in psychiatric illness. *Hum Psychopharmacol* 16:229–236.

Rao, M.L., Hawellek, B., Papassotiropoulos, A., Deister, A., Frahnert, C. 1998. Upregulation of the platelet serotonin2A receptor and low blood serotonin in suicidal psychiatric patients. *Neuropsychobiology* 38:84–89.

Rausch, J.L., Janowsky, D.S., Risch, S.C., Huey, L.Y. 1986. A kinetic analysis and replication of decreased platelet serotonin uptake in depressed patients. *Psychiatry Res* 19:105–112.

Roggenbach, J., Muller-Oerlinghausen, B., Franke, L. 2002. Suicidality, impulsivity and aggression—Is there a link to 5HIAA concentration in the cerebrospinal fluid? *Psychiatry Res* 113:193–206.

Roggenbach, J., Muller-Oerlinghausen, B., Franke, L., Uebelhack, R., Blank, S., Ahrens, B. 2007. Peripheral serotonergic markers in acutely suicidal patients. 1. Comparison of serotonergic platelet measures between suicidal individuals, nonsuicidal patients with major depression and healthy subjects. *J Neural Transm* 114:479–487.

Rosel, P., Arranz, B., San, L., Vallejo, J., Crespo, J.M., Urretavizcaya, M., Navarro, M.A. 2000. Altered 5-HT(2A) binding sites and second messenger inositol trisphosphate (IP(3)) levels in hippocampus but not in frontal cortex from depressed suicide victims. *Psychiatry Res* 99:173–181.

Rosel, P., Arranz, B., Vallejo, J., Alvarez, P., Menchon, J.M., Palencia, T., Navarro, M.A. 1999. Altered [^{3}H]imipramine and 5-HT$_2$ but not [^{3}H]paroxetine binding sites in platelets from depressed patients. *J Affect Disord* 52:225–233.

Roy, A., De Jong, J., Linnoila, M. 1989. Cerebrospinal fluid monoamine metabolites and suicidal behavior in depressed patients. A 5-year follow-up study. *Arch Gen Psychiatry* 46:609–612.

Roy, A., Ninan, P., Mazonson, A., Pickar, D., Van Kammen, D., Linnoila, M., Paul, S.M. 1985. CSF monoamine metabolites in chronic schizophrenic patients who attempt suicide. *Psychol Med* 15:335–340.

Roy-Byrne, P., Post, R.M., Rubinow, D.R., Linnoila, M., Savard, R., Davis, D. 1983. CSF 5HIAA and personal and family history of suicide in affectively ill patients: A negative study. *Psychiatry Res* 10:263–274.

Secunda, S.K., Cross, C.K., Koslow, S., Katz, M.M., Kocsis, J.H., Maas, J.W. 1986. Studies of amine metabolites in depressed patients. Relationship to suicidal behavior. *Ann N Y Acad Sci* 487:231–242.

Sheline, Y.I., Bardgett, M.E., Jackson, J.L., Newcomer, J.W., Csernansky, J.G. 1995a. Platelet serotonin markers and depressive symptomatology. *Biol Psychiatry* 37:442–447.

Sheline, Y.I., Black, K.J., Bardgett, M.E., Csernansky, J.G. 1995b. Platelet binding characteristics distinguish placebo responders from nonresponders in depression. *Neuropsychopharmacology* 12:315–322.

Sher, L., Carballo, J.J., Grunebaum, M.F., Burke, A.K., Zalsman, G., Huang, Y.Y., Mann, J.J., Oquendo, M.A. 2006. A prospective study of the association of cerebrospinal fluid monoamine metabolite levels with lethality of suicide attempts in patients with bipolar disorder. *Bipolar Disord* 8:543–550.

Spreux-Varoquaux, O., Alvarez, J.C., Berlin, I., Batista, G., Despierre, P.G., Gilton, A., Cremniter, D. 2001. Differential abnormalities in plasma 5-HIAA and platelet serotonin concentrations in violent suicide attempters: Relationships with impulsivity and depression. *Life Sci* 69:647–657.

Stanley, M., Mann, J.J. 1983. Increased serotonin-2 binding sites in frontal cortex of suicide victims. *Lancet* 1:214–216.

Stanley, M., Traskman-Bendz, L., Dorovini-Zis, K. 1985. Correlations between aminergic metabolites simultaneously obtained from human CSF and brain. *Life Sci* 37:1279–1286.

Stanley, M., Virgilio, J., Gershon, S. 1982. Tritiated imipramine binding sites are decreased in the frontal cortex of suicides. *Science* 216:1337–1339.

Stockmeier, C.A. 2003. Involvement of serotonin in depression: Evidence from postmortem and imaging studies of serotonin receptors and the serotonin transporter. *J Psychiatr Res* 37:357–373.

Stockmeier, C.A., Dilley, G.E., Shapiro, L.A., Overholser, J.C., Thompson, P.A., Meltzer, H.Y. 1997. Serotonin receptors in suicide victims with major depression. *Neuropsychopharmacology* 16:162–173.

Targum, S.D., Rosen, L., Capodanno, A.E. 1983. The dexamethasone suppression test in suicidal patients with unipolar depression. *Am J Psychiatry* 140:877–879.

Teitler, M., Herrick-Davis, K. 1994. Multiple serotonin receptor subtypes: Molecular cloning and functional expression. *Crit Rev Neurobiol* 8:175–188.

Traskman-Bendz, L., Asberg, M., Bertilsson, L., Thoren, P. 1984. CSF monoamine metabolites of depressed patients during illness and after recovery. *Acta Psychiatr Scand* 69:333–342.

Tripodianakis, J., Markianos, M., Sarantidis, D., Agouridaki, M. 2002. Biogenic amine turnover and serum cholesterol in suicide attempt. *Eur Arch Psychiatry Clin Neurosci* 252:38–43.

Tuomisto, J., Tukiainen, E. 1976. Decreased uptake of 5-hydroxytryptamine in blood platelets from depressed patients. *Nature* 262:596–598.

Tyano, S., Zalsman, G., Ofek, H., Blum, I., Apter, A., Wolovik, L., Sher, L., Sommerfeld, E., Harell, D., Weizman, A. 2006. Plasma serotonin levels and suicidal behavior in adolescents. *Eur Neuropsychopharmacol* 16:49–57.

van Heeringen, K., Audenaert, K., Van de Wiele, L., Verstraete, A. 2000. Cortisol in violent suicidal behaviour: Association with personality and monoaminergic activity. *J Affect Disord* 60:181–189.

Yerevanian, B.I., Feusner, J.D., Koek, R.J., Mintz, J. 2004. The dexamethasone suppression test as a predictor of suicidal behavior in unipolar depression. *J Affect Disord* 83:103–108.

21 Medication in Suicide Prevention

Insights from Neurobiology of Suicidal Behavior

J. John Mann and Dianne Currier

CONTENTS

Suicidal behavior occurs in the context of a diathesis or predisposition that is characterized by traits in multiple domains: behavioral, clinical, personality, biological, and cognitive (see [1] for an overview). These traits likely have their origins in combinations of genetic and early-life experiences during critical formative periods of development. Thirty years of research have yielded considerable insight into neurobiological dysfunction associated with suicidal behavior as a phenotype and with a number of traits belonging to the diathesis such as impulsive aggression, deficits in executive function, negative or rigid cognitive processes, and recurrent mood disorders. The major systems where abnormalities have been observed in suicide and nonfatal suicide attempts are the serotonergic system and the stress response systems of the noradrenergic system and the hypothalamic–pituitary–adrenal (HPA) axis.

While the complexity and variability of suicidal behavior clearly requires a multifaceted prevention approach, the identification of neurobiological dysfunction suggests targets for pharmacological intervention that may have protective effects against suicidal behavior.

In this chapter, we outline the main alterations in neurobiological function that have been documented in suicide attempters and individuals who died by suicide and describe putative mechanisms of action whereby different classes of medication might help preventing suicidal behavior.

21.1 LOW SEROTONERGIC ACTIVITY IS A BIOLOGICAL TRAIT RELATED TO SUICIDAL BEHAVIOR

More than 30 years ago, Åsberg et al. observed that depressed individuals who had either attempted suicide by violent means or subsequently died by suicide were more likely to have lower cerebrospinal fluid 5-hydroxyindoleacetic acid (CSF 5-HIAA) levels indicative of less serotonin release [2]. Since that time, the function of the serotonergic system in suicide and attempted suicide has been examined in many paradigms, and while not all studies agree there is substantial consensus that individuals who die by suicide, or make serious nonfatal suicide attempts, exhibit a deficiency in regional brain serotonin neurotransmission such as in the orbital prefrontal cortex.

Evidence of serotonin impairment or hypofunction comes from CSF studies in living patients and from postmortem brain tissue studies. 5-HIAA is the major metabolite of serotonin and the level of CSF 5-HIAA is a guide to serotonin activity in parts of the brain including the prefrontal cortex. There have been over 20 studies of CSF 5-HIAA and suicidal behavior in mood disorder patients, and a meta-analysis of prospective studies of 5-HIAA found that in mood disorders lower CSF 5-HIAA increased the chance of death by suicide by about 4.6-fold over follow-up periods of between 1 and 14 years [3]. Lower concentration of CSF 5-HIAA has also been reported in patients with other psychiatric disorders such as schizophrenia, bipolar disorder, and personality disorders who have made suicide attempts compared with nonattempter psychiatric controls. Some report that low CSF 5-HIAA characterizes those who use violent methods to suicide or who make nonfatal attempts of higher lethality [4–6]. These results are consistent with reports of low levels of 5-HT and/or 5-HIAA in the brainstem of suicides that is similar in magnitude across diagnostic categories [7]. Finding the same serotonin system deficit in several psychiatric disorders sharing the common feature of a suicide or suicide attempt indicates that this biological abnormality is related to the suicidal behavior and not to a specific psychiatric disorder [7]. Once this was understood, many studies have since addressed the state of the serotonin system in suicidal behavior and how it may relate to clinical and cognitive aspects of the diathesis for suicidal behavior.

Suicides have localized lower serotonin transporter (SERT) binding in the ventromedial prefrontal cortex and anterior cingulate, which might reflect reduced serotonin input to these brain areas that are known to be involved in willed action or decision making [8,9]. Findings in suicide of more serotonin neurons, greater tryptophan hydroxylase 2 gene expression and protein, lower SERT expression and protein, and more 5-HT_{1A} inhibitory autoreceptors all favor enhancement of serotonin transmission [7] and have been postulated to represent a homeostatic response to lower levels of serotonin and/or 5-HIAA in brainstem of suicides and lower 5-HIAA in CSF of serious suicide attempters [7]. For example, morphometric analysis of serotonin neurons in the brainstem demonstrates greater neuronal density in depressed suicides compared with nonsuicides [8], suggesting that reduction in serotonin activity is associated with dysfunctional cells and not with fewer neurons.

Neuroendocrine challenge studies using fenfluramine provide further evidence of impaired serotonergic function associated with suicidal behavior. Fenfluramine is a serotonin-releasing agent and a reuptake inhibitor that may also directly stimulate postsynaptic 5-HT receptors. The release of serotonin by fenfluramine causes an increase in serum prolactin levels and is an indirect index of central serotonergic responsiveness. In depressed patients, those with a history of suicide attempts have a more blunted prolactin response to fenfluramine challenge compared with nonattempters, and there is some evidence that the effect is more strongly related to seriousness of past attempt [10].

Lower serotonergic transmission in the central nervous system (CNS) may be accompanied by a compensatory upregulation of some serotonergic postsynaptic receptors such as the 5-HT_{1A} and 5-HT_{2A} and a decrease in the number of serotonin reuptake sites [11]. In depressed and nondepressed suicides, there is evidence that 5-HT_{2A} receptors are upregulated in the dorsal prefrontal cortex but not in the rostral prefrontal cortex [12]. This increased binding is also reflected in more protein and may be due to elevated gene expression observed in youth suicides [13]. Elevated 5-HT_{2A} binding has also been reported in the amygdala in depressed suicides [14]. The emerging picture from postmortem studies of greater 5-HT_{2A} receptor binding in the frontal cortex of depressed individuals who die by suicide, fewer brainstem 5-HT_{1A} autoreceptors, and fewer SERTs in the cortex, as well as findings of greater tryptophan hydroxylase (the rate-limiting step in serotonin synthesis) immunoreactivity in serotonin nuclei in the brainstem [15] all point to homeostatic changes in response to deficient serotonergic transmission evidenced by low 5-HIAA in CSF and brain, low 5-HT and 5-HIAA in brainstem, and blunted prolactin response to fenfluramine challenge.

Platelet studies examine 5-HT_{2A} receptors in living subjects with respect to nonfatal suicide attempt. 5-HT_{2A} receptors, serotonin reuptake sites, and serotonin second messenger systems are present in blood platelets and changes in these platelet measurements may reflect similar changes in the CNS. Several studies have reported higher platelet 5-HT_{2A} receptor numbers in suicide attempters compared with nonattempters and healthy controls [16]. Greater 5-HT_{2A} receptor binding appears offset by impaired signal transduction in the prefrontal cortex of suicides [17] and in platelets from patients with major depression who have made high-lethality suicide attempts compared to depressed patients who have made low-lethality suicide attempts [18].

21.1.1 Serotonergic Dysfunction and Suicide Endophenotypes

Impulsive aggressive traits are postulated to be part of the diathesis for suicidal behavior [19]. Increased aggression has been associated with suicide, and more highly lethal suicide attempts and impulsivity have shown a stronger relationship to nonfatal suicide attempts [20]. Reduced activity of the serotonin system has been implicated in impulsive violence and aggression in studies of several serotonergic indices including low CSF 5-HIAA in individuals with a lifetime history of aggressive behavior with personality and other psychiatric disorders [21,22], a blunted prolactin response to serotonin-releasing agent fenfluramine in personality disorder patients [23,24], and greater platelet 5-HT_{2A} binding correlated with aggressive behavior in personality and other psychiatric disorder patients [25,26]. In a postmortem study of aggression, suicidal behavior, and serotonergic function, a positive correlation was observed between lifetime history of aggression scores and 5-HT_{2A} binding in several regions of prefrontal cortex of suicides [27]. A role for serotonergic function in aggressive and, to a lesser extent, impulsive behavior is well documented [28,29], consistent with observations that low SERT binding associated with suicide appears to be concentrated in the ventromedial PFC and anterior cingulate regions, which play a role in mediating inhibition and restraint [8,9]. Moreover, the association observed between prefrontal hypofunction and impaired serotonergic responsivity and lethality of nonfatal suicide attempts and the linkage in familial transmission of suicide attempts and aggressive traits [30] suggest that aggressive traits may be an intermediate clinical phenotype linking serotonergic dysfunction and suicidal behavior [31].

Positron emission tomography (PET) studies have shown a deficient response to serotonergic challenge in the orbitofrontal cortex, medial frontal, and cingulate regions in individuals with impulsive aggression compared to controls [32,33] and lower SERT binding in the anterior cingulate cortex in impulsive aggressive individuals compared to healthy controls [34]. The prefrontal cortex is important for the inhibitory control of behavior, including impulsive and aggressive behavior [35]. Thus, aggressive/impulsive traits, related to serotonergic dysfunction, are potentially an aspect of the diathesis for suicidal behavior whereby aggressive/suicidal behaviors is manifested in response to stressful circumstances or powerful emotions. This tendency might be conceived of as a diminution in natural inhibitory circuits, or as a volatile cognitive decision style. Lower C-α-methyl-L-tryptophan trapping in the orbital cortex and ventromedial prefrontal cortex was found in suicide attempters who made high-intent attempts [36], and 5-HT_{2A} binding negatively correlated with levels of hopelessness, a correlate of suicide and suicide attempt [37]. In response to the administration of the serotonin agonist fenfluramine, depressed high-lethality suicide attempters had lower fluorodeoxyglucose (^{18}F) regional cerebral metabolism of glucose (rCMRGlu) in anterior cingulate and superior frontal gyri, compared with depressed low-lethality attempters [31]. In that study, lethality of the most serious lifetime suicide attempt correlated negatively with rCMRGlu in the anterior cingulate, right superior frontal, and right medial frontal gyri consistent with prefrontal cortex hypofunction in high-lethality depressed suicide attempters [31].

21.2 NORADRENERGIC SYSTEM DYSFUNCTION IN SUICIDAL BEHAVIOR

Postmortem studies in depressed suicides report fewer noradrenergic neurons in the locus coeruleus [38], higher α_2-adrenergic binding in the frontal cortex of suicide victims [39–41], α_1 receptor binding in layers IV and V of PFC [42], α_2 receptor binding in the locus coeruleus [43], and greater tyrosine hydroxylase immunoreactivity in the locus coeruleus [44]. These changes suggest a pattern of cortical noradrenergic overactivity that may be the outcome of excessive NE release with stress leading to depletion due to the smaller population of NE neurons found in suicides [42,45]. The hypothesized excessive norepinephrine release when stressed in adulthood may result from sensitization of the system as a result of childhood adversity [46].

CSF 3-methoxy-4-hydroxyphenylglycol (MHPG) is a metabolite of norepinephrine and studied as an indicator of brain noradrenergic activity. Cross-sectional studies of CSF MHPG and past suicide attempt in samples of varied psychiatric diagnoses have been largely negative (reviewed in [47]), with two studies reporting significantly lower CSF MHPG in suicide attempters compared to nonattempters in major depression [48] and offenders [49]. Cross-sectional studies have also examined the relationship between violence and/or lethality of suicide attempts and CSF MHPG levels, again with negative results reported in various diagnostic groups [50–54]. Given that the noradrenergic system is stress-responsive, these generally negative findings may reflect the inability of cross-sectional and retrospective study designs to adequately assess associations between state-dependent biological alterations and past suicidal behavior. State-dependent biological abnormalities are best detected in closer time proximity to the behavior of interest and a short term. Prospective study may be better able to detect these effects; however, there have been few prospective studies of CSF MHPG and suicide and suicide attempt. One prospective study of major depressive disorder found that 20% of low CSF MHPG patients made suicide attempts versus 5% of high MHPG patients in follow-up. Most of the attempts were in the first 3 months after the initial evaluation, and among the 15 individuals who made a suicide attempter during the follow-up period, the lethality of suicide attempts in low MHPG group was greater than in high MHPG group with very little overlap [55]. Another study of hospitalized suicide attempters observed that those who made a further suicide attempt or died by suicide in the year following their index attempt were more likely to have CSF MHPG levels above the median [56]; however, that result was not tested statistically. Studies based on death records 1–12 years after index admission for a suicide attempt report no differences in baseline CSF MHPG between those who did or did not eventually die by suicide during follow-up [57–59]. In other studies, no difference in CSF MHPG levels was observed in suicide attempters during an 11 year follow-up in schizophrenia [4], or in a 2 year follow-up of bipolar disorder [6]. In the later study, CSF MHPG level was negatively correlated with maximum lethality of follow-up attempt raising the possibility that the relationship of low CSF MHPG to future suicide attempts might be stronger in those making more lethal suicide attempts.

Noradrenergic response to stress in adulthood appears to be greater in those reporting an abusive experience in childhood [46,60]. In remitted depressed individuals treated with norepinephrine reuptake inhibitors (NRIs), acute catecholamine depletion resulted in an increase in hopelessness or depression [61], consistent with preclinical studies suggesting that raising NE transmission can reduce hopelessness or depression [62].

21.3 HYPOTHALAMIC–PITUITARY–ADRENAL AXIS STRESS RESPONSE SYSTEM IS RELATED TO SUICIDAL BEHAVIOR

The HPA axis is the other major stress response system. A number of indicators of HPA axis hyperactivity of the HPA axis have been associated with suicide death including failure to suppress cortisol secretion following the administration of dexamethasone (DST), higher levels of plasma and urinary cortisol, and loss of diurnal variation due to less afternoon decline in cortisol levels [63]. In other indices of HPA axis function, larger pituitary and larger adrenal gland volumes are reported in depressed suicides [64,65], and lower corticotrophin-releasing hormone (CRH) receptor binding was observed in the frontal cortex in suicides [66] and higher CSF CRH was also reported [67]. These latter findings suggest a state of excessive CRH release downregulating receptors [68]. In terms of nonfatal suicidal behavior, although there is a more than fourfold risk of dying by suicide in depressed individuals who are cortisol nonsuppressors in DST challenge [3], DST nonsuppression is not clearly related to future nonfatal suicide attempt, although some studies find an association with serious or violent attempts (see [69] for a review). In other indices of HPA axis function, nonfatal suicide attempt is associated with lower CSF CRH but no difference in plasma CRH or plasma cortisol [70], higher urinary cortisol in violent attempters [71], and higher serum cortisol after 5-hydroxytryptophan (5-HT) challenge [72].

The HPA axis is complexly interrelated with the serotonergic and noradrenergic systems. The HPA axis and serotonergic system have a bidirectional relationship [73]. CRH neurons project to the serotonin raphe nuclei, and projections from the raphe nuclei extend to various brain regions that contain CRH and are involved in stress response [74]. The hyperactivity of the HPA axis observed in suicidal patients may mediate or moderate some of the serotonin abnormalities also seen in these patients [75], and modulation of serotonin receptors by corticosteroids in response to stress may have important implications for the pathophysiology of suicide [75,76]. There is also a bidirectional relationship between the HPA axis and the NE system. Stress activates both the HPA axis and *locus coeruleus* (LC), the major source of NE neurons in the brain [77] leading to increased NE release during stress. LC neurons influence neuroendocrine stress response system through their broad innervation of the paraventricular nucleus (PVN) projection pathways, and these reciprocal interactions connect cerebral NE and CRH systems and may generate a feed-forward loop [78]. Severe anxiety in response to stress may be related to NE overactivity and hyperactivity of the HPA axis and thereby contribute to suicide risk [79]. Thus, there are multiple potential pathways through which stress may be involved in the biological anomalies observed in suicidal behavior: directly through dysfunction of the HPA axis and the noradrenergic system and interactions between those two systems, as well as indirectly through downstream effects on serotonergic system function.

As with the noradrenergic system, early-life adversity appears to have lasting effects on stress response in the HPA axis in adulthood. Abnormalities in HPA axis function have been implicated in poor response to antidepressant treatment, and greater likelihood of relapse in major depression, both of which increase the risk for suicidal acts [69].

21.4 OTHER BIOLOGICAL ABNORMALITIES ASSOCIATED WITH SUICIDAL BEHAVIOR

Suicide is more common in groups with very low cholesterol levels or after cholesterol lowering by diet (see [80] for a review). This relationship between cholesterol and suicide may be mediated by serotonergic function, as studies of nonhuman primates on a low-fat diet found lower serotonergic activity and increased aggressive behaviors [81]. Long-chain polyunsaturated fatty acids, particularly omega-3, may also be a mediating factor in the relationship between low cholesterol and increased risk for depression and suicide [82]. A 2 year follow-up study of depressed patients found that lower docosahexaenoic acid percentage of total plasma polyunsaturated fatty acids and a higher omega-6/omega-3 ratio predicted suicide attempt in follow-up [83]. Lower eicosapentaenoic acid was found in red blood cells of suicide attempters compared to controls [84].

21.5 MEDICATION IN SUICIDE PREVENTION

21.5.1 Targets of Action of Antidepressant Agents

Most effective antidepressants amplify serotonin or norepinephrine signaling by inhibiting reuptake at the synaptic cleft; these include selective serotonin reuptake inhibitors (SSRIs), norepinephrine reuptake inhibitors (NRIs), dual-action agents that inhibit uptake of serotonin and norepinephrine (SNRIs), and tricyclic antidepressants (TCAs). All SRRIs selectively inhibit serotonin reuptake; however, the potency and selectivity of reuptake inhibition differ between individual agents. For example, paroxetine is the most potent while citalopram has the highest selectivity for 5-HT over NE [85]. Fluoxetine's active metabolite has a longer half-life compared to other SSRIs [86]. There are also differences between SSRIs in their secondary pharmacological actions; for example, paroxetine and sertraline in higher doses also block dopamine reuptake (see [85] and [87] for a review). Monoamine oxidase inhibitors (MAOIs) inhibit monoamine degradation by monoamine oxidase A or B [88]. Other antidepressant agents antagonize α_2-adrenergic autoreceptors with resultant increased release of norepinephrine, and/or antagonize 5-HT_{2A} receptors [88]. For example, mirtazapine enhances norepinephrine release by blocking α_2-adrenergic autoreceptors and blocks serotonin 5-HT_{2A}, 5-HT_3, and histamine H_1 receptors. It has comparable efficacy to TCAs and SSRIs [89]. Nefazodone blocks the 5-HT_{2A} serotonin receptor and serotonin reuptake and has comparable antidepressant efficacy to SSRIs [90,91]. Other agents used in the treatment of depression do not act directly on the serotonin system; for example, bupropion inhibits both norepinephrine and dopamine reuptake and generally has comparable efficacy to TCAs [92], and mifepristone, a glucocorticoid antagonist, has been used to treat delusional depression [93].

Given that most antidepressant agents enhance serotonergic function and/or noradrenergic function and that deficient function in those systems is widely reported in suicidal behavior, one would expect that these agents would have some anti-suicidal effect. Direct evidence of this however is relatively rare, and indeed there is widespread debate that these agents may have a pro-suicidal effect in younger adults, adolescents, and children, for which direct evidence is also sparse.

21.5.2 Antidepressant Agents and Suicidal Behavior

Experimental studies provide the most robust evidence regarding the effects of antidepressant agents; however, there are considerable barriers to investigating the effect of antidepressant agents on suicide and suicide attempt within an experimental design. One difficulty is that as suicide death and suicide attempt are rare outcomes, very large sample sizes and long observation periods are required to detect effects. Another is that it would be ethically untenable to have a placebo arm in such high-risk studies. Thus, much of the experimental data on the relation of antidepressants to suicidality currently available is derived from secondary analysis of treatment efficacy or effectiveness trials, which are generally not designed with suicidal behavior as a specific outcome. There are a number of limitations to this approach. These studies are largely uninformative about suicide death, as to maximize the number of observable outcome events suicidal ideation is generally taken as the outcome measure, or a composite outcome of "suicidality" comprising ideation and attempts is used. Moreover, in such studies, observation periods are generally short, actively suicidal subjects are often excluded, and suicidal outcomes may be poorly defined, measured, and not systematically assessed (see [94]). Methodological issues such as these make interpreting the findings of secondary analyses of clinical trial data difficult. Despite these limitations, such analyses have played a prominent role in the debate and regulatory decision making around the issue of suicidality and antidepressant use in children, youth, and young adults.

Evidence for a pro-suicidal effect of antidepressants was reported by the FDA in retrospective review of clinical data from 24 RCTs of SSRIs and other newer antidepressants in pediatric patients, which found more medication-related adverse event reports of suicidal ideation or acts (but no suicide deaths) in the active treatment groups (4%) compared to the placebo group (2%) [95]. When suicidal outcome was assessed using the suicide item on the depression rating scale that was administered regularly throughout the course of treatment, there was no difference in suicidality between drug and placebo groups. In a subsequent further meta-analysis by the FDA of adverse event reports in 372 antidepressant trials comprising 99,231 patients of all ages, 77,383 of whom had psychiatric diagnoses, young adults aged 18–24 years had increased risk for suicidal behavior on antidepressants compared to placebo [96]. However, there was a protective effect on suicidality (ideation or attempt combined) in all adults 25 years and older, across all psychiatric diagnosis for antidepressants compared to placebo, and most significantly in the older age groups [96]. Other meta-analyses note a beneficial effect of antidepressant use for suicidal ideation in adults [97,98]. Subsequent meta-analysis of clinical trial data has not found an increased risk in young patients. In a more recent pediatric antidepressant RCT meta-analysis, higher suicidal ideation/attempt scores for antidepressants versus placebo were more modest than the first FDA study and were only significant statistically when data were pooled across indications, but no such increase was observed when each drug was analyzed separately [99]. What is puzzling about these results is that given the deficits in serotonin and noradrenergic function in suicidal behavior, why would antidepressants such as SSRIs or noradrenergic-enhancing drugs not reduce suicidal ideation and behavior. There has been much debate over the FDA's use of suicide-related adverse event reports as the outcome measure and the advantages of a systematically administered rating scale such that one needs to examine other data sets to see if findings from other types of studies support the enhanced risk of suicide-related symptoms being more severe on antidepressants compared with placebo (see Gibbons and Mann [100] for a review). Moreover, the unavoidable methodological limitations of secondary analysis of efficacy trial data require consideration of observational and ecological studies as another source of information on the effects of antidepressant agents on suicidal behavior.

21.5.3 Nonexperimental Studies of Antidepressant Use and Suicidal Outcomes

One set of observational studies that is suggestive of a protective effect of antidepressant use on both suicide death and nonfatal suicide attempt are national population level ecological studies. Studies from various countries including Sweden, Norway, Denmark, Finland [101], Japan [102], and Italy [103] have demonstrated that increasing use of antidepressants population-wide correlates with a decline in population suicide rates (see Ludwig and Marcotte [104] for a review). Others have not found this correlation [105,106], although in the case of Iceland this may have been due to the very low national rater to begin with, while in Italy it was found observed in women but not men and may have reflected the segment of the population where the greatest increase in antidepressant use occurred. Another set of studies evaluated smaller communities where programs to improve diagnosis and treatment of depression, generally in the elderly, have produced a robust decrease in suicide rates in that segment of the population. This finding has been documented in Sweden [107], Hungary [108], and Japan [109,110]. For nonfatal suicide attempt, there is evidence from both community-level studies and quasi-experimental studies of a benefit for suicide attempt with the use of antidepressants. In Denmark and Germany, community-level interventions to place suicide attempters into treatment and retain them in treatment saw a decrease in reattempt compared to nonintervention comparison regions [111,112].

Cohort studies in health care have shown declines in suicidal behavior following the initiation of antidepressant treatment. Gibbons et al. in a study of veterans found that the highest rate of suicide attempt was immediately prior to treatment initiation, which subsequently declined dramatically [113]. Prescription of antidepressant was associated with a decline in suicide rates but it was also noteworthy that the pretreatment suicide rates were quite different between the treated and untreated groups and need to be considered to fully understand the impact of antidepressants on suicide rates. Likewise, Simon et al. reported that the highest rate of suicide attempts was in the month prior to the initiation of treatment in adults [114] and in adolescents and young adults [115] and that the rate declined in the months following treatment initiation. In a large record-based case–control study, Olfson et al. reported an increased odds of children and adolescents who made a suicide attempt having an antidepressant prescribed at the time of the attempt compared to nonattempter controls (95% CI 1.12–2.07, $p = 0.007$) [116]. By drug class, the odds were nonsignificant for SSRIs but significantly increased for tricyclics and venlafaxine. In such studies, the cases are a high-risk group and, thus it is difficult to rule out the possibility that illness severity, including two important risk factors for a suicide attempt, namely, suicidal ideation and behavior, triggered the prescription of an antidepressant, and that explains the observed association between prescription of antidepressant and suicide attempt.

Clearly, such studies cannot establish causality, nor be informative on the molecular mechanisms by which antidepressants might mitigate suicidal ideation or behavior, but given the low rates of suicide and serious suicide attempts such studies are needed given the limited capacity to do a large enough RCT to study effects on suicide attempts. In keeping with the methodological difficulties in conducting experimental trials for suicidal behavior outcomes, RCTs most often use suicidal ideation as the primary outcome, and less frequently use suicide attempt. While less is known of the neurobiological underpinnings of suicidal ideation, it is a known predictor of future suicide attempt and death by suicide in community and clinical studies. A large national epidemiological study found that 90% of unplanned and 60% of planned first suicide attempts occurred within 1 year of onset of ideation [117], and prospective studies in mood disorder patients [20,118] found that severity of suicidal ideation predicted risk of future suicide attempt and suicide death. The most severe lifetime suicidal ideation was the best predictor of suicide.

21.5.4 Comparator Studies with Suicidal Outcomes

Given that drugs from different classes with different targets of action have antidepressant efficacy [89–92,119–128], studies comparing their effect on suicidal outcomes provide insight into the most salient neurobiological actions that affect suicidal ideation and behavior. In eight studies, drugs that act predominantly on the serotonergic system in comparison with drugs that act primarily on the noradrenergic system had greater effect in terms of improving suicidal ideation [129–136], while four studies found no superior benefit [137–140]. Only one study found the noradrenergic drug more effective [141]. Differences in study durations, patient samples, and treatment protocols make it difficult to draw firm conclusions from this evidence, but it seems to indicate that the serotonergic system is a key target for pharmacotherapeutic approaches to suicide prevention.

21.5.5 Antidepressant Action on the HPA Axis

Studies in rodents have demonstrated effects of different classes of antidepressant agents, including MOAIs, SSRIs NRIs, on the mineralocorticoid and glucocorticoid receptors in the hippocampus and other limbic and cortical brain areas (see Schule [142] for a review). In humans, improvement in HPA axis function is linked to the therapeutic action of antidepressants. For example, remitted MDD patients who had higher cortisol levels following DEX/CRH challenge were more likely to relapse [143,144], and that those with persistent cortisol hypersecretion during the DEX/CRH test despite clinical improvement are also at increased risk of relapse in the following 6 months [145]. HPA axis abnormalities predict the risk for future suicide in MDDs and postulated mechanisms include relapse into an MDE or effects of chronically elevated steroids on the brain and cognitive rigidity.

21.5.6 Other Psychotropic Drugs and Suicidal Behavior

Other psychopharmacological agents that are not classed primarily as antidepressants but have some action on the same neurobiological targets as antidepressants or even some secondary antidepressant action, such as lithium, may also have anti-suicidal properties. Two classes of drugs that have been studied in this regard are mood stabilizers, most notably lithium, which is used as a first-line treatment for bipolar disorder and an augmenting agent for treatment-resistant depression, and antipsychotic agents, principally clozapine, used in the treatment for schizophrenia spectrum disorders. One of the mechanisms proposed for lithium's antidepressant effect is that it enhances serotonin neurotransmission [146]. Several studies have shown that lithium increases serotonin release in certain brain regions [147–149], and in challenge studies the increase in plasma prolactin following the administration of L-tryptophan (the serotonin precursor) was further enhanced by lithium [150,151] similar to the effect seen with tricyclic and SSRI antidepressants [152]. There is a large body of evidence from meta-analyses and observational studies that lithium reduces the risk for suicide death and suicide attempt (for a review see [153]). It has been proposed that one of the pathways through which lithium decreases risk for suicide is that it reduces aggressive and impulsive behaviors. Aggression and impulsivity have been associated with suicide and suicide attempt [154], both are associated with impaired serotonergic function [28,29], and there is evidence from human and animal studies that lithium reduced both aggression and impulsivity (see [155,156] for reviews). Lithium also augments antidepressant effect and that may reduce suicide risk.

Clozapine an antipsychotic agent has also been shown to have a protective effect on suicide death in schizophrenia (see [157] for a review). A large randomized clinical trial designed specifically to assess suicide outcomes using olanzapine as a comparator found a 25% reduction in risk for suicide attempt in the clozapine group [158]. Clozapine blocks the serotonin 2A receptor [159]. Higher serotonin 2A receptor binding has been reported in suicide [12,14,160] and suicide attempt [16,18]. Higher serotonin 2A receptor binding has also been associated with aggression [25–27] suggesting that blocking this receptor may reduce aggression and be one mechanism through which clozapine exerts an anti-suicidal effect. Other atypical antipsychotics that also block the 5-HT_{2A} receptor are being evaluated in terms of suicide risk.

21.5.7 Role of Medication in Suicide Prevention Going Forward

Much has been learned and written about other neurobiological mechanisms that play a role in suicidal behavior and whether pharmacological agents that act on those systems can contribute to reducing the risk for suicide and suicide attempt. Antiepileptic drugs were also found by the FDA in a meta-analysis to be associated with greater risk for ideation [161] but others found the opposite [162]. Antiepileptic drugs act mostly to enhance GABAergic inhibitory function in the cortex and that raises a question about the impact of GABA on suicide risk as an area for future research. PUFA profile can predict suicide attempt risk and perhaps omega-3 supplements may help reduce risk. Ketaminc is very rapidly acting antidepressant and is reported to also have a profound therapeutic benefit for suicidal ideation. It is an AMPA receptor antagonist and raises GABA levels and perhaps those properties contribute to its rapid and profound reduction in suicidal ideation. Other new agents targeting other neurobiological mechanisms related to suicidal behavior beyond the major neurotransmitters that are currently being examined, including the corticotrophin-releasing factor 1 receptor, neuropeptide Y, vasopressin V1b, *N*-methyl-D-aspartate, nicotinic acetylcholine, dopamine D1, glucocorticoid system, δ-opioid, cannabinoid and cytokine receptors, γ-amino butyric acid (GABA), intracellular messenger systems, transcription, and neuroprotective and neurogenic factors [163]. Moreover, emerging data from pharmacogenetic studies showing gene effects on treatment response will provide additional tools for developing more personalized pharmacotherapies for suicidal behavior [164].

REFERENCES

1. Mann JJ, Currier D: Suicide and attempted suicide, in *The Medical Basis of Psychiatry*. Fatemi SH, Clayton PJ, eds. Philadelphia, PA: Humana Press, 2008, pp. 561–576.
2. Åsberg M, Thorén P, Träskman L, Bertilsson L, Ringberger V: "Serotonin depression"—A biochemical subgroup within the affective disorders? *Science* 1976; 191:478–480.
3. Mann JJ, Currier D, Stanley B, Oquendo MA, Amsel LV, Ellis SP: Can biological tests assist prediction of suicide in mood disorders? *Int J Neuropsychopharmacol* 2006; 9(4):465–474.
4. Cooper SJ, Kelly CB, King DJ: 5-Hydroxyindoleacetic acid in cerebrospinal fluid and prediction of suicidal behaviour in schizophrenia. *Lancet* 1992; 340:940–941.
5. Mann JJ, Malone KM: Cerebrospinal fluid amines and higher-lethality suicide attempts in depressed inpatients. *Biol Psychiatry* 1997; 41(2):162–171.
6. Sher L, Carballo JJ, Grunebaum MF, Burke AK, Zalsman G, Huang YY, Mann JJ, Oquendo MA: A prospective study of the association of cerebrospinal fluid monoamine metabolite levels with lethality of suicide attempts in patients with bipolar disorder. *Bipolar Disord* 2006; 8(5 Pt 2):543–550.
7. Mann JJ, Brent DA, Arango V: The neurobiology and genetics of suicide and attempted suicide: A focus on the serotonergic system. *Neuropsychopharmacology* 2001; 24(5):467–477.
8. Arango V, Underwood MD, Gubbi AV, Mann JJ: Localized alterations in pre- and post-synaptic serotonin binding sites in the ventrolateral prefrontal cortex of suicide victims. *Brain Res* 1995; 688(1–2):121–133.
9. Mann JJ, Huang YY, Underwood MD, Kassir SA, Oppenheim S, Kelly TM, Dwork AJ, Arango V: A serotonin transporter gene promoter polymorphism (5-HTTLPR) and prefrontal cortical binding in major depression and suicide. *Arch Gen Psychiatry* 2000; 57(8):729–738.
10. Kamali M, Oquendo MA, Mann JJ: Understanding the neurobiology of suicidal behavior. *Depress Anxiety* 2001; 14(3):164–176.
11. Mann JJ, Underwood MD, Arango V: Postmortem studies of suicide victims, in *Biology of Schizophrenia and Affective Disease*. Watson SJ, ed. Washington, DC: American Psychiatric Press, Inc., 1996, pp. 197–220.

12. Stockmeier CA: Involvement of serotonin in depression: Evidence from postmortem and imaging studies of serotonin receptors and the serotonin transporter. *J Psychiatr Res* 2003; 37(5):357–373.
13. Pandey GN, Dwivedi Y, Rizavi HS, Ren X, Pandey SC, Pesold C, Roberts RC, Conley RR, Tamminga CA: Higher expression of serotonin 5-HT(2A) receptors in the postmortem brains of teenage suicide victims. *Am J Psychiatry* 2002; 159(3):419–429.
14. Hrdina PD, Demeter E, Vu TB, Sótónyi P, Palkovits M: 5-HT uptake sites and 5-HT_2 receptors in brain of antidepressant-free suicide victims/depressives: Increase in 5-HT_2 sites in cortex and amygdala. *Brain Res* 1993; 614:37–44.
15. Boldrini M, Underwood MD, Mann JJ, Arango V: More tryptophan hydroxylase in the brainstem dorsal raphe nucleus in depressed suicides. *Brain Res* 2005; 1041(1):19–28.
16. Pandey GN: Altered serotonin function in suicide. Evidence from platelet and neuroendocrine studies. *Ann N Y Acad Sci* 1997; 836:182–200.
17. Pandey GN, Dwivedi Y, Pandey SC, Teas SS, Conley RR, Roberts RC, Tamminga CA: Low phosphoinositide-specific phospholipase C activity and expression of phospholipase C beta1 protein in the prefrontal cortex of teenage suicide subjects. *Am J Psychiatry* 1999; 156(12):1895–1901.
18. Malone KM, Ellis SP, Currier D, John MJ: Platelet 5-HT_{2A} receptor subresponsivity and lethality of attempted suicide in depressed in-patients. *Int J Neuropsychopharmacol* 2006;1–9.
19. Mann JJ, Waternaux C, Haas GL, Malone KM: Toward a clinical model of suicidal behavior in psychiatric patients. *Am J Psychiatry* 1999; 156(2):181–189.
20. Oquendo MA, Galfalvy H, Russo S, Ellis SP, Grunebaum MF, Burke A, Mann JJ: Prospective study of clinical predictors of suicidal acts after a major depressive episode in patients with major depressive disorder or bipolar disorder. *Am J Psychiatry* 2004; 161(8):1433–1441.
21. Brown GL, Goodwin FK: Cerebrospinal fluid correlates of suicide attempts and aggression. *Ann N Y Acad Sci* 1986; 487:175–188.
22. Stanley B, Molcho A, Stanley M, Winchel R, Gameroff MJ, Parsons B, Mann JJ: Association of aggressive behavior with altered serotonergic function in patients who are not suicidal. *Am J Psychiatry* 2000; 157(4):609–614.
23. Coccaro EF, Siever LJ, Klar HM, Maurer G, Cochrane K, Cooper TB, Mohs RC, Davis KL: Serotonergic studies in patients with affective and personality disorders. Correlates with suicidal and impulsive aggressive behavior. *Arch Gen Psychiatry* 1989; 46:587–599.
24. New AS, Trestman RF, Mitropoulou V, Goodman M, Koenigsberg HH, Silverman J, Siever LJ: Low prolactin response to fenfluramine in impulsive aggression. *J Psychiatr Res* 2004; 38(3):223–230.
25. Coccaro EF, Kavoussi RJ, Sheline YI, Berman ME, Csernansky JG: Impulsive aggression in personality disorder correlates with platelet 5-HT_{2A} receptor binding. *Neuropsychopharmacology* 1997; 16(3):211–216.
26. McBride PA, Brown RP, DeMeo M, Keilp JG, Mieczkowski T, Mann JJ: The relationship of platelet 5-HT_2 receptor indices to major depressive disorder, personality traits, and suicidal behavior. *Biol Psychiatry* 1994; 35:295–308.
27. Oquendo MA, Russo SA, Underwood MD, Kassir SA, Ellis SP, Mann JJ, Arango V: Higher postmortem prefrontal 5-HT_{2A} receptor binding correlates with lifetime aggression in suicide. *Biol Psychiatry* 2006; 59(3):235–243.
28. Ryding E, Lindstrom M, Traskman-Bendz L: The role of dopamine and serotonin in suicidal behaviour and aggression. *Prog Brain Res* 2008; 172:307–315.
29. Congdon E, Canli T: A neurogenetic approach to impulsivity. *J Pers* 2008; 76(6): 1447–1483.

30. Brent DA, Oquendo MA, Birmaher B, Greenhill L, Kolko D, Stanley B, Zelazny J, Brodsky B, Bridge J, Ellis S, Salazar JO, Mann JJ: Familial pathways to early-onset suicide attempt: Risk for suicidal behavior in offspring of mood-disordered suicide attempters. *Arch Gen Psychiatry* 2002; 59(9):801–807.
31. Oquendo MA, Placidi GP, Malone KM, Campbell C, Keilp J, Brodsky B, Kegeles LS, Cooper TB, Parsey RV, Van Heertum RL, Mann JJ: Positron emission tomography of regional brain metabolic responses to a serotonergic challenge and lethality of suicide attempts in major depression. *Arch Gen Psychiatry* 2003; 60(1):14–22.
32. Siever LJ, Buchsbaum MS, New AS, Spiegel-Cohen J, Wei T, Hazlett EA, Sevin E, Nunn M, Mitropoulou V: D,l-Fenfluramine response in impulsive personality disorder assessed with [18F]fluorodeoxyglucose positron emission tomography. *Neuropsychopharmacology* 1999; 20(5):413–423.
33. New AS, Hazlett EA, Buchsbaum MS, Goodman M, Reynolds D, Mitropoulou V, Sprung L, Shaw RB, Jr., Koenigsberg H, Platholi J, Silverman J, Siever LJ: Blunted prefrontal cortical 18fluorodeoxyglucose positron emission tomography response to meta-chlorophenylpiperazine in impulsive aggression. *Arch Gen Psychiatry* 2002; 59(7):621–629.
34. Frankle WG, Lombardo I, New AS, Goodman M, Talbot PS, Huang Y, Hwang DR, Slifstein M, Curry S, Abi-Dargham A, Laruelle M, Siever LJ: Brain serotonin transporter distribution in subjects with impulsive aggressivity: A positron emission study with [11C]McN 5652. *Am J Psychiatry* 2005; 162(5):915–923.
35. de Almeida RM, Rosa MM, Santos DM, Saft DM, Benini Q, Miczek KA: 5-HT(1B) receptors, ventral orbitofrontal cortex, and aggressive behavior in mice. *Psychopharmacology (Berl)* 2006; 185(4):441–450.
36. Leyton M, Paquette V, Gravel P, Rosa-Neto P, Weston F, Diksic M, Benkelfat C: Alpha-[11C]methyl-L-tryptophan trapping in the orbital and ventral medial prefrontal cortex of suicide attempters. *Eur Neuropsychopharmacol* 2006; 16(3):220–223.
37. van Heeringen C, Audenaert K, Van Laere K, Dumont F, Slegers G, Mertens J, Dierckx RA: Prefrontal 5-HT_{2a} receptor binding index, hopelessness and personality characteristics in attempted suicide. *J Affect Disord* 2003; 74(2):149–158.
38. Arango V, Underwood MD, Mann JJ: Fewer pigmented locus coeruleus neurons in suicide victims: Preliminary results. *Biol Psychiatry* 1996; 39:112–120.
39. Arango V, Ernsberger P, Marzuk PM, Chen JS, Tierney H, Stanley M, Reis DJ, Mann JJ: Autoradiographic demonstration of increased serotonin 5-HT_2 and beta-adrenergic receptor binding sites in the brain of suicide victims. *Arch Gen Psychiatry* 1990; 47(11):1038–1047.
40. Biegon A, Israeli M: Regionally selective increases in β-adrenergic receptor density in the brains of suicide victims. *Brain Res* 1988; 442:199–203.
41. Mann JJ, Stanley M, McBride PA, McEwen BS: Increased $serotonin_2$ and β-adrenergic receptor binding in the frontal cortices of suicide victims. *Arch Gen Psychiatry* 1986; 43:954–959.
42. Arango V, Ernsberger P, Sved AF, Mann JJ: Quantitative autoradiography of α_1- and α_2-adrenergic receptors in the cerebral cortex of controls and suicide victims. *Brain Res* 1993; 630:271–282.
43. Ordway GA, Widdowson PS, Smith KS, Halaris A: Agonist binding to α_2-adrenoceptors is elevated in the locus coeruleus from victims of suicide. *J Neurochem* 1994; 63:617–624.
44. Ordway GA, Smith KS, Haycock JW: Elevated tyrosine hydroxylase in the locus coeruleus of suicide victims. *J Neurochem* 1994; 62:680–685.
45. Mann JJ: Neurobiology of suicidal behaviour. *Nat Rev Neurosci* 2003; 4(10):819–828.
46. Heim C, Nemeroff CB: The role of childhood trauma in the neurobiology of mood and anxiety disorders: Preclinical and clinical studies. *Biol Psychiatry* 2001; 49(12):1023–1039.

47. Lester D: The concentration of neurotransmitter metabolites in the cerebrospinal fluid of suicidal individuals: A meta-analysis. *Pharmacopsychiatry* 1995; 28(2):45–50.
48. Agren H, Niklasson F: Suicidal potential in depression: Focus on CSF monoamine and purine metabolites. *Psychopharmacol Bull* 1986; 22(3):656–660.
49. Virkkunen M, De Jong J, Bartko J, Linnoila M: Psychobiological concomitants of history of suicide attempts among violent offenders and impulsive fire setters. *Arch Gen Psychiatry* 1989; 46(7):604–606.
50. Sher L, Oquendo MA, Grunebaum MF, Burke AK, Huang YY, Mann JJ: CSF monoamine metabolites and lethality of suicide attempts in depressed patients with alcohol dependence. *Eur Neuropsychopharmacol* 2007; 17(1):12–15.
51. Mann JJ, Malone KM, Sweeney JA, Brown RP, Linnoila M, Stanley B, Stanley M: Attempted suicide characteristics and cerebrospinal fluid amine metabolites in depressed inpatients. *Neuropsychopharmacology* 1996; 15:576–586.
52. Placidi GP, Oquendo MA, Malone KM, Huang YY, Ellis SP, Mann JJ: Aggressivity, suicide attempts, and depression: Relationship to cerebrospinal fluid monoamine metabolite levels. *Biol Psychiatry* 2001; 50(10):783–791.
53. Roy A, Ninan PT, Mazonson A, Pickar D, van Kammen DP, Linnoila M, Paul SM: CSF monoamine metabolites in chronic schizophrenic patients who attempt suicide. *Psychol Med* 1985; 15:335–340.
54. Träskman L, Åsberg M, Bertilsson L, Sjöstrand L: Monoamine metabolites in CSF and suicidal behavior. *Arch Gen Psychiatry* 1981; 38:631–636.
55. Galfalvy H, Currier D, Oquendo MA, Sullivan G, Huang YY, John MJ: Lower CSF MHPG predicts short-term risk for suicide attempt. *Int J Neuropsychopharmacol* 2009; 12(10):1327–1335.
56. Traskman-Bendz L, Alling C, Oreland L, Regnell G, Vinge E, Ohman R: Prediction of suicidal behavior from biologic tests. *J Clin Psychopharmacol* 1992; 12(2 Suppl):21S–26S.
57. Nordström P, Samuelsson M, Åsberg M, Träskman-Bendz L, Aberg-Wistedt A, Nordin C, Bertilsson L: CSF 5-HIAA predicts suicide risk after attempted suicide. *Suicide Life Threat Behav* 1994; 24(1):1–9.
58. Engstrom G, Alling C, Blennow K, Regnell G, Traskman-Bendz L: Reduced cerebrospinal HVA concentrations and HVA/5-HIAA ratios in suicide attempters. Monoamine metabolites in 120 suicide attempters and 47 controls. *Eur Neuropsychopharmacol* 1999; 9(5):399–405.
59. Sunnqvist C, Westrin A, Traskman-Bendz L: Suicide attempters: Biological stressmarkers and adverse life events. *Eur Arch Psychiatry Clin Neurosci* 2008; 258(8):456–462.
60. Morilak DA, Barrera G, Echevarria DJ, Garcia AS, Hernandez A, Ma S, Petre CO: Role of brain norepinephrine in the behavioral response to stress. *Prog Neuropsychopharmacol Biol Psychiatry* 2005; 29(8):1214–1224.
61. Miller HL, Delgado PL, Salomon RM, Berman R, Krystal JH, Heninger GR, Charney DS: Clinical and biochemical effects of catecholamine depletion on antidepressant-induced remission of depression. *Arch Gen Psychiatry* 1996; 53(2):117–128.
62. Henn FA, Vollmayr B: Stress models of depression: Forming genetically vulnerable strains. *Neurosci Biobehav Rev* 2005; 29(4–5):799–804.
63. Coryell W, Schlesser M: The dexamethasone suppression test and suicide prediction. *Am J Psychiatry* 2001; 158(5):748–753.
64. Szigethy E, Conwell Y, Forbes NT, Cox C, Caine ED: Adrenal weight and morphology in victims of completed suicide. *Biol Psychiatry* 1994; 36(6):374–380.
65. Dumser T, Barocka A, Schubert E: Weight of adrenal glands may be increased in persons who commit suicide. *Am J Forensic Med Pathol* 1998; 19(1):72–76.
66. Nemeroff CB: The neurobiology of aging and the neurobiology of depression: Is there a relationship? *Neurobiol Aging* 1988; 9:120–122.

67. Arato M, Banki CM, Bissette G, Nemeroff CB: Elevated CSF CRF in suicide victims. *Biol Psychiatry* 1989; 25(3):355–359.
68. Nemeroff CB, Owens MJ, Bissette G, Andorn AC, Stanley M: Reduced corticotropin releasing factor binding sites in the frontal cortex of suicide victims. *Arch Gen Psychiatry* 1988; 45:577–579.
69. Mann JJ, Currier D: A review of prospective studies of biologic predictors of suicidal behavior in mood disorders. *Arch Suicide Res* 2007; 11(1):3–16.
70. Brunner J, Stalla GK, Stalla J, Uhr M, Grabner A, Wetter TC, Bronisch T: Decreased corticotropin-releasing hormone (CRH) concentrations in the cerebrospinal fluid of eucortisolemic suicide attempters. *J Psychiatr Res* 2001; 35(1):1–9.
71. van Heeringen K, Audenaert K, Van de WL, Verstraete A: Cortisol in violent suicidal behaviour: Association with personality and monoaminergic activity. *J Affect Disord* 2000; 60(3):181–189.
72. Meltzer HY, Perline R, Tricou BJ, Lowy M, Robertson A: Effect of 5-hydroxytryptophan on serum cortisol levels in major affective disorders. II. Relation to suicide, psychosis, and depressive symptoms. *Arch Gen Psychiatry* 1984; 41(4): 379–387.
73. Meijer OC, De Kloet ER: Corticosterone and serotonergic neurotransmission in the hippocampus: Functional implications of central corticosteroid receptor diversity. *Crit Rev Neurobiol* 1998; 12(1–2):1–20.
74. Owens MJ, Nemeroff CB: Physiology and pharmacology of corticotropin-releasing factor. *Pharmacol Rev* 1991; 43:425–473.
75. Lopez JF, Vazquez DM, Chalmers DT, Watson SJ: Regulation of 5-HT receptors and the hypothalamic–pituitary–adrenal axis. Implications for the neurobiology of suicide. *Ann N Y Acad Sci* 1997; 836:106–134.
76. Stoff DM, Mann JJ: Suicide research. Overview and introduction. *Ann N Y Acad Sci* 1997; 836:1–11.
77. Aston-Jones G, Shipley MT, Chouvet G, Ennis M, van BE, Pieribone V, Shiekhattar R, Akaoka H, Drolet G, Astier B: Afferent regulation of locus coeruleus neurons: Anatomy, physiology and pharmacology. *Prog Brain Res* 1991; 88:47–75.
78. Dunn AJ, Swiergiel AH, Palamarchouk V: Brain circuits involved in corticotropin-releasing factor-norepinephrine interactions during stress. *Ann N Y Acad Sci* 2004; 1018:25–34.
79. Brown RP, Stoll PM, Stokes PE, Frances A, Sweeney J, Kocsis JH, Mann JJ: Adrenocortical hyperactivity in depression: Effects of agitation, delusions, melancholia, and other illness variables. *Psychiatry Res* 1988; 23(2):167–178.
80. Golomb BA: Cholesterol and violence: Is there a connection? *Ann Intern Med* 1998; 128:478–487.
81. Muldoon MF, Rossouw JE, Manuck SB, Glueck CJ, Kaplan JR, Kaufmann PG: Low or lowered cholesterol and risk of death from suicide and trauma. *Metabolism* 1993; 42(Suppl):145–156.
82. Brunner J, Parhofer KG, Schwandt P, Bronisch T: Cholesterol, essential fatty acids, and suicide. *Pharmacopsychiatry* 2002; 35(1):1–5.
83. Sublette ME, Hibbeln JR, Galfalvy H, Oquendo MA, Mann JJ: Omega-3 polyunsaturated essential fatty acid status as a predictor of future suicide risk. *Am J Psychiatry* 2006; 163(6):1100–1102.
84. Huan M, Hamazaki K, Sun Y, Itomura M, Liu H, Kang W, Watanabe S, Terasawa K, Hamazaki T: Suicide attempt and *n*–3 fatty acid levels in red blood cells: A case control study in China. *Biol Psychiatry* 2004; 56(7):490–496.
85. Wong DT, Bymaster FP: Development of antidepressant drugs. Fluoxetine (Prozac) and other selective serotonin uptake inhibitors. *Adv Exp Med Biol* 1995; 363:77–96.

86. Rosenbaum JF, Fava M, Hoog SL, Ascroft RC, Krebs WB: Selective serotonin reuptake inhibitor discontinuation syndrome: A randomized clinical trial. *Biol Psychiatry* 1998; 44(2):77–87.
87. Nemeroff CB, Owens MJ: Pharmacologic differences among the SSRIs: Focus on monoamine transporters and the HPA axis. *CNS Spectr* 2004; 9(6 Suppl 4):23–31.
88. Westenberg HG: Pharmacology of antidepressants: Selectivity or multiplicity? *J Clin Psychiatry* 1999; 60(Suppl):174–178.
89. Benkert O, Muller M, Szegedi A: An overview of the clinical efficacy of mirtazapine. *Hum Psychopharmacol* 2002; 17(Suppl 1):S23–S26.
90. Feiger A, Kiev A, Shrivastava RK, Wisselink PG, Wilcox CS: Nefazodone versus sertraline in outpatients with major depression: Focus on efficacy, tolerability, and effects on sexual function and satisfaction. *J Clin Psychiatry* 1996; 57(Suppl 2):53–62.
91. Rush AJ, Armitage R, Gillin JC, Yonkers KA, Winokur A, Moldofsky H, Vogel GW, Kaplita SB, Fleming JB, Montplaisir J, Erman MK, Albala BJ, McQuade RD: Comparative effects of nefazodone and fluoxetine on sleep in outpatients with major depressive disorder. *Biol Psychiatry* 1998; 44(1):3–14.
92. Agency for Health Care Policy and Research: *Evidence Report on Treatment of Depression: Newer Pharmacotherapies*. Washington, DC: AHCPR, 1999.
93. Stahl SM: Placebo-controlled comparison of the selective serotonin reuptake inhibitors citalopram and sertraline. *Biol Psychiatry* 2000; 48(9):894–901.
94. Mann JJ, Emslie G, Baldessarini RJ, Beardslee W, Fawcett JA, Goodwin FK, Leon AC, Meltzer HY, Ryan ND, Shaffer D, Wagner KD: ACNP Task Force report on SSRIs and suicidal behavior in youth. *Neuropsychopharmacology* 2006; 31(3):473–492.
95. Hammad TA, Laughren T, Racoosin J: Suicidality in pediatric patients treated with antidepressant drugs. *Arch Gen Psychiatry* 2006; 63(3):332–339.
96. Levinson M, Hollanc: Antidepressants and suicidality in adults: Statistical evaluation. http://www.fda.gov/ohrms/dockets/ac/06/slides/2006-4272s1-04-FDA_files/frame.htm. 2006. 7/1/2009. Accessed December 18, 2010.
97. Beasley CM, Jr., Ball SG, Nilsson ME, Polzer J, Tauscher-Wisniewski S, Plewes J, Acharya N: Fluoxetine and adult suicidality revisited: An updated meta-analysis using expanded data sources from placebo-controlled trials. *J Clin Psychopharmacol* 2007; 27(6):682–686.
98. Acharya N, Rosen AS, Polzer JP, D'Souza DN, Perahia DG, Cavazzoni PA, Baldessarini RJ: Duloxetine: Meta-analyses of suicidal behaviors and ideation in clinical trials for major depressive disorder. *J Clin Psychopharmacol* 2006; 26(6):587–594.
99. Bridge JA, Iyengar S, Salary CB, Barbe RP, Birmaher B, Pincus HA, Ren L, Brent DA: Clinical response and risk for reported suicidal ideation and suicide attempts in pediatric antidepressant treatment: A meta-analysis of randomized controlled trials. *J Am Med Assoc* 2007; 297(15):1683–1696.
100. Gibbons R, Mann JJ: Proper studies of selective serotonin reuptake inhibitors are needed for youth with depression. *CMAJ* 2009; 180(3):270–271.
101. Isacsson G: Suicide prevention—A medical breakthrough? *Acta Psychiatr Scand* 2000; 102:113–117.
102. Nakagawa A, Grunebaum MF, Ellis SP, Oquendo MA, Kashima H, Gibbons RD, Mann JJ: Association of suicide and antidepressant prescription rates in Japan, 1999–2003. *J Clin Psychiatry* 2007; 68(6):908–916.
103. Castelpietra G, Morsanutto A, Pascolo-Fabrici E, Isacsson G: Antidepressant use and suicide prevention: A prescription database study in the region Friuli Venezia Giulia, Italy. *Acta Psychiatr Scand* 2008; 118(5):382–388.
104. Ludwig J, Marcotte DE, Norberg K: Anti-depressants and suicide. *J Health Econ* 2009; 28(3):659–676.

105. Barbui C, Campomori A, D'Avanzo B, Negri E, Garattini S: Antidepressant drug use in Italy since the introduction of SSRIs: National trends, regional differences and impact on suicide rates. *Soc Psychiatry Psychiatr Epidemiol* 1999; 34:152–156.
106. Helgason T, Tomasson H, Zoega T: Antidepressants and public health in Iceland. Time series analysis of national data. *Br J Psychiatry* 2004; 184:157–162.
107. Rutz W: Preventing suicide and premature death by education and treatment. *J Affect Disord* 2001; 62:123–129.
108. Rihmer Z, Belso N, Kalmar S: Antidepressants and suicide prevention in Hungary. *Acta Psychiatr Scand* 2001; 103(3):238–239.
109. Oyama H, Fujita M, Goto M, Shibuya H, Sakashita T: Outcomes of community-based screening for depression and suicide prevention among Japanese elders. *Gerontologist* 2006; 46(6):821–826.
110. Oyama H, Goto M, Fujita M, Shibuya H, Sakashita T: Preventing elderly suicide through primary care by community-based screening for depression in rural Japan. *Crisis* 2006; 27(2):58–65.
111. Hvid M, Vangborg K, Sorensen HJ, Nielsen IK, Stenborg JM, Wang AG: Preventing repetition of attempted suicide-II. The Amager Project, a randomized controlled trial. *Nord J Psychiatry* 2011; 65(5):292–298.
112. Hegerl U, Althaus D, Schmidtke A, Niklewski G: The alliance against depression: 2-Year evaluation of a community-based intervention to reduce suicidality. *Psychol Med* 2006; 36(9):1225–1233.
113. Gibbons RD, Brown CH, Hur K, Marcus SM, Bhaumik DK, Mann JJ: Relationship between antidepressants and suicide attempts: An analysis of the Veterans Health Administration data sets. *Am J Psychiatry* 2007; 164(7):1044–1049.
114. Simon GE, Savarino J, Operskalski B, Wang PS: Suicide risk during antidepressant treatment. *Am J Psychiatry* 2006; 163(1):41–47.
115. Simon GE, Savarino J: Suicide attempts among patients starting depression treatment with medications or psychotherapy. *Am J Psychiatry* 2007; 164(7):1029–1034.
116. Olfson M, Marcus SC, Shaffer D: Antidepressant drug therapy and suicide in severely depressed children and adults: A case–control study. *Arch Gen Psychiatry* 2006; 63(8):865–872.
117. Kessler RC, Borges G, Walters EE: Prevalence of and risk factors for lifetime suicide attempts in the National Comorbidity Survey. *Arch Gen Psychiatry* 1999; 56(7):617–626.
118. Fawcett J, Scheftner WA, Fogg L, Clark DC, Young MA, Hedeker D, Gibbons R: Time-related predictors of suicide in major affective disorder. *Am J Psychiatry* 1990; 147:1189–1194.
119. Anderson IM: Meta-analytical studies on new antidepressants. *Br Med Bull* 2001; 57:161–178.
120. Stahl SM, Entsuah R, Rudolph RL: Comparative efficacy between venlafaxine and SSRIs: A pooled analysis of patients with depression. *Biol Psychiatry* 2002; 52(12):1166–1174.
121. Thase ME, Entsuah AR, Rudolph RL: Remission rates during treatment with venlafaxine or selective serotonin reuptake inhibitors. *Br J Psychiatry* 2001; 178:234–241.
122. Smith D, Dempster C, Glanville J, Freemantle N, Anderson I: Efficacy and tolerability of venlafaxine compared with selective serotonin reuptake inhibitors and other antidepressants: A meta-analysis. *Br J Psychiatry* 2002; 180:396–404.
123. Hirschfeld RM: Efficacy of SSRIs and newer antidepressants in severe depression: Comparison with TCAs. *J Clin Psychiatry* 1999; 60(5):326–335.
124. Detke MJ, Wiltse CG, Mallinckrodt CH, McNamara RK, Demitrack MA, Bitter I: Duloxetine in the acute and long-term treatment of major depressive disorder: A placebo- and paroxetine-controlled trial. *Eur Neuropsychopharmacol* 2004; 14(6):457–470.
125. Nieuwstraten CE, Dolovich LR: Bupropion versus selective serotonin-reuptake inhibitors for treatment of depression. *Ann Pharmacother* 2001; 35(12):1608–1613.

126. Thase ME, Trivedi MH, Rush AJ: MAOIs in the contemporary treatment of depression. *Neuropsychopharmacology* 1995; 12(3):185–219.
127. Lotufo-Neto F, Trivedi M, Thase ME: Meta-analysis of the reversible inhibitors of monoamine oxidase type A moclobemide and brofaromine for the treatment of depression. *Neuropsychopharmacology* 1999; 20(3):226–247.
128. Cipriani A, Brambilla P, Furukawa T, Geddes J, Gregis M, Hotopf M, Malvini L, Barbui C: Fluoxetine versus other types of pharmacotherapy for depression. *Cochrane Database Syst Rev* 2005;(4):CD004185.
129. Filteau MJ, Lapierre YD, Bakish D, Blanchard A: Reduction in suicidal ideation with SSRIs: A review of 459 depressed patients. *J Psychiatry Neurosci* 1993; 18(3):114–119.
130. Eker SS, Kirli S, Akkaya C, Cangur S, Sarandol A: Are there differences between serotonergic, noradrenergic and dual acting antidepressants in the treatment of depressed women? *World J Biol Psychiatry* 2009; 10(4 Pt 2):400–408.
131. Gonella G, Baignoli G, Ecari U: Fluvoxamine and imipramine in the treatment of depressive patients: A double-blind controlled study. *Curr Med Res Opin* 1990; 12(3):177–184.
132. Kasper S, Moller HJ, Montgomery SA, Zondag E: Antidepressant efficacy in relation to item analysis and severity of depression: A placebo-controlled trial of fluvoxamine versus imipramine. *Int Clin Psychopharmacol* 1995; 9(Suppl 4):3–12.
133. Montgomery S, Cronholm B, Asberg M, Montgomery DB: Differential effects on suicidal ideation of mianserin, maprotiline and amitriptyline. *Br J Clin Pharmacol* 1978; 5(Suppl 1):77S–80S.
134. Mahapatra SN, Hackett D: A randomised, double-blind, parallel-group comparison of venlafaxine and dothiepin in geriatric patients with major depression. *Int J Clin Pract* 1997; 51(4):209–213.
135. Sacchetti E, Vita A, Guarneri L, Cornacchia M: The effectiveness of fluoxetine, clomipramine, nortriptyline and desipramine in major depressives with suicidal behavior: Preliminary findings, in *Serotonin-Related Psychiatric Syndromes: Clinical and Therapeutic Links*. Proceedings of an international meeting held under the auspices of the Italian Society of neurosciences and sponsored by Eli Lilly Italia. Cassano GB, Aksiskal HS, eds. London, U.K.: Royal Society of Medicine Services, 1991, pp. 47–53.
136. Grunebaum MF, Ellis SP, Duan N, Burke AK, Oquendo MA, Mann JJ. A pilot randomized controlled clinical trial of an SSRI versus bupropion on suicidal behavior and ideation in at-risk depressed patients. *Neuropsychopharmacology* 2012; 37(3): 697–706.
137. Lapierre YD: Controlling acute episodes of depression. *Int Clin Psychopharm* 1991; 6(Suppl):23–35.
138. Judd FK, Moore K, Norman TR, Burrows GD, Gupta RK, Parker G: A multicentre double blind trial of fluoxetine versus amitriptyline in the treatment of depressive illness. *Aust N Z J Psychiatry* 1993; 27:49–55.
139. Möller HJ, Steinmeyer EM: Are serotonergic reuptake inhibitors more potent in reducing suicidality? An empirical study on paroxetine. *Eur Neuropsychopharmacol* 1994; 455–459.
140. Tollefson GD, Greist JH, Jefferson JW, Heiligenstein JH, Sayler ME, Tollefson SL, Koback K: Is baseline agitation a relative contraindication for a selective serotonin reuptake inhibitor: A comparative trial of fluoxetine versus imipramine. *J Clin Psychopharmacol* 1994; 14(6):385–391.
141. Marchesi C, Ceccherininelli A, Rossi A, Maggini C: Is anxious-agitated major depression responsive to fluoxetine? A double-blind comparison with amitriptyline. *Pharmacopsychiatry* 1998; 31(6):216–221.
142. Schule C: Neuroendocrinological mechanisms of actions of antidepressant drugs. *J Neuroendocrinol* 2007; 19(3):213–226.

143. Aubry JM, Gervasoni N, Osiek C, Perret G, Rossier MF, Bertschy G, Bondolfi G: The DEX/CRH neuroendocrine test and the prediction of depressive relapse in remitted depressed outpatients. *J Psychiatr Res* 2007; 41(3–4):290–294.
144. Appelhof BC, Huyser J, Verweij M, Brouwer JP, Van DR, Fliers E, Hoogendijk WJ, Tijssen JG, Wiersinga WM, Schene AH: Glucocorticoids and relapse of major depression (dexamethasone/corticotropin-releasing hormone test in relation to relapse of major depression). *Biol Psychiatry* 2006; 59(8):696–701.
145. Zobel AW, Nickel T, Sonntag A, Uhr M, Holsboer F, Ising M: Cortisol response in the combined dexamethasone/CRH test as predictor of relapse in patients with remitted depression. A prospective study. *J Psychiatr Res* 2001; 35(2):83–94.
146. Scheuch K, Holtje M, Budde H, Lautenschlager M, Heinz A, hnert-Hilger G, Priller J: Lithium modulates tryptophan hydroxylase 2 gene expression and serotonin release in primary cultures of serotonergic raphe neurons. *Brain Res* 2010; 1307:14–21.
147. Kitaichi Y, Inoue T, Nakagawa S, Izumi T, Koyama T: Effect of co-administration of lithium and reboxetine on extracellular monoamine concentrations in rats. *Eur J Pharmacol* 2004; 489(3):187–191.
148. Wegener G, Linnet K, Rosenberg R, Mork A: The effect of acute citalopram on extracellular 5-HT levels is not augmented by lithium: An in vivo microdialysis study. *Brain Res* 2000; 871(2):338–342.
149. Treiser SL, Cascio CS, O'Donohue TL, Thoa NB, Jacobowitz DM, Kellar KJ: Lithium increases serotonin release and decreases serotonin receptors in the hippocampus. *Science* 1981; 213(4515):1529–1531.
150. Cowen PJ, McCance SL, Ware CJ, Cohen PR, Chalmers JS, Julier DL: Lithium in tricyclic-resistant depression. Correlation of increased brain 5-HT function with clinical outcome. *Br J Psychiatry* 1991; 159:341–346.
151. Cowen PJ, McCance SL, Cohen PR, Julier DL: Lithium increases 5-HT-mediated neuroendocrine responses in tricyclic resistant depression. *Psychopharmacology (Berl)* 1989; 99(2):230–232.
152. Delgado PL, Miller HL, Salomon RM, Licinio J, Krystal JH, Moreno FA, Heninger GR, Charney DS: Tryptophan-depletion challenge in depressed patients treated with desipramine or fluoxetine: Implications for the role of serotonin in the mechanism of antidepressant action. *Biol Psychiatry* 1999; 46(2):212–220.
153. Tondo L, Baldessarini RJ: Long-term lithium treatment in the prevention of suicidal behavior in bipolar disorder patients. *Epidemiol Psychiatr Soc* 2009; 18(3):179–183.
154. Oquendo MA, Currier D, Mann JJ: Prospective studies of suicidal behavior in major depressive and bipolar disorders: What is the evidence for predictive risk factors? *Acta Psychiatr Scand* 2006; 114(3):151–158.
155. Kovacsics CE, Gottesman II, Gould TD: Lithium's antisuicidal efficacy: Elucidation of neurobiological targets using endophenotype strategies. *Annu Rev Pharmacol Toxicol* 2009; 49:175–198.
156. Muller-Oerlinghausen B, Lewitzka U: Lithium reduces pathological aggression and suicidality: A mini-review. *Neuropsychobiology* 2010; 62(1):43–49.
157. Meltzer HY: Suicide in schizophrenia, clozapine, and adoption of evidence-based medicine. *J Clin Psychiatry* 2005; 66(4):530–533.
158. Meltzer HY, Alphs L, Green AI, Altamura AC, Anand R, Bertoldi A, Bourgeois M, Chouinard G, Islam MZ, Kane J, Krishnan R, Lindenmayer JP, Potkin S: Clozapine treatment for suicidality in schizophrenia: International Suicide Prevention Trial (InterSePT). *Arch Gen Psychiatry* 2003; 60(1):82–91.
159. Meltzer HY, Li Z, Kaneda Y, Ichikawa J: Serotonin receptors: Their key role in drugs to treat schizophrenia. *Prog Neuropsychopharmacol Biol Psychiatry* 2003; 27(7):1159–1172.

160. Stanley M, Mann JJ: Increased serotonin-2 binding sites in frontal cortex of suicide victims. *Lancet* 1983; 321:i214–i216.
161. Bell GS, Mula M, Sander JW: Suicidality in people taking antiepileptic drugs: What is the evidence? *CNS Drugs* 2009; 23(4):281–292.
162. Gibbons RD, Hur K, Brown CH, Mann JJ: Relationship between antiepileptic drugs and suicide attempts in patients with bipolar disorder. *Arch Gen Psychiatry* 2009; 66(12):1354–1360.
163. Pacher P, Kecskemeti V: Trends in the development of new antidepressants. Is there a light at the end of the tunnel? *Curr Med Chem* 2004; 11(7):925–943.
164. Brent D, Melhem N, Turecki G: Pharmacogenomics of suicidal events. *Pharmacogenomics* 2010; 11(6):793–807.

Index

C

D

E

N

P

R

S

T

U

W

Y

Printed in the United States
By Bookmasters

Printed in the United States
by Baker & Taylor Publisher Services